Fire Effects on Soils and Restoration Strategies

Land Reconstruction and Management
Vol. 5, 2009

Land Reconstruction and Management Series

Series Editor: Martin J. Haigh

Vol. 1: Reclaimed Land, Erosion Control, Soils and Ecology
Vol. 2: Ecological Effects of Roads
Vol. 3: Stone Deterioration in Polluted Urban Environments
Vol. 4: Better Land Husbandry
Vol. 5: Fire Effects on Soils and Restoration Strategies

Fire Effects on Soils and Restoration Strategies

Volume Editors

Artemi Cerdà

Departament de Geografia
Universitat de València
Spain

Peter R. Robichaud

U.S. Department of Agriculture, Forest Service
Rocky Mountain Research Station
Moscow, Idaho
USA

Volume 5 of Series:
Land Reconstruction and Management

Series Editor
MARTIN J. HAIGH
Oxford Brookes University
Oxford, UK

Science Publishers
Enfield (NH) Jersey Plymouth

Science Publishers
234 May Street
Post Office Box 699
Enfield, New Hampshire 03748
United States of America

www.scipub.net

General enquiries : *info@scipub.net*
Editorial enquiries : *editor@scipub.net*
Sales enquiries : *sales@scipub.net*

Published by Science Publishers, Enfield, NH, USA
An imprint of Edenbridge Ltd., British Channel Islands
Printed in India

ISBN 978-1-57808-526-2

Library of Congress Cataloging-in-Publication Data

Fire effects on soils and restoration strategies / Artemio Cerda, Peter Robichaud [editors]. -- 1st ed.
p. cm. -- (Land reconstruction and management series ; v. 5)
Includes bibliographical references and index.
ISBN 978-1-57808-526-2 (hardcover)
1. Forest fires. 2. Fire ecology. 3. Soil ecology. 4. Forest restoration. I. Cerda, Artemio. II. Robichaud, Peter R.
SD421.F514 2008
577.2'4--dc22

2008032400

Preface

Heat, oxygen and fuel are required for fire ignition. Oxygen is always present in the earth's troposphere and fuel is abundant on vegetated landscapes. The heat needed for burning to occur can come from lighting or human sources, such as intentional ignition (prescribed fire or arson), or accidental ignition. Once ignition occurs, fires may burn on continuous vegetation for days, weeks, and even months. Fire is a natural process within forest, scrubland, and grassland ecosystems, and is a determining factor of landscape processes and features. High frequency fires are essential in maintaining many ecosystems, such as grasslands and savannahs. The post-fire influx of fire-dependent and fire-resistant species in many different ecosystems demonstrates the integral role of fire in the natural succession of these ecosystems.

Since the earliest times, but especially after the development of agriculture, fire has been a useful tool for humankind to control and modify nature. The shift to agriculture-based economies increased fire recurrence. In the past 100 years, forest fire suppression efforts have decreased fire recurrence, resulting in changes of the forest structure in much of the western United States (Agee 1993). Forest fire suppression has resulted in larger-than-average fuel loads, which, according to many researchers, have contributed to the large, high-intensity forest fires of the past 10 to 15 years in the American West. Other regions, such as the northern Mediterranean, have had increased fire occurrence since the 1970's due to land abandonment and propitious natural climatic and cultural conditions. Climate change is also increasing forest fire occurrence in many ecosystems, but especially in the boreal forests. These and other human-induced changes in natural fire cycles have increased the frequency, size, and severity of fires in recent years and significantly impacted human and natural resources as well as ecosystem functions.

Fires can kill, char, and/or consume living vegetation and non-living organic material normally found in various stages of decomposition on the soil surface or within the upper layer of the mineral soil. In addition, fire can change the physical, chemical, mineralogical, and biological soil properties (Certini 2005). Soil has been described as a chemical reactor powered by living matter that modifies the earth surface. Fire-induced changes in soil properties impact soil functions, such as water movement, nutrient flows, and habitat for soil-dwelling organisms, which are essential and interdependent components of the ecosystem. For example, fire-induced changes in soil structure and water repellency can change the hydrologic response of soil, resulting in decreased infiltration, increased overland flow, and soil erosion. The degree of fire-induced changes in soil properties and the length of time needed for soil to recover to its pre-fire functionality are dependent on the severity of the fire. Consequently, after severe fires, such as wildfires, land managers often decide to use stabilization and rehabilitation strategies to reduce erosion and sedimentation and to protect important downstream structures and resources as well as the soil resource itself.

Fire Effects on Soils and Restoration Strategies is being published a decade after *Fire's Effects on Ecosystems* by DeBano, Neary, and Folliott (1998), and builds on their foundation to update knowledge on natural post-fire processes and describe the use and effectiveness of various restoration strategies that may be applied when human intervention is warranted. The chapters in this book, written by leading scientists, have been compiled to provide relevant and accessible information to students, land managers, and policy-makers as well as other scientists.

When choosing where and when to apply post-fire restoration treatments, it is important to consider: 1) the effects of fire on the ecosystem; 2) the potential threats, such as increased flooding, erosion, and invasive weeds, that may result from the fire-induced changes; 3) the potential effectiveness of the treatments being considered; and 4) the sequence and time frame of natural recovery. Consequently, this book is organized to provide current information on all these topics. It begins (Chapter 1) with an historical examination of the role of forest fires as seen through the fossil record. *Section I: Fire Effects on Soil* (Chapters 2 to 9) presents current knowledge of the geological, hydrological, geomorphologic, and biological effects of fire on soil, especially as these relate to the use of post-fire rehabilitation and restoration strategies. In *Section II: Rehabilitation and Restoration Strategies* (Chapters 10 to 14), the contributions of several authors combine to provide a comprehensive review of common post-fire restoration strategies. The research studies described in these chapters reveal key parameters of the most successful strategies. In *Section III: Regional Strategies* (Chapters 15 to 20), authors from Australia, Canada, Chile, Portugal, and two areas of the western United States describe how knowledge of wildfire effects and post-fire restoration strategies have been applied within their regions. The final chapter (21) provides a summary of the book and some thoughts on future research.

Fire Effects on Soils and Restoration Strategies is the result of a long and productive collaboration between the editors and has grown to include scientists from across the Americas, Europe, and Australia. The publication of this book reflects a decade of international scientific exchange among many scientists, some of whom are chapter authors, who have made important contributions through journal articles, conference presentations, and extensive field site visits. The concept of producing a book such as *Fire Effects on Soils and Restoration Strategies* with an international perspective was first discussed at the 2000 American Geophysical Union meeting special session, Wildfire and Surficial Processes (see special issue of Hydrological Processes, Elsenbeer and Robichaud (eds.) 2001) and again at the 2002 Geological Society of America meeting. Scientific exchanges continued at European Geoscience Union meetings that occurred in 2003 to 2006 where many of the authors were able to come together and discuss their work on post-fire effects and rehabilitation techniques. The positive effects of extensive field excursions, with international participation, deserve special mention. In April 2003, Artemi Cerdà organized and escorted 30 participants on a trip along the eastern Iberian Peninsula coast of Spain to observe and discuss the post-fire research in progress at several sites. Many of these same participants travelled together on another excursion, organized by John Moody and Deborah Martin, through post-fire research sites in Colorado with an added week in New Mexico leaded by Craig Allen. A third, short but very useful, excursion was organized by António Ferreira in Portugal in 2005. Field excursion participants had extensive opportunities to share information, discuss research in progress, and develop joint projects. Most of the authors of this book continue to meet at international conferences to exchange new information and techniques and to maintain professional and personal affiliations. We hope that *Fire Effects on Soils and Restoration Strategies* reflects the valuable collaboration among scientists that has produced our current understanding of post-fire effects and restoration, and that these dynamic international efforts continue to be nurtured and productive as we face important global issues in the coming years.

We wish to thank the many authors for their contributions and enthusiastic collaboration and Louise Ashmun for her technical and editorial assistance throughout the book. In addition, we would like to acknowledge the encouragement received from Dr. Samran Sambatpanit, the past-president of the World Association of Soil and Water Conservation, and thank him for his part in moving this project forward.

Artemi Cerdà and Peter R. Robichaud

Literature cited

Agee, J.K. 1993. *Fire Ecology of Pacific Northwest Forests.* Island Press, Washington, D.C., USA.

Certini, G. 2005. Effects of fire on properties of forest soils: a review. *Ocologia,* 143: 1-10. DOI 10.1007/s00442-004-1788-8.

DeBano, L.F., D.G. Neary, and P.F. Folliott. 1998. *Fire's Effects on Ecosystems.* John Wiley & Sons, New York, New York, USA.

Elsenbeer, H. and P.R. Robichaud [eds.]. 2001. Wildfire and Surficial Processes. *Special Issue of Hydrological Processes,* 15: 2865-3092.

Contents

Section III—Regional Strategies

List of Contributors

Alloza J.A.
Centro de Estudios Ambientales del Mediterráneo (CEAM) Foundation, Parque Tecnológico, Ch. Darwin 14, 46980 Paterna, Valencia, Spain
Armesto J.J.
Center for Advanced Studies in Ecology and Biodiversity, Pontificia Universidad Católica de Chile, Alameda 340, P.C. 6513677, Santiago, Chile
Baeza M.J.
Centro de Estudios Ambientales del Mediterráneo (CEAM) Foundation, Parque Tecnológico, Ch. Darwin 14, 46980 Paterna, Valencia, Spain
Bárcenas G.M.
Grupo de Edafología Ambiental (GEA), Departamento de Agroquímica y Medio Ambiente, Universidad Miguel Hernández, Avenida de la Universidad s/n, 03202-Elche, Alicante, Spain
Bautista S.
Departament de Ecologia Apdo. 99, Universidad de Alicante, E-03080, Alicante, Spain
Berg S.
School of Geography and Environmental Science, Clayton Campus, Monash University, Victoria 3800, Australia
Beyers J.L.
U.S. Department of Agriculture, Forest Service, Pacific Southwest Research Station, 4955 Canyon Crest Dr., Riverside, California 92507 USA
Bladé C.
Centro de Estudios Ambientales del Mediterráneo (CEAM) Foundation, Parque Tecnológico, Ch. Darwin 14, 46980 Paterna, Valencia, Spain
Boulet A-K.
Centro de Estudos do Ambiente e do Mar (CESAM), Departamento de Ambíente e Ordenamento, Universidade de Aveiro, P3810-193 Aveiro, Portugal
Bustamante-Sánchez M.A.
Center for Advanced Studies in Ecology and Biodiversity, Pontificia Universidad Católica de Chile, Alameda 340, P.C. 6513677, Santiago, Chile

Cerdà A.
Departament de Geografia, Universitat de València, Blasco Ibàñez, 28, 46010, València, Spain
Chirino E.
Centro de Estudios Ambientales del Mediterráneo (CEAM) Foundation, Parque Tecnológico, Ch. Darwin 14, 46980 Paterna, Valencia, Spain
Coelho C.
Centro de Estudos do Ambiente e do Mar (CESAM), Departamento de Ambíente e Ordenamento, Universidade de Aveiro, P3810-193 Aveiro, Portugal
Curran M.P.
British Columbia Ministry of Forests and Range (BC Forest Service) Kootenay Lake Forestry Centre, 1907 Ridgewood Rd, Nelson, British Columbia VIL6KI, Canada
Díaz M.F.
Center for Advanced Studies in Ecology and Biodiversity, Pontificia Universidad Católica de Chile, Alameda 340, P.C. 6513677, Santiago, Chile
Doerr S.H.
Institute of Environmental Sustainability, School of the Environment and Society, Swansea University, Singleton Park, Swansea, SA2 8PP, UK
Duguy B.
Centro de Estudios Ambientales del Mediterráneo (CEAM) Foundation, Parque Tecnológico, Ch. Darwin 14, 46980 Paterna, Valencia, Spain
Dunkerley D.
School of Geography and Environmental Science, Clayton Campus, Monash University, Victoria 3800, Australia
Ferguson R.
Melbourne Water, PO Box 4342, Melbourne, Victoria 3001, Australia
Ferreira A.J.D.
Centro de Estudos de Recursos Naturais (CERNAS), Escola Superior Agrária de Coimbra, Bencanta, P-3040-316 Coimbra, Portugal
Fuentes D.
Centro de Estudios Ambientales del Mediterráneo (CEAM) Foundation, Parque Tecnológico, Ch. Darwin 14, 46980 Paterna, Valencia, Spain
García-Orenes F.
Grupo de Edafología Ambiental (GEA), Departamento de Agroquímica y Medio Ambiente, Universidad Miguel Hernández, Avenida de la Universidad s/n, 03202-Elche, Alicante, Spain
González M.E.
Forest Ecosystem Services to Aquatic Ecosystems under Climatic Change (FORECOS), Universidad Austral de Chile, Casilla 567, Valdivia, Chile

Guerrero C.
Grupo de Edafología Ambiental (GEA), Departamento de Agroquímica y Medio Ambiente, Universidad Miguel Hernández, Avenida de la Universidad s/n, 03202-Elche, Alicante, Spain
Holz A.
Department of Geography, University of Colorado, Boulder, Colorado, 80309, USA
Hubbert K.R.
Hubbert and Associates, Apple Valley, California, 92307, USA
Jacobsen K.L.S.
Commonwealth Scientific and Industrial Research Organisation (CSIRO) Sustainable Ecosystems, P.O. Box E4008, Kingston, ACT 2604, Australia
Keizer J.J.
Centro de Estudos do Ambiente e do Mar (CESAM), Departamento de Ambíente e Ordenamento, Universidade de Aveiro, P3810-193 Aveiro, Portugal
Khanna P.K.
Commonwealth Scientific and Industrial Research Organisation (CSIRO) Sustainable Ecosystems, P.O. Box E4008, Kingston, ACT 2604, Australia
Larsen I.J.
Department of Forest, Rangeland, and Watershed Stewardship, Colorado State University, Fort Collins, Colorado 80523-1472, USA
Lloret F.
Center for Ecological Research and Forestry Applications (CREAF) – Departament de Biologia Animal, Biologia Vegetal i Ecologia, Universitat Autònoma de Barcelona, 08193 Bellaterra, Barcelona, Spain
MacDonald L.H.
Department of Forest, Rangeland, and Watershed Stewardship, Colorado State University, Fort Collins, Colorado 80523-1472, USA
Martin D.A.
U.S. Geological Survey, 3215 Marine Street, Suite E-127 Boulder, Colorado 80303, USA
Martin N.
North East Catchment Management Authority, PO Box 616, Wodonga, Victoria 3689, Australia
Mataix-Solera J.
Grupo de Edafología Ambiental (GEA), Departamento de Agroquímica y Medio Ambiente, Universidad Miguel Hernández, Avenida de la Universidad s/n, 03202-Elche, Alicante, Spain
Moody J.A
U.S. Geological Survey, 3215 Marine Street, Suite E-127, Boulder, Colorado 80303, USA

Nuñez-Avila M.
Forest Ecosystem Services to Aquatic Ecosystems under Climatic Change (FORECOS), Universidad Austral de Chile, Independcia 641, Valdivia, Chile
Outeiro L.R.
Grup de Recerca Ambiental Mediterrània (GRAM), Department of Physical Geography, Universitat de Barcelona, Montalegre, 6. 08001 Barcelona, Spain
Pausas J.G.
Centro de Estudios Ambientales del Mediterráneo (CEAM) Foundation, Parque Tecnológico, Ch. Darwin 14, 46980 Paterna, Valencia, Spain
Raison R.J.
Commonwealth Scientific and Industrial Research Organisation (CSIRO) Sustainable Ecosystems, P.O. Box E4008, Kingston, ACT 2604, Australia
Robichaud P.R.
U.S. Department of Agriculture, Forest Service, Rocky Mountain Research Station, 1221 S. Main St, Moscow, Idaho 83843, USA
Romanya J.
Department of Natural Products, Plant Biology and Soil Science, Universitat de Barcelona, Av. Joan XXIII s/n, 08028 Barcelona, Spain
Scott A.C.
Geology Department, Royal Holloway University of London, Egham, Surrey, TW20 0EX, UK
Scott D.F.
Watershed Management, The University of British Columbia Okanagan, Kelowna, British Columbia, V1V1V7, Canada
Serrasolses I.
Unitat d'Ecologia, Facultad de Biosciences, Universitat Autònoma de Barcelona, 08193 Bellaterra, Barcelona, Spain
Shakesby R.A.
Institute of Environmental Sustainability, School of the Environment and Society, Swansea University, Singleton Park, Swansea, SA2 8PP, UK
Silva J.S.
Centro de Estudos de Recursos Naturais (CERNAS), Escola Superior Agrária de Coimbra, Bencanta, P-3040-316 Coimbra, Portugal
Smith-Ramírez C.
Center for Advanced Studies in Ecology and Biodiversity, Pontificia Universidad Católica de Chile, Alameda 340, P.C. 6513677, Santiago, Chile
Torres M.P.
Departamento de Biología Aplicada, Universidad Miguel Hernández, Avenida de la Universidad s/n, 03202-Elche, Alicante, Spain
Úbeda X.
Grup de Recerca Ambiental Mediterrània (GRAM), Department of Physical Geography, Universitat de Barcelona, Montalegre 6. 08001, Barcelona, Spain

Valdecantos A.
Centro de Estudios Ambientales del Mediterráneo (CEAM) Foundation, Parque Tecnológico, Ch. Darwin 14, 46980 Paterna, Valencia, Spain
Vallejo V.R.
Centro de Estudios Ambientales del Mediterráneo (CEAM) Foundation, Parque Tecnológico, Ch. Darwin 14, 46980 Paterna, Valencia, Spain
Vilagrosa A.
Centro de Estudios Ambientales del Mediterráneo (CEAM) Foundation, Parque Tecnológico, Ch. Darwin 14, 46980 Paterna, Valencia, Spain
Wagenbrenner J.W.
U.S. Department of Agriculture, Forest Service, Rocky Mountain Research Station, 1221 S. Main Street, Moscow, Idaho 83843, USA
Wohlgemuth P.M.
U.S. Department of Agriculture, Forest Service, Pacific Southwest Research Station, 4955 Canyon Crest Drive, Riverside, California 92507, USA
Zedler P.H.
Nelson Institute for Environmental Studies and the Arboretum, University of Wisconsin-Madison, 550 N. Park St, Madison, Wisconsin 53706, USA

1

Forest Fire in the Fossil Record

Andrew C. Scott

Abstract

Fire may not only be seen as a destructive force but also as a preservational mechanism. Charcoal, a fire residue, preserves anatomy allowing for the identification of plants, is inert, and may survive transport, burial and diagenesis. Evidence of fires in deep time, before man, comes predominantly from macroscopic charcoal deposits. Our earliest records of wildfire come from the late Silurian and early Devonian (420 to 400 million years ago) but evidence of fire through the Devonian (400 to 350 my) is rare, possibly because of low atmospheric oxygen. Atmospheric oxygen levels are thought to have risen rapidly through the Carboniferous and Permian (350 to 250 my) and this coincides with the spread of fire into a range of environments from lowland tropical mires to floodplains and in to upland regions. Sedimentological evidence suggests that post-fire erosion-depositional systems are more widespread in the fossil record than has been previously thought. Studies of charcoalified plants not only provide data on the evolution of plants but also of fire-prone vegetation. Fire has played an important role in the Earth system processes for over 400 million years impacting on the atmosphere, climate and the evolution of terrestrial ecosystems. It is in this context that we should see the Earth as a 'fire planet' (S.J. Pyne).

INTRODUCTION

Fire has been called a 'global herbivore' (Bond and Keeley 2005) and it has been shown that fire has shaped global ecosystems for over 350 million years (Scott 2000a, Bond and Keeley 2005). However, fire may also be considered as a major Earth system process interacting with climate and the atmosphere as well as providing a possible driving force for the evolution and spread of new plants and biomes (Scott and Glasspool 2006, Berner et al. 2003, Osborne and Beerling 2006). In addition, as is obvious from other papers presented in this volume, post-fire erosion-deposition is a factor that has recently come to the

Andrew C. Scott: Department of Earth Sciences, Royal Holloway University of London, Egham, Surrey, TW20 0EX, UK. Tel: 44+(0)1784 443608, Fax: 44+(0)1784 471780, e-mail: a.scott@gl.rhul.ac.uk

fore yet it has not been widely recognized until now in the fossil record. A major problem is a lack of awareness of the phenomenon for many Earth scientists but a recent paper by Shakesby and Doerr (2006) may help address this.

The main source of data on the fossil record of fires comes from charcoal, although the presence of soot and polyaromatic hydrocarbons (PAH) have also been used (Scott 1989, 2000a, 2001, 2002, Jones et al. 1997, Wolbach et al. 1990, Finklestein et al. 2005). In studies of more recent fires, fire scars on trees have also been widely used (Westerling et al. 2006).

Wildfires produce a number of carbonaceous residues, which have a broad range of terminology (Jones et al. 1997). These include combustion products such as smoke, gases and soot as well as residues such as ash and charcoal (Fig. 1). The range of methods for detecting such products and the confusion in their definition has hindered rather than helped a broader discussion on the history of wildfire (see Jones et al. 1997, Forbes et al. 2006, Scott and Glasspool 2007).

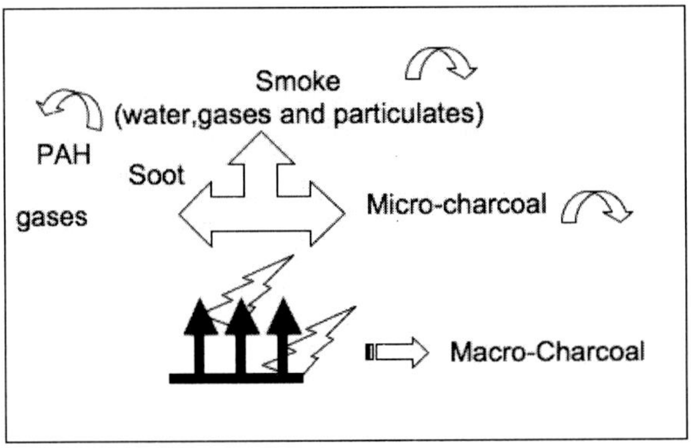

Fig. 1 Transport of fire combustion products and residues. Particulates and gases are transported by wind and most macroscopic charcoal is produced from the burning of surface vegetation and litter and is transported by water.

In this chapter, a brief account of the geological history of wildfire and its significance in the Earth system will be provided and followed by the ways in which combustion products are formed, recognized and used.

THE RECOGNITION OF FIRE IN THE FOSSIL RECORD

Wildfires may produce both solid residues and gases (Novakov et al. 1997). Both solid residues and gases may combine together to form smoke which can cause significant environmental damage, in particular to human health as well as causing the build-up of greenhouse gases which may affect climate (Pyne et al. 1996, Page et al. 2002, Flannigan et al. 2000). Most geologists are concerned with the solid particulate residues from wildfire, which is predominantly charcoal (Scott 2000a), also known in the fossil record as fusain (Scott 1989). However, the wide use of terms such as elemental carbon, black carbon etc. and the overlap in their recognition and definition has created some confusion (Forbes et al. 2006, Bird 1997, Schmidt and Noack 2000). Some of the residues in smoke include soot, which is formed by the agglomeration of condensed hydrocarbons and tars (Jones et al. 1997). However, soot may not only form as a result of biomass burning but also for coal and oil so that presence in the fossil record cannot be used to signal exclusively the occurrence of vegetation fires (Belcher et al. 2005).

The largest of the combustion residues comprise charcoal – that is charcoalified and partially charred plants, all of which have the potential of being incorporated into the fossil record (Scott 2003). In the fossil record, such material has been described under the term 'fusain' and has been identified by coal petrologists under the group of macerals known as inertinites (Scott 2002, Scott and Glasspool 2007).

Charcoal may be recognized in hand specimen as being black with a silky sheen and showing anatomical structure under the hand lens, as well as having a black streak. Charcoals have also been characterized using a range of chemical and physical methods.

CHARCOAL

Chemical changes

Charcoals have been characterized chemically using a range of techniques. Most woods, for example, comprise around 70 percent cellulose and 30 percent lignin but during the charcoalification process these products break down with the cellulose being less stable than the lignin. It has been well established that the increase in temperature is the main driving force of chemical change, which causes the breaking of particular chemical bonds and changes in functional group chemistry (Bustin and Guo 1999).

When wood is heated rapidly by an ignition source such as lightning, volatile gases are produced which mix with atmospheric oxygen resulting in combustion. This combustion, then provides the heat for pyrolysis to occur which provides more flammable gases and leaves a charcoal residue. With sufficient time and oxygen supply, all the wood will be combusted but this is often not the case and the pyrolysis residue, or charcoal, remains. The initial temperature rise drives off moisture (up to 110°C) before cellulose decomposition is initiated between 110 to 270°C. At this stage both gases and tars may be produced. This breakdown continues up to 400°C when lignin also breaks down and the formation of charcoal is completed around 500°C when little tar or volatiles are driven off (Pyne et al. 1996).

The transformation of wood to charcoal has been studied using a range of techniques including electron spin resonance (Cope 1980), infra-red spectroscopy (Guo and Bustin 1998) and 13C nuclear magnetic resonance (Jones 1993). All these methods show the change in chemical structure with increasing temperature. It has also been shown that, in general, plants retain their stable carbon isotopic signature through the charring process, although some exceptions have been reported (Jones 1994, McParland et al. 2007).

Physical changes

Charcoal is characterized by exceptional anatomical preservation (Fig. 3; Scott 2001) and together with the fact that it is relatively inert, means that it has a rich information content that is only now being widely exploited. Whilst charcoal may appear as small poorly preserved black fragments (Fig. 2), a study even under a hand lens will reveal anatomical as well as morphological preservation. The use of scanning electron microscopy reveals this anatomical preservation to its full potential (Figs. 3 and 5; Scott 2001). The anatomical data may allow the identification of the plants as well as characters such as stomata on leaves (Fig. 5b, d) and growth rings in woods, which can be used as palaeoatmospheric and palaeoclimate proxies. Many new groups of plants have been identified from fossil charcoal deposits (Scott 2000a). Even such delicate structures as flowers (Fig. 3d) and glandular hairs on ovules (Fig. 5g) may be preserved.

Whilst there may be some shrinkage during the charcoalification process (Prior and Alvin 1983, Lupia 1995) anatomical data survives. Cell wells, however, may typically be homogenized between 300 to 325°C.

When studied in polished blocks under oil in reflected light, charcoal shows characteristic high reflectance (Figs. 3e). Experiments have demonstrated that there is an increase in reflectance with temperature (Fig. 7b; Jones et al. 1991, Scott and Jones 1991, Guo and Bustin 1998, Bustin and Guo 1999, Scott and Glasspool 2007, McParland et al. 2007) and with time (Fig. 7a; Bustin and Guo 1998, Guo and Bustin 1999, Scott and Glasspool 2005). Charcoal reflectance may be used, therefore, to provide an estimate of minimum fire temperatures, an area that has yet to be widely exploited (Scott and Jones 1994, McParland et al. 2007).

Soot

Soot is often defined as a black carbonaceous substance, which is produced during the imperfect combustion of plant material but also of coal and oil. However, as pointed out by Jones et al. (1997), it has been variously and widely used in the literature normally comprising material formed via a gas-phase condensation process. The mixing of terms such as soot and black/elemental carbon creates significant difficulty, especially as much black carbon has been thought to be fossil charcoal (Forbes et al. 2006). Jones et al. (1997) proposed that soot be defined as "particles emitted with smoke and formed via gas phase processes gas-to-particle conversion. Particle sizes range from sub-micrometer (mainly) up to less than 1 μm".

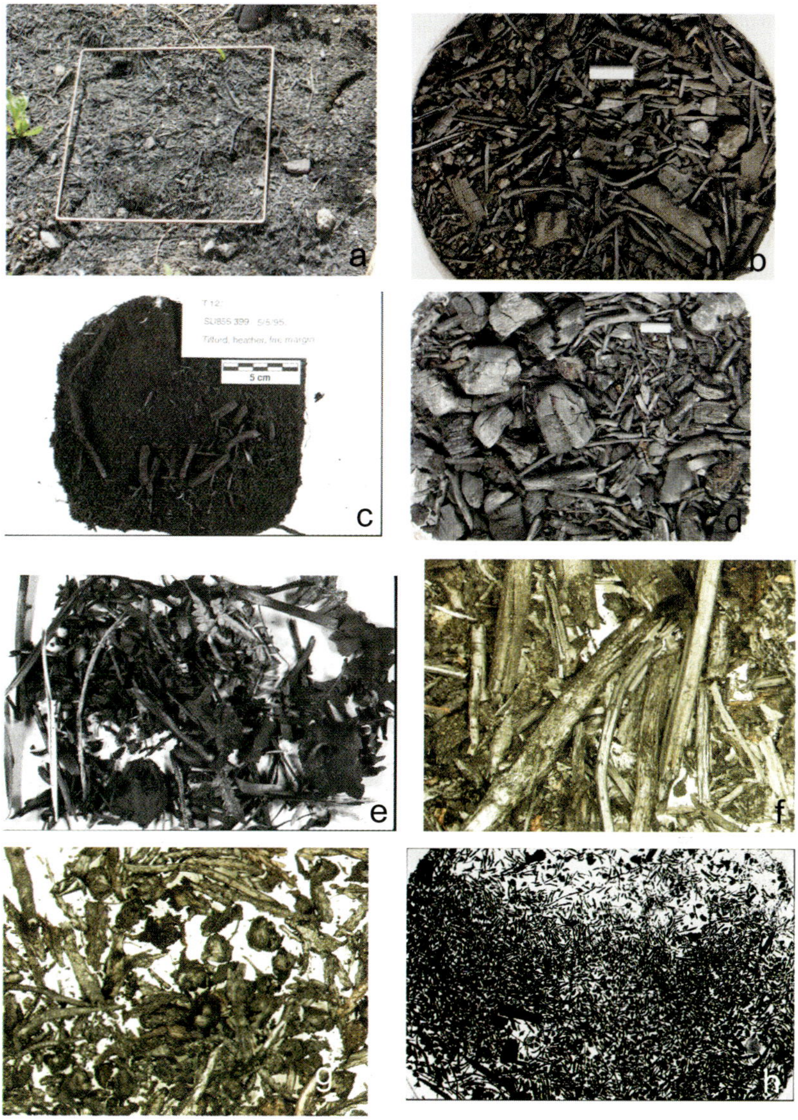

Fig. 2 Modern charcoal assemblages: a) charred surface litter from the 2002 Hayman Fire, Colorado, USA; b) detail of material collected in a; c) charred heathland surface from the 1995 Frensham Fire, Surrey, UK; d) charcoal, mainly wood, from the 1995 Frensham Fire, Surrey, UK; e) charcoal from the 1995 Frensham Fire, Surrey, UK showing charred fern litter; f) heather dominated charcoal from the 2006 Thursley Fire, Surrey, UK; g) charcoal from the 1995 Frensham Fire, Surrey, UK showing wind concentrations of *Calluna* flowers; h) heather dominated charcoal from the 2006 Thursley Fire, Surrey, UK showing wind concentrated of *Calluna* flowers.

The recognition of soot in the fossil record has remained controversial. Wolbach et al. (1990) reported soot from the days of the Cretaceous-Tertiary boundary, which they identified both chemically and morphologically, having characteristic 'grape bunch' morphology. However, soot is particularly difficult to extract and identify and may be derived not only from vegetation fires but also the burning of coal and oil deposits (Harvey et al. 2008). Such material may not be the best, therefore, to identify biomass burning in the fossil record (Belcher et al. 2005).

Polyaromatic Hydrocarbons (PAH)

One group of chemicals that has been strongly linked to vegetation fires is polyaromatic hydrocarbons (PAH) (Zepp and Macko 1997). Many of these chemicals with high molecular weights are resistant to degradation and hence may be found in the fossil record (Killops and Massoud 1992). It is also possible to infer a vegetation source from the range of PAHs in a sample, as many are strongly linked to the burning of certain plants (Simoneit 2002). In addition, it is possible to distinguish PAHs derived from biomass burning from those derived from fossil fuel combustion (Zepp and Macko 1997). Additional data on compound specific isotopes have also shown potential of identifying biomass source (Zepp and Macko 1997).

In the fossil record the occurrence of PAHs to infer fire history has not been widely needed but they may be particularly useful in the absence of charcoal to infer the presence of fires. However, the combined use of charcoal analysis and chemistry is always advocated (Finklestein et al. 2005).

Fire Scars

When trees are burned by a fire they may be charred yet survive. Often this charring results in the presence of a fire scar. The occurrence of fire scars to interpret fire history is of major use in the historic past where growth ring studies can help date fire events (Agee 1990, Whitlock and Millspaugh 1995). Many recent fire records from North America have used a combination of charcoal occurrence in lakes and peats together with fire scar data to obtain data on fire frequency and climate which have helped in modelling the relationship between fire and climate (Westerling et al. 2006).

It is not possible to use fire scars in deep time to interpret fire history but rare examples have been reported from fossil woods (Dechamps 1984, Putz and Taylor 1996).

MACRO- AND MICRO-CHARCOAL

Origin

During fires there are many potential sources of charcoal and each may yield different information. One of the most obvious sources is small charcoal particles that are lofted in smoke by the wind (Komarek et al. 1973). For example, leaves and fine twigs may be combusted in crown fires and the small (often less than 1 mm) charcoal fragments may be transported by the wind away from the fire (Clark 1988). It has been shown (Clark and Patterson 1997) that heavier particles are deposited from the air sooner than smaller particles. Particles <63 μm may be transported many kilometers away from the fire site. It is these particles that have been predominantly used for fire frequency analysis and have been used as evidence of background fires (Clark and Patterson 1997). However, microscopic charcoals may also be produced from surface fires burning shrubs, herbs as well as litter.

Most macroscopic charcoals (>1 mm often up to 1 cm) are not formed by burning of crowns of trees but from the burning of shrubs, herbs and dead trees and litter (Fig. 2). A prerequisite of the formation of large quantities of macroscopic charcoal is a sufficient build-up of litter. In this case, surface fires will burn and char the litter (Fig. 2a). Only the smaller charcoal fragments, which may include small charcoalified twigs, leaves and flowers may be wind blown (Fig. 2g). Most of the macroscopic charcoal is transported first by overland flow (Figs. 8a, b) following rainstorms after the fire, then as suspended load sediment in rivers and finally by bed load transport after becoming waterlogged.

Experiments have shown that larger charcoal fragments take longer to waterlog (Nichols et al. 2000) (Fig. 9d) and hence may be transported further. Charcoal may be transported into the ocean and be deposited in a range of marine environments. It is unlikely that all occurrences of larger charcoal fragments indicate local fires as has been claimed by some authors (Clark and Patterson 1997).

A third source of charcoal is from the burning of peat (Cohen 1974). However, there have been few instances where this has been shown to be significant in the fossil record despite such material being petrographically characteristic (Peterson 1998).

Whilst botanical identification of microscopic charcoals <63 μm is not generally possible, it is often possible to identify the botanical origin of larger charcoals and hence their greater significance in fossil sediments.

Transport and Deposition of Charcoal

Charcoal may be transported by either wind or water or a combination of these and deposited in a wide range of environments from lakes and mires to rivers, estuaries and into the ocean. Charcoal may be transported during the initial phases of a fire mainly in smoke and by wind and post-fire by wind or water. There have been extensive studies, from both natural and experimental burns, that have tracked the movement of microscopic charcoal by wind (Clark and Patterson 1997) and these and other authors show that the particle size of windblown charcoal decreases away from the fire. It has also been shown (Clark et al. 1997, Lynch et al. 2004) that little macroscopic charcoal is moved by wind (Ohlson and Tryerud 2000). These observations have led to the view that the record of microscopic charcoal provides data on regional or background fires whereas macroscopic charcoal provides evidence of local fires (see Clark and Patterson 1997). However, observational data from modern wildfires, from laboratory experiments and from the distribution of charcoal in ancient sediments suggests that this is not the case and that macroscopic charcoals may be widely transported by water and deposited into most depositional settings (Nichols et al. 2000, Scott 2000a, Scott et al. 2000b).

Considerable amounts of macroscopic charcoals may be formed from surface fires charring plant litter and shrub or herbaceous vegetation (Fig. 2; Scott et al. 2000b). Some of this macroscopic charcoal may be selected and concentrated by wind after the fire has passed (e.g., charcoalified flowers concentrating in wind ripples; Fig. 2g; Scott et al. 2000a) most is moved initially be water often in a slurry by overland flow (Fig. 8a, b). Large quantities of charcoal and sediment may move in this way and eventually will flow into moving or standing water. In some cases, this initial movement will cause a sorting of the charcoal by size and origin (Scott et al. 2000b). Once in water charcoal will remain buoyant. Laboratory experiments have shown that smaller charcoal particles sink more quickly than larger particles which would allow larger particles to be transported further (Fig. 9). There is also a difference in the hydrodynamic properties of wood charcoals formed at different temperatures (Fig. 10; Vaughn and Nichols 1995) and of charcoals formed by different plant organs (Fig. 10; Nichols et al. 2000). This work indicates that charcoal may be transported considerable distances in suspension and help explain their abundance in some near-shore marine sediments in the fossil record (Nichols and Jones 1992, Falcon-Lang 1998, Scott 2000a).

Once waterlogged, charcoal may be transported as part of a bedload (Fig. 11; Nichols et al. 2000). Despite being fragile, tumbler experiments have shown that the charcoal may be moved many tens of kilometers without appreciable damage. However, to be incorporated into bedload sediments it has been shown by flume-tank experiments that only moderate currents will permit this to happen (Fig. 11; Nichols et al. 2000). Often, however, the charcoal may be transported into a lake, lagoon, onto a floodplain or into a mire and settle out of suspension.

Only by analyzing the botanical origin of a charcoal assemblage can an assessment be made as to what vegetation it has come from. Rather than microscopic charcoal being used to indicate regional fires and macroscopic charcoals to represent local fires, a mixture of microscopic and macroscopic charcoal indicates local fires, whereas well sorted macroscopic charcoal assemblages may imply distance transport from the fire site (see Collinson et al. 2007).

CAN WE RECOGNIZE POST-FIRE EROSION IN THE FOSSIL RECORD?

A major feature of modern fire is that of post-fire erosion, as indicated by chapters in this volume (see also Moody and Martin 2001, Shakesby and Doerr 2006). However, the nature of the fossil record is that it records the history of sedimentation in depositional settings rather than in erosive upland settings. To identify the occurrence of post-fire erosion/deposition cycles we need evidence firstly of a fire, from the occurrence of abundant charcoal and an unusual sediment in terms of thickness or grain size. This phenomenon has until now rarely been recognized in deep time. Indeed, fire as a whole, and its impact, has been rarely considered by sedimentologists working on ancient sediments.

However, a number of deposits may be identified as being related to post-fire erosion. In the Carboniferous, coastal sediments contain abundant charcoal with coarse angular sand grains, which suggest rapid deposition (Nichols and Jones 1992, Falcon-Lang 1998) and were almost certainly the result of post-fire erosion/deposition following major wildfires (Fig. 8a). Evidence is also found within Carboniferous and Permian peats (coals), where thick and sudden influxes of sediment and charcoal is common. This can be seen both in coals but also in premineralized peats. Finally, a number of coarse sand deposits with charcoal from the Jurassic and Cretaceous may be formed in a similar manner (Scott 2000a, Collinson et al. 2000, Jones 1997).

Post-fire erosion depositional systems in deep time may be recognized by: a sharp or erosionally-based sedimentary unit; a thicker than normal sediment-rich horizon; a coarser sediment; immature sediments; charcoal with a range of clast sizes (possibly with a range of origins) and commonly with non-charred plants in addition. However, it is only a combination of charcoal and sediment characteristics that may indicate that a deposit is derived from post-fire erosion.

QUATERNARY AND PRE-QUATERNARY METHODOLOGIES

There has been a significant dichotomy of methodologies for those studying the history of fire in the recent past and for those researching fire regimes in deep time. Most papers on recent and Holocene fire histories use the microscopic charcoal record which may be linked to fire scar data, varve data or historical records (Bradbury 1996, Clark and Patterson 1997, Edwards and Whittington 2000). However, the questions being addressed for those interested in the deep time record may be different. In such cases it is the larger or macroscopic charcoal record that is used, as it provides evidence of the vegetation, which was burned as well as data on fire temperature (Scott 2000a, 2001).

The Use of Charcoal Size Categories

Fire may produce a wide range of charcoal particles from around 1 cm to less than 1 µm (Clark and Patterson 1997). Many smaller particles less than 125 µm are regularly wind blown although some may also be water transported (Fig. 1). Small particles may be prepared as part of palynological preparations and particles can be counted at the same time as pollen and spores. Microscopic charcoal counts through lake or peat sequences have been used to interpret fire history (Campbell and Campbell 2000). Reworking of microscopic charcoals may cause some problems (Bradbury 1996, Edwards and Whittington 2000).

Some authors prefer chemical methods to identify charcoal concentrations (Winkler 1985) but to what extent all the charcoal is measured as opposed to high temperature charcoals alone may be debated (see Schmidt and Noack 2003, Forbes et al. 2006, Scott and Glasspool 2007).

Larger charcoal fragments have often been studied by those interested in fire in deep time. There are several reasons for this: the ability to recognize charcoal (Figs. 2 and 4; Scott 1989); the anatomical preservation hence allowing the identity of the plant to be ascertained (Figs. 3 and 5; Scott 2001) and the ability to derive reflectance data for fire temperature interpretations (Fig. 7; Scott and Jones 1994, Scott and Glasspool 2005, McParland et al. 2007).

WHAT CAN WE LEARN FROM FOSSIL CHARCOAL?

The excellent preservation of charcoalified plants (Figs. 2 and 4) allows a wide range of morphological and anatomical data to be derived allowing not only organ and species identification but also information concerning growth and ecology to be gained (Scott 2001). However, plants may be exposed to a wide range of temperatures during charring and experiments have allowed relationships between physical and chemical characteristics caused by the charring process to be identified (Scott 1989, Cope 1980, 1981, Scott and Jones 1991, Jones et al. 1991, Jones 1993).

Experimental Charcoalification Experiments

In the fossil record, charcoal has been called fusain (Scott 1989). Some researchers doubted that fusain represented charcoal and was formed by fire and some researchers persist in that view (see Scott 1989, 2002, Scott and Glasspool 2007). Much effort has gone into demonstrating the similarities, both chemical and physical, between charcoal and fusain, and experimental charcoalification has played a major role in this discussion (see Cope 1980, 1981, Jones et al. 1991, Scott and Jones 1991, Bustin and Guo 1998, Guo and Bustin 1999, Scott 2000, Scott and Glasspool 2005, 2007, McParland et al. 2007).

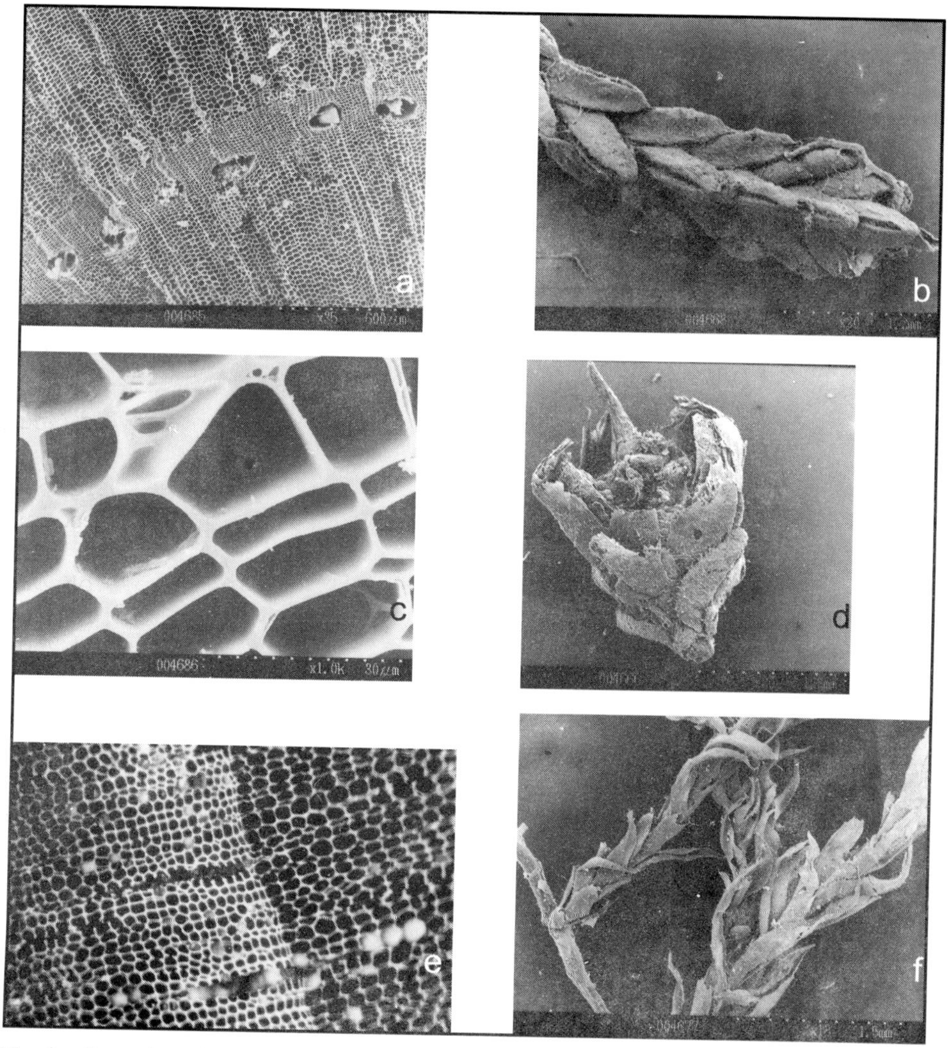

Fig. 3 Scanning electron microscopy of modern charcoalified plants – charcoal from the 1995 Frensham Fire, Surrey, UK (after Scott et al. 2000b): a) wood of *Pinus*; b) leaves of *Calluna*; c) detail of *Pinus* wood showing fusion of the middle lamella; d) *Calluna* flower; e) reflected light photo of charred *Pinus*; f) charred moss.

Controlled charring experiments have shown that the reflectance of charcoal as seen in polished blocks under oil (measured at a wavelength of 546 nm) increases with increasing temperature (Fig. 7b). This gave rise to the idea that charcoal reflectance might be used to interpret fire temperature (Jones et al. 1991, Scott and Jones 1994). Additional studies (Guo and Bustin 1998, Bustin and Guo 1999, Scott and Glasspool 2005, 2007) have shown that time also played a role. However, it was also shown in these studies that

Fig. 4 Fossil charcoal assemblages: a) wood charcoal in fluvial Jurassic sandstone, Yorkshire, UK; b) macerated limestone with charcoal fragments and uncharred megaspores from the Mississippian of Scotland; c) small charcoal fragments from a Palaeocene silty-mudstone, Colorado, USA; d) detail of c showing charcoalified wood fragment; e) maceration of Upper Cretaceous clastic sediment yielding charcoal from Japan; f) maceration from Upper Cretaceous muddy siltstone yielding charcoalified flowers, Sweden.

reflectance stabilizes after a few hours (Fig. 7a) and thus if a piece of wood charcoal had a reflectance of 2.0 R_o then it probably experienced temperatures over 400°C (Fig. 7a). This provides the idea of a minimum temperature estimate of charcoal formation and provides some idea of likely fire temperatures (Fig. 7b). Reflectance data appears to correlate with known fire temperatures (Scott and Jones 1994, Scott et al. 2000b).

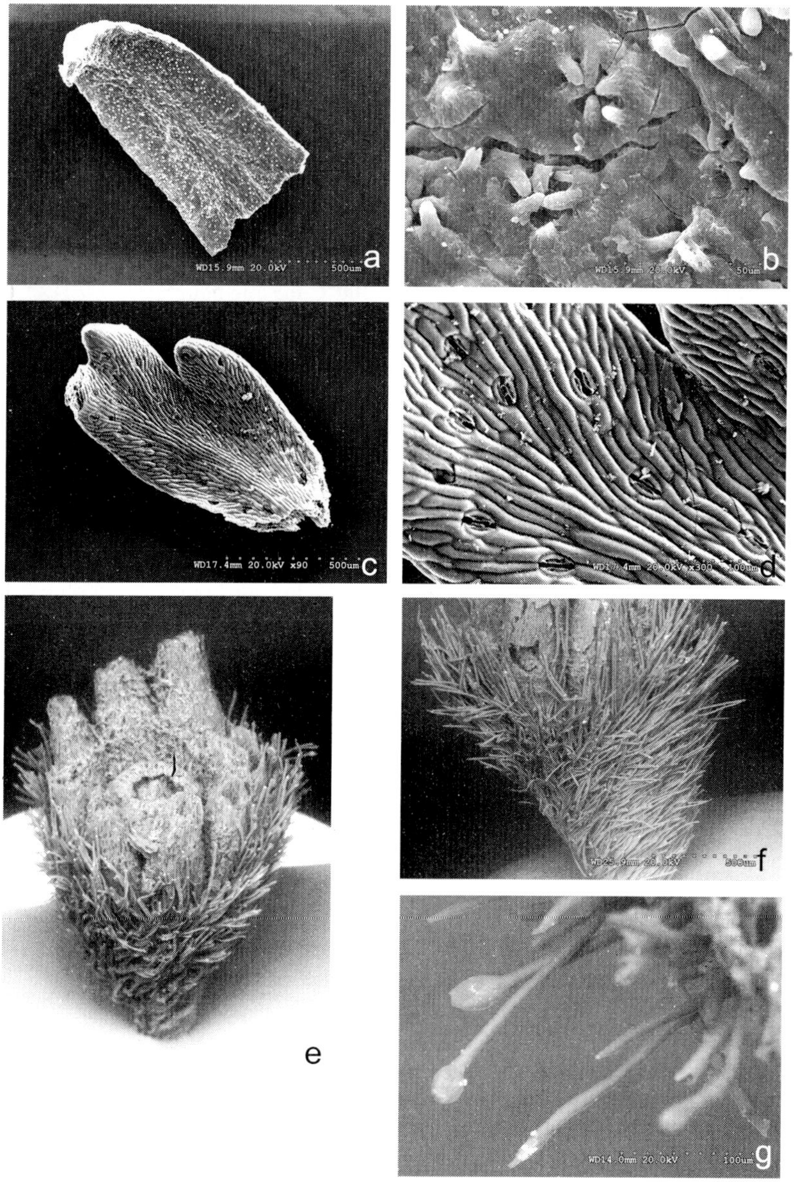

Fig. 5 Scanning electron microscopy of fossil charcoalified plants: a) charcoalified conifer from the early Pennsylvanian, Illinois, USA; b) detail showing stomata with over-arching papillae; c) charcoalified pteridosperm leaf from the late Mississippian of Scotland; d) detail of c showing the leaf surface with stomata; e-g) charcoalified ovule from the Mississippian of Scotland; e) whole ovule showing ovule lobes and glandular hairs; f) showing spirally arranged glandular hairs; g) detail of glandular hairs.

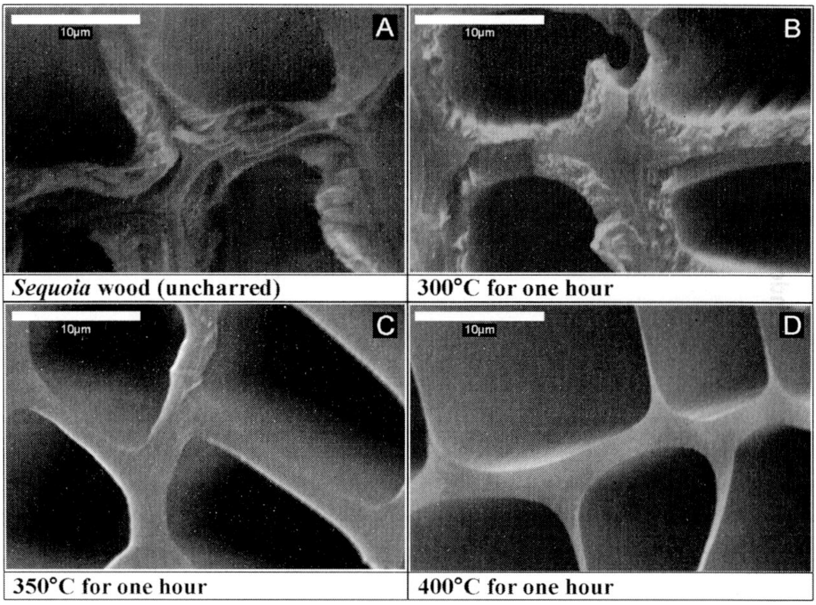

Fig. 6 Charring of *Sequoia* wood showing the homogenization of the cell wall: a) uncharred showing middle lamella; b) 350°C showing presence of middle lamella; c) 350°C showing the homogenization of the cell wall; d) 400°C showing the homogenization of the cell wall (from McParland et al. 2007).

New research on a variety of plant and other materials shows that this increase in reflectance with temperature is common. For example, in addition to wood, fern tissues and fungal material also show this pattern (Scott and Glasspool 2007, McParland et al. 2007). The use of charcoal reflectance as a fire temperature proxy has yet to be fully exploited.

Charcoal and the Carbon Cycle

A significant problem is any consideration of charcoal and the carbon cycle comes from the diversity of combustion products and residues from vegetation fires (Jones et al. 1997, Bird 1997, Schmidt and Noak 2000, Forbes et al. 2006) and a general lack of agreed nomenclature and methods of study. These authors suggest a good starting point into the literature and only some brief comments will be provided.

In many marine sediments black/elemental carbon has been documented (Goldberg 1985, Smith et al. 1973) and this has also been found in more ancient deep water sediments (Herring 1985). However, some particles can be identified microscopically, as small charcoal fragments based upon characteristic morphology (Griffin and Goldberg 1979, Goldberg 1985), others may represent soot or other carbonaceous particles.

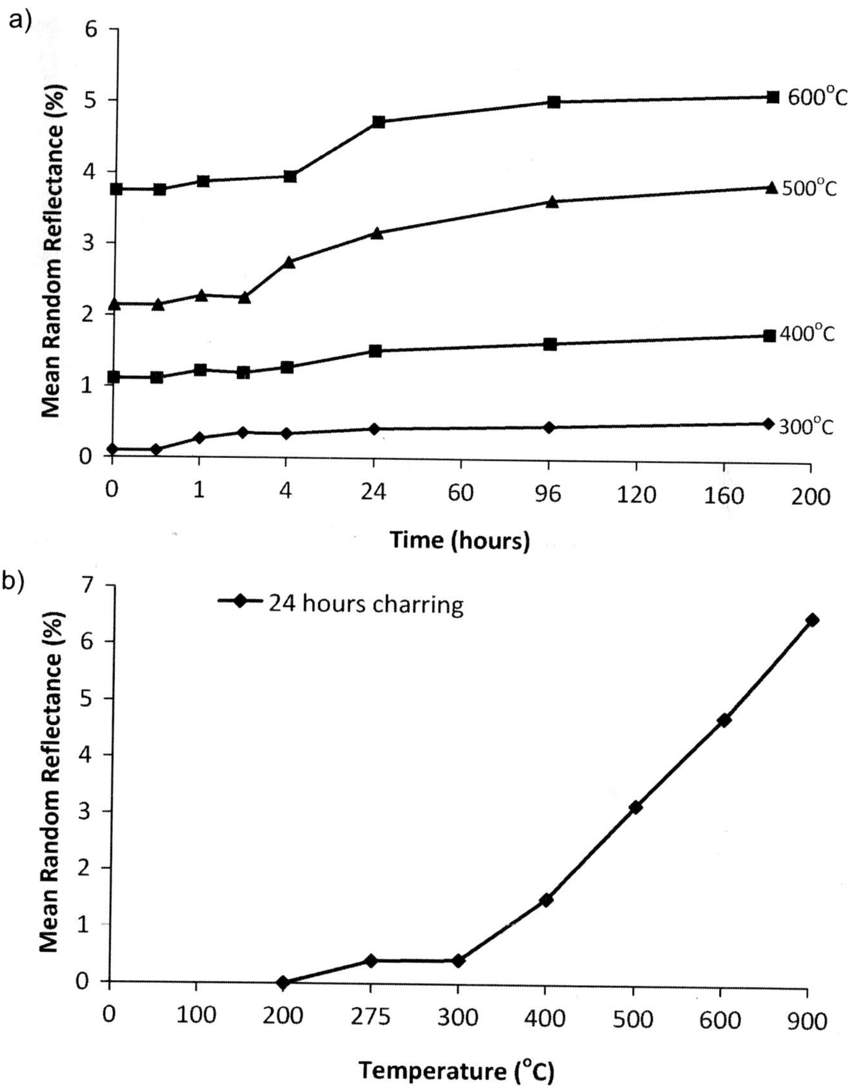

Fig. 7 a) Reflectance of experimentally charred *Sequoia* wood charred for different temperatures and times (after Scott and Glasspool 2007); b) Reflectance of experimentally charred *Sequoia* at different temperatures for 24 hours. This data can give the minimum fire temperatures for wildfire charcoals.

Analysis of the fossil record does show that large quantities of charcoal may be buried in a wide range of sediments including in peat mires and as it is relatively inert (not necessarily pure carbon) it survives and hence contributes to the carbon sink.

Extensive fires may generate considerable CO_2 from biomass burning but in the long term it is the conversion of wood to charcoal, which provides a mechanism for increased carbon burial. As will be discussed later, this has implications for both long-term atmospheric composition and climate change.

Soot, Carbon, PAH – How Reliable as Indicators of Ancient Forest Fires?

If there is debate about modern fire combustion products and residues and their characterization, there is also uncertainty for some in identifying such materials in deep time. In many cases, rocks need to be dissolved to release their organic materials and the separation of the different organic fractions with different origins may be problematic. Four main categories of combustion products and residues have been identified in sediments from deep time.

The most easily extractable and recognizable of the combustion products is charcoal, often called fusain in ancient sedimentary rocks. This is easily released by simple sieving (often using a 200 μm sieve; Figs. 4b, e, f). However, occurrences in shales or siltstones/sandstones (Figs. 4a, c, d and 8c, d, e) involve the use of hydrofluoric acid (HF), for limestones the use of diluted hydrochloric acid (HCl) and coals the use of nitric acid (HNO_3) (see Pearson and Scott 1999 for techniques of maceration). Many charcoal fragments are easily identifiable, some specimens may be only partially charred (Jones et al. 1993). The occurrence of macroscopic charcoal in ancient sediments is a reliable indicator of wildfire. Only with the analysis of the plant organ and identity is it possible to identify a forest fire, for example.

Charcoal is very brittle and may be easily crushed upon burial in sediment. Micro-charcoal is more problematic in that it may represent wind-blown fire charcoal from a contemporary fire, waterborne charcoal and in some cases reworked charcoal. Micro-charcoal has been widely used in Quaternary studies (Patterson et al. 1987, Clark and Patterson 1997) but it may be more difficult to identify in some more ancient sediments (Highton et al. 1991). This is because organic material becomes coalified and blacker during burial diagenesis. Separating on a palynology slide, black coaly particles from black charcoal particles may be very difficult, especially if they are small (<63 μm).

Soot is very difficult to extract from sediments and identify. It is different from charcoal in that it is a product formed by combustion and comprises small carbon particles. When dissolving sediments to study the organic matter it is possible to generate organic precipitates which may be difficult to distinguish from soot (Belcher et al. 2005). Soot has been reported in some sediments (Wolbach et al. 1990), it has not been widely reported and it may, in

Fig. 8 Transport and deposition of charcoal: a, b) from a forest fire by overland flow, California, USA (Photo D. Neary); c-d) fossil sediments yielding charcoal; c) Mississippian near-shore marine sediments with abundant charcoal from the Mississippian of Donegal, Ireland; d) sandy siltstones with charcoalified conifer needles from the early Pennsylvanian, Illinois, USA; e) silty mudstones with abundant charred ferns, Lower Cretaceous of the Isle of Wight, UK.

addition, be derived from a number of sources – biomass burning as well as from natural coal and oil fires (Harvey et al. 2008).

Polyaromatic hydrocarbons (PAH) have only recently been used to interpret fire signals (Killops and Massoud 1992, Finklestein et al. 2005) in the fossil record. It is not entirely clear if such compounds may migrate in sediments or could be formed by some diagenetic processes but some characteristic compounds may be useful as a fire signal in the absence of charcoal, but its use is strengthened when combined with charcoal occurrence (Finklestein et al. 2005).

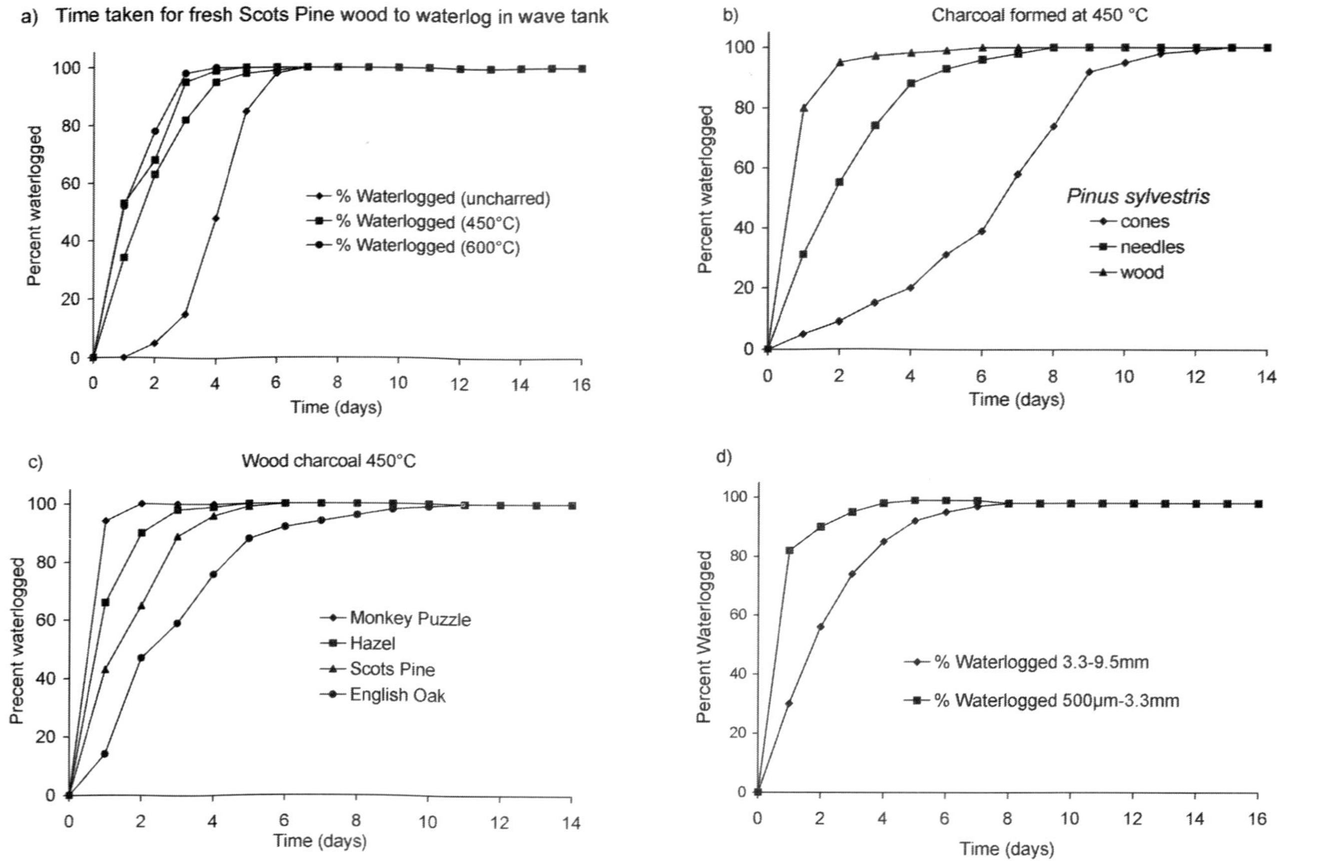

Fig. 9 Waterlogging rates for different charcoalified plants and plant organs (after Scott 2000a): a) *Pinus* charred at different temperatures; b) different organs of charred *Pinus*; c) different charred wood species; d) different size classes of wood charcoal.

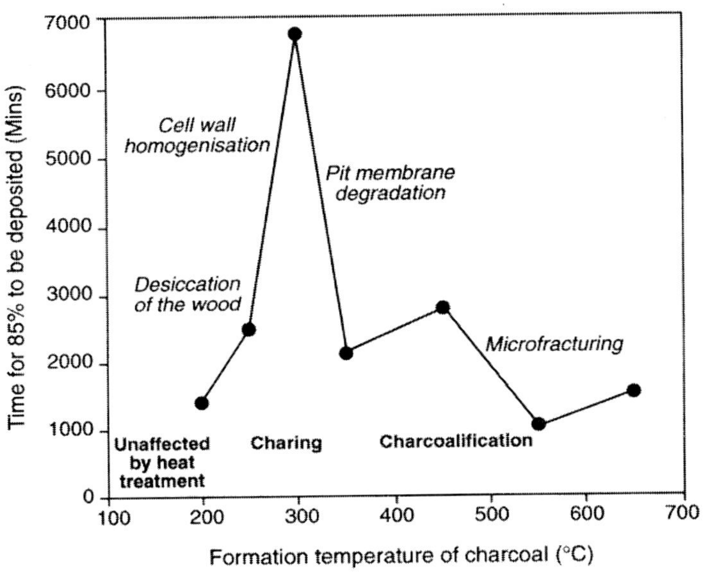

Fig. 10 The relationship between temperature of charcoal formation and time taken to sink (after Vaughn and Nichols 1995).

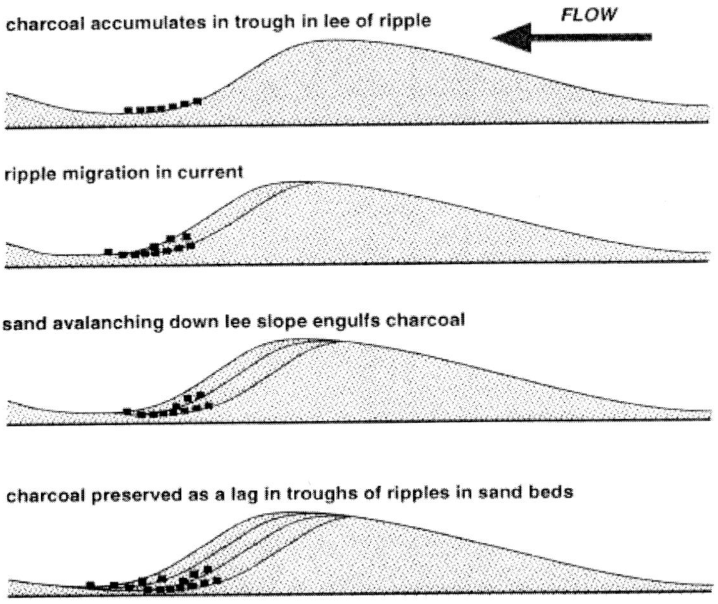

Fig. 11 Experimental flume results incorporating charcoal into bedload sands (after Nichols et al. 2000).

FIRE AND CLIMATE

There is a strong link between fire and climate (Terasmae and Weeks 1979, Fosberg et al. 1993, Carcaillet et al. 2001) and this interest and relationship has been heightened with discussion concerning global warming (Page et al. 2002, Westerling et al. 2006). It is clear that climate is likely to be a major driver of fire, especially catastrophic fire. In areas where there have been few fires, fuel build-up may occur. If there are increased periods of drought then a fire may spread, burn very hot and prove catastrophic both to the vegetation and also may lead to major post-fire erosion (Robichaud and Elsenbeer 2001).

Unfortunately, since it is possible to interpret fire and climate on a decadal or even millennial scale, understanding the relationship on a million year timescale is much more difficult. In addition, there is the added complication of atmospheric oxygen changes that are likely to affect fire intensity and fire frequency. We might assume that fires would be less frequent during cooler and wetter periods in Earth's history, and there has been an attempt to examine this on a tens of millions of years time scale. It is possible to note that climate may have an effect on fire systems at some intervals in Earth's history. It has been observed that fires were common through the Paleocene (65 to 55 my). Charcoal beds are common, for example, in southern England (Collinson et al. 2007). However at the Paleocene-Eocene boundary, there was a period of rapid global warming which had a major effect upon terrestrial systems (Gingerich 2006). An examination of the distribution of Eocene charcoal indicates that fires were not common (Scott 2000a). In the section in southern England, which crosses the Paleocene-Eocene boundary (55.8 my), charcoal becomes rare or absent after the onset of the Palaeocene-Eocene Thermal Maximum (PETM), and Collinson et al. (2007) and Steart et al. (2007) argue that this is caused by a significant change in rainfall.

A much more comprehensive documentation of charcoal records in deep time are needed before a full analysis of the long-term relationship between fire and climate can be addressed.

FIRE AND ATMOSPHERE

The interaction between fire and the atmosphere is widely recognized. However, the importance of fire in affecting atmospheric composition over geological time has only recently been discussed (e.g., Lenton 2003, Berner et al. 2003).

Oxygen

While charcoal forms in the absence of oxygen, fire can only be sustained in its presence. It has been demonstrated that a fire cannot be sustained when

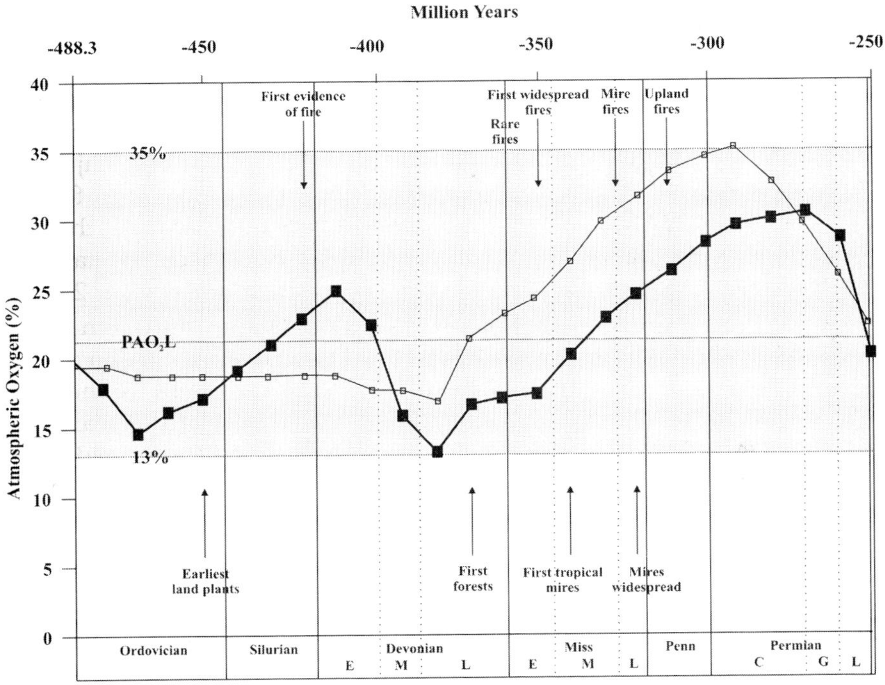

Fig. 12 The occurrence of charcoal and interpreted fire systems in relation to modelled atmospheric oxygen for the Paleozoic (after Scott and Glasspool 2006a).

there is less than 13 percent atmospheric oxygen (today's value is 21 percent – PAL, the present atmospheric level) (Chaloner 1989, Cope and Chaloner 1980, 1981). Experiments have indicated that with increasing oxygen, wetter plants may burn (Watson et al. 1978, Wildman et al. 2004) and an upper threshold of 35 percent oxygen provides the basis of a 'fire window' (Jones and Chaloner 1991). A range of experiments suggests that below 13 percent O_2 no fire would spread and there would be an absence of charcoal in the fossil record, between 13 to 16 percent fires would be rare and only very dry plant material would burn. This would mean that only vegetation, which was liable to dry out, would burn. Between 18 to 23 percent fire occurrences would be similar to those under the present atmospheric level of 21 percent O_2. At >25 percent O_2 fires would become more widespread and at >30 percent fire would be frequent in all environments (Watson et al. 1978, Wildman et al. 2004, Scott and Glasspool 2006).

All major models of atmospheric oxygen show variation through Phanerozoic time, especially in the last 400 million years (Berner et al. 2003, Lenton 2003, Berner 2006). Most models agree with a period of high oxygen during the Carboniferous from 350 to 300 million years. Scott and Glasspool (2006) have shown that this rise in oxygen may be seen in the charcoal record whereas oxygen rose fires became more widespread and common in all environments and even thick Permian coals may be composed of over 70 percent charcoal.

Modelling suggests a fall in oxygen at the end of the Permian around 250 million years ago to below the PAL in the Triassic (Berner 2006). However, models that suggest very low oxygen levels in the Jurassic, below 13 percent, are not in agreement with the charcoal record. The Cretaceous (120 to 65 my) is of particular importance. It is during this period that flowering plants evolved. Many flowers were found to be preserved as charcoal (Friis et al. 2006). Numerous models have indicated that oxygen levels were elevated above PAL during this period (Berner et al. 2003, Lenton 2003). However, a recent model (Berner 2006) suggests an oxygen level of <PAL. The abundance of charcoal through the Cretaceous is a wide range of environments (Scott 2000a, Collinson et al. 2000, Scott and Stea 2002, Falcon-Lang et al. 2001) suggest a higher level. It is clear, however, that there is significant feedback between atmospheric oxygen and fire (Fig. 13)

Carbon Dioxide

There is an interesting relationship between fire and atmospheric CO_2. It has been recently suggested that there is a direct link (Carcaillet et al. 2002), while

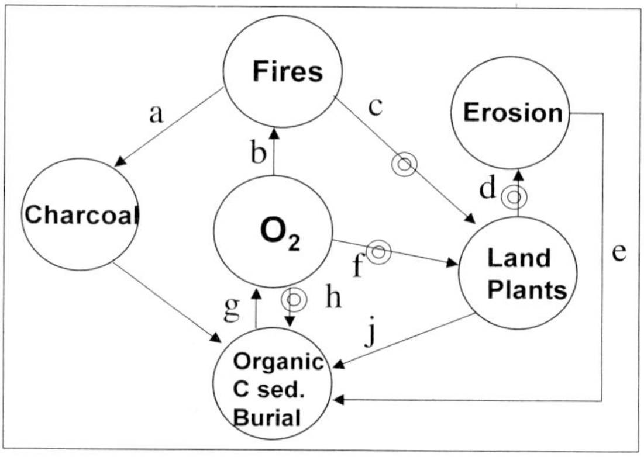

Fig. 13 Systems analysis showing the feedbacks between fire and atmospheric oxygen.

Arrows originate with causes and end at effects. Plain arrows indicate direct responses, and arrows marked with bull's-eyes show inverse responses. Closed loops with an even number of bull's-eyed arrows or solely plain arrows are positive feedbacks, and those with an odd number of arrows with bull's-eyes are negative feedbacks. Straight arrows lead to a positive response (e.g., oxygen increases, fires increase) and arrows with bull's-eyes are negative responses (e.g., fires increase, vegetation decreases). A closed loop with an odd number of bull's-eyes leads to negative feedback and stability. An even number or no bull's-eyes leads to positive feedback and enhancement (but not always destabilization as can be shown mathematically) (after Berner et al. 2003).

systems analysis indicates a strong fire feedback and that CO_2 levels may be at least indirectly linked to fire (Lenton 2003, Berner et al. 2003). Berner et al. (2003) in their systems analysis (Fig. 13) show that with increasing oxygen there is increasing fire, which in turn causes a decrease in land biomass (the path b-c-j-g in Fig. 13). They suggest this leads to less organic matter burial, less oxygen production and thus is a negative feedback. They also point out that fire may also lead to a positive feedback with regard to atmospheric oxygen exemplified with the production of charcoal (Berner et al. 2003). As charcoal is inert and tends not to biodegrade its burial leads to organic matter preservation, which in turn leads to increased oxygen production, a positive feedback represented in the path b-a-i-g (Fig. 13). In addition, increased burning may give rise to increased erosion and more transport of sediment and charcoal to the sea, which enhances organic matter burial and raises oxygen levels (path b-c-d-e-g in Fig. 13).

While burial of carbon may lead to a rise in oxygen it may also lead to a fall in carbon dioxide level. This is well illustrated by models of CO_2 through time, which shows in the late Paleozoic there is a rapid fall in CO_2 levels as O_2 levels rise. This fall in CO_2 provides the stimulus for the onset of the late Paleozoic ice age.

FIRE IN DIFFERENT ECOLOGICAL AND GEOGRAPHIC SETTINGS

In today's world, fire systems vary in relation to vegetation type, climate, latitude and geomorphic setting. However, fire is an integral part of many terrestrial ecosystems, from the northern coniferous forests to the equatorial mires and including grasslands and forests (Pyne et al. 1996, Agee 1990, Cohen 1984, Crutzen and Andrae 1990, Johnson 1984, Sandford et al. 1985, Stocks and Kauffman 1997).

In the fossil record, evidence of fire has been found in most sedimentary environments from temperate to tropical areas and from uplands to lowland mires (Scott 2000a). Charcoal derived from upland fires have been reported from the Carboniferous (Scott 1974, Scott and Chaloner 1983, Falcon-Lang and Scott 2000) and from the Triassic/Jurassic (Harris 1958). Charcoal occurs commonly in lowland fluvial sediments from bed-load sandstones or the Jurassic and Cretaceous (Cope 1993, Scott 2000a, Collinson et al. 2000) to overbank floodplain fires of the Carboniferous (Scott 1978) and Cretaceous (Scott and Stea 2002, Friis et al. 1988). Charcoal is particularly common in peat deposits (coals) and occur abundantly in all ages (Scott 2000a), but are particularly abundant in the Carboniferous, Permian and Cretaceous (Scott and Glasspool 2006a, Scott and Stea 2002).

Charcoal also gets washed into both shallow and deep marine sediments. Extensive seashore marine charcoal beds have been reported for the Carboniferous (Nichols and Jones 1992, Falcon-Lang 1998) and have also been reported as a significant component of late Devonian marine black shales

(Rimmer et al. 2004). There has been a significant reporting of charcoal deposits since the year 2000 when a volume of fire and the paleoenvironment was published (Scott et al. 2000a) and continued reporting of charcoal in deep time will allow the unravelling of ancient fire systems more completely than is currently now possible.

THE GEOLOGICAL HISTORY OF FIRE

The First Fires

The two basic criteria for the occurrence of fire in the fossil record is the evolution of a land vegetation and sufficient atmospheric oxygen for fire to be sustained and spread (Fig. 14). In addition, there also needs to be a build-up of fuel. Vascular plants first evolved on land probably in the Silurian (430 my) (Scott 2000b). These plants were relatively small being no more than 10 cm tall (Edwards and Wellman 2001). We have evidence of charcoalified plants from the latest Silurian (400 my) (Glasspool et al. 2004) but records are scarce. Charcoalified plants have also been reported from the earliest Devonian (Edwards and Axe 2004), so small wildfires must have occurred at that time.

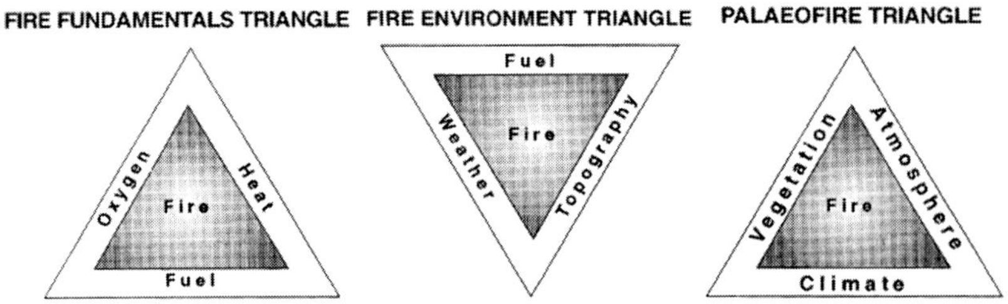

Fig. 14 Fire fundamentals triangles: fire fundamentals triangle (after Pyne et al. 1996); fire environment triangle (after Pyne et al. 1996); paleofire triangle (after Scott 2000a).

While land plants continued to evolve and diversify and become larger (Kenrick and Crane 1997) through the Devonian, we have little evidence of fire from the fossil record of charcoal (the charcoal gap; Scott and Glasspool 2006). This may be because atmospheric oxygen levels fell at this time (Scott and Glasspool 2006).

Evolution of Widespread Fire Systems

Trees first evolved during the middle of the Devonian period (380 my) and by the end of the Devonian there were extensive coastal forests dominated by the tree *Archaeopteris/Callixlon* (Meyer-Berthaud et al. 1999). Despite this, we have little evidence of extensive forest fires. Modelled atmospheric oxygen is shown to increase in the late Devonian so that fires may be sustained (Fig. 12; Berner 2006, Scott and Glasspool 2006).

We have only scattered records of charcoal in late Devonian terrestrial sediments and the vegetation being burned appears to be dominated by fern-like shrubby plants rather than trees indicating the occurrence of surface fires. Microscopic charcoal from such fires may be transported by wind into the ocean. Petrographic studies of late Devonian black shales (Rimmer et al. 2004) have shown that there is a rapid rise in the concentration of charcoal in these sediments at the end of the Devonian which indicates a rise in the spread and frequency at this time (Scott et al. 2006, Scott 2006) and this may reflect rising atmospheric oxygen levels (Scott and Glasspool 2006).

By the Devonian/Carboniferous boundary extensive charcoal deposits are recognized and the sedimentary environments in which charcoal is found and the vegetation being burned suggests the spread of fires into a wide range of environments through the Mississippian (350 to 320 my) (Scott and Glasspool 2006). Extensive charcoal beds are found throughout the Mississippian and evidence from the charred plants and sediments suggests (Fig. 8c) very widespread fires that had a significant impact on the environment (Nichols and Jones 1982, Falcon-Lang 1998, 2000, Scott 2000a).

Diversification of Fire Systems

It has been suggested from atmospheric models that oxygen concentrations rose through the Carboniferous and Permian to significantly above the present atmospheric level (21 percent) to 30 to 35 percent (Fig. 12; Berner et al. 2003, Berner 2006), and at these levels fires may be maintained in much wetter vegetation (Wildman et al. 2004). Wetland peat-forming mires also evolved during the late Mississippian and at this time evidence of fires are found abundantly in coals. Charcoal, which represents the inertinite group of macerals (Scott and Glasspool 2007), is common throughout the Pennsylvanian and Permian (Scott and Glasspool 2006) in coal seams to indicate frequent fires in wetland mire systems (Scott and Jones 1994). A high level of atmospheric oxygen may have allowed wetter vegetation to burn (Scott and Glasspool 2006).

Evidence of fires in other lowland ecosystems is shown by the presence of charcoal in floodplain shales (Scott and Jones 1994). In addition, charcoal is often found in fluvial sandstones which have drained burned upland vegetation (Falcon-Lang and Scott 2000) and we have evidence of charcoals from early Pennsylvanian cave deposits of fires through conifer and cordaite vegetation living in drier upland habitats (Figs. 5a, b and 8d; Plotnick et al. 2008).

While evidence of widespread fire is found throughout the late Palaeozoic (see Scott 2000a, Scott and Glasspool 2006) there is much less evidence of widespread fire systems in the succeeding Triassic. It is possibly related to a major fall in atmospheric oxygen at this time (Berner 2006).

In the Mesozoic charcoal is abundant in a wide variety of sedimentary systems. Charcoal becomes frequent in the Middle Jurassic (Fig. 4a) and there is evidence of fires related to drier intervals (Cope 1993, Morgans et al. 1999). Charcoal is especially abundant in the Cretaceous (Fig. 8e; Collinson et al. 2000, Scott 2000a, Scott and Stea 2002). It is during this period that flowering plants (angiosperms) first evolved. Many of the early flowering plants were small herbaceous herbs and often their flowers are preserved as charcoal (Friis et al. 2006). It is possible that regular fires through angiosperm dominated vegetation aided in the diversification and success of this group of plants.

Fires were abundant throughout the Cretaceous and the earliest Tertiary but claims of a global wildfire at the Cretaceous-Tertiary boundary (Wolbach et al. 1990) have not been supported from charcoal studies (Belcher et al. 2005).

A major innovation in vegetation biomes was the development of grasslands, in particular savannas dominated by C4 grasses. Bond and Keeley (2005) have suggested that it is fire that maintains this savanna. This would imply a long interaction between fire and grasslands. Further, Osborne and Beerling (2006) have suggested that fire was the stimulating force in the evolution of C4 grasslands.

There is no question of the importance of fire in shaping not only the biosphere but also the significance of fire feedbacks in the evolution of the Earth system as a whole. All of this happened before man appeared on the planet.

IMPACT OF FIRE ON THE ENVIRONMENT

The impact of fire upon the environment is well understood (Crutzen and Goldammer 1993, Rundel 1981, Cypert 1973). However, less is known in a fossil context and many issues that form the basis for intense discussion concerning modern fire systems have not been addressed. In addition, time adds an extra important dimension.

Four areas that are of current concern are discussed for those studying ancient fire systems: impact on plant evolution; impact on fossil preservation; impact on plants; and post-fire erosion.

Impact of Fire on Plant Evolution

While it may be accepted that fire may play a role in shaping plant communities – giving rise to fire-prone vegetation, there has been less consideration of how fire may have helped in plant evolution/diversification. Cause and effect is, however, very difficult to unravel in the fossil record, thus a few observations and suggestions may be made at this time, awaiting further study.

Many early ferns (late Devonian/Early Carboniferous) are often found as charcoal (Scott and Galtier 1985) and burned ferns are frequently encountered in periods such as the Cretaceous (Harris 1981). This has led to idea of fern 'prairies' being maintained by fire (Collinson et al. 2000). However, as ferns may survive fire and regenerate very quickly, it is possible that fire regimes of the early Carboniferous encouraged their spread and diversification.

Likewise, many early conifers are also preserved as charcoal (Scott 1974, Scott and Chaloner 1983, Scott 2000a) and fire may have played a role in their diversification, as they were some of the first plants to have lived in drier uplands.

Most records of early angiosperm flowers in the Cretaceous come from charcoal deposits. Most of these appear to come from small herbaceous forms and fire may have maintained an open habitat in which they could thrive.

Fire has also been invoked as a mechanism to create the savannah biome and in the evolution of C4 grasses (Osborn and Beerling 2006,) but more research is needed to prove these hypotheses.

Impact of Fire on Fossil Preservation

It is intuitive to believe that fire is a destructive force rather than a preservational mechanism. However, the rapid heating of pyrolysis caused by a fire, and conversion of plant tissues to charcoal, has the effect of preserving both the morphology and anatomy of the plants as well as converting cell walls to carbon-rich and non-biodegradable materials (Scott 2001). The charcoalification process may cause some shrinking of plant tissues (Prior and Alvin 1981, Lupia 1995) but the overall relative dimension of the plants are often not changed. Other effects include the homogenization of the cell walls with the loss of the middle lamella but this does not affect botanical anatomical data (Jones 1993).

Studies using scanning electron microscopy on charcoalified plants following recent fires, show that a wide range of plants and plant organs may be beautifully preserved such as wood, leaves, flowers, and seeds (Figs. 3 and 5). Flowers may show exquisite preservation (Fig. 3d) and despite being fragile may be both wind and water transported (Scott et al. 2000b). Likewise in the fossil record a wide variety of charcoalified plant organs have been reported, not only flowers but even ovules with glandular hairs (Figs. 5e, f, g). Often charcoalified plants look fragmentary, black, and uninteresting (Figs. 4c, d), but initial observations under a hand lens and subsequent microscopical studies using scanning electron microscopy reveal a wealth of important data that may be used to unravel plant history, ecology, and evolution.

Impact of Fire on Plants

There are many vegetation types, for example in Australia and Africa that have evolved in response to fire. In some cases plants will not reproduce without fire. This leads to the conclusion that fire must have played an important role in the evolution of some biomes. It has become well established that savannas are kept open by fire (Bond and Keeley 2006) and others have indicated that the evolution of C4 grasslands may have been a response to fire regimes (Osborne and Beerling 2006). It is possible that fire may have also played a role in the rise and spread of other plants and plant communities such as ferns and early angiosperms, as many of these are found as charcoal and are associated with frequent fires (Scott and Galtier 1985, Friis et al. 2006). As the significance of fire in Earth systems processes is more fully understood then the role of fire in stimulating plant evolution may be addressed more widely.

Burial and Preservation by Post-fire Erosion

Over the past 20 years, the scale and significance of post-fire erosion and deposition has been more fully appreciated. As demonstrated by several authors in this volume, fire may not only destroy or alter surface litter but also affects the soil so that subsequent rainstorms may cause extensive erosion and sediment transport initially by surface flow (Figs. 8a, b). The mix of sediment and charcoal may be transported down hillslopes and even across floodplains before being deposited in rivers or lakes. The phenomenon of post-fire erosion is now considered a major sedimentological process (Cannon 2001, Cannon et al. 2001, Moody and Martin 2001, this volume). However, post-fire erosion has not been widely recognized in the Pre-Quaternary fossil record. It has been recognized that many alluvial fan sequences may be triggered as a result of post-fire erosion processes (Meyer et al. 1992).

Examination of the fossil record indicates that post-fire erosion and deposition is a more important phenomenon that has been previously realized. There are numerous examples where sudden influx of sediment with abundant charcoal is found, for example, in the early Carboniferous of Ireland, in the Triassic of New Mexico, in the Jurassic of Yorkshire, the Cretaceous of the Isle of Wight (Nichols and Jones 1992, Falcon-Lang 1998, Collinson et al. 2000, Scott 2000a, Zeigler 2003).

The rapid erosion and burial of sediments may also have had an effect on the global atmosphere. It is often not appreciated that many thick coals from the Permian of India or the Early Cretaceous of Canada are deposits with over 70 percent charcoal and probably were the result of post-fire erosion/depositional processes.

HOW CAN THE PAST HELP US UNDERSTAND THE PRESENT AND FUTURE OF WILDFIRE?

While modern fire systems can help us understand ancient fires, the geological record can offer the perspective of time, where there have been changes in atmospheric composition, changes in climate including periods of rapid global warming. In each case it is possible to see effects that can only be modelled into the future.

The fossil record offers a glimpse of the rare event or a large event, so that, for example, the impact of a large fire had on estuarine fish communities in the Carboniferous may offer a new perspective to fire scientists studying only their fire area.

What the fossil record does show, however, is that major wildfires were an integral part of the Earth system for hundreds of millions of years before man and we may regard man as an additional complicating factor.

Our understanding of ancient fire systems is based mainly on studies of fossil charcoal. However, charcoals are rarely studied by those interested in modern fires. Perhaps some of the data and techniques used for the study of macroscopic charcoals may be of use in the current debates concerning the recognition and definition of fire intensity and burn severity.

FUTURE DIRECTIONS

Our understanding of ancient fire systems is relatively poor. The significance of fires in the Pre-Quaternary, before the advent of man, has only recently been appreciated. Likewise, the significance of charcoal as a major data source is still not fully appreciated. Charcoal studies may contribute to not only understanding past fire systems but may also help in the interpretation of both fire and post-fire processes. Several important questions need resolution:

- Can charcoal provide data on the question of fire intensity and burn severity?
- Are all charcoal assemblages representative of the vegetation being burned ?
- Can we use charcoal assemblages to recognize different vegetation types?
- Can we use charcoals to interpret fire type?
- Are most macroscopic charcoal assemblages derived from plant litter and surface plants?
- How might sedimentary processes bias charcoal assemblages?
- Can we use charcoal to identify sedimentary deposits formed by post-fire erosion/depositional processes?

In relation to our understanding of ancient fire systems many of the above questions are still relevant. We might add a time dimension:

- What is the role of fire in 'Earth systems processes'?
- How has the evolution of the atmosphere been affected by or have an influence on the diversification of fire systems?
- What is the role of charcoal in long-term carbon storage?
- How does fire influence plant evolution?
- How might some biomes be a direct result of fire activity?
- What are the major drivers of fire systems in deep time – oxygen levels, climate or both?

It is clear that we are only just scratching the surface of charcoal studies and the importance of ancient fire systems is only just being appreciated by the wider science community.

CONCLUSIONS

Fire products and residues include gases, organic compounds such as polyaromatic hydrocarbons (PAH), soot and charcoal. Whilst soot and PAHs have been described from ancient sediments, it is the occurrence of charcoal that provides the best evidence of ancient forest fires.

Charcoal preserves plant anatomy and allows the identification of plant organ and species. Plants charred at increasing temperatures show increased cell wall reflectance when studied in polished blocks under oil. This data may help in the interpretation of fire type and fire temperatures as well as indicating fire intensity and burn severity.

Evidence of the first fires comes from the late Silurian/early Devonian but it is not until the Carboniferous that fires became diverse and widespread. This may be in part due to a rise in atmospheric oxygen concentration at this time. The suggestion of high oxygen concentrations in the Carboniferous and Permian (over 30 percent) may be supported by the occurrence and abundance of charcoal in coals at this time.

Fires may have stimulated the evolution of a range of plants and may have helped in the evolutions and maintenance of the savanna biome.

Major fires from the Carboniferous onwards may have generated major sedimentary sequences as post-fire erosion.

Strong feedback between fire and the atmosphere have been identified and especially strong links have been suggested between the importance of fire with increasing oxygen and the link between rising oxygen and falling carbon dioxide levels and the erosion of rocks.

Fire has played an important role in the Earth system for at least 400 million years.

Acknowledgments

I thank Artemi Cerdà and Peter Robichaud for inviting me to contribute to this volume. Writing this chapter has been made possible by grants from the Royal Holloway University of London Research Strategy Fund and Yale University, who made my sabbatical at Yale possible. I thank Derek Briggs, Leo Hickey and Bob Berner for their support and encouragement. I thank Deborah Martin and John Moody for inviting me to join their wildfire workshops. I thank in particular Ian Glasspool and Margaret Collinson for continuing collaboration of the study of modern and ancient charcoals. I acknowledge the receipt of research grants from the Leverhulme Trust and the Royal Society. I thank Bill

Chaloner for sparking my interest in ancient fires in 1973 while a research student and for his continued support and my many research students including Tim Jones, Ian Glasspool and Howard Falcon-Lang for stimulating discussions. I thank Neil Holloway, Sharon Gibbons and Pat Goggin for their technical support. Dan Neary kindly allowed me to use the photos in Figs. 8a, b.

References

Agee, J.K. 1990. The historical role of fire in Pacific Northwest forests. pp. 25-38. *In* J.D. Walstad, S.R. Radosevich and D.V. Sandberg [eds.]. *Natural and Prescribed Fire in Pacific Northwest Forests.* Oregon State University Press, Corvallis, Oregon, USA.

Belcher, C.M., M.E. Collinson, and A.C. Scott. 2005. Constraints on the thermal energy released from the Chicxulub impactor: New evidence from multi-method charcoal analysis. *Journal of the Geological Society, London,* 162: 591-602.

Berner, R.A. 2006. A combined model for Phanerozoic atmospheric O_2 and CO_2. *Geochemica et Cosmochimica Acta,* 70: 5653-5664.

Berner R.A., D.J. Beerling, R. Dudley, J.M. Robinson, and R.A. Wildman. 2003. Phanerozoic atmospheric oxygen. *Annual Review of Earth and Planetary Sciences,* 31: 105-134.

Bird, M.I. 1997. Fire in the Earth Sciences. *Episodes,* 20: 223-226.

Bond, W.J. and J.E. Keeley. 2005. Fire as a global 'herbivore': the ecology and evolution of flammable ecosystems. *TRENDS in Ecology and Evolution,* 20: 387-394.

Bradbury, J.P. 1996. Charcoal deposition and redeposition in Elk Lake, Minnesota, USA. *The Holocene,* 6: 339-344.

Brown, R., A.C. Scott, and T.P. Jones. 1994. Taphonomy of fossil plants from the Viséan of East Kirkton, West Lothian, Scotland. *Transactions of the Royal Society of Edinburgh: Earth Sciences,* 84: 267-274.

Bustin, R.M and Y. Guo. 1999. Abrupt changes (jumps) in reflectance values and chemical compositions of artificial charcoals and inertinite in coals. *International Journal of Coal Geology,* 38: 237-260.

Campbell, I.D. and C. Campbell. 2000. Late Holocene vegetation and fire history at the southern boreal forest margin in Alberta, Canada. *Palaeogeography, Palaeoclimatology, Palaeoecology,* 164: 263-280.

Cannon, S.H. 2001. Debris flow generation from recently burned watersheds. *Environmental and Engineering Geoscience,* 7: 321-341.

Cannon, S.H., E.R. Bigio and E. Mine. 2001. A process for fire-related debris flow initiation, Cerro Grande fire, New Mexico. *Hydrological Processes,* 15: 3011-3023.

Carcaillet, C., H. Almquist, H. Asnong, et al. 2002. Holocene biomass burning and global dynamics of the carbon cycle. *Chemosphere,* 49: 845-863.

Carcaillet, C., Y. Bergeron, P.J.H. Richard, B. Fréchette, S. Gauthier, and Y.T. Prairie. 2001. Change in fire frequency in the eastern Canadian boreal forests during the Holocene: does vegetation composition or climate trigger the fire regime? *Journal of Ecology,* 89: 930-946.

Chaloner, W.G. 1989. Fossil charcoal as an indicator of palaeo-atmospheric oxygen level. *Journal of the Geological Society, London,* 146: 171-174.

Clark, J.S. 1988. Particle motion and the theory of charcoal analysis: Source area, transport, deposition and sampling. *Quaternary Research,* 30: 67-80.

Clark, J.S. and W.A. Patterson III. 1997. Background and local charcoal in sediments: scales of fire evidence in the palaeo record. pp. 23-48. *In* J.S. Clark, H. Cachier, J.C. Goldammer, and B. Stocks [eds.]. *Sediment Records of Biomass Burning and Global Change.* NATO Advanced Science Institutes Series I, Volume 51. Springer-Verlag, Berlin, Germany.

Clark, J.S., H. Cachier, J.C. Goldammer, and B. Stocks [eds.]. 1997. *Sediment Records of Biomass Burning and Global Change.* NATO Advanced Science Institutes Series I, Volume 51. Springer-Verlag, Berlin, Germany.

Cohen, A.D. 1984. Evidence of fires in the ancient Everglades and coastal swamps of Southern Florida. Pages 459-464. *In* P.J. Gleason [ed.]. *Environments of South Florida: present and past.* Miami Geological Society, Coral Gables, Florida, USA.

Collinson, M.E., C. Featherstone, J.A. Cripps, G.J., Nichols, and A.C. Scott. 2000. Charcoal-rich plant debris accumulations in the Lower Cretaceous of the Isle of Wight, England. *Acta Palaeobotanica Supplement* 2 (for 1999): 93-105.

Collinson, M.E., D. Steart, A.C. Scott, I.J. Glasspool, and J.J. Hooker. 2007a. Fire and episodic runoff and deposition at the Paleocene-Eocene boundary. *Journal of the Geological Society, London,* 167: 87-97.

Cope, M.J. 1980. Physical and chemical properties of coalified and charcoalified phtyoclasts from some British Mesozoic sediments: An organic geochemical approach to palaeobotany. pp. 663-677. *In* A.G. Douglas and J.R. Maxwell [eds.]. Advances in organic geochemistry. Pergamon Press, Oxford, UK.

Cope, M.J. 1981. Products of natural burning as a component of the dispersed organic matter of sedimentary rocks. pp. 89-109. *In* J. Brooks [ed.]. *Organic Maturation Studies and Fossil Fuel Exploration.* Academic Press, London, UK.

Cope, M.J. 1993. A preliminary study of charcoalfield plant fossils from the Middle Jurrassic Scalby Formation of North Yorkshire. *Special Papers in Palaeontology,* 49: 101-111.

Cope, M.J. and W.G. Chaloner. 1980. Fossil charcoal as evidence of past atmospheric composition. *Nature,* 283: 647-649.

Cope, M.J. and W.G. Chaloner. 1981. Reply. *Nature,* 290: 428.

Crutzen, P.J. and M. Andrae. 1990. Biomass burning in the tropics: impact on atmospheric chemistry and biogeochemical cycles. *Science,* 250: 1669-1678.

Crutzen, P.J. and J.G. Goldammer [eds.]. 1993. *Fire in the environment: The ecological, atmospheric and climatic importance of vegetation fires. Dahlem workshop reports.* Environmental Science Research Reports No. 13. J. Wiley and Sons Ltd., Chichester, UK.

Cypert, E. 1973. Plant succession on burned areas in Okefenokee swamp following the fires of 1954 and 1955. *Proceedings of the Tall Timbers Fire Ecology Conference,* 12: 199-217.

Dechamps, R. 1984. Evidence of bush fires during the Plio-Pleistocene in Africa (Omo and Sahabi) with the aid of fossil woods. *Palaeoecology of Africa,* 16: 291-294.

Edwards, D. and C.H. Wellman. 2001. Embryophytes on land: the Ordovician to Lochkovian (Lower Devonian) record. pp. 3-28. *In* P.G. Gensel and D. Edwards [eds.]. *Plants Invade the Land: Evolutionary and Environmental Perspectives.* Columbia University Press, New York, New York, USA.

Edwards, D. and L. Axe. 2004. Anatomical evidence in the detection of the earliest wildfires. *Palaios,* 19: 113-128.

Edwards, K.J. and G. Whittington. 2000. Multiple charcoal profiles in a Scottish lake: a study in taphonomy. *Palaeogeography, Palaeoclimatology, Palaeoecology*, 164: 67-86.

Elliott, J.G. and R.S. Parker. 2001. Developing a post-fire chronology and recurrence probability from alluvial stratigraphy in the Buffalo Creek Watershed, Colorado, USA. *Hydrological Processes*, 15: 3039-3052.

Falcon-Lang, H.J. 1998. The impact of wildfire on an Early Carboniferous coastal system, North Mayo, Ireland. *Palaeogeography, Palaeoclimatology, Palaeoecology*, 139: 121-138.

Falcon-Lang, H.J. 2000. Fire ecology of the Carboniferous tropical zone. *Palaeogeography, Palaeoclimatology, Palaeoecology*, 164: 339-355.

Falcon-Lang, H. and A.C. Scott. 2000. Upland ecology of some Late Carboniferous Cordaitalean trees from Nova Scotia and England. *Palaeogeography, Palaeoclimatology, Palaeoecology*, 156: 225-242.

Falcon-Lang, H.J., J. Kvacek, and D. Ulicny. 2001. Fire-prone plant communities and palaeoclimate of a Late Cretaceous fluvial to estuarine environment, Pecinov Quarry, Czech Republic. *Geological Magazine*, 138: 563-576.

Finklestein, D.B., L.M. Pratt, T.M. Curtin, and S.C. Brassell. 2005. Wildfires and seasonal aridity recorded in Late Cretaceous strata from south-eastern Arizona, USA. *Sedimentology*, 52: 587–599.

Flannigan, M.D., B.J. Stocks, and B.M. Wotton. 2000. Climate change and forest fires. *The Science of the Total Environment*, 262: 221-229.

Forbes, M.S., R.J. Raison, and J.O. Skjemstad. 2006. Formation, transformation and transport of black carbon (charcoal) in terrestrial and aquatic ecosystems. *Science of the Total Environment*, 370: 190-296.

Fosberg, M.A., L.O. Mearns, and C. Price. 1993. Climate change – fire interactions at the global scale: predictions and limitations of methods. pp. 123-137. *In* P.J. Crutzen and J.G. Goldammer [eds.]. *Fire in the Environment. The Ecological, Atmospheric, and Climatic Importance of Vegetation Fires.* J. Wiley and Sons Ltd., Chichester, UK.

Friis, E.M., P.R. Crane, and K.R. Pedersen. 1988. Reproductive structures of Cretaceous Platanaceae. *Det Kongelige Danske Videnskaberns Selskab Biologiske Skrifter*, 31: 1-55.

Friis, E.M., K.R. Pedersen, and P.R. Crane. 2006. Cretaceous angiosperm flowers: Innovation and evolution in plant reproduction. *Palaeogeography, Palaeoclimatology, Palaeoecology*, 232: 251-293.

Gingerich, P.D. 2006. Environment and evolution through the Paleocene–Eocene thermal maximum. *TRENDS in Ecology and Evolution*, 21: 246-253.

Glasspool, I.J., D. Edwards, and L. Axe. 2004. Charcoal in the Silurian as evidence for the earliest wildfire. *Geology*, 32: 381-383.

Griffin, J.J. and E.D. Goldberg. 1979. Morphologies and origin of elemental carbon in the environment. *Science*, 206: 563-565.

Goldberg, E.D. 1985. *Black Carbon in the Environment*. J. Wiley and Sons, New York, USA.

Guo, Y. and R.M. Bustin. 1998. FTIR spectroscopy and reflectance of modern charcoals and fungal decayed woods: implications for studies of inertinite in coals. *International Journal of Coal Geology*, 37: 29-53.

Harris, T.M. 1958. Forest fire in the Mesozoic. *Journal of Ecology*, 46: 447-453.

Harris, T.M. 1981. Burnt ferns from the English Wealden. *Proceedings of the Geologists Association*, 92: 47-58.

Harvey, M.C., S.C. Brassell, C.M. Belcher, and A. Montanari. 2008. Combustion of fossil organic matter at the Cretaceous-Paleocene (K-P) boundary. *Geology,* 36: 355-358.

Herring, J.R. 1985. Charcoal fluxes into sediments of the North Pacific Ocean: the Cenozoic record of burning. *In* The Carbon Cycle and Atmospheric CO_2: Natural Variations Archean to Present. *Geophysical Monographs,* 32: 419-442.

Highton, P.J.C., A. Pearson, and A.C. Scott. 1991. Palynofacies and Palynodebris and their use in coal measures palaeoecology and palaeoenvironmental analysis. *Neus Jarbuch fur Geologie und Paläontologie Abhandlungen,* 183: 135-169.

Johnson, B. 1984. The great fire of Borneo. Report of a visit to Kalimantan-Timur a year later, May 1984. World Wild Life Fund. Godalming, Surrey, UK.

Jones, T.P. 1993. New morphological and chemical evidence for a wildfire origin for fusain from comparisons with modern charcoal. *Special Papers in Palaeontology,* 49: 113-123.

Jones, T.P. 1994. 13C enriched Lower Carboniferous fossil plants from Donegal, Ireland. Carbon isotope constraints on taphonomy, diagenesis and Palaeoenvironment. *Review of Palaeobotany and Palynology,* 18: 53-64.

Jones, T.P. 1997. Fusain in Late Jurassic sediments from the Witch Ground Graben, North Sea, UK. *Medelingen Nederlands Instituut voor Toegepaste geowetenschappen. TNO,* 58: 93-103.

Jones, T.P. and W.G. Chaloner. 1991. Fossil charcoal, its recognition and palaeoatmospheric significance. *Palaeogeography, Palaeoclimatology, Palaeoecology (Global and Planetary Change Section),* 97: 39-50.

Jones, T.P., A.C. Scott, and M. Cope. 1991. Reflectance measurements against temperature of formation for modern charcoals and their implications for the study of fusain. *Bulletin de la Société Géologique de France,* 162: 193-200.

Jones, T.P., A.C. Scott, and D.P. Mattey. 1993. Investigations of "fusain transition fossils" from the Lower Carboniferous: comparisons with modern partially charred wood. *International Journal of Coal Geology,* 22: 37-59.

Jones, T.P., W.G. Chaloner, and T.A.J. Kuhlbusch. 1997. Proposed bio-geological and chemical based terminology for fire-altered plant matter. pp. 9-22. *In* J.S. Clark, H. Cachier, J.C. Goldammer, and B. Stocks et al. [eds.]. *Sediment Records of Biomass Burning and Global Change.* NATO Advanced Science Institutes Series I, Volume 51, Springer-Verlag, Berlin, Germany.

Katz, M.E., D.K. Pak, G.R. Dickens, and K.G. Miller. 1999. The source and fate of massive carbon input during the latest Paleocene thermal maximum. *Science,* 286: 1531-1533.

Kenrick, P. and P.R. Crane. 1997. The origin and early evolution of plants on land. *Nature,* 389: 33-39.

Killops, S.D. and M.S. Massoud. 1992. Polyeyclic aromatic hydrocarbons of pyrolitic origin in ancient sediments: evidence for vegetation fires. *Organic Geochemistry,* 18: 1-7.

Komarek, E.V., B.B. Komarek, and T.C. Carlysle. 1973. The ecology of smoke particulates and charcoal residue from forest and grassland fires: a preliminary atlas. Miscellaneous Publication 3. Tall Timbers Research Station. Tallahassee, Florida, USA.

Lenton, T.M. 2003. The role of land plants, phosphorous weathering and fire in the rise and regulation of atmospheric oxygen. *Global Change Biology,* 7: 613-629.

Lupia, R. 1995. Palaeobotanical data from fossil charcoal: an actualistic study of seed plant reproductive structures. *Palaios,* 10: 465-477.

McParland, L., M.E. Collinson, A.C. Scott, and D. Steart. 2007. Ferns and fires: Experimental charring of ferns compared to wood and implications for paleobiology, coal petrology and isotope geochemistry. *Palaios,* 22: 528-538.

Meyer-Berthaud, B., S.E. Scheckler, and J. Wendt. 1999. *Archaeopteris* is the earliest known modern tree. *Nature,* 398: 700-701.

Meyer, G.A., S.G. Wells, R.C. Balling, and A.J.T. Jull. 1992. Response of alluvial systems to fire and climate change in Yellowstone National Park. *Nature,* 357: 147-150.

Millspaugh, S.H. and C. Whitlock. 1995. A 750-year fire history baes on lake sediment records in central Yellowstone National Park, USA. *The Holocene,* 5: 283-282.

Moody, J.A. and D.A. Martin. 2001. Hydrological and sedimentologic response of two burned watersheds. Water-Resources Investigation Report, Report Number 01-4122. U.S. Geological Survey, Denver, Colorado, USA.

Morgans, H.S., S.P. Hesselbo, and R.A. Spicer. 1999. The seasonal climate of the Early-Middle Jurassic Cleveland Basin, England. *Palaios,* 14: 261-272.

Nichols, G.J., J. Cripps, M.E. Collinson, and A.C. Scott. 2000. Experiments in waterlogging and sedimentology of charcoal: results and implications. *Palaeogeography, Palaeoclimatology Palaeoecology,* 164: 43-56.

Nichols, G.J and T.P. Jones, 1992. Fusain in Carboniferous shallow marine sediments, Donegal, Ireland: The sedimentological effects of wildfire. *Sedimentology,* 39: 487-502.

Osborne, C.P. and D.J. Beerling. 2006. Nature's green revolution: the remarkable evolutionary rise of C4 plants. *Philosophical Transactions of the Royal Society,* B 361: 173-194.

Page, S.E., F. Siegert, J. Rieley, H.D.V. Boehm, A. Jaya, and S. Limin. 2002. The amount of carbon released from peat and forest fires in Indonesia during 1977. *Nature,* 420: 61-65.

Patterson III, W.A., K.J. Edwards, and D.J. Maguire. 1987. Microscopic charcoal as an indicator of fire. *Quaternary Science Reviews,* 6: 3-23.

Pearson, T. and A.C. Scott. 1999. Large palynomorphs and debris. pp. 20-25. *In* T.P. Jones and N.P. Rowe [eds.]. *Fossil plants and spores: Modern techniques.* Geological Society, London, UK.

Petersen, H.I. 1998. Morphology, formation and palaeoenvironmental implications of naturally formed char particles in coals and carbonaceous mudstones. *Fuel,* 77: 1177-1183.

Plotnick, R.E., F. Keing, A. Scott, and I. Glasspool. 2008. Exceptionally well-preserved paleokarst and Pennsylvanian cavefills. pp. 79-87. *In* B.B. Curry [ed.]. *Deglacial History and Paleoenvironments of Northern Illinois.* ISGS Open File Series 2008-1. Illinois State Geological Survey, Champaign, Illinois, USA.

Prior, J. and K.L. Alvin. 1983. Structural changes on charring woods of *Dichrostachys* and *Salix* from Southern Africa. *International Association of Wood Anatomists Bulletin,* 4: 197-206.

Putz, M.K. and E.L. Taylor. 1996. Wound response in fossil trees from Antarctica and its potential as palaeoenvironmental indicators. *International Association of Wood Anatomists Journal,* 17: 77-88.

Pyne, S.J., P.L. Andrews, and R.D. Laven. 1996. *Introduction to Wildland Fire.* J. Wiley and Sons, New York, New York, USA.

Rimmer, S.M., J.A. Thompson, S.A. Goodnight, and T.L. Robl. 2004. Multiple controls on the preservation of organic matter in Devonian–Mississippian marine black shales: geochemical and petrographic evidence: *Palaeogeography, Palaeoclimatology, Palaeoecology,* 215: 125-154.

Robichaud, P.R. and H. Elsenbeer. 2001. Wildfire and surficial processes. *Hydrological Processes,* 15: 2865-3091.

Rundel, P.W. 1981. Fire as an ecological factor. pp. 501-538. *In* O.L. Lange, P.S. Nobel, C.B. Osmond and H. Ziegler [eds.]. *Physiological Plant Ecology I, Response to the Physical Environment.* Springer-Verlag, Berlin, Germany.

Sanford, R.L. Jr., J. Saldarriage, K.E. Clark, C. Uhl, and R. Herra. 1985. Amazon rainforest fires. *Science,* 227: 53-55.

Schmidt, M.W.I. and A.G. Noack. 2000. Black carbon in soils and sediments: analysis, distribution, implications, and current challenges. *Global Biogeochemical Cycles,* 14: 777-793.

Scott, A. 1974. The earliest conifer. *Nature,* 251: 707-708.

Scott, A.C. 1978. Sedimentological and ecological control of Westphalian B plant assemblages from West Yorkshire. *Proceedings of the Yorkshire Geological Society,* 41: 461-508.

Scott, A.C. 1989. Observations on the nature and origin of fusain. *International Journal of Coal Geology,* 12: 443-475.

Scott, A.C. 2000a. The Pre-Quaternary history of fire. *Palaeogeography, Palaeoclimatology, Palaeoecology,* 164: 281-329.

Scott, A.C. 2000b. Fossil Plants. pp. 364-371. *In* P.L. Hancock and B.J. Skinner [eds.]. *The Oxford Companion to the Earth.* Oxford University Press, Oxford, UK.

Scott, A.C. 2001. Preservation by fire. pp. 277-280. *In* D.E.G. Briggs and P.J. Crowther [eds.]. *Palaeobiology II.* Blackwells, Oxford, UK.

Scott, A.C. 2002. Coal petrology and the origin of coal macerals: a way ahead? *International Journal of Coal Geology,* 50: 119-134.

Scott, A.C. 2003. Charcoal in sediments. pp. 121-123. *In* G.V. Middleton [ed.]. *Encyclopaedia of Sediments and Sedimentary Rocks.* Kluwer Academic Publishers, Dordrecht, The Netherlands.

Scott, A.C. and W.G. Chaloner. 1983. The earliest fossil conifer from the Westphalian B of Yorkshire. *Proceedings of the Royal Society of London B.* 220: 163-182.

Scott, A.C., J. Cripps, G. Nichols, and M.E. Collinson. 2000b. The taphonomy of charcoal following a recent heathland fire and some implications for the interpretation of fossil charcoal deposits. *Palaeogeography, Palaeoclimatology, Palaeoecology,* 164: 1-31.

Scott, A.C. and J. Galtier. 1985. The distribution and ecology of early ferns. *Proceedings of the Royal Society of Edinburgh B,* 86: 141-149.

Scott, A.C. and I.J. Glasspool. 2005. Charcoal reflectance as a proxy for the emplacement temperature of pyroclastic flow deposits. *Geology,* 33: 589-592.

Scott, A.C. and I.J. Glasspool. 2006. The diversification of Paleozoic fire systems and fluctuations in atmospheric oxygen concentration. Proceedings of the National Academy of Sciences, USA, 103: 10861-10865.

Scott, A.C. and I.J. Glasspool. 2007. Observations and experiments on the origin and formation of inertinite group macerals. *International Journal of Coal Geology,* 70: 55-66.

Scott, A.C., I.J. Glasspool, S.M. Rimmer, and W.L. Cressler III. 2006. The origin of widespread fire systems and their wider impact. Paper 138-7. Geological Society of America *Abstracts with Programs,* 38(7): 341. *online at* http://gsa.confex.com/gsa/2006AM/finalprogram/abstract_111956.htm [accessed 27 Aug. 2008].

Scott, A.C. and T.P. Jones. 1991. Microscopical observations of recent and fossil charcoal. *Microscopy and Analysis,* 25: 13-15.

Scott, A.C. and T.P. Jones. 1994. The nature and influence of fire in Carboniferous ecosystems. *Palaeogeography, Palaeoclimatology, Palaeoecology,* 106: 91-112.

Scott, A.C. and R. Stea. 2002. Fires sweep across the Mid-Cretaceous landscape of Nova Scotia. *Geoscientist,* 12(1): 4-6.

Scott, A.C., J. Moore, and B. Brayshay [eds.]. 2000a. Fire and the Palaeoenvironment. *Palaeogeography, Palaeoclimatology, Palaeoecology,* 164: 1-412.

Shakesby, R.A. and S.H. Doerr. 2006. Wildfire as a hydrological and geomorphological agent. *Earth Science Reviews,* 74: 269-307.

Simoneit, B.R.T. 2002. Biomass burning – a review of organic tracers for smoke from incomplete combustion. *Applied Geochemistry,* 17: 129-162.

Smith, D.M., J.J. Griffin, and E.D. Goldberg. 1973. Elemental carbon in marine sediments: a baseline for burning. *Nature,* 241: 268-270.

Steart, D., M.E. Collinson, A.C. Scott, I.J. Glasspool, and J.J. Hooker. 2007. Palaeobotany as revealed by petrology of two contrasting lignites on either side of the Paleocene-Eocene Thermal Maximum. *Acta Palaeobotanica,* 47(1): 109-125.

Stocks, B.J. and J.B. Kauffman. 1997. Biomass consumption and behaviour of wildland fires in boreal, temperate and tropical ecosystems: parameters necessary to interpret historic fire regimes and future fire scenarios. pp. 169-188. *In* J.S. Clark, H. Cachier, J.C. Goldammer, and B. Stocks [eds.]. *Sediment Records of Biomass Burning and Global Change.* NATO Advanced Science Institutes Series I, Volume 51. Springer-Verlag, Berlin, Germany.

Terasmae, J. and N.C. Weeks. 1979. Natural fires as an index of Palaeoclimate. *Canadian Field-Naturalist,* 93: 116-125.

Vaughan, A. and G.J. Nichols. 1995. Controls on the deposition of charcoal: implications for sedimentary accumulations of fusain. *Journal of Sedimentary Research* A, 65(1): 129-135.

Watson, A.J., J.E. Lovelock, and L. Margulis. 1978. Methanogenesis, fires and the regulation of atmospheric oxygen. *Biosystems,* 10: 293-298.

Westerling, A.L., H.G. Hidalgo, D.R. Cayan, and T.W. Swetnam. 2006. Warming and earlier spring increase western U.S. forest wildfire activity. *Science,* 313: 940-943.

Wildman, R.A., L.J. Hickey, M.B. Dickinson, R.A. Berner, J.M. Robinson, M. Dietrich, R.H. Essenhigh, and C.B. Wildman. 2004. Burning of forest materials under late Paleozoic high atmospheric oxygen levels. *Geology,* 32: 457-460.

Wolbach, W.S., I. Gilmour, and E. Anders. 1990. Major wildfires at the Cretaceous/ Tertiary Boundary. *Geological Society of America Special Paper,* 247: 391-400.

Whitlock, C. and S.H. Millspaugh. 1996. Testing the assumptions of fire-history studies: an examination of modern charcoal accumulation in Yellowstone National Park, USA. *The Holocene,* 6: 7-15.

Winkler, M.G. 1985. Charcoal analysis for paleoenvironmental interpretation: A chemical assay. *Quaternary Research,* 23: 313-326.

Zeigler, K.E. 2003. Taphonomic analysis of the Snyder Quarry; a fire-related Upper Triassic vertebrate fossil assemblage from north-central New Mexico. *Bulletin - New Mexico Museum of Natural History and Science,* 24: 49-62.

Zepp, R.G., and S.A. Macko. 1997. Polycyclic Aromatic Hydrocarbons in sedimentary records of biomass burning. pp. 145-166. *In* J.S. Clark, H. Cachier, J.C. Goldammer, and B. Stocks [eds.]. *Sediment Records of Biomass Burning and Global Change.* NATO Advanced Science Institutes Series I, Volume 51. Springer-Verlag, Berlin, Germany.

Section I
Fire Effects on Soil

2

Forest Fire Effects on Geomorphic Processes

John A. Moody[1]* and *Deborah A. Martin*[1]

Abstract

Landscapes comprise a variety of geomorphic features including bedrock either bare or draped with a mantle of soil and sediment. Within the landscape some features represent reservoirs for the storage of sediment. Three fundamental geomorphic processes (erosion, transport, and deposition) act on sediment particles in these reservoirs to change geomorphic features. These processes require available sediment and a driver such as wind, rainfall, runoff, or gravity. They are components of any process that transfers sediment between different sediment-storage reservoirs. These transfer processes can be viewed as complex non-linear systems characterized by power laws, thresholds, feedbacks, and sensitivity, which can produce a response to the driver that is inherently unstable and, under certain conditions, extreme.

Forest fires produce high soil temperatures that vary widely over a burned landscape because the spatial distribution of vegetation is patchy. Elevated soil temperatures increase the availability of sediment for erosion by lowering thresholds and removing the protective cover over the soil. This increased sediment availability is patchy and temporary. If a fire is soon followed by a storm with localized cells of high-intensity rainfall, and these cells are superimposed over large patches of available sediment, then the response can be a catastrophic flood that erodes hillslopes, incises channels, and deposits sediment. The resulting erosional and depositional features are localized in space across the burned landscape. These features become legacies, a form of landscape memory, and part of the feedback process for the next fire-flood cycle.

INTRODUCTION

Accurate predictions of floods and erosion from burned landscapes are a major unsolved problem. One watershed can produce catastrophic floods or

[1] *U.S. Geological Survey, Boulder, Colorado, USA.*
* *Corresponding author: John A. Moody, U.S. Geological Survey, 3215 Marine Street, Suite E-127, Boulder, Colorado 80303, USA. Tel: 303 541 3011, Fax: 303 541 3084, e-mail: jamoody@usgs.gov*

debris flow while its neighbor produces only a trickle. To approach this problem it is necessary to take the approach Lewis Thomas (1979) suggests "if you want to fix something you are first obliged to understand, in detail, the whole system." In the case of burned watersheds, the something that needs to be 'fixed' may be the determination of what action, if any, should be taken to mitigate the short- and long-term effects of the fire. This chapter provides some insights into the reasons for bewildering post-fire responses, and we hope these insights will provide some initial understanding of what is required to develop predictive models of these responses.

Geomorphology is the study of the creation and evolution of landscapes. Any landscape is a collection of historical legacies in the form of geomorphic features molded by climate, animals, and catastrophic and uniform geomorphic processes. The complex fire history of a landscape (Chapter 1) is an important predictor of the response of today's landscapes (Schumm 1991) to a disturbance such as another forest fire. Fires affect landscapes repeatedly over geological time (Chapter 1), and the geomorphic response for watersheds with similar physical characteristics depends on the joint probability of three factors: 1) the probability of a fire, 2) the probability of a significant rainstorm, and 3) the probability that the rainstorm follows soon after the fire (Elliott and Parker 2001). This joint probability will be less than the probability of the fire itself, because a catastrophic geomorphic response does not necessarily follow after every fire. In this chapter, we do not consider these long-term probabilistic effects of multiple fire-flood cycles over geological time but only the possible short-term geomorphic responses that follow a single fire.

Landscapes consist of a variety of geomorphic features surfaced by a veneer of soil and sediment. Burned landscapes can be viewed as one type of a geomorphic system having two components: 1) sediment-storage reservoirs (Dietrich and Dunne 1978, Dietrich et al. 1982) and 2) different transfer processes that redistribute the sediment between storage reservoirs (Fig. 1). Sediment particles spend different residence times in each storage reservoir and are not necessarily the same age. Hillslope colluvium, sediment behind logs and rocks, rill levees, unchannelized drainages, hollows, alluvial fans, terraces, floodplains, and stream channels with alternate bars are examples of some possible temporary sediment storage reservoirs.

Erosion, transport, and deposition of sediment are some of the short-term fundamental geomorphic processes that change geomorphic features after fire. Erosion consists of the separation, detachment or initiation of motion of soil aggregates or particles from among neighboring aggregates or particles and entrainment into the fluid. Transport of the eroded material can be as dissolved load, suspended load, bed load, or gravity flows, and certainly as a combination of these processes. Sediment transport is maintained if the fluid's shear stress on the boundary (force per unit area exerted on the sediment surface beneath the flow) or the equivalent shear velocity is greater than the critical shear velocity necessary for the initiation of motion or entrainment of

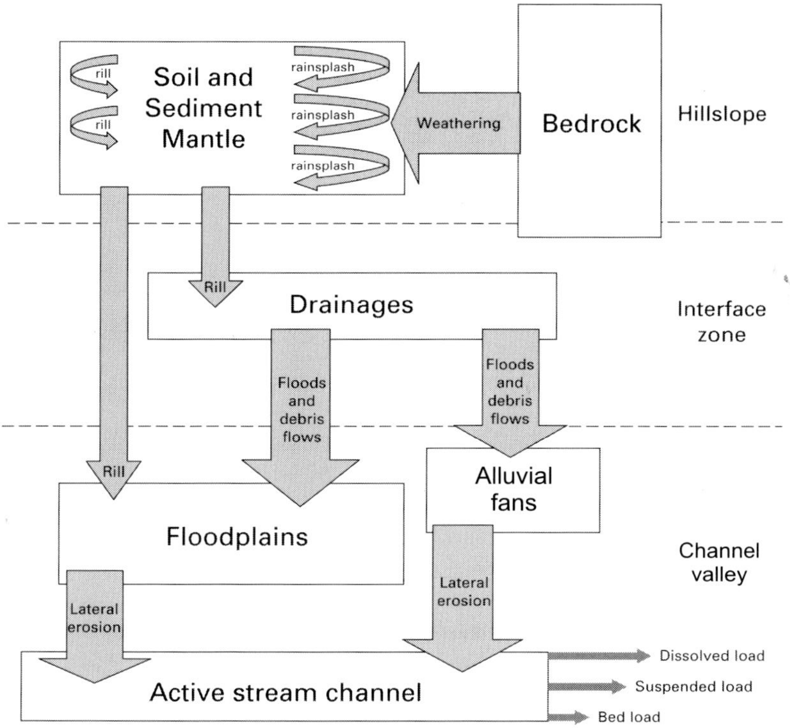

Fig. 1 Components of a burned landscape. Sediment storage reservoirs are shown as rectangles and the transfer processes as gray arrows. The areas of the rectangles are roughly proportional to the residence times and the widths of the gray arrows are roughly proportional to the flux of sediment between storage reservoirs. The interface zone is the area between the channels and the hillslope, which can be described as unchannelized drainages.

sediment particles (Middleton and Southard 1984, Wiberg and Smith 1987). Some net deposition of sediment closest to the boundary begins as the shear velocity falls below the critical shear velocity.

These three fundamental processes are components of any transfer processes. Wind, rainsplash, soil creep, biogenic transport, overland flow, channel flow, flash floods, landslides, and debris flows are examples of transfer processes. Many more sediment particles are rearranged within a storage reservoir than are transferred out of the storage reservoir to another one. All these transfer processes are not simultaneous in time nor are they spread uniformly across a watershed. Transfer processes are frequently limited to narrow spatial corridors such as channels and they are not steady in time but are frequently episodic, moving sediment between storage reservoirs in short bursts of energy. Transport processes must conserve mass. The transport of sediment into a storage reservoir, I, and out of a storage reservoir, O, over an interval of time can be expressed as a discrete form of the conservation of mass equation:

$$I + \Delta V = O \quad (1)$$

where ΔV is the change in mass or volume (either positive or negative) of the storage reservoir.

The geomorphic transfer processes require two elements: available sediment and a driver. The driver provides the energy for doing work and comes in the form of gravity, wind, rain, flowing water, freezing/thawing, lava flows, and animals (Peters and Havstad 2006). Tectonic energy is also a driver but is beyond the scope of this chapter. The book *Active Tectonics and Alluvial Rivers* (Schumm et al. 2000) can be referred to for a thorough discussion.

Sediment Availability

While landscapes are composed of soil, sediment, and bedrock, not all of these materials are available to be eroded or transported. Soil comprises inorganic sediment particles, organic matter, liquid, and gases and its production involves long-term processes. In this chapter we focus on sediment and on the short-term processes that affect sediment availability. The availability of sediment depends on three variables: 1) erodibility, 2) vegetation protection, and 3) volume or mass of sediment in storage.

Erodibility has been expressed as a ratio of the mass of sediment eroded from the soil per unit area per unit time per unit flow variable. Some flow variables are kinetic energy per unit area (Poesen and Savat 1981), rain intensity raised to a power (Flanagan and Nearing 1995), rainfall erosivity index (Renard et al. 1997), unit stream power (Rose et al. 1983, Hairsine 1988), and boundary shear stress (Elliot et al. 1989). A frequently used metric for erodibility throughout the world is the USLE K-factor (Fig. 2) and values are available in the United States at a resolution of 1 km^2.

Vegetation is a critical variable that affects the rates of many geomorphic transfer processes. The live canopy layer when viewed from above has a complex patchy distribution based on the morphological characteristics of trees, shrubs, forbs, and grasses. The surface litter layer and lower duff layer on the soil surface prevent erosion by wind, raindrop impacts, and overland flow. The natural spatial distribution of these layers is extremely patchy with intervening bare ground in some places (Cerdà 1997, Davenport et al. 1998). When these patchy layers are burned by a fire in an erratic pattern there will be a more complex pattern of burn effects on the underlying soil. This pattern of burn effects may or may not coincide with the spatial distribution of the erodibility of sediment storage reservoirs.

The volume of sediment stored on the landscape depends on the location of the storage reservoirs, how rapidly new soil or sediment is produced from bedrock, and how quickly storage reservoirs are refilled by transfer processes.

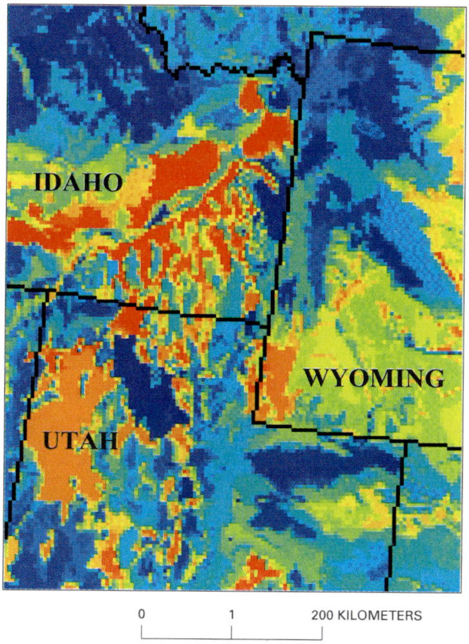

Fig. 2 Spatial variability of soil erodibility. An example of the degree of spatial variability of the erodibility factor K is shown for a portion of the western United States (southern Idaho, western Wyoming and northeastern Utah) on a scale of 1 km^2. The red color indicates the most erodible soil and the blue color indicates the least erodible soil. The soil erodibility (USLE K-factor) was taken from the STATSGO database and the data were analyzed and processed by David Mixon.

The distribution of these sediment storage reservoirs is not uniformly spread across the landscape and the transfer of sediment between storage reservoirs depends on the size of the watershed. As the drainage area increases, the sediment transfer or yield (mass per unit area) decreases because large watersheds have more potential storage reservoirs (Walling 1983, Walling and Webb 1996). Thus, low-order channels within a watershed will often yield the most sediment per unit area and this eroded sediment is stored downstream in higher-order channels. We have not included in this chapter a compilation of such sediment yields after fire (see Moody and Martin in press). Shakesby and Doerr (2006) provide a detailed discussion of the caveats in interpreting the diverse methods and correspondingly diverse units used to measure sediment yield over different scales.

Driver

To improve predictions of floods and erosion after fire, we must learn more about the quantitative characteristics of the drivers such as rainfall. Rainfall is variable in time and space. However, determining the short-time scale characteristics such as the size and shape of a rainstorm, as well as its duration and intensity are essential to predicting short-duration floods typical of the geomorphic response that follows fire. Rainfall is notoriously episodic with discrete intervals of rain followed in most cases by long periods with no rain and defining a storm is not trivial. Storms do not drop all their rain in one place but move across the landscape and evolve with time. Land management agencies frequently use a 'design storm' with a specified recurrence interval, duration, and associated rainfall intensity, (for example, 25-year, 1-hour storm, 48 mm h^{-1}) to determine post-fire restoration measures. A 'design storm' represents a hypothetical storm with uniform steady rainfall (Sivapalan and Wood 1987) that is convenient for computer models. Real storms rarely fit into such rigid models. The practice of using such uniform models for a single storm rather than modeling a series of storms may partially explain why predicted responses of burned watersheds often differ substantially from actual outcomes.

A storm's momentary spatial footprint on the landscape has several structural scales. This mesoscale footprint consists of smaller scale rain bands with even smaller-scale embedded rain cells. This hierarchy of spatial scales is usually measured at widely separated points on the ground with a relatively limited number of rain gages. In the last 10 years Doppler radar has given us better resolution (~1 km^2) of the spatially variable areal rainfall over large areas (~10,000 km^2), provided the radar is not blocked by topography and the area is within range of the radar (Fig. 3a). This radar-based areal rainfall is an indirect measurement of rain and is not the same as a point-rainfall value measured by rain gages (Sivapalan and Blöschl 1998).

While radar can resolve the spatial variability over ~10,000 km^2 it can not resolve the fine-scale spatial variability critical for predicting runoff from individual burned watersheds at scales <1 km^2. Data from a network of gages show spatial variability at a smaller scale (Fig. 3b) as indicated by the large rainfall-intensity gradient (35 mm h^{-1} km^{-1}) across Rendija Canyon and the even larger gradient (40 mm h^{-1} km^{-1}) across a small subwatershed such as the one in the southwest corner of Rendija Canyon. This indicates that spatial changes can be on the order of 0.1 km. The spatial variability within the footprint of the storm is also shown by the recurrence intervals associated with the rainfall intensity based on data compiled by Hershfield (1961). These range from 10 to 100 years and illustrate that a storm cannot be represented by a single recurrence interval.

The rainfall pattern is superimposed on the landscape topography only momentarily. The spatial pattern of rainfall changes frequently and suddenly. The topography focuses the wide-spread but spatially variable rainwater into relatively narrow and steep drainage features, which represent a small fraction of the total landscape. The spatial and temporal variability of the rainfall can produce a spatially and temporally variable transient surface flow

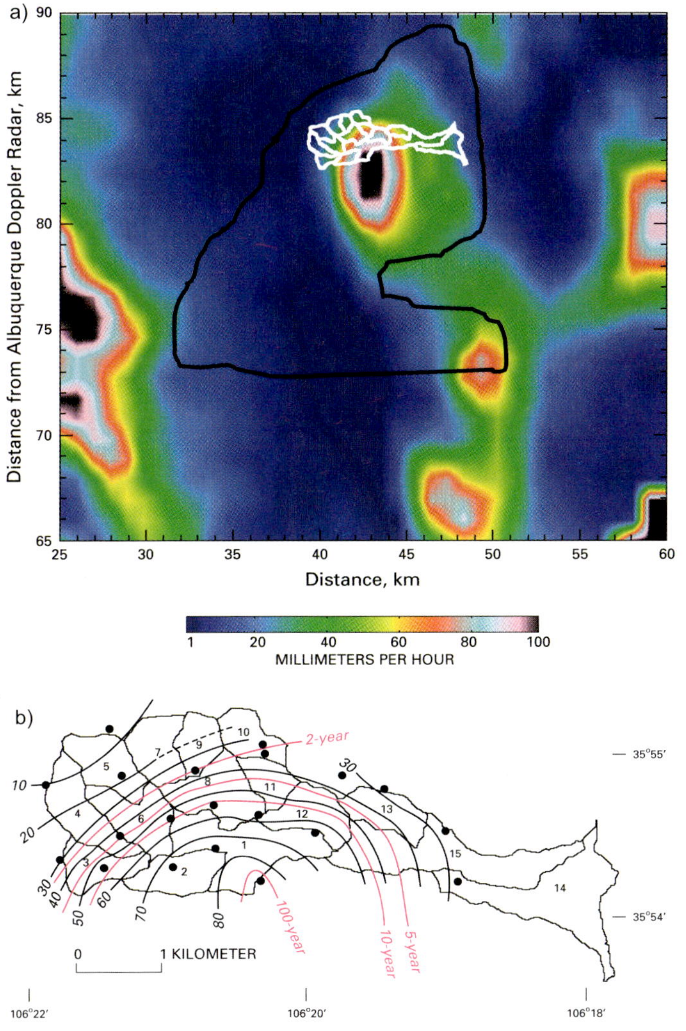

Fig. 3 a) An example of spatial variability of areal rainfall over the area burned by the Cerro Grande Fire near Los Alamos, New Mexico, USA. Color contours are values of the maximum 30-minute intensity during an approximately 8-hour time interval. Several different size rain cells (~10 km^2) are embedded in two rain bands of a convective storm over the area burned by the 2000 Cerro Grande Fire. The outline of the fire perimeter near Los Alamos, New Mexico, USA, is in black. The white outline is the upper part of Rendija Canyon (7.9 km^2). The Doppler radar was located near Albuquerque, New Mexico, and the resolution of the rainfall was 1 km × 1 km. Data were processed and analyzed by Andrea Williams. b) Spatial variability based on point rainfall measured by using a rain gage network (~ 3 gages per 1 km^2; locations are shown by black dots) in the upper part of Rendija Canyon during the same storm shown in A above. Contours show both the rainfall intensity and the recurrence intervals in years associated with the rainfall intensity.

in low order drainages. Under certain conditions this surface flow can produce flash floods. These floods are non-uniform and unsteady. That is, the flow characteristics (depth, velocity, and discharge) change with distance (non-uniform) downstream and with time (unsteady) making this type of runoff more difficult to model than steady flow.

NON-LINEAR CHARACTER OF GEOMORPHIC TRANSFER PROCESSES

More often than not, geomorphic processes are non-linear. This means that: 1) process outputs are not proportional to the process inputs over the entire range of inputs, 2) the process is sensitive to initial and boundary conditions, and 3) the effects of disturbances or perturbations tend to grow and produce disproportionately large and erratic responses (Phillips 2003). As a consequence, a geomorphic process cannot be reduced into separate smaller processes and then reconstructed as a linear combination of solutions of the smaller processes. Being able to linearly combine solutions is the distinct advantage of a linear system and the motivation to try to linearize a non-linear process. Rice (1982) stated in his paper on sedimentation in the chaparral that hydrologists often pray that "the world be linear and Gaussian," but cautioned that this "prayer is rarely answered."

A non-linear response is common in a system where variables such as rainfall may produce more than one effect. In the short term, rainfall can increase runoff but in the long term it also can increase vegetation density, which decreases runoff. Thus, the non-linear character of geomorphic processes makes predictions like interpolation and especially extrapolation beyond measured responses difficult (Schumm 1991) and riddled with uncertainty. These uncertainties are seldom communicated adequately in the transfer of technology information to land managers (Peters and Havstad 2006). Some characteristics of non-linear processes are: 1) power law functions, 2) thresholds, 3) feedback mechanisms, and 4) sensitivity, and these characteristics can explain some of the disproportionate responses and 'strange behavior' of geomorphic processes after fire.

Power Law Functions

If a geomorphic process is a function of a variable raised to a power, then small changes in the variable can produce disproportionately large changes in the process. This was noted by Brown (1972) in Australia after fire when he wrote "Since the carrying capacity is not a linear function of the discharge the overall effect is that enormous quantities of sediment can be moved." Thus, power law functions are part of the explanation for the disproportionate response of burned watersheds.

Thresholds

Another reason for the disproportionate response is that many geomorphic processes depend on thresholds. This is a form of instability where a small increase in the independent variable beyond the threshold value results in an abrupt change in process response (Schumm 1973, Schumm 1979, Phillips 2003). A geomorphic response, R, with a threshold typically has the functional form, f

$$R = f(x - x_0) \quad \text{or} \quad f\left(\frac{1}{x - x_0}\right) \tag{2}$$

where x is the independent variable and x_0 is the threshold value. When $x < x_0$, there is no response. Only when $x > x_0$ does a response begin and if the response has the second functional form in equation (2), then the response can be extremely large as x approaches x_0. In reality it is usually a range of values near x_0, rather than the single value of x_0, that produces rapid changes in the geomorphic response. Most thresholds are not fixed but can be changed by disturbances to a geomorphic system.

Fires change the density of vegetation and vegetation density thresholds have been shown to affect sediment availability. Results from a study in a piñon-juniper vegetation ecosystem in the semi-arid climate of New Mexico are directly applicable to vegetation burned by fire. Davenport et al. (1998) found a threshold for the percent of area occupied by patches of ground cover to be near 10 percent. A small 2 percent decrease (from 11 percent to 9 percent) in ground cover resulted in an abrupt 21 percent increase in the area of patches generating runoff and erosion. Fires also change the infiltration properties of the soil (see Chapters 3, 4, and 7), which control the generation of runoff from rainfall. One property is the critical moisture threshold (Dekker et al. 2001), below which water repellency (Chapter 7, Doerr et al. 2000) and the surface runoff increase. Critical moisture thresholds for water repellency, θ_{cr}, are typically about 0.18 $cm^3\ cm^{-3}$ for dune sands (Dekker et al. 2001) and were inferred to be about 0.062 $cm^3\ cm^{-3}$ after the 2002 Bobcat Fire near Ft. Collins, Colorado, USA (MacDonald and Huffman 2004).

Feedback Mechanisms

A feedback mechanism is an internal interaction between two or more variables of a geomorphic system that promotes the growth and development of the response without an outside force (positive feedback) or acts to self-regulate the system (negative feedback) (Phillips 2003). A familiar example of a negative feedback mechanism is the formation of ripples and dunes by wind or water transport of sediment. As these ripples or dunes grow their form alters the flow, which in turn alters the sediment transport process, eventually producing a stable feature (Nelson and Smith 1989).

An interesting example of a positive feedback mechanism is found in permafrost regions, which after fire can accelerate the evolution of thermokarst topography. Here the permafrost supports the overlying soil. A critical thickness of the litter and duff layers to maintain the permafrost is 0.07-0.12 m (Yoshikawa et al. 2003) and when the thickness becomes less than this threshold, the permafrost begins to melt. If the permafrost should begin to melt, because fire has removed the insulating litter and duff layers, then voids are created under the soil. These voids decrease the support of the overlying soil and, if they become large enough, the soil subsides and cracks. Rain and runoff can flow into the cracks causing more permafrost to melt thus creating more voids, subsidence, and cracking (Yoshikawa and Hinzman 2003). The result of this feedback mechanism is subsidence (Viereck 1973), accelerated erosion, solifluction (Swanson 1981), detachment failures in the form of translational land slides (Lewkowicz and Harris 2005), and in one case the creation of a new drainage network on a burned hillslope (Brown 1983).

The piñon-juniper vegetation ecosystem example described above also illustrates a positive feedback mechanism that is applicable to burned landscapes. Once the threshold is crossed a decrease in ground cover increases the number of patches generating runoff. The increased runoff also carries off litter protecting the surface, decreasing the ground cover even more. In addition, it erodes available sediment, in some cases from around the ground cover leaving it "pedestalled and isolated, in a harsher microclimate, and less able to capture runon" (Davenport et al. 1998). The plants may die, which exposes more sediment to erosion. As the ground cover decreases, evaporation increases, drying the soil and reducing the amount of water available to the remaining plants. This feedback mechanism involves multiple processes, the system is unstable, and the erosion continues to accelerate and produces disproportionate amounts of runoff containing water and sediment.

Sensitivity

Non-linear systems are frequently so sensitive to small perturbations that responses to 'minuscule' variations in properties such as the initial rainfall intensity or soil hydrologic properties result in 'much larger' variations in runoff (Phillips 1992). This supports observations by Gupta et al. (2002) in the semi-arid southwestern United States, where they state "in this area small variations in precipitation can result in huge fractional changes in run-off and recharge."

FIRE EFFECTS ON GEOMORPHIC TRANSFER PROCESSES

Fire differs fundamentally from other disturbances because its energy source is the combustion of vegetation and it transforms vegetation from one organic state to another (Swanson 1981, Bond and Keeley 2005). Other disturbances such as hurricanes, ice storms, and floods are fueled by different sources of energy, and vegetation plays a passive role in those disturbances. Fire interacts dynamically with the landscape (Peterson 2002) and depends on the spatial configuration of the landscape. Fire increases the hydraulic connectivity of the landscape by burning and removing live and dead vegetation barriers.

Fire effects on geomorphic processes are transient because of the relatively rapid regrowth of vegetation after a fire. The initial response phase has a reaction time (Fig. 4) where in process rates usually increase disproportionately after the first post-fire rainstorm and flood (Moody and Martin 2001b). This is followed by a recovery phase when process rates generally decrease to pre-fire levels. The total time (reaction plus recovery) is the relaxation time defined to be the time period for the features in the landscape to attain a new character-

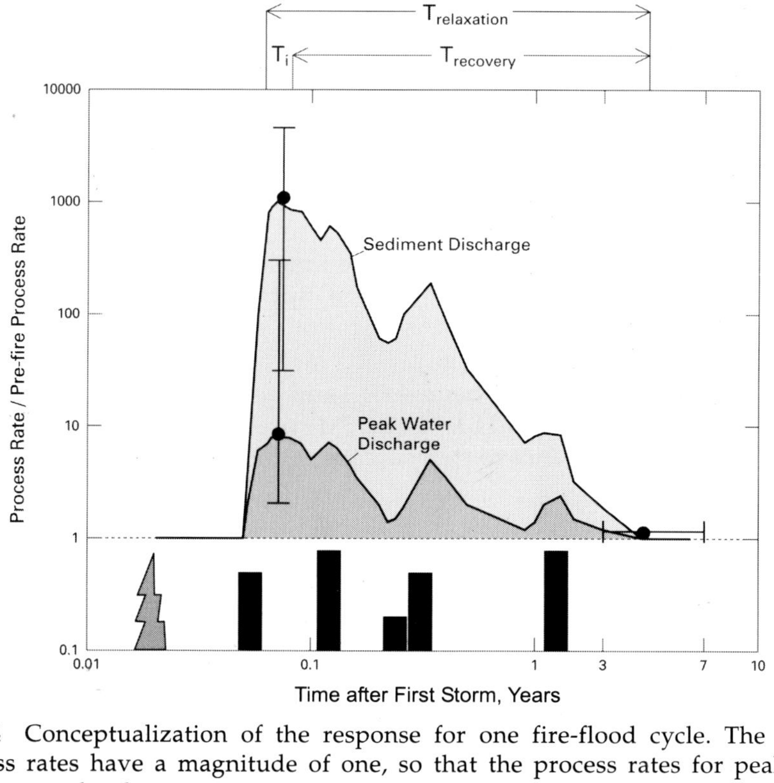

Fig. 4 Conceptualization of the response for one fire-flood cycle. The pre-fire process rates have a magnitude of one, so that the process rates for peak water discharge and sediment transport are relative rates. These relative rates are based on data published by Shakesby and Doerr (2006) and represent median values. The bars indicate the maximum and minimum range for their data. T_i is the reaction time for the initial response phase (measured after the first flood); $T_{recovery}$ is the recovery time when process rates decrease to pre-fire levels. The time of the fire is shown by the 'lightning bolt.' The vertical black bars represent rainfall intensity for five storms. The horizontal bar represents estimates of the relaxation time (Moody and Martin 2001b) for a study site in Colorado Front Range Mountains, USA.

istic form (Brunsden and Thornes 1979), and is on the order of three to seven years (Moody and Martin 2001b) for semi-arid mountainous terrain in Colorado.

Forest fires are frequently described by their recurrence intervals, fire size, and burn severity (Long 2003). We will discuss only the burn severity here because it is a factor in the short-term response while the recurrence interval is a factor in the long-term effects on geomorphic processes.

Burn Severity

Fires increase soil temperatures, which alters substantially the biota, chemistry, and physical properties of the soil. These alterations depend upon the magnitude of the temperature increase (Table 1). Fire intensity is quantified by scorch height, flame length, heat per unit length per unit time or heat per unit area (Romme 1980, Ryan and Noste 1985, Flannigan et al. 2000). Burn severity or fire severity generally refers to effects of fire on components of an ecosystem and both terms have been used to describe the response of the hydrologic component of burned watersheds (Neary et al. 2005, Moody et al. 2007).

Fires leave a mosaic pattern of burn severity. The exact pattern depends on the distribution of the vegetation that burns, the surface soil grain sizes, the vagaries of the wind, the humidity, the time of day when the vegetation burns, and the topography of the landscape. Burn severity has been recently quantified by using the normalized burn ratio (Key and Benson 2006).

The ratio is derived from remote sensing measurements of earth radiation. Raw digital values for spectral band 4 (R_4) and band 7 (R_7) measured by the Landsat Thematic Mapper and Landsat ETM+ are converted to radiance and then to at-satellite reflectance to compute the normalized burn ratio:

$$NBR = 1000(R_4 - R_7)/(R_4 + R_7) \tag{3}$$

The *NBR* represents the change detected by two spectral bands sensitive to fire effects. Band 4 in the near infrared ($0.76\text{-}0.90 \times 10^{-6}$ m) measures the reflected radiation from vegetation (which typically decreases as a consequence of fire) and band 7 in the short-wave infrared ($1.0\text{-}3.0 \times 10^{-6}$ m) measures the reflected radiation from bare soil. The amount of bare soil typically increases as a consequence of fire and the runoff and erosion depend directly on the amount of bare soil (Benavides-Solorio and MacDonald 2005). To calculate the change in the normalized burn ratio, ΔNBR, the *NBR* for an image obtained before the fire is subtracted from the *NBR* for an image obtained after the fire. Each image should have similar phenology (Key and Benson 2006). The ΔNBR values for an area burned by the 2000 Cerro Grande Fire near Los Alamos, New Mexico show a mosaic pattern at the 30-m × 30-m scale (Fig. 5), at several other scales (Fig. 5), and has been reported at the 5-m × 5-m scale for fires in Australia (Price et al. 2003).

Table 1 Temperature effects on soil biota, chemistry, and physical properties

Temperature, °C		Effect	Reference
Minimum	Maximum		
40	70	Plant tissue death	Ryan 2002
60	100	Water loss	Ryan 2002
50	120	Seed death	Ryan 2002
50	160	Bacteria and fungi death	Ryan 2002
175	200	Development of intense water repellency	DeBano 2000
175	275	Maximum critical shear stress (~1-2 N m^{-2}) for initiation of particle motion	Moody et al. 2005
200	900	Decrease in the modulus of elasticity of rocks	Goudie et al. 1992
200	315	Organic matter destructively distilled	Ryan 2002
~220		Organometallic cements destroyed	Giovannini et al. 1988
		Decrease in silt and clay fraction possibly by fusion of finer particles resulting in an increase in sand fraction	Dyrness and Youngberg 1957, Giovannini et al. 1988, Duriscoe and Wells 1982
>275		Decrease in critical shear stress to ~0.5 N m^{-2}	Moody et al. 2005
280	400	Destruction of water repellency	DeBano 2000
350+	450	Organic matter charred	Ryan 2002
400		Fusion temperature of colloidal particles	Puri and Asghar 1940
460	700	Loss of OH groups from clay minerals, and loss of plasticity and elasticity of soil;	Giovannini et al. 1988
774+		Phosphorus volatilized	Ryan 2002
800+		Sulfur volatilized	Ryan 2002

The spatial sequence of the small-scale burn severity can control the runoff along hillslope flow paths. If runoff is generated in severely burned patches on a ridge (see black areas in Fig. 6) but flows downhill onto patches with decreasing burn severity (see brown and green areas in Fig. 6), then much of the runon may infiltrate and never reach the channel. If the sequence of patches is reversed (low burn severity at the ridge but increasing burn severity downhill) then less runon will infiltrate and more runoff will continue to the channel. Not only is the magnitude of the burn severity in each patch important, but some measure of the connectivity and order of the patches is important in predicting runoff from burned hillslopes (Moody et al. 2008).

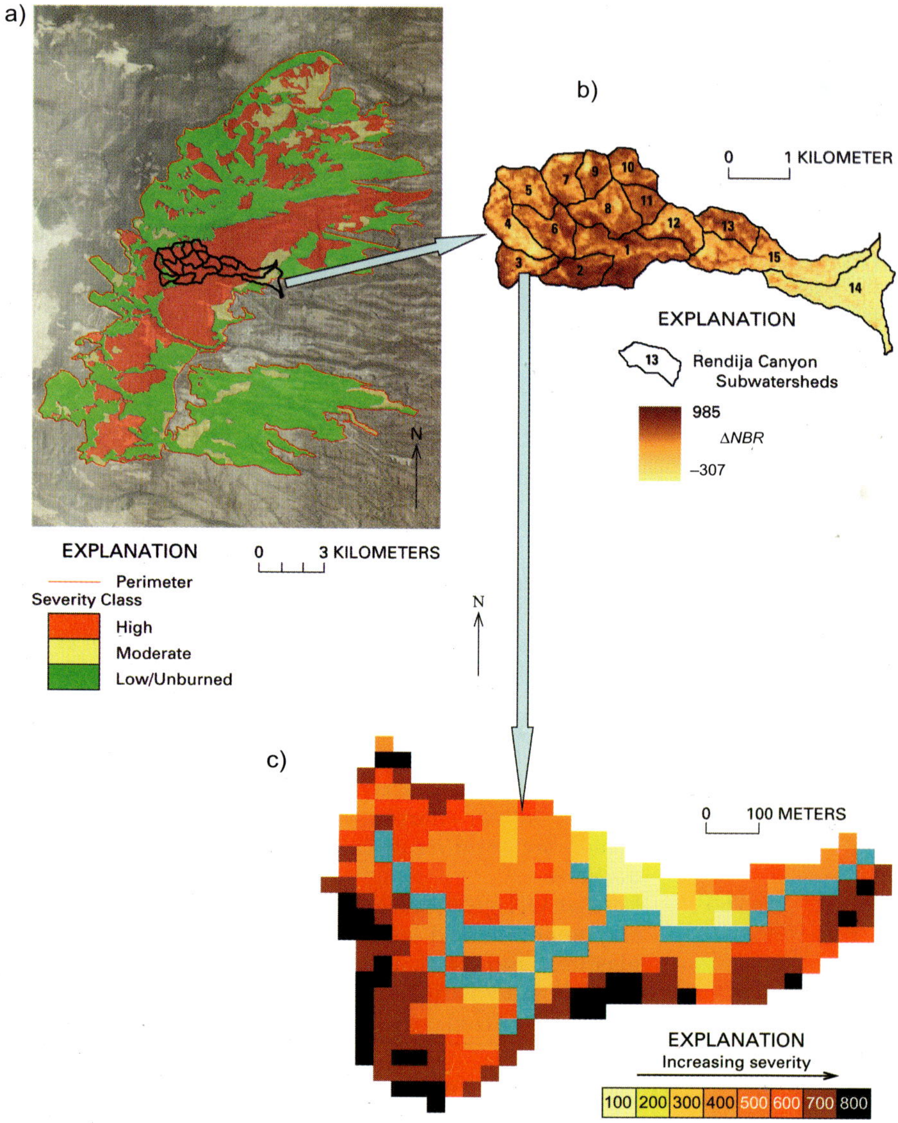

Fig. 5 Burn severity at different scales. a) Burn severity map generated by the Burned Area Emergency Response team with three burn severity categories (BAER 2000) for the 2000 Cerro Grande Fire near Los Alamos, New Mexico, USA. b) The spatial pattern of burn severity is expressed as the change in the normalized burn ratio (ΔNBR) in subwatersheds in the upper part of Rendija Canyon within the area burned by the Cerro Grande Fire. This measure of burn severity incorporates the spatial variability of all hydrologic response units, averaged within a 30-m × 30-m pixel. The ΔNBR values for unburned areas vary among different burns, but often fall between –150 and +150. c) Burn severity (ΔNBR) for one subwatershed in the southwest corner of the upper part of Rendija Canyon. The blue pixels are stream channels.

Fig. 6 Photograph of the pattern of burn severity in the 2000 Bobcat Fire located near Ft. Collins, Colorado, USA. The black areas indicated the highest burn severity. The brown areas (mostly dead trees with brown needles) are often classified as moderate severity burn. The understory vegetation is usually burned under the green areas adjacent to the brown and black areas.

Sediment Availability

Fire removes much of the protective litter and duff layer covering the mineral soil. It produces patches with a thin layer of ash, partially burned organic matter, or bare soil. Once the ash is removed by wind and overland flow, the bare patches represent potential sites of available sediment. Benavides-Solorio and MacDonald (2005) found the percent of bare patches within an area was one of the best predictors of sediment erosion. Our discussion here is limited to a few examples to illustrate why post-fire responses can be so variable in space and time. A more detailed discussion of actual transfer processes has been given by Shakesby and Doerr (2006) and in Chapter 7.

Direct effects

Fire temperatures can directly alter the particle-size distribution in soils and affect erodibility. The content of clay-size particles in soils has been measured in the laboratory to decrease rapidly with an increase in temperature up to about 400°C with a corresponding increase in the cement-like cohesion properties (Puri and Asghar 1940). This decrease in silt and clay-sized particles and a corresponding increase in stable composite particles in the sand-size range have been reported by several investigators (Dyrness and

Youngberg 1957, Duriscoe and Wells 1982). The implication is that surface soils are more erodible because the cohesive influence of the clay particles has been removed. At a threshold temperature of around 220°C, the organometallic cements are destroyed (Table 1; Giovannini et al. 1988) and soils become almost cohesionless and thus readily available for transport. This issue of sediment availability is complex because some evidence suggests that aggregate stability increases in some soils after fire with a corresponding decrease in erodibility (Mataix-Solera and Doerr 2004). Both situations are probably present to some degree and may explain some of the disparate erosional responses after fire.

A quantitative measure of soil erodibility is the critical boundary shear stress or the equivalent critical shear velocity of water. The critical shear velocity is measured at essentially the surface of the soil or sediment and it changes when subjected to the temperatures typical of forest fires. It represents a threshold for the entrainment or initiation of motion of soil or sediment particles and applies to both cohesive and non-cohesive particles. This threshold is a function of temperature (Fig. 7), and critical shear velocity for soils in unburned conditions ranges from 0.022 to 0.19 m s^{-1} (Elliot et al. 1989, Zhu et al. 2001, Moody et al. 2005).

The relation between critical shear velocity and temperature can be separated into three different temperature ranges (Moody et al. 2005). For temperatures less than 175°C, the critical shear velocity is essentially constant for a given type of soil. However, for temperatures between 175° and 275°C the critical shear velocity was a maximum (>0.063 m s^{-1}). At these temperatures the soils were indurated, failed to wet, and would either erode as one or two large chunks representing at least 30 percent of the sample or would fail to erode at the maximum velocity possible in the flume (Moody et al. 2005). This induration of soils was measured by a penetrometer by Wahlenberg et al. (1939) who showed that burned soils could be as much as five times as hard as unburned soils. For temperatures >275°C the critical shear velocity decreases rapidly with an increase in temperature to a value that is essentially a constant (0.016 to 0.025 m s^{-1}). The similarities in the changes with temperature of both critical shear velocity and water repellency (Table 1; DeBano 2000) suggest that the two soil properties may be linked. This link may be through the types of cementation processes exhibited by cohesive mixed-grain soils at different temperatures (Table 1; Giovannini et al. 1988).

The interesting implication is that the erodibility component of sediment availability in burned watersheds can be modeled more simply than erodibility in unburned watersheds. Forest fires reduce the complex spatial pattern of soil erodibility associated with unburned watersheds by reducing the range of critical shear velocity. Thus, the erodibility of soil after fire has two thresholds separating three different temperature ranges so that predicting the erosional response after a fire depends primarily on determining the distribution of the maximum soil temperature.

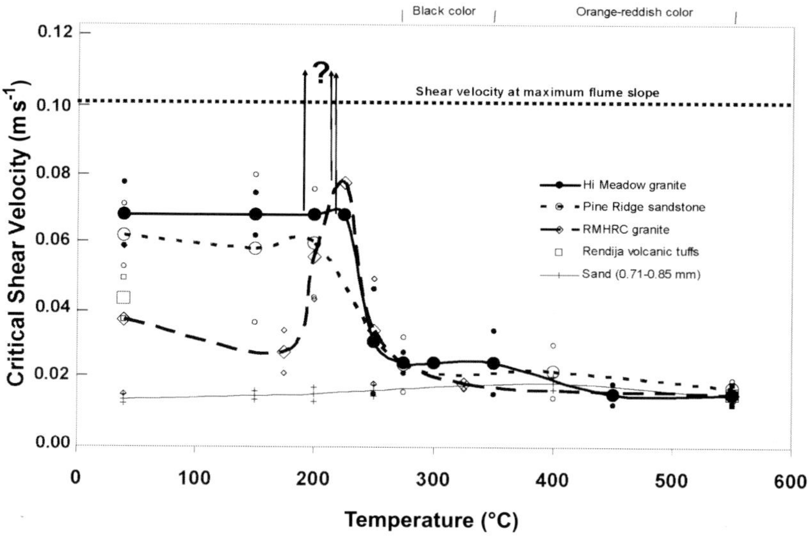

Fig. 7 Critical shear velocity for the initiation of erosion of soils subjected to temperatures typical of forest fires. Three types of forest soils (granite, sandstone, and volcanic tuffs) were used. Measurements were made in a recirculating flume, 1.84 m long by 0.13 m wide, where the slope of the flume could be changed to alter the velocity at the bed. Smaller symbols indicate the 95 percent confidence limits for the mean of replicate samples shown by the larger symbol (some limits do not show because larger symbols are superimposed). The ? indicates that the critical shear velocity measurements are a minimum and are certainly much greater than the upper limit of 0.10 m s^{-1} shown on the graph (Moody et al. 2005).

Indirect effects

The canopy and understory vegetation are reduced by fire, the frictional drag caused by vegetation is less, and the wind blows more readily through a more open forest. Wind erosion after fire has been measured in rangeland ecosystem (Blaisdel 1953, Hinds 1976), and in a shrub-land ecosystem in the Chihuahuan Desert in New Mexico, USA. Here Whicker et al. (2002) measured 70 times more transport of 0.010-mm particles (critical velocity was 7 m s^{-1} at 3 m above the ground) from a burned site than from a nearby unburned site. In a forested ecosystem (burned by the 2000 Cerro Grande Fire near Los Alamos, New Mexico), Whicker et al. (2001) found that wind speeds (1 m above the ground) in a severely burned area were greater than in an unburned area. Initially wind-driven dust flux from burned sites was more than one order of magnitude greater than from unburned sites. Surprisingly, this flux increased two to three times by the end of the first year after the fire and remained fairly constant for the second and third years after the fire (Whicker et al. 2006).

In addition, to transport via wind, the loss of the canopy, litter, and duff layers to fire makes the sediment more available to transport by rainsplash. The rainsplash transport process can couple non-linearly with the shallow overland-flow transport process to produce a more efficient transport process called "rain-flow" (Moss and Green 1983). It reaches a maximum transport rate when the water depth (~2 mm) equals about two to three raindrop diameters (Moss 1988). At this threshold depth raindrops can still penetrate the shallow water and dislodge particles, which are immediately entrained in the overland flow. Shallower depths have less transport capacity and larger depths prevent the raindrops from impacting and disrupting the soil surface.

Raindrop impacts and the transport of fine particles such as ash and sediment can seal pores at the surface of bare soil (Mallik et al. 1984, Martin and Moody 2001). The sealing process after a fire is an example of a positive feedback mechanism. Sealing of pores decreases the infiltration of water into the soil. This increases the runoff, and the detachment and transport of more sediment particles fill more pores. However, if runoff produces enough shear velocity on the soil surface, then it may erode the sealed layer (Poesen 1993) and suddenly alter the runoff. Sealing may happen on one patch of ground producing substantially more runoff than from another patch where conditions for sealing are not sufficient. Rock fragments on the soil surface after a fire can confound the runoff process. Poesen (1993) found that "rock fragments at the soil surface have an ambivalent effect on overland-flow production. On the one hand, rocks prevent direct infiltration of intercepted raindrop water into the soil. On the other hand, rock fragments prevent soil-surface sealing by protecting the soil surface against raindrop impact forces and, therefore, they have a positive effect on water intake rate."

Fires remove vegetation barriers like grass clumps and stem clusters that hold sediment on hillslopes. These vegetation barriers are a kind of threshold. When they are destroyed by fire, sediment is suddenly released from one storage reservoir on the hillslope and transported: 1) down the hillslope but still within the same storage reservoir, 2) to reservoirs in some unchannelized drainages, or 3) to reservoirs in channels. Where sediment is not stored on the hillslope behind vegetation barriers, it lies on the open slope at the angle of repose. The angle of repose is a threshold. This angle is greater when sediment is wet than when it is dry, because of the added cohesion of the soil moisture (Burkalow 1945). Thus, as the sediment on a hillslope dries, the sediment becomes unstable and it slides downhill when disturbed by "a breeze that shook a bush," a lizard, a deer, or even the "vibrations of an airplane" (Anderson et al. 1959). Gravity is the driver in this transfer process termed dry ravel. Dry ravel is the dominant transfer process in southern California after fire and in some places in Oregon (Anderson et al. 1959, Rice 1982, Wells 1987, Gabet 2003, Roering and Gerber 2005, Shakesby and Doerr 2006).

Fires remove vegetation barriers and increase the connectivity of patches on the hillslopes. The vegetation barriers can be considered to be surface roughness elements with various heights above the surface, which act to decrease the velocity of runoff on the soil surface and impede transfer processes. The height and density of surface roughness elements on a hillslope depends on the burn severity (Fig. 8). Surface roughness can be parameterized by the Darcy-Weisbach friction factor, f, where runoff velocity decreases as the f increases (Moore and Foster 1990). This friction factor on unburned slopes with grass, woody brush, and dense brush ranges from 50 to 140 (Weltz et al. 1992), on desert slopes with less vegetation it ranges from 1 to 15 (Abrahams et al. 1986), and on bare soil it can be less than 1 (Weltz et al. 1992). Lavee et al. (1995) found on Mt. Carmel in northwestern Israel that high intensity fire produced a smooth, homogenous surface. On the relatively smooth patches, they measured a "drastic increase in runoff". The decrease in friction also decreases the time for water to reach the outlet of a watershed and, thus, the resulting flood hydrograph reaches a higher peak discharge sooner than flow delayed by obstructions. Lower burn severity leaves the surface rougher with more obstructions. Therefore, a burned watershed has a mosaic-like surface that contains both rough and smooth patches of ground (Lavee et al. 1995). And the situation becomes more complex when "needle casts" from partially burned conifer trees later fall on the smooth surface (Pannkuk and Robichaud 2003). In addition to increasing the potential for

Fig. 8 Comparison of slopes with woody vegetation and grass and bare slopes after a forest fire.

surface water storage, the "needle casts" increase the friction for surface runoff.

As connectivity of these burned patches may be important, so Moody et al. (2008) defined a new variable, the functional hydraulic connectivity, to link burn severity and runoff. The hydraulic functional connectivity incorporates the spatial sequence of the burn severity of patches that compose a hillslope flow path. The burn severity affects hydraulic variables like the friction factor, infiltration rates, flow depth, and flow velocities along a hillslope flow path. Runoff from a hillslope will be greater if the sequence of burn severity associated with patches comprising a flow path increase (on average) downhill rather than if the sequence of burn severity decrease downhill toward the channel (Fig. 9).

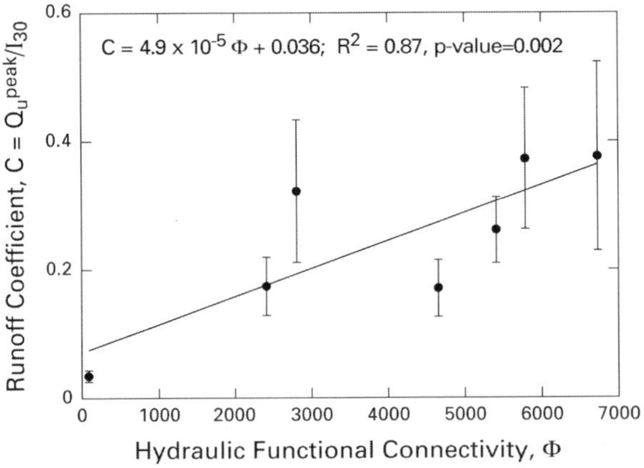

Fig. 9 Relation between runoff coefficient and hydraulic functional connectivity. The runoff coefficient is a non-dimensional variable and is the ratio of the peak discharge per unit drainage area divided by the maximum 30-minute intensity for the associated rainstorm. The hydraulic functional connectivity is the average of the burn severity (ΔNBR) for each pixel composing a flow path (in a 30-m × 30-m digital elevation model) weighted by the uphill contributing area and the local slope of the pixel. The starting point of each flow path was selected randomly along the boundaries of the watershed. (Detail in Moody et al. (2008)).

Driver

Fire has both direct effects and indirect effects on the drivers of geomorphic processes. Most indirect effects relate to changes in soil hydrologic properties (Neary et al. 2005, Chapters 4 and 6), infiltration (Chapter 3), as well as effects on soil (Chapter 5) and nutrients (Chapter 8). As these topics are covered in other chapters, we again limit our discussion to a few examples to illustrate the reasons why responses can be so variable in space and time.

Direct effects

Winds generated by forest fires certainly have enough energy to transport ash, fine silt and clay and even larger sediment sizes. These winds have caused blow-downs (Swanson 1981) and have sheared off trees (Fig. 10). Wind speeds are similar to those in tornados (Graham 1955, Moran and Stieglitz 1983, Brown 2006), and average wind speeds of 17.7 m s^{-1} and maximum 10-minute gusts on the order of 27 m s^{-1} were measured in a forest fire in northeastern Portugal (Viegas 2005). The critical shear velocity to initiate the movement of a relatively large 1.0-mm particle in the air is 0.46 m s^{-1} (Bagnold 1954). This translates into a critical wind speed at a height of 4 m above the surface equal to 13 m s^{-1}. Thus, winds generated during some fires can transport any particles less than 1.0 mm in diameter, in potentially any direction – even uphill. Project Flambeau measured maximum winds of 56 m s^{-1}, "which scoured the ground clean" of all mineral and fine fuel smaller than about 1 cm diameter (Palmer 1981).

Fig. 10 Trees sheared off during the 2002 Missionary Ridge Fire near Durango, Colorado, USA. This fire generated three fire-vortices or tornados.

Indirect effects

Fire-induced water repellency is a well-known post-fire phenomenon. The direct effect is on the infiltrability and the indirect effect is on the runoff. For convenience of discussion, we have separated water repellency into two processes – one process is chemical and one process is physical. The chemical process of coating particles with hydrophobic compounds is well reported in the literature (DeBano 2000, Doerr et al. 2000, Shakesby and Doerr 2006; Chapters 3, 4, and 7). This fire-induced water repellency is in addition to the natural water repellency present in most soils before a fire (Doerr and Thomas 2000, Shakesby and Doerr 2006).

The physical process of heating during a fire removes water from a thin layer of soil at the surface. This thin layer controls the subsequent movement of water into the soil. Temperatures during a fire (up to about 900°C; DeBano 1981) are hot enough to decrease the soil moisture well below the critical soil moisture threshold (Chapter 4). These temperatures can remove even the tenacious film of adsorbed and capillary-bound water coating soil particles (Bachmann and van der Ploeg 2002) that is normally present at the wilting point (critical soil moisture available for plant use) and even at the lower soil moistures associated with the residual water content. Puri and Asghar (1940) found that the hygroscopicity of soils (with ~5 percent clay content) decreased with increase in temperature above 400°C. These soils adsorbed about 0.01 cm^3 cm^{-3} at 10 percent humidity and the humidity may be even lower immediately after a fire. Soil moisture measured for samples subjected to 550°C in the laboratory had values less than 0.001 cm^3 cm^{-3} (Moody et al., unpublished data).

A second moisture threshold has been measured in aggregates with dual porosity composed of intra-aggregate (pores between individual particles of the aggregate) and inter-aggregate (pores between soil aggregates) pores. This is the hydraulic critical soil moisture, θ_{ch}, where the soil matrix pressure changes rapidly by about three orders of magnitude with essentially no change in the soil moisture (Blonquist et al. 2006). Thus, the first step that must begin after a fire is the physical process of re-adsorption and reforming of capillary-bound water or rewetting of the soil. The time lag in rewetting the soil may partially explain why disproportionate overland flow is frequently observed during the first rainstorms after a fire before the rewetting process is complete.

Infiltration-excess overland flow (Hortonian runoff) is rarely observed in forested watersheds but is common in burned watersheds (Shakesby and Doerr 2006). This type of runoff depends on another threshold. This threshold corresponds to the limiting infiltration rate of the soil or infiltrability (Horton 1933, Smith and Goodrich 2005). Fires change the soil properties and thus the infiltrability. A rainfall-intensity threshold has also been identified for this type of flow in burned watersheds (Table 2) and may represent a threshold equal to the spatial average of the infiltrability of some fundamental patch size within the watershed.

Table 2 Rainfall intensity thresholds

Rainfall intensity (mm h^{-1})	Location	Country	Reference
13	San Gabriel Mountains in California	USA	Doehring 1968
10	Mount Carmel	Israel	Inbar et al. 1998
10-20	Bega Batholith, New South Wales	Australia	Mackay and Cornish 1982
10	2002 Bobcat Fire near Ft. Collins, Colorado	USA	Kunze and Stednick 2006
10	1996 Buffalo Creek Fire near Denver, Colorado	USA	Moody and Martin 2001c
8.5	2000 Cerro Grande Fire near Los Alamos, New Mexico	USA	Moody et al. 2008

Another reason for a rainfall-intensity threshold could be the nature of the friction factor on burned hillslopes. The height of surface roughness elements, k, on the hillslope produces friction that controls the velocity of shallow overland flow. Recent studies by Lawrence (1997) and Nikora et al. (2001) have shown that for the special conditions when the flow depth, h, is less than k, the frictional factor, f, increases with an increase in flow depth, h, or

$$f \sim h \quad \text{for} \quad < k \tag{4}$$

Only when the surface roughness elements are completely inundated does the frictional factor begin to decrease with an increase in flow depth (Manning's equation; Moore and Foster 1990), which can be written as a power law:

$$f \sim h^{-0.33} \quad \text{for} \quad h > k \tag{5}$$

This non-linear change in the nature of the frictional factor implies the existence of a critical threshold near $h = k$. Therefore if rainfall intensities are large enough to cause the overland flow depths to completely cover the surface roughness elements there could be a sudden increase in the runoff from a burned hillslope.

As vegetation regrowth continues after a forest fire, the surface roughness would increase with perhaps a corresponding increase in the rainfall-intensity threshold. This would slowly raise the threshold and eventually it could reach a large value typical of forested watershed where infiltration-excess overland flow is not the dominant process. Four years of rainfall-runoff data after the 1996 Buffalo Creek Fire in Colorado (Moody and Martin 2001b) suggested that the rainfall threshold may have increased with time after the fire with an increase from 10 mm h^{-1} in 1996-97 to about 16 mm h^{-1} in 1999. Rainfall-runoff data collected after the 2000 Cerro Grande Fire also indicated an increase in the rainfall-intensity threshold from 7.6 mm h^{-1} in 2001 to 11.1 mm h^{-1} in 2002 (Moody et al. 2008). However, in both cases the conditions after a fire are continually changing and the number of measured storms was insufficient each year to establish this trend with statistical confidence.

Linking Sediment Availability and Driver

A general effect of fires on geomorphic processes is to increase the connectivity of the landscape. Specific areas of a burned landscape that will produce the largest response can be identified by superimposing the spatial distribution of the driver over the spatial distribution of sediment availability. The actual magnitude of the response is some non-linear function of the sediment availability, the driver, and some measure of the degree of connectivity.

Superposition

In Colorado, two high-intensity rainstorms with different sizes and shapes caused floods that produced different erosional responses from the same burned area. The first substantial flood after the 1996 Buffalo Creek Fire in the Pikes Peak batholith in Colorado (Moody and Marin 2001a) was on 12 July 1996. The footprint of the rain storm based on radar data (Yates et al. 2000) was nearly identical to the burned area (Fig. 11). Maximum intensity (based on rain gages and paleohydrology) was estimated to be about 90 mm h^{-1} (Jarrett and Browning 1999) and the response produced the erosion and deposition of about 1.1×10^6 m^3 of sediment. This can be expressed as a yield per burned area to compare with other erosion amounts in the literature. The yield is 40,000 Mg km^{-2} y^{-1} and is similar to the yield (43,000 Mg km^{-2} y^{-1}) reported by Meyer and Pierce (2003) after a fire in the Idaho granitic batholith. It was estimated that 80 percent of the sediment was eroded from the channel area in the Pikes Peak batholith (Moody and Martin 2001b) and at least 40 percent was eroded from the channel area in the Idaho batholith (J. Pierce, personal communication 2006). Using the channel area rather than the burned area gives a more meaningful yield of 8×10^8 Mg km^{-2} y^{-1} in the Pikes Peak batholith. A second storm on 13 August 2001 had a small footprint (Fig. 11) with a maximum 30-minute rainfall intensity of 66 mm h^{-1} measured by a recording rain gage near the center of the storm. The erosional response was confined to about 700 m adjacent to the main channel of Spring Creek and the flood produced about 300 Mg km^{-2} y^{-1} or an equivalent yield from the tributary channel areas of 17,000 Mg km^{-2} y^{-1}. While some of the differences in sediment yield from the two floods could be attributed to the lower rainfall intensity, it is clear that the relative size of the storm compared to the size of the watershed is the primary reason for the difference.

Quite different responses from the same watershed also could be produced by two rainstorms with the same size, shape, and average rainfall intensity, but with different spatial distributions of rain cells. The different spatial distribution of rainfall means that a different combination of hillslope flow paths would be affected within the watershed. And because the average

a)

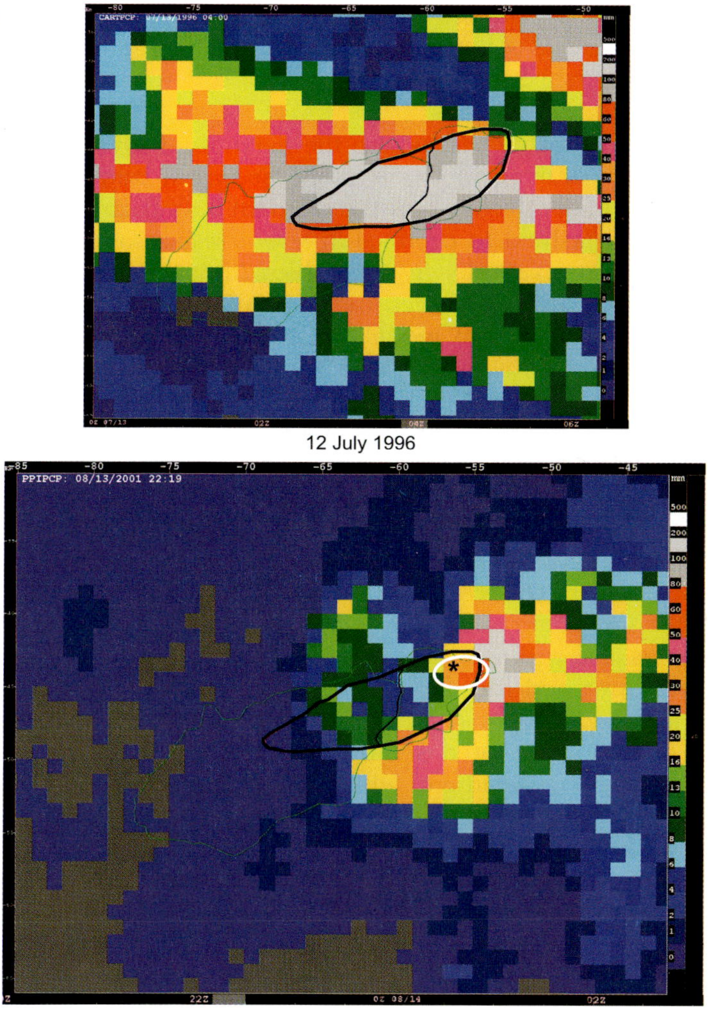

12 July 1996

b)

13 August 2001

Fig. 11 Relative size of rainstorm and watershed. a) Estimates of the rainfall (based on radar reflectivity within 1-km^2 pixels) for the rain storm on 12 July 1996 are shown superimposed over the boundaries of Spring Creek where 79 percent of the drainage area was burned by the 1996 Buffalo Creek Fire. The fire perimeter is the thick black cigar-shape. The thinner irregular line separates the Buffalo Creek watershed to the left from the Spring Creek watershed to the right. The majority of the rainfall accumulated during two hours (Yates et al. 2000). b) Estimates of the rainfall (based on radar reflectivity) for the rainstorm on 13 August 2001 are shown superimposed over the boundaries of Spring Creek. Rainfall accumulation was for two hours. The asterisk shows the location of a recording rain gage. The white oval indicates the 700 m reach of Spring Creek where alluvial fans were deposited at the mouths of tributaries to Spring Creek in response to the rain on 13 August 2001. (Images prepared by David Yates, National Center for Atmospheric Research).

hydraulic functional connectivity is not necessarily the same for the two different combinations of flow paths, then quite different responses can result for the same watershed for identical storms at different times (Moody et al. 2008).

It is highly probable that where the maxima of several superimposed variables such as drainage network, slope, sediment storage volumes, burn severity, and rainfall intensity are all coincident the variables will combine in some non-linear fashion to produce different responses from neighboring watersheds. As an example, five variables (topography, slope, sediment storage volumes, burn severity, and rainfall intensity) are shown in a simplistic and imaginary box model of a burned landscape (Fig. 12). All are spatially variable and the maxima are not coincident. It is assumed that no rain fell outside the zero contour and therefore no water can flow from upstream (outside the box) through the two tributary watersheds shown inside the box.

The interfluve area labeled 'A' between the two channels has the highest burn severity. Here the critical shear velocity for the entrainment of sediment is probably the lowest because the high burn severity indicates soil temperatures may have exceeded 300°C (Fig. 7) and any vegetation that once protected the soil from raindrop impact probably has been removed (Fig. 8). The area 'A' is also within a cell of high rainfall intensity but no sediment is stored in this area and the slope is flat so that there will be essentially no sediment transport response from area 'A'. Area 'B' has a patch with a 'moderate' volume of stored sediment and is located in an area with a relatively strong slope gradient, which would increase sediment transport. The rainfall intensity is moderate (between 50 and 100 units) but most of this area did not burn (except for the moderate burn severity at the small downstream tip of the sediment storage reservoir) so that vegetation still covers most of the sediment storage reservoir and the critical shear velocity would be that of an unburned soil. Thus, while the potential exists for substantial sediment transfer down the channel from area 'B', the response would be little or nothing. Area 'C' has a large volume of stored sediment. While the burn severity is moderate probably most of the protective vegetation has been burned and the critical shear velocity in some portions of the sediment storage reservoir may be quite high if the soil temperatures were between 175° and 275°C during the fire (Fig. 7). The slope gradient is moderate but the rainfall intensity in the cell centered over the storage reservoir is the highest within this burned landscape. Therefore, this watershed would probably respond and produce much more water and sediment than the other watershed.

The examples in this section have illustrated the importance of the spatial superposition of the variables controlling sediment availability and the driver. The response cannot be predicted as a linear combination of each essential variable because of the non-linear interaction of the variable. Therefore in

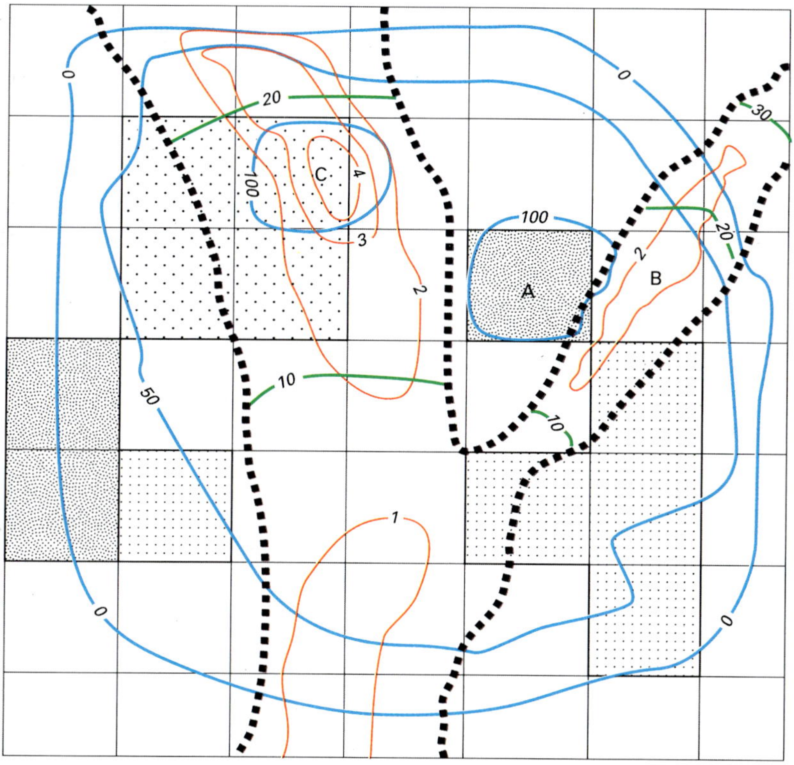

Fig. 12 Superposition of variables that contribute to sediment transport in a simplistic and imaginary box model of a burned landscape. Two converging watersheds are shown as hatched lines with flow visualized coming from the top to bottom. The green lines are contours of equal slope gradient in the channel shown with arbitrary units. The red lines represent volume of stored sediment with arbitrary units. The stippled squares represent the burn severity as might be quantified using Δ*NBR* (for example, see Figure 5). Denser stippling represents higher burn severity. The blue contours represent the rainfall intensity using arbitrary units and the footprint of the rainstorm is defined by the zero contour. No rain fell outside this contour and therefore no runoff enters the two watersheds from outside the box.

some cases the greatest response may not correspond to the maxima of all variables but some optimal combination (Wolman and Miller 1960).

Storm sequence

The effectiveness of floods of a given magnitude will vary significantly depending on the time sequence of storms (Beven 1981). This is illustrated in Fig. 4 where the smallest rainstorm had almost no effect on the peak water discharge or sediment transport. This small storm primed the hillslope by leaving high antecedent soil moisture behind such that the following moderate storm produced a response nearly equal to the preceding larger rainstorm. Germanoski et al. (2002) found that in the Central Great Basin of Nevada two similar watersheds (Crow and Wall Canyons) had substantially different responses after fire because of the sequence of storms. After the fire in Crow Canyon, the precipitation was twice the annual rainfall and the average incision was 1.9 m, while after the fire in Wall Canyon, the precipitation was less than the annual rainfall and no incision was observed.

Burned watersheds in Colorado (Spring Creek) and New Mexico (Rendija Canyon) also had a different sequence of storms, flood magnitude, and sediment transport response. The first storm to impact the Spring Creek watershed (granitic terrain) was two months after the 1996 Buffalo Creek Fire and had the most intense rainfall (I_{30} ~90 mm h^{-1}; ~100-year recurrence interval; Miller et al. 1973) of all the storms measured after the fire (Fig. 13). In contrast, the first storm to impact the Rendija Canyon watershed (volcanic terrain) was one month after the 2000 Çerro Grande Fire but was one of the smallest storms (I_{30} ~17 mm h^{-1}; ~1-year recurrence interval; Reneau et al. 2003). The most intense storm (I_{30} ~62 mm h^{-1}) in the Rendija Canyon

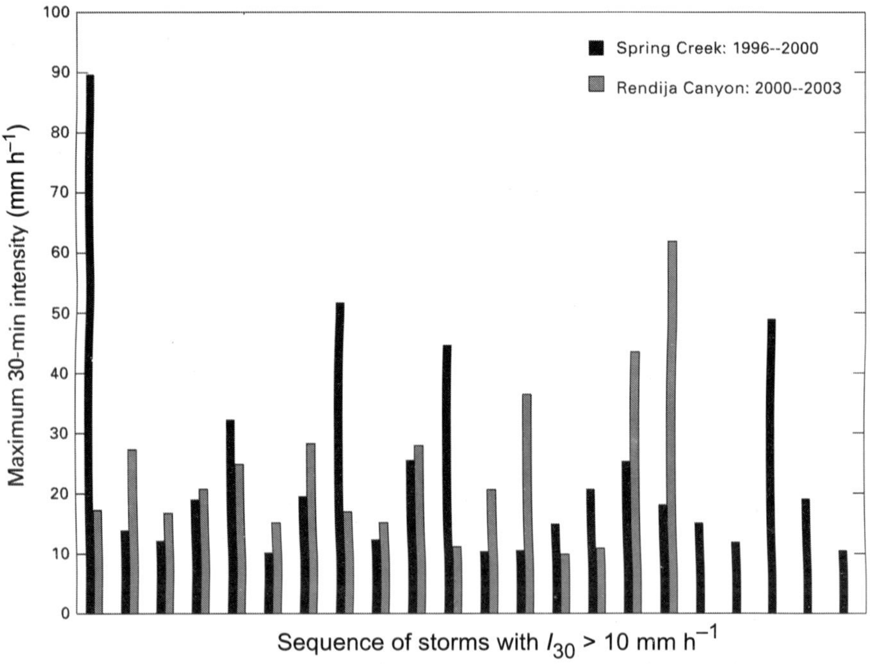

Fig. 13 Magnitude of a sequence of storms (only those with maximum 30-minute rainfall intensity, I_{30} > 10 mm h^{-1}) in the Spring Creek and Rendija Canyon watersheds. The time interval between storms is different for each pair of storms and for each watershed.

watershed was three years after the fire and this had about a 25- to 50-year recurrence interval (Reneau et al. 2003). However, the regrowth of vegetation was substantial and the sediment transfer response was damped.

Geomorphic Transfer Processes

Complex response

Schumm (1991) observed that "when dealing with complex systems, a single explanation in most cases will not be sufficient." Another insight for planning management strategies of burned landscapes is that "the complex system when interfered with or modified is unable to adjust in a progressive and systematic fashion, and its response can be complex" (Schumm 1991).

One type of complex response is the changes in the transport process. Fire plays a role in these changes. Fires indirectly lower the threshold rate for infiltration (see discussion above and Chapters 4 and 7), and this changes the runoff process from saturated-excess overland flow (typical of forested watersheds) to infiltration-excess overland flow (typical of burned watersheds). In some landscapes, the decay of roots of burned trees after four to ten years can change the process again from infiltration-excess overland-flow immediately after the fire to the saturated landslide process (Meyer and Pierce 2003).

Post-fire erosion after the first rainstorm depends on the stream order of the channel. In low-order channels where sediment is frequently stored, these sediment storage reservoirs are flushed out and the system switches from a transport-limited to a supply-limited system (Keller et al. 1997). In burned watersheds in Colorado, the sediment eroded soon after the fire from low-order channels was deposited in higher-order channels and the system shifted from a supply-limited to a transport-limited system (Moody 2001). Not only was there a change in transport processes but the response in the low-order channels and drainages (erosion) was different than the response in the high-order channels (deposition).

Sediment transport processes within a short reach of channel can also change. Three sediment transport processes were observed in a 5000 m reach of the main channel in the Spring Creek watershed burned by the 1996 Buffalo Creek Fire (Moody 2001). These were a uniform, discontinuous, and unsteady process at different places along the channel and at different times after the fire. Initially, the first flood on 12 July 1996 deposited a large quantity of sediment up to 3 m thick in the main channel with alluvial fans at the mouths of tributaries. The uniform transport process was usually in the spring and early summer when the discharge was steady. The flow incised some reaches, formed a channel 1 to 2 m wide and typically 0.5 m deep, but downstream from the incision, the flow formed a braided channel about 20 to 40 m wide. This alternating pattern of incised and braided reaches was observed along the entire 5000 m length of channel.

The discontinuous transport process was usually in the late summer when the base flow had decreased. Surface water disappeared into the bed, depositing the bedload, and creating in-channel fans (Fig. 14). These fans moved slowly upstream as additional sediment was deposited over time at the upstream edge where the water disappeared and eroded from the downstream edge where the water re-emerged. The unsteady process was episodic flash floods. In the first few years after the fire, rainfall greater than the threshold of 10 mm h^{-1} usually generated a flash flood of some magnitude. Much of the sediment, which moved as bedload (1 to 32 mm) during the other transport processes was transported as suspended load during these flash floods. These floods caused an aggradation of the channel, and usually left a relatively smooth surface on which the other two transport processes began again.

Fig. 14 View of an in-channel fan looking upstream in the main channel of Spring Creek. The man is about 1.8 m tall. The photo was taken 1 year after the fire and about 10 months after the flood on 12 July 1996 that deposited sediment in the main channel up to 3 m thick in some places. The green vegetation is herbaceous ground cover composed of fire-adapted plants (mostly leafy spurge, *Euphorbia esula*; and dog bane, *Apocynum androsaemfolium*). They covered some slopes near the channel but the slopes in the background are bare. Water percolated into the bed at the upstream edge of the fan, deposited sediment, and the fan form slowly moved upstream. Water emerged at the downstream edge of the fan (near the feet of the man), easily eroded non-cohesive coarse sand and gravel (0.5 to 2 mm in diameter), leaving small terraces, and producing a braided stream.

SUMMARY

Geomorphic systems are usually non-linear with special characteristics (power laws, threshold, feedback mechanisms, and sensitivity) that partially explain the contrasting responses from two similar watersheds. Power laws may be the consequence of multiple variables controlling a complex system. The response of a system can be disproportionate when the exponent is greater than one or when thresholds are exceeded. Positive feedback mechanisms cause the response to increase dramatically and sensitivity to small perturbations can cause the response to grow to a point way beyond what would be predicted for a linear system.

The two elements of any geomorphic transfer process are the availability of sediment and a driver, such as rainfall, to provide the energy to transport the sediment. These two elements will have a complex spatial distribution within and over a burned landscape. Fires, in general, increase the connectivity of the landscape but produce a patchy distribution of burn severity. The soil burn severity has both direct and indirect effects on the variables that control the sediment availability and it affects the variables that influence the rainfall-runoff response.

In most burned landscapes, the final response is a superposition of the patchy distribution of the sediment availability and rainfall. The superposition is not a linear combination of variables but a complex non-linear combination of the variables. Some optimal combination of the sediment availability, rainfall, and a measure of connectivity can produce quite disparate responses from neighboring watersheds that appear quite similar.

The disparate responses are not just a function of the spatial variability. The sequence of rainstorms determines the geomorphic response in the short time of the response and recovery after a forest fire disturbance. If a 'large' rainstorm follows soon after the fire, then the response can be catastrophic, while nothing may happen if the first storm is small. However, it is possible that several small storms with short interstorm intervals may 'prime' the landscape to respond disproportionately to what is considered a 'moderate' storm.

In this chapter, we have taken a short-term viewpoint and have focused on a single fire's effects on geomorphic processes. The long-term viewpoint is that fire effects on geomorphic processes are the cumulative legacies from several fire-flood cycles. Each cycle will leave behind a legacy of geomorphic features such as eroded hillslopes, incised channels, alluvial fans, aggraded floodplains, and small terraces that shape the landscape. These landscape features will also affect the behavior of the next fire and the possible subsequent geomorphic responses (Swanson 1981). Thus, landscape evolution and fire are forever linked.

Acknowledgments

David Mixon provided extensive assistance in GIS analysis of landscapes and Andrea Williams unraveled the complexities of transforming radar reflectivities into meaningful estimates of rainfall. Bob Meade provided excellent perspective and polish. Jennifer Pierce made many sound technical suggestions to improve this chapter.

References

Abrahams, A.D., A.J. Parsons, and S.H. Luk. 1986. Resistance to overland flow on desert hillslopes. *Journal of Hydrology*, 88: 343-363.

Anderson, H.W., G.B. Coleman, and P.J. Zinke. 1959. Summer slides and winter scour – dry-wet erosion in southern California mountains. Technical Paper 36, U.S. Department of Agriculture, Forest Service, Pacific Southwest Forest and Range Experiment Station, Berkeley, California, USA.

Bachmann, J., and R.R. van der Ploeg. 2002. A review on recent developments in soil water retention theory: interfacial tension and temperature effects. *Journal of Plant Nutrition and Soil Science*, 165(4): 468-478.

BAER. 2000. Cerro Grande Fire, Burned Area Emergency Rehabilitation BAER plan. Interagency BAER Team, U.S. Department of Agriculture, Forest Service, Region 3, Albuquerque, New Mexico, USA.

Bagnold, R.A. 1954. *The Physics of Blown Sand and Desert Dunes*. Methuen & Co., London, UK.

Benavides-Solorio, J. and L.H. MacDonald. 2005. Measurement and prediction of post-fire erosion at the hillslope scale, Colorado Front Range. *International Journal of Wildland Fire*, 14(5): 1-18.

Beven, K. 1981. The effect of ordering on the geomorphic effectiveness of hydrologic events, pp. 510-526. *In* T.R.H. Davies and A.J. Pearce [eds.]. Symposium on Erosion and Sediment Transport in Pacific Rim Steeplands, Christchurch, New Zealand, IAHS-AISH Publication Number 132. International Association of Hydrological Sciences, Washington, D.C., USA.

Blaisdell, J.P. 1953. Ecological effects of planned burning of sagebrush-grass range on the Upper Snake River Plains. Technical Bulletin 1075: 1-39. U.S. Department of Agriculture, Washington, D.C., USA.

Blonquist, J.M. Jr., S.B. Jones, I. Lebron, and D.A. Robinson. 2006. Microstructural and phase configurational effects determining water content: Dielectric relationships of aggregated porous media. *Water Resources Research*, 42(5): WO5424, doi: 10. 1029/2005WR004418.

Bond, W.J. and J.E. Keeley. 2005. Fire as a global 'herbivore': the ecology and evolution of flammable ecosystems. *Trends in Ecology and Evolution*, 20(7): 387-394.

Brown, D.J. 2006. *Under a Flaming Sky – The Great Hinckley Firestorm of 1894*. The Lyons Press, Guilford, Connecticut, USA.

Brown, J.A.H. 1972. Hydrologic effects of a brushfire in a catchment in south-eastern New South Wales. *Journal of Hydrology*, 15: 77-96.

Brown, R.J.E. 1983. Effects of fire on the permafrost ground thermal regime, 18: 97-110, 317. *In* R.W. Wein, and D.A. MacLean [eds.]. *The Role of Fire in Northern Circumpolar Ecosystems.* John Wiley & Sons, SCOPE (Scientific Committee on Problems of the Environment), Chichester, UK.

Brunsden, D. and J.B. Thornes. 1979. Landscape sensitivity and change. *Transactions of the Institute of British Geographers,* 4(4): 463-484.

Burkalow, A. van. 1945. Angle of repose and angle of sliding friction: an experimental study. *Bulletin of the Geology Society of America,* 56: 669-707.

Cerdá, A. 1997. The effect of patchy distribution of *Stipa tenacissima* L. on runoff and erosion. *Journal of Arid Environments,* 36: 37-51.

Davenport, D.W., D.D. Breshears, B.P. Wilcox, and C.D. Allen. 1998. Viewpoint: Sustainability of pinyon-juniper ecosystems: a unifying perspective of soil erosion thresholds. *Journal of Range Management,* 51(2): 231-240.

DeBano, L.F. 1981. Water repellent soils: a state-of-the-art. General Technical Report PSW-46. U.S. Department of Agriculture, Forest Service, Pacific Southwest Forest Range Experiment Station, Berkeley, California, USA.

DeBano, L.F. 2000. The role of fire and soil heating on water repellency in wildland environments: a review. *Journal of Hydrology,* 231-232: 195-206.

Dekker, L.W., S.H. Doerr, K. Oostindie, A.K. Ziogas, and C.J. Ritsema. 2001. Actual water repellency and critical soil water content in a dune sand. *Soil Science Society of America Journal,* 65: 1667-1675.

Dietrich, W.E. and T. Dunne. 1978. Sediment budget for a small catchment in mountainous terrain. *Zeitschrift für Geomorphology Neues Folga, v. Supplement Band* 29: 191-206.

Dietrich, W.E., T. Dunne, N.F. Humphrey, and L.M. Reid. 1982. Construction of sediment budgets for drainage basins. *In* F.J. Swanson, R.J. Janda, T. Dunne and D.N. Swanston [tech. eds.]. Workshop on Sediment Budgets and Routing in Forested Drainage Basins: Proceedings, General Technical Report PNW-141. U.S. Department of Agriculture, Forest Service, Pacific Northwest Forest and Range Experiment Station, Portland, Oregon, USA.

Doehring, D.O. 1968. The effect of fire on geomorphic processes in the San Gabriel Mountains, California. *Contributions to Geology,* 7: 43-65.

Doerr, S.H., R.A. Shakesby, and R.P.D. Walsh. 2000. Soil water repellency: its causes, characteristics and hydro-geomorphological significance. *Earth-Science Reviews,* 51: 33-65.

Doerr, S.H. and A.D. Thomas. 2000. The role of soil moisture in controlling water repellency: new evidence from forest soils in Portugal. *Journal of Hydrology,* 231-232: 134-147.

Duriscoe, D.M. and W.G. Wells II. 1982. Effects of fire on certain physical properties of selected chaparral soils. *In* C.E. Conrad and W.C. Oechel [tech. coord.]. Proceedings of the Symposium on Dynamics and Management of Mediterranean-type Ecosystems, San Diego, California. General Technical Report PSW-58. U.S. Department of Agriculture, Forest Service, Pacific Southwest Forest and Range Experiment Station, Berkeley, California, USA.

Dyrness, C.T. and C.T. Youngberg. 1957. The effect of logging and slash-burning on soil structure. *Soil Science Society of America Proceedings,* 21: 444-447.

Elliot, W.J., A.M. Liebenow, J.M. Laflen and K.D. Kohl. 1989. A Compendium of Soil Erodibility Data From WEPP Cropland Soil Field Erodibility Experiments 1987 &

1988, NSERL Report Number 3. U.S. Department of Agriculture, Agricultural Research Service, National Soil Erosion Research Laboratory, Purdue University, West Lafayette, Indiana, USA.

Elliott, J.G. and R.S. Parker. 2001. Developing a post-fire flood chronology and recurrence probability from alluvial stratigraphy in the Buffalo Creek watershed, Colorado, USA. *Hydrological Processes*, 15(15): 3039-3051.

Flanagan, D.C. and M.A. Nearing [eds.]. 1995. USDA-Water Erosion Prediction Project (WEPP): Hillslope profile and watershed model documentation, NSERL Report 10, U.S. Department of Agriculture, Agricultural Research Service, National Soil Erosion Research Laboratory, Purdue University, West Lafayette, Indiana, USA.

Flannigan, M.D., B.J. Stocks, and B.M. Wotton. 2000. Climate change and forest fires. *The Science of the Total Environment*, 262: 221-229.

Gabet, E.J. 2003. Post-fire thin debris flows; sediment transport and numerical modeling. *Earth Surface Processes and Landforms*, 28(12): 1341-1348.

Germanoski, D., J.R. Miller, and D. Latham. 2002. The importance of event sequencing on the geomorphic impact of wildfire in the central Great Basin. Geological Society of America Annual Meeting, Denver, Colorado. Abstracts, 34(6): 319. *Online at:* http://gsa.confex.com/gsa/2002AM/finalprogram/abstract_41994.htm [accessed 2 Sep. 2008]

Giovannini, G., S. Lucchesi, and M. Giachetti. 1988. Effect of heating on some physical and chemical parameters related to soil aggregation and erodibility. *Soil Science*, 146(4): 255-261.

Goudie, A.S., R.J. Allison, and S.J. McLaren. 1992. The relations between modulus of elasticity and temperature in the context of the experimental simulation of rock weathering by fire. *Earth Surface Processes and Landforms*, 17(6): 605-615.

Graham, H.E. 1955. Fire whirlwinds. *Bulletin of the American Meteorological Society*, 36(3): 99-103.

Gupta, H., S. Sorooshian, X. Gao, B. Imam, K.L. Hsu, L. Bastida, J.Li, and S. Mahami. 2002. The challenge of predicting flash floods from thunderstorm rainfall. *Philosophical Transactions of the Royal Society*, 360: 1363-1371.

Hairsine, P.B. 1988. A physically based model of the erosion of cohesive soils, Ph.D. thesis, Griffith University, Division of Australian Environmental Studies, Brisbane, Queensland, Australia.

Hershfield, D.M. 1961. Rainfall frequency atlas of the United States for duration from 30 minutes to 24 hours and return periods from 1 to 100 Years, U.S. Department of Commerce, Washington, D.C., USA.

Hinds, W.T. 1976. Range fires can be bad medicine on some soils. p.56. *In* Abstracts 29th Annual Meeting of the Society for Range Management, Wheat Ridge, Colorado, USA.

Horton, R.E. 1933. The role of infiltration in the hydrologic cycle. *Transactions, American Geophysical Union*, 14: 446-460.

Inbar M., M. Tamir and L. Wittenberg. 1998. Runoff and erosion processes after a forest fire in Mount Carmel, a Mediterranean area. *Geomorphology*, 24: 17-33.

Jarrett, R.D. and T.W. Browning. 1999. Geomorphic estimates of rainfall, floods, and sediment runoff: Applied to the 1996 wildfire, Buffalo Creek, Colorado, Abstract, Geological Society of America, Denver, Colorado, 31(7): 313.

Keller, E.A., D.W. Valentine, and D.R. Gibbs. 1997. Hydrological response of small watersheds following the southern California Painted Cave Fire of June 1990. *Hydrological Processes*, 11(4): 401-414.

Key, C.H. and N.C. Benson. 2006. Landscape Assessment: Remote sensing of severity, the normalized burn ratio; and ground measure of severity, the composite burn index. *In* D.C. Lutes, R.E. Keane, J.F. Caratti, et al. [eds.]. FIREMON: Fire Effects Monitoring and Inventory System, General Technical Report RMRS-GTR-164-CD. U.S. Department of Agriculture, Forest Service, Rocky Mountain Research Station, Ogden, Utah, USA.

Kunze, M.D. and J.D. Stednick. 2006. Streamflow and suspended sediment yield following the 2000 Bobcat Fire, Colorado. *Hydrological Processes*, 208: 1661-1681.

Lavee, H., P. Kutiel, M. Segev, and Y. Benyamini. 1995. Effect of surface roughness on runoff and erosion in a Mediterranean ecosystem: the role of fire. *Geomorphology*, 11(3): 227-234.

Lawrence, D.S.L. 1997. Macroscale surface roughness and frictional resistance in overland flow. *Earth Surface Processes and Landforms*, 22: 365-382.

Lewkowicz, A.G. and C. Harris. 2005. Frequency and magnitude of active-layer detachment failures in discontinuous and continuous permafrost, northern Canada. *Permafrost and Periglacial Processes*, 16: 115-130.

Long, J.N. 2003. Diversity complexity and interactions an overview of Rocky Mountain forest ecosystems. *Tree Physiology*, 23: 1091-1099.

MacDonald, L.H. and E.L. Huffman. 2004. Post-fire soil water repellency: Persistence and soil moisture thresholds. *Soil Science Society of America Journal*, 68: 1729-1734.

Mackay, S.M. and P.M. Cornish. 1982. Effects of wildfire and logging on the hydrology of small catchments near Eden, N.S.W. pp. 111-117. *In* E.M. O'Loughlin and L.J. Bren [eds.]. *Preprints of Papers for the First National Symposium on forest Hydrology*. Institute of Engineers, Barton, Australian Capital Territories, Australia.

Mallik, A.U., C.H. Gimingham, and A.A. Rahman. 1984. Ecological effects of heather burning, I. Water infiltration, moisture retention and porosity of surface soil. *Journal of Ecology*, 72: 767-776.

Martin, D.A. and J.A. Moody. 2001. Comparison of soil infiltration rates in burned and unburned mountainous watersheds. *Hydrological Processes*, 15: 2893-2903.

Mataix-Solera, J. and S.H. Doerr. 2004. Hydrophobicity and aggregate stability in calcareous topsoils from fire-affected pine forests in southeastern Spain. *Geoderma*, 118: 77-88.

Meyer, G.A. and J.L. Pierce. 2003. Climatic controls on fire-induced sediment pulses in Yellowstone National Park and Central Idaho: a long-term perspective. *Forest Ecology and Management*, 178: 89-104.

Middleton, G.V. and J.B. Southard. 1984. Mechanics of Sediment Movement, Lecture notes for Short Course No. 3, Eastern Section of the Society of Economic Paleontologists and Mineralogists, Providence, Rhodes Island, USA.

Miller, J.F., R.H. Frederick, and R.J. Tracey. 1973. Precipitation-frequency atlas of the western United States, Volume III – Colorado, National Oceanic and Atmospheric Administration, National Weather Service, Silver Springs, Maryland, USA.

Moody, J.A. 2001. Sediment transport regimes after a wildfire in steep mountainous terrain. Volume 2, pp. X-41 - X-48. *In* Proceedings of the Seventh Federal Interagency Sedimentation Conferences 1947 to 2006. U.S. Geological Survey,

Denver, Colorado, USA. *Online at:* http://pubs.usgs.gov/misc_reports/ FISC_1947-2006/pdf/1st-7thFISCs-CD/7thFISC/7Fisc-V2/7FSC2-10.pdf [accessed 2 Sep. 2008]

Moody, J.A. and D.A. Martin. 2001a. Hydrological and Sedimentologic Response of Two Burned Watersheds in Colorado. Water-Resources Investigations Report 01-4122, U.S. Geological Survey, Denver, Colorado, USA.

Moody, J.A. and D.A. Martin. 2001b. Initial hydrologic and geomorphic response following a wildfire in the Colorado Front Range. *Earth Surface Processes and Landforms*, 26(10): 1049-1070.

Moody, J.A. and D.A. Martin. 2001c. Post-fire, rainfall intensity – peak discharge relations for three mountainous watersheds in the western USA. *Hydrological Processes*, 15(15): 2981-2993.

Moody, J.A. and D.A. Martin. 2008. Synthesis of sediment yields after wildland fire in different rainfall regimes in the western United States. *International Journal of Wildland Fire* (in press).

Moody, J.A., D.J. Smith, and B.W. Ragan. 2005. Critical shear stress for erosion of cohesive soils subjected to temperature typical of wildfires. *Journal of Geophysical Research*, 110: 1-13.

Moody, J.A., D.A. Martin, S.L. Haire, and D.A. Kinner. 2008. Linking runoff response to burn severity after wildfire, *Hydrological Processes*, 22(13): 2063-2074.

Moore, I.D. and G.R. Foster. 1990. Hydraulics, and overland flow. *In* M.G. Anderson and T.P. Burt [eds.]. *Process Studies in Hillslope Hydrology*. John Wiley & Sons Ltd., Chichester, UK.

Moran, J.M. and R.D. Stieglitz. 1983. Tornadoes of fire: the tragic story of Williamsville, Wisconsin, 8 October, 1871. *Weatherwise*, 36(6): 298-300.

Moss, A.J. 1988. Effects of flow-velocity variation on rain-driven transportation and the role of rain impact in the movement of solids. *Australian Journal of Soil Research*, 26: 443-450.

Moss, A.J. and P. Green. 1983. Movement of solids in air and water by raindrop impact. Effects of drop-size and water-depth variations. *Australian Journal of Soil Research*, 21: 257-269.

Neary, D.G., K.C. Ryan, and L.F. DeBano. 2005 (revised 2008). Wildland fire in ecosystems: effects of fire on soils and water. General Technical Report. RMRS-GTR-42-vol. 4. U.S. Department of Agriculture, Forest Service, Rocky Mountain Research Station, Ogden, Utah, USA.

Nelson, J.M. and J.D. Smith. 1989. Mechanics of flow over ripples and dunes. *Journal of Geophysical Research*, 94 (C6): 8146-8162.

Nikora, V., D. Goring, I. McEwan, and G. Griffiths. 2001. Spatially averaged open-channel flow over rough bed. *Journal of Hydraulic Engineering*, 127(2): 123-133.

Palmer, T.Y. 1981. Large fire winds, gases and smoke. *Atmospheric Environment*, 15(10/11): 2079-2090.

Pannkuk, C.D. and P.R. Robichaud. 2003. Effectiveness of needle cast at reducing erosion after forest fires. *Water Resources Research*, 39(12): 1333, doi: 10.1029/2003WR002318.

Peters, D.P.C. and K.M. Havstad. 2006. Nonlinear dynamics in arid and semi-arid systems: Interactions among drivers and processes across scales. *Journal of Arid Environments*, 65: 196-206.

Peterson, G.D. 2002. Contagious disturbance, ecological memory, and the emergence of landscape pattern. *Ecosystems*, 5: 329-338.

Phillips, J.D. 1992. Deterministic chaos in surface runoff. *In* A.J. Parson and A.D. Abrahams [eds.]. *Overland Flow: Hydraulic and Erosion Mechanics*. University College Press, London, UK.

Phillips, J.D. 2003. Sources of nonlinearity and complexity in geomorphic systems. *Progress in Physical Geography*, 27(1): 1-23.

Poesen, J. and J. Savat. 1981. Detachment and transportation of loose sediments by raindrop splash, Part II. Detachability and transportability measurements. *Catena*, 8: 19-41.

Poesen, J.W.A. 1993. Mechanisms of overland flow generation and sediment production on loamy and sandy soils with and without rock fragments. *In* A.J. Parsons and A.D. Abrahams [eds.]. *Overland Flow; Hydraulics and Erosion Methods*, University College Press, London, UK.

Price, O., J. Russell-Smith, and A. Edwards. 2003. Fine-scale patchiness of different fire intensities in sandstone heath vegetation in northern Australia. *International Journal of Wildland Fire*, 12: 227-236.

Puri, A.N. and A.G. Asghar. 1940. Physical characteristics of soils: VII. Effect of ignition. *Soil Science*, 49: 369-378.

Renard, K.G., G.R. Foster, G.A. Weesies, and D.K. McCool. 1997. Predicting soil erosion by water: a guide to conservation planning with the Revised Universal Soil Loss Equation (RUSLE), Agriculture Handbook 703, U.S. Department of Agriculture, Washington, D.C., USA.

Reneau, S.L., G.A. Kuyumjian, D.V. Malmon, and M.F. Tardiff. 2003. Precipitation-Frequency Relations on the Pararito Plateau and in the Eastern Jemez Mountains, New Mexico, and Examples of Extreme or Flood-Producing Storms. Los Alamos National Laboratory Report LA-UR-03-6484, Los Alamos, New Mexico, USA.

Rice, R.M. 1982. Sedimentation in the chaparral: How do you handle unusual events? *In* F.J. Swanson, R.J. Janda, T. Dunne, and others [tech. coords.]. Workshop on Sediment Budgets and Routing in Forested Drainage Basins: Proceedings. General Technical Report PNW-141. U.S. Department of Agriculture, Forest Service, Pacific Northwest Forest and Range Experiment Station, Portland, Oregon, USA.

Roering, J.J. and M. Gerber. 2005. Fire and the evolution of steep, soil-mantled landscapes. *Geology*, 33(5): 349-352.

Romme, W.H. 1980. Fire history terminology: Report of the ad hoc Committee. *In* M.A. Stokes and J.H. Dieterich [tech. coords.]. Proceedings of the Fire History Workshop, October 20-24, Tucson, Arizona. General Technical Report RM-81. U.S. Department of Agriculture, Forest Service, Rocky Mountain Research Station, Fort Collins, Colorado, USA.

Rose, C.W., J.R. Williams, G.C. Sander, and D.A. Barry. 1983. A mathematical model of soil erosion and depositions processes: I. Theory for a plane land element. *Soil Science Society of America Journal*, 47: 991-995.

Ryan, K.C. and N. Noste. 1985. Evaluating pescribed fires. pp. 230-238. *In* J.E. Lotan, B.M. Kilgore, W.C. Fischer, and R.W. Mutch [tech. coords.]. Symposium and Workshop on Wilderness Fire, General Technical Report INT-182. U.S. Department of Agriculture, Forest Service, Intermountain Research Station, Ogden, Utah, USA.

Ryan, K.C. 2002. Dynamic interactions between forest structure and fire behavior in boreal ecosystems, *Silva Fennica*, 36(1): 13-39.

Schumm, S.A. 1973. Geomorphic thresholds and complex response of drainage systems. pp. 299-310. *In* M. Morisawa [ed.]. *Fluvial Geomorphology,* A proceedings volume of the Fourth Annual Geomorphology Symposia Series, New York State University Publications in Geomorphology, Binghamton, New York, USA.

Schumm, S.A. 1979. Geomorphic thresholds: the concept and its applications. *Transactions of the Institute of British Geographers,* 4(4): 485-515.

Schumm, S.A. 1991. *To Interpret the Earth, Ten Ways to be Wrong*. Cambridge University Press, New York, New York, USA.

Schumm, S.A., J.F. Dumont, and J.M. Holbrook. 2000. *Active Tectonics and Alluvial Rivers*. Cambridge University Press, New York, New York, USA.

Shakesby, R.A. and S.H. Doerr. 2006. Wildfire as a hydrological and geomorphological agent. *Earth-Science Reviews,* 74: 269-307.

Sivapalan, M. and E.F. Wood. 1987. A multidimensional model of nonstationary space-time rainfall at the catchment scale. *Water Resources Research,* 23(7): 1289-1299.

Sivapalan, M. and G. Blöschl. 1998. Transformation of point rainfall to areal rainfall: intensity-duration-frequency curves. *Journal of Hydrology,* 204: 150-167.

Smith, R.E. and D.C. Goodrich. 2005. Rainfall excess overland flow, pp. 1707-1718. *In* M.G. Anderson [ed.]. *Encyclopedia of Hydrological Sciences*. John Wiley & Sons, Ltd., Hoboken, New Jersey, USA.

Swanson, F.J. 1981. Fire and geomorphic processes. *In* H.A. Mooney, T.M. Bonnicksen, N.L. Christensen, J.E. Lotan, and W.A. Reiners [tech. coords.]. Fire Regimes and Ecosystem Processes, Proceedings of the Conference, Honolulu, Hawaii, General Technical Report WO-26. U.S. Department of Agriculture, Forest Service, Washington, D.C., USA.

Thomas, L. 1995. *The Medusa and the Snail: More Notes of a Biology Watcher*. Penguin Books, New York, New York, USA.

Viegas, D.X. 2005. A mathematical model for forest fires blowup. *Combustion Science and Technology,* 177: 27-51.

Viereck, L.A. 1973. Wildfire in the taiga of Alaska. *Quaternary Research,* 3(3): 465-495.

Wahlenberg, W.G., S.W. Greene, and H.R. Reed. 1939. Effects of fire and cattle grazing on longleaf pine lands as studied at McNeill, Mississippi, Technical Bulletin, 683: 1-52. U.S. Department of Agriculture, Washington, D.C., USA.

Walling, D.E. 1983. The sediment delivery problem. *Journal of Hydrology,* 65: 209-237.

Walling, D.E. and B.W. Webb [eds.]. 1996. Erosion and sediment yield: Global and regional perspectives, International Association of Hydrological Sciences Publication no. 236, University of Exeter, Exeter, UK.

Wells, W.G. II. 1987. The effect of fire on the generation of debris flows in southern California. *In* J.E. Costa and G.F. Wieczorek [eds.]. Debris Flows/Avalanches. Geological Society of America Reviews in Engineering Geology VII: 105-114.

Weltz, M.A., A.B. Arslan, and L.J. Lane. 1992. Hydraulic roughness coefficients for native rangelands: *Journal of Irrigation and Drainage Engineering,* 118(5): 776-789.

Whicker, J.J., J. Pinder, III, D.D. Breshears, F.W. Whicker, C. Eberhart, and J.C. Rodgers. 2001. Wildfire effects on contaminant transport through wind erosion, Los Alamos National Laboratory Report LA-UR-01-6243. U.S. Department of Energy, Los Alamos National Laboratory, Los Alamos, New Mexico, USA.

Whicker, J.J., D.D. Breshears, P.T. Wasiolek, and K.N. Baker. 2002. Temporal and spatial variation of episodic wind erosion in unburned and burned semiarid shrubland: *Journal of Environmental Quality,* 31: 599-612.

Whicker, J.J., J.E. Pinder, III, and D.D. Breshears. 2006. Increased wind erosion from forest wildfire: Implications for contaminant-related risks. *Journal of Environmental Quality*, 35: 468-478.

Wiberg, P. and J.D. Smith. 1987. Calculations of critical shear stress for motion of uniform and heterogeneous sediments. *Water Resources Research*, 23(8): 1471-1480.

Wolman, M.G. and J.P. Miller. 1960. Magnitude and frequency of forces in geomorphic processes. *Journal of Geology*, 68: 54-74.

Yates, D.N., T.T. Warner, and G.H. Leavesley. 2000. Prediction of a flash flood in a complex terrain. Part 2: A comparison of flood discharge simulations using rainfall input from radar, a dynamic model, and an automated algorithmic system. *Journal of Applied Meteorology*, 39(6): 815-825.

Yoshikawa, K. and L.D. Hinzman. 2003. Shrinking thermokarst ponds and groundwater dynamics in discontinuous permafrost near Council, Alaska. *Permafrost and Periglacial Processes*, 14: 151-160.

Yoshikawa, K., W.R. Bolton, V.E. Romanovsky, M. Fukuda, and L.D. Hinzman. 2003. Impacts of wildfire on the permafrost in the boreal forests of interior Alaska. *Journal of Geophysical Research*, 108 (D1): 1-14.

Zhu, J.C., C.J. Gantzer, S.H. Anderson, R.L. Peyton, and E.E. Alberts. 2001. Comparison of concentrated-flow detachment equations at low shear stress. *Soil and Tillage Research*, 61: 203-212.

3

Fire Effects on Soil Infiltration

Artemi Cerdà[1*] and *Peter R. Robichaud*[2]

Abstract

Within the hydrological cycle, infiltration plays the key role in partitioning rainfall into surface runoff and subsurface flows. Fire modifies the soil as well as the vegetation/litter cover on the soil, thus altering the infiltration process on burned lands. Fire effects on the infiltration process and infiltration rates are described and illustrated with data from studies where infiltration was measured before and after wildfires or compared in paired (burned and unburned) plots. Some research shows that immediately after a fire, ash absorbs the rainfall and briefly enhances the infiltration into the mineral soil. As the ash is washed downslope due to raindrop impact and surface runoff, it clogs the soil pores causing surface sealing, and infiltration reaches its lowest values. The extent and degree of fire-induced or enhanced soil water repellency is directly related to the reduction in infiltration. As vegetation recovery enhances the hydrological functioning of the soil and soil water repellency decreases, infiltration recovers approaching its pre-fire infiltration values within a few years.

INTRODUCTION

Some of the water that falls on land as rain, snow, or hail moves into the soil (Dunne and Leopold 1978). The term infiltration refers to this movement of water into the soil (Hillel 1971). The infiltration rate (volume of water infiltrated per time) usually declines rapidly during the early portion of a rain event and reaches a constant value after several minutes of rainfall (Dunin 1976). The water that does not infiltrate becomes surface runoff and may cause erosion (Kirkby 1978). Forest fires burn the vegetation and litter, induce physical, chemical and biological effects on soil, and thus, disturb the

[1] *Departament de Geografia, Universitat de València, València, Spain.*
[2] *U.S. Department of Agriculture, Forest Service, Rocky Mountain Research Station, Moscow, Idaho, USA.*
* *Corresponding author: Artemi Cerdà, Departament de Geografia, Universitat de València, Blasco Ibàñez, 28, 46010, València, Spain. Tel: +34(9) 6 3864882; Fax: +34(9) 6 3983054; E-mail: artemio.cerda@uv.es*

infiltration process (Cerdà 1995, Robichaud 2000). However, post-fire infiltration changes do not follow a simple temporal model of progressive recovery where infiltration rates return to pre-fire levels over some timeframe. Instead a highly dynamic (both temporal and spatial) post-fire infiltration system results from seasonal variation of soil water content, the surface ash layer developed immediately after the fire, the fire-induced water repellent soil conditions, and the post-fire changes in vegetation. This chapter will review current scientific knowledge related to the effects of fire on water infiltration into soil. Research data provide examples of post-fire changes in infiltration rates and illustrate temporal (seasonal and general post-fire recovery) and spatial variability of fire effects on infiltration.

The Infiltration Process

The volume of water that infiltrates per unit of time is the infiltration rate, and is influenced by rainfall intensity and water acceptance by the soil (Douglas 1977). The maximum rate that water can be accepted by a soil is the infiltration capacity (Knapp 1978). The water enters the soil in response to water potential and gravitational gradients (Hendricks 1990). Water potential is governed by the soil water content and pore structure, which combine to determine the capillary action and adhesive forces of the soil particle surfaces and is often expressed as the sorptivity factor. The gravity potential depends on soil pore size, continuity, and distribution, and describes the rate of water flow through soil under the influence of gravity. The initial water infiltration rate is largely governed by the sorptivity forces of the dry soil, and is replaced, once the soil wets up by the gravitational force (Hillel 1982).

During a single rainfall event, the decreasing infiltration rate is due to the filling of fine soil pores with water that reduces the capillary forces, the swelling of soil particles that reduces the size of soil pores, and the raindrop impact that splash and seal the soil surface (Chorley 1969). After fires, increased sediment and ash in the runoff and lack of vegetation cover contribute to these processes (Shakesby and Doerr 2006). Generally, during a rainfall event, infiltration declines (exponentially and asymptotically) to a constant value called saturated hydraulic conductivity or the steady-state infiltration rate (Kirkby 1978). Several models have been developed that predict this behavior (Green and Ampt 1911, Kostiakov 1932, Horton 1940, Philip 1957). However, in fire-affected soils, where fire-induced or fire enhanced soil water repellency has occurred, these models do not fit (Robichaud 2000). In fact, under water repellent conditions, infiltration rates are initially low and increase after soil wetting occurs.

Infiltration is a key process of the hydrological cycle as most rainfall infiltrates into the soil. Soil infiltration determines the proportion of water that moves by overland flow versus subsurface flow (Burt 1987), and changes in water infiltration behavior result in changes in surface runoff and then in the stream flow (Emblenton and Thornes 1979). The direct connection between infiltration and runoff generation is demonstrated by the Hortonian (Horton 1933), variable-source (Bernier 1985), and partial-source (Hewlett and Hibbert 1967) models. Infiltration rates also determine the amount of water stored between the aquifer and the surface, which is the water available to vegetation (Ponce 1989). Given that fire often consumes forest floor material, it can reduce water storage and greatly influence the partitioning of water into subsurface and surface water, which can increase the risk of flooding (Summerfield 1991). Usually fires reduce the infiltration capacity of soils, causing more surface flow and decreasing the subsurface flow, but the infiltration capacity of burned soils is highly dynamic in time and space.

Factors Affecting the Infiltration Process

Soil properties, such as grain size, porosity, cracks or surface crusts, influence the infiltration process and the amount of water absorbed (Dunne and Dietrich 1980). For example, clay soils absorb water at a slower rate than sandy soils (Abrahams et al. 1988). Land use and management influence the vegetation cover and the soil characteristics, which have various effects on infiltration rates. Biological activity usually enhances soil porosity and structure, increasing the infiltration capacity of soils. The burrowing of worms and penetration of plant roots can increase the size and number of macro- and micro-pores within the soil, which allow water to infiltrate the soil. Water infiltration also depends on the surface roughness, which influences runoff velocity and surface storage capacity. Generally, as surface slope increases, the amount of ponding decreases and runoff increases; however, complex relationships have been found between sediment and water output related to slope angle (Neal 1938, Poesen 1984, Abrahams and Parsons 1991).

Although many variables control infiltration rates (Parr and Bertrand 1960, Nassif and Wilson 1975, Agassi and Levy 1991), vegetation has the largest impact (Meewing 1969, Simanton and Renard 1982, Soni et al. 1985, Faulkner 1990, Wood and Luk 1990, Cerdà 1999a). Vegetation increases litter and organic material that improve soil structure and porosity, allowing the soil to hold larger quantities of water, and as a consequence, improve infiltration rates (Munn et al. 1973, Berndtsson and Larson 1987). Vegetation protects the soil from raindrop impact and increases soil porosity as the roots develop macropores. Vegetation results in a deeper litter layer, which encourages biological activity, water storage, soil aggregation, and micro- and macro-pore development. Forest fires burn the vegetation and litter, which cause ash to cover the soil surface and heats the surface layers of the soil (Robichaud and Hungerford 2000). Thus, fire reduces the vegetation, which reduces infiltration and increases runoff. Vegetation recovery begins quickly after fires in some ecosystems, such as the Mediterranean and temperate forests of the United States, and the vegetation cover is often reestablished within several years (Traboud 1981, Naveh 1990); however, full recovery of the pre-fire ecosystem is a long process, which takes more time than the initial recovery of vegetation cover. Vegetation recovery enhances water infiltration during the post-fire recovery period.

SOIL INFILTRATION RESEARCH METHODS

Measurement of Infiltration Rates

Ring infiltrometers are the most widely used technique to measure soil infiltration rates in flat agriculture fields (Hills 1970); however, in areas affected by wildfires, slopes are seldom flat enough to use a ring infiltrometer. Some researchers have used cylinder infiltrometers to measure post-fire changes in infiltration. During a post-fire study in Portugal, Shakesby et al. (1993) had difficulties maintaining a seal around the perimeter of the infiltrometer rings when inserted into the stony soils – typical of forest soils but rarely a problem on plowed agricultural land (Cerdà 1995). Additionally, ring infiltrometers reproduce the infiltration process under ponding conditions, which is uncommon in rangeland and forest areas where most post-fire research is done (Amerman 1983). For these reasons, rainfall simulation is one of the most reliable methods to assess infiltration rates on rangeland (Cerdà, 1998a, 1999b, Pierson et al. 2001) and in forest environments (Robichaud 2000).

The high spatial and temporal variability of natural rainfall encourage the development and use of field rainfall simulators that standardize the rainfall and provide more accurate measurements of rainfall and runoff. Removing the variability of natural rainfall, allows rainfall and runoff measurements to be related to infiltration and erosion rates with more confidence (Nichols and Sexton 1932, Neal 1938, Beutner et al. 1940, Amerman et al. 1970). Most research on fire effects and soil infiltration use rainfall simulators, which can produce selected rain events at the time and place needed for the study. To obtain the same results using natural rainfall would require many years of measurements and, due to natural recovery processes, make variable control very difficult. Thus, simulated rainfall experiments have been widely used to measure the infiltration rates of burned soils throughout the post-fire recovery period (Cerdà 1998a, Robichaud 2000).

Infiltration can be indirectly measured on a gaged watershed where the difference between rainfall and runoff measurements can provide information about the infiltration capacity of the watershed under natural rainfall (Robichaud 2005). Using natural or simulated rainfall, rainfall and runoff measurements from plots of different sizes can be used to determine the effects of scale on infiltration rates (Robichaud et al. 2008a). Although simulated rainfall can be used on different plot sizes, small plots that minimize runoff delay provide more accurate infiltration measurements (Amerman 1983, Cerdà 1996).

Use of Neutron probes or Time-domain Reflectometry (TDR) meters to measure changes in soil water content also provide indirect measurement of water infiltration into soil. Other measurements related to infiltration are being carried out using disc permeameters, mini-disk infiltrometers (Robichaud et al. 2008b), and other tools that allow hydraulic conductivity to be determined for a range of negative potentials.

Comparing Burned and Unburned Sites

To understand the effect of wildfire on soil infiltration, many researchers have studied plots or watersheds under burned and unburned conditions. They generally compare measurements from paired plots (burned and unburned) and/or from repeated measurements made before and after the fire.

FIRE EFFECTS ON SOIL INFILTRATION

General Post-fire Trends

The earliest studies related to fire effects on soil infiltration and runoff production were done in the USA. Most researchers have reported an increase in surface runoff due to a reduction in soil infiltration and water storage capacity (Arend 1941, Hendricks and Johnson 1944, Anderson 1949, Burgy and Scott 1952, Scott 1956, Musgrave and Holtan 1964, Imeson 1971, Zwolinski 1971, Brown 1972). Other studies reported an order of magnitude increase in peak flows and surface runoff after fire as compared to little or no runoff before the fire, which is attributed to the reduction of the soil infiltration and surface storage (Wright et al. 1978, 1982, Helvey 1980, Knight et al. 1983, Garza and Blackburn 1985, Hibbert 1985, Steichen et al. 1987, Heede et al. 1988, Scott and van Wyk 1990, Link et al. 1990). During the 1990's and early 2000's research continued to confirm that fire generally reduces infiltration rates and to determine the factors that influence the magnitude of the effect (Cerdà and Calvo 1991, Shahlaee et al. 1991, Emmerich and Cox 1992, 1994, Imeson et al. 1992, Marques and Mora 1992, Soler and Sala 1992, Scott 1993, Calvo and Cerdà 1994, Robichaud and Waldrop 1994, Soto et al. 1994, Sánchez et al. 1994, Cerdà et al. 1995, Lavee et al. 1995, Hester et al. 1997, Inbar et al. 1997, 1998, Rubio et al. 1997, Robichaud and Miller 1999, Gimeno et al. 2000, Marcos et al. 2000, Robichaud 2000, Pierson et al. 2001, Andreu et al. 2001, Martin and Moody 2001, Moody and Martin 2001a, 2001b, Meyer et al. 2001, Wondzell and King 2003). However, due to the high spatial and temporal variability of rainfall, the measurements carried out under natural rainfall have sometimes shown a decrease in runoff production (Dunne et al. 1991, Sánchez et al. 1994).

Usually infiltration rates are greater on unburned soil and decrease with increasing burn severity on burned soil (Cerdà 1995). In a wet Mediterranean environment in Portugal, the lowest infiltration rates were found on the recently burned sites and the greatest in the unburned sites (Shakesby et al. 1993); Campbell et al. (1977) also demonstrated that the removal of litter by fire resulted in more frequent soil freezing, which may reduce infiltration rates. As the vegetation recovers, the infiltration rates return to pre-burn conditions (Cerdà 1998a, 1998b). However, because infiltration rates are dependent on rainfall characteristics, infiltration rates can be greater after a fire than before. Bautista (1999) found the greatest runoff rates the fifth year after a fire due to an intense rainfall event. Robichaud et al. (2008c) found rainfall intensity was the driving factor in post-fire runoff and peak flow responses at five of six sites in the western United States. Cerdà and Lasanta (2005) found significant differences in post-fire runoff and erosion as well as recovery time among two similarly burned plots. These differences may be due to small differences in soil organic matter and texture, rainfall after the fire, and vegetation recovery or plant distribution on the plots.

Soil burn severity and post-fire infiltration

Soil burn severity describes, in qualitative terms, the fire-induced changes in the soil properties that impact hydrological and biological soil functions (see Chapter 4). Several classification systems have been developed to describe the condition of the post-fire soil in terms of high, moderate, and low soil burn severity (e.g., Hungerford 1996, DeBano et al. 1998). Some fire effects resulting from the consumption of organic material near and on the soil surface lead to changes in soil properties that impact soil infiltration, such as: 1) the formation or enhancement of soil water repellency; 2) change in soil structure (e.g., consumption of fine roots increases micro- and macro-pores); and 3) change in bulk density (e.g., due to collapse of aggregates and clogging of voids by ash; Certini 2005). After prescribed burns in western Montana, Robichaud (2000) compared the effects of soil burn severity on infiltration under simulated rainfall. Infiltration rates of 60 to 80 mm h^{-1} were measured on low soil burn severity plots and decreased as soil burn severity increased. Moderate soil burn severity plots had infiltration rates of 30 to 84 mm h^{-1} the infiltration on high burn severity plots was only 23 to 55 mm h^{-1} (Fig. 1).

The Effects of Ash on Soil Infiltration

During the 1990's, research results showed that ash on the soil surface increased the infiltration capacity of burned soils (Cerdà 1995, Cerdà 1998a), which surprised researchers and others. The expected post-fire increase in runoff and reduction in infiltration was not immediately observed. Most post-fire studies miss the first days or weeks after a fire, and do not measure the influence of the dry, high porosity fire ash. Ash is easily affected by wind and water erosion, and after several weeks of wind/rain events, ash often fills soil pores sealing the upper soil horizon with the remainder being carried to valley bottoms. The soil sealing by ash may reduce infiltration and enhance

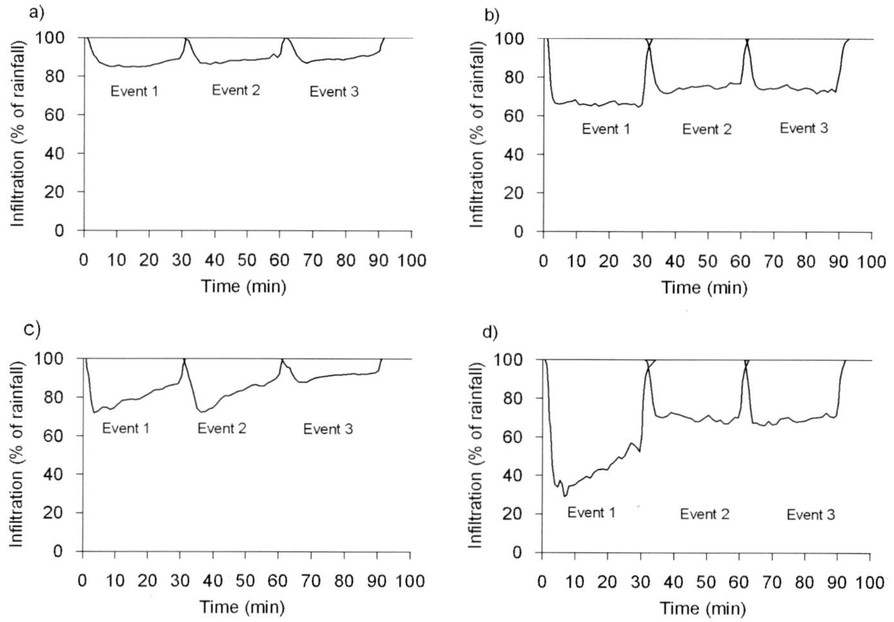

Fig. 1 Following a prescribed fire in western Montana, USA, mean infiltration (% of rainfall) was measured on 14 plots (1 m^2) with varying soil burn severity and soil water repellency. Examples are for: a) unburned-undisturbed plots; b) low soil burn severity; c) high soil burn severity; and d) high soil burn severity with water repellent soils. Each plot received three 30-min rainfall (mean intensity 94 mm h^{-1}) events: Event 1 conducted with antecedent soil moisture; Event 2 conducted the following day; and Event 3 conducted 30 min later (adapted from Robichaud 2000).

runoff and sediment transport. However, immediately after the fire, the infiltration rate of the ash is high (Cerdà 1998a), and the ash layer may be thick enough to absorb the rainfall and avoid runoff. Additionally, the ash may increase infiltration into the mineral soil. This ephemeral effect of ash is often missed in post-fire studies. For example, in the Central Spanish Pyrenees, Cerdà and Lasanta (2005) and Lasanta and Cerdà (2005) measured prescribed fire effects on infiltration and runoff from two burned and one unburned plot. They observed an increase in runoff on both burned plots with seasonal variations, due to differences in rainfall characteristics, during the seven-year recovery period. Nonetheless, recent data have confirmed the effect of ash on reducing runoff immediately after fires (Cerdà and Doerr 2007).

Most post-fire experiments take place after the ash has moved and/or soil surface crusting has occurred (Burch et al. 1989, Shakesby et al. 1993, Cerdà et al. 1995, Inbar et al. 1998, Prosser and Williams 1998, Shakesby et al. 2000, Ferreira et al. 2000, 2005, Benavides-Solorio and MacDonald 2001, Huffman et al. 2001, Martin and Moody 2001a, Meyer et al. 2001). The post-fire decrease in infiltration that is often observed after fires is likely due to the development of water repellent soil conditions near the soil surface (DeBano 1971, 1981), the lack of litter and vegetation to store water and wet the soil surface, and/or the clogging of soil pores by ash which results in the soil surface sealing.

The persistence of ash on the soil surface depends on the post-fire rain and wind characteristics. In two post-fire studies carried out at the Pedralba site in eastern Spain (Central Valencia Province), the 0.25 m^2 study plots had ash on the soil surface for nine weeks after the fire (wildfire 1 August 1990; seasonal rains began second week of October). Using simulated rainfall of 60 mm h^{-1}, infiltration rates of about 59 mm h^{-1} (negligible runoff) were measured on these ash covered plots immediately after the fire (Cerdà 1998b; Table 1; Fig. 2). Those high infiltration rates were similar to the infiltration rates measured in a nearby unburned oak (*Quercus ilex*) Mediterranean forest or in the dense cover of Matorral (*Quercus coccifera* and *Pistacea lentiscus*) shrubland (Cerdà 1996, 1997a). However, four months after the fire (December 1990), the soil surface no longer had any ash cover and the infiltration rates had decreased to extremely low values (17 mm h^{-1}), similar to the infiltration rates measured in badland areas (Cerdà 1999b) or on bare road embankments (Table 1; Fig. 2; Cerdà 2007). In another study at the Naquera site (east of Valencia province), a fire took place on 12 August 2004 and the first post-fire rain event (115 mm in 3 hours) occurred on 8 September 2004, which removed the ash with surface runoff (Fig. 3). Also noted in this study, the ash-enhanced infiltration rates had little spatial variability due to the homogeneity of the ash cover on burned soils. Post-fire ash-enhanced infiltration was examined in another experiment carried out in the weeks after a wildfire and before the first rainfall on 30 plots: 15 with ash as natural surface and 15 with the ash cover removed (Fig. 4). The infiltration was three times greater in the ash covered plots (95 percent of the simulated rainfall infiltrated) compared to the ash-removed plots (30 percent of the simulated rainfall infiltrated) (Table 2). In all of these studies high infiltration rates were observed immediately after the fire due to the presence of ash on dry soil; and, after ash removal and soil sealing by the raindrop impact, infiltration rates decreased significantly. The

Fig. 2 Rainfall simulation plots from the Pedralba site, Spain. From left to right: a) two days after fire with ash cover; b) during the first simulated rainfall experiment; c) 4 months later with sealed surface; and d) 8 months later with the *Ulex parviflorus* and *Pinus halepensis* saplings.

Fig. 3 The Naquera study sites, Spain, before and after the first intense post-fire rain event. The ash was removed by 15 mm rainfall in three hours. a-b) ash cover and c-d) eroded surfaces after a thunderstorm.

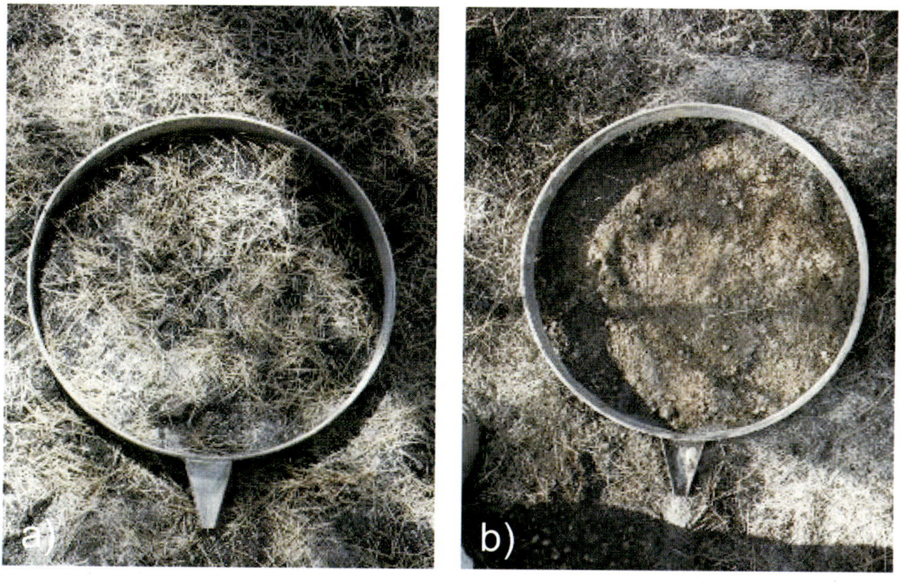

Fig. 4 Naquera study site immediately after fire. a) plot with ash cover; b) bare plot after removal of ash by hand.

Table 1 Comparison of the infiltration rates from simulated rainfall (60 mm h^{-1} for 1 h) applied two days after the fire and four months later (after ash-removing autumn rains). I_{15} is the infiltrated depth of water after 15 minutes and I_{60} is infiltrated depth of water after 60 minutes.

Plots	I_{15} 2 d after fire (mm)	I_{15} 4 mon after fire (mm)	I_{60} 2 d after fire (mm)	I_{60} 4 mon after fire (mm)
1	15.0	4.4	60.0	9.9
2	15.0	6.0	60.0	16.9
3	15.0	7.8	60.0	12.5
4	15.0	8.4	60.0	18.7
5	15.0	8.5	56.8	27.1
Average	**15.0**	**7.0**	**59.4**	**17.0**
Std	**0.0**	**1.8**	**1.5**	**6.6**
Vc (%)	**0.0**	**25.5**	**2.5**	**38.8**

length of time that ash-enhanced infiltration occurs depends on the time that the ash remains on the soil surface and is highly variable. Thus, the infiltration capacity of soils affected by fire can change from extremely high

Table 2 Infiltration rates (percent of total rainfall) on two sets of 15 plots, one with ash remaining and with ash removed by hand, immediately after the August 2004 Naquera Fire in eastern Spain.

Plots	With Ash (%)	Ash Removed (%)
1	100	56.2
2	100	50.5
3	100	48.4
4	100	44.0
5	98.0	40.9
6	98.3	36.1
7	98.0	33.3
8	97.3	30.8
9	96.4	29.4
10	94.3	26.5
11	93.0	21.1
12	91.0	20.1
13	90.0	15.0
14	86.3	10.8
15	83.5	5.5
Average	**95.1**	**31.2**
Std	**5.3**	**15.1**
Vc (%)	**5.5**	**48.2**

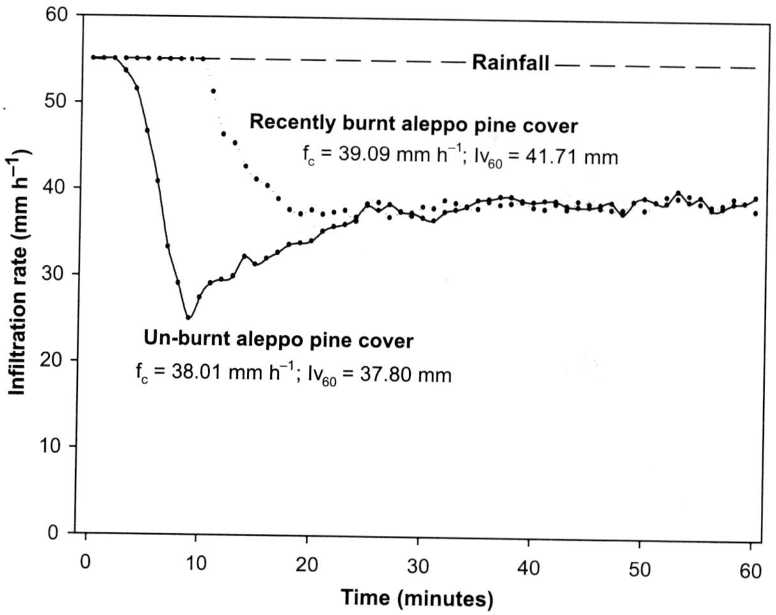

Fig. 5 A comparison between a water repellent versus a hydrophilic infiltration curve observed on the aleppo pine woodland in eastern Spain during summer. The dry and hot soil conditions enhance the water repellent response on the aleppo pine cover. The hydrophobic response is removed by the effect of fire as shown the fire affected burned soil.

(such as those found in natural conifer forests) to the extremely low (such as those found in bare badlands) in a very short time.

Fire Effects on Infiltration due to Soil Water Repellency

Another key factor affecting post-fire soil hydrology is the fire-induced or fire-enhanced soil water repellency (DeBano 2000, Robichaud and Hungerford 2000, Doerr et al. 2006). Given that fire effects on soil water repellency are described in Chapter 7, the discussion here will be restricted to its relationship to soil infiltration. Infiltration rates can be slightly increased (Cerdà and Doerr 2005, 2007) or reduced (Robichaud 2000, Benavides-Solorio and MacDonald 2001) depending on the degree of fire-induced or fire-modified soil water repellency; the degree and extent of soil water repellency is generally related to soil burn severity (Robichaud 2000, Robichaud and Hungerford 2000). Robichaud (2000) showed that the most apparent hydrological effect of water repellent soil conditions is reduction of infiltration (Fig. 1d), which results in greater surface runoff and soil erosion. Infiltration rates from an unburned soil during the dry summer months were compared to the infiltration rates of the same soil that had been burned (Fig. 5; Cerdà, 1998a). The steady-state infiltration rates and total amount of water infiltrated in one hour are similar for both soils; however, during the first 20 minutes of simulation, the infiltration rate of the burned soil was significantly less than the unburned soil. Given that short rain events of 10 to 20 minutes occur frequently, these results suggest that under natural rainfall conditions, lower infiltration rates would be frequently observed on burned, water- repellent soil conditions, especially in the summer dry season (Fig. 5).

In areas of moderate and high soil burn severity, fire-induced or enhanced water repellency is most often detected at 1 to 3 cm below the surface and the surface soil (immediately below the ash layer) is often highly absorbent (Robichaud et al. 2007). In fact, on forest soils that are naturally water repellent on the surface, high severity fires may destroy the surface water repellency and increase the subsurface water repellency (Doerr et al. 2006, Robichaud et al. 2007). Water rapidly infiltrates into this highly absorbent surface layer; however, water does not readily infiltrate into the water repellent subsurface layer, and the wettable topsoil layer can quickly become saturated. Once saturated, the infiltration rate rapidly decreases and the soil material within the wettable layer is easily entrained in the subsequent overland flow and carried downslope (DeBano 1998, Doerr et al. 2006).

As alluded to above, not all soil water repellency is fire-induced. In a western Montana forest under simulated rainfall, Brady et al. (2001) found infiltration rates were lower in root mat-removed (unburned, mineral soil exposed) plots than in either the burned or undisturbed (organic forest floor layer intact on surface) plots. These results were due to dry antecedent soil moisture conditions and the naturally-occurring soil water repellency in volcanic ash soils with high organic matter content. Although fire usually encourages the formation and/or translocation of soil water repellency, burning can destroy water repellency, as found by Doerr et al. (2004) in sandstone tablelands near Sydney, or reduce the water repellency, as found by Cerdà and Doerr (2005) in a forest in eastern Spain.

Variation in Post-fire Infiltration Rates

Infiltration rates have large spatial variability that varies over time (Fig. 6). If the fuel loading and antecedent soil moisture conditions are similar, the post-fire ash cover and the initial infiltration rates will have low spatial variability; however after several weeks, the coefficients of variation are close to 40 percent. The spatial variability of infiltration is high due to the anisotropy of the soils and variability in soil burn severity due to differences in pre-fire fuel loading, forest floor thickness, and moisture content of the soil (Springer and Gifford 1980, Starr 1990); these factors affect the peak and duration of high soil temperatures during burning (Robichaud and Hungerford 2000).

Spatial variation

The spatial variability of the infiltration rates observed in many post-fire research studies are often due to the patchy distribution of pre-fire vegetation

(Seyfred 1991; Dunkerly and Brown 1995; Cerdà 1997b; Reid et al. 1999). When a fire burns a patchy (mosaic or tussock) vegetation cover, the bare soil and the soil in the vegetated patch are heated differentially. After the Denio Fire in the rangeland of northwestern Nevada, USA, Pierson et al. (2001, 2006) used simulated rainfall experiments to determine that the coppice microsites (area directly under shrubs) had reduced infiltration rates and increased overland flow due to soil water repellency, while the bare soil between the vegetation was not water repellent and had greater infiltration rates. This is also related to the micro- and macro-pores caused by the vegetation which influence the infiltration capacity of soils. Hubbert et al. (2006) found in chaparral landscapes of California, that fire-induced soil water repellency was preferentially produced under *Ceanothus crassifolius* cover, and this would likely result in variable infiltration rates under natural rainfall. The post-fire recovery of vegetation corresponds to an increase in the heterogeneity of infiltration has been demonstrated in a number of studies (Dunkerly and Brown 1995, Cerdà 1997b, Cerdà and Doerr 2005, Cerdà and Doerr 2007).

Post-fire infiltration rate recovery – temporal variation

After fire occurs and the soil is covered with ash, the next change in the soil surface is ash redistribution by raindrop splash, surface runoff, and wind. The first rain events easily remove the ash cover, clog the pores with small particles of ash and sediment, and seal the soil surface. These changes reduce infiltration rates to the lowest post-fire values. However, the same post-fire rainfall events that move ash downslope provide needed moisture for vegetation recovery. As vegetation re-establishes, it protects the soil from raindrop impact and begins to develop the litter layer. These changes, due to increased vegetation, increase infiltration and reduce surface runoff.

The studies in eastern Spain have shown it takes 2 to 4 years under dry (summer) conditions and 4 to 6 years in wet (winter) conditions for infiltration to return to pre-fire conditions (Cerdà 1998a, 1998b). In addition, infiltration rates vary seasonally where there is large variation in seasonal climatic conditions (soil moisure), such as in the Mediterranean (Cerdà 1996). Increases in soil sorptivity values and steady state infiltration rates are observed during summer droughts (Simmanton and Renard 1982, Blackburn et al. 1990, Cerdà 1995). The Pedralba sites were monitored from 1990 (beginning immediately after the August fire) until 1995. Infiltration rates were lowest during the first post-fire autumn rains, after the ash layer was washed downslope. The recovery of vegetation increased the infiltration rates over subsequent post-fire recovery years. The seasonal oscillations of infiltration rates were superimposed on the temporal post-fire recovery of infiltration (Fig. 6). The same trends in post-fire infiltration rates over time were observed at the Bolbens en La Costera District study site in eastern Spain following an August 1989 wildfire. Low infiltration rates (45 percent of rainfall) occurred during the first post-fire winter with an increase to 95 percent of rainfall five years

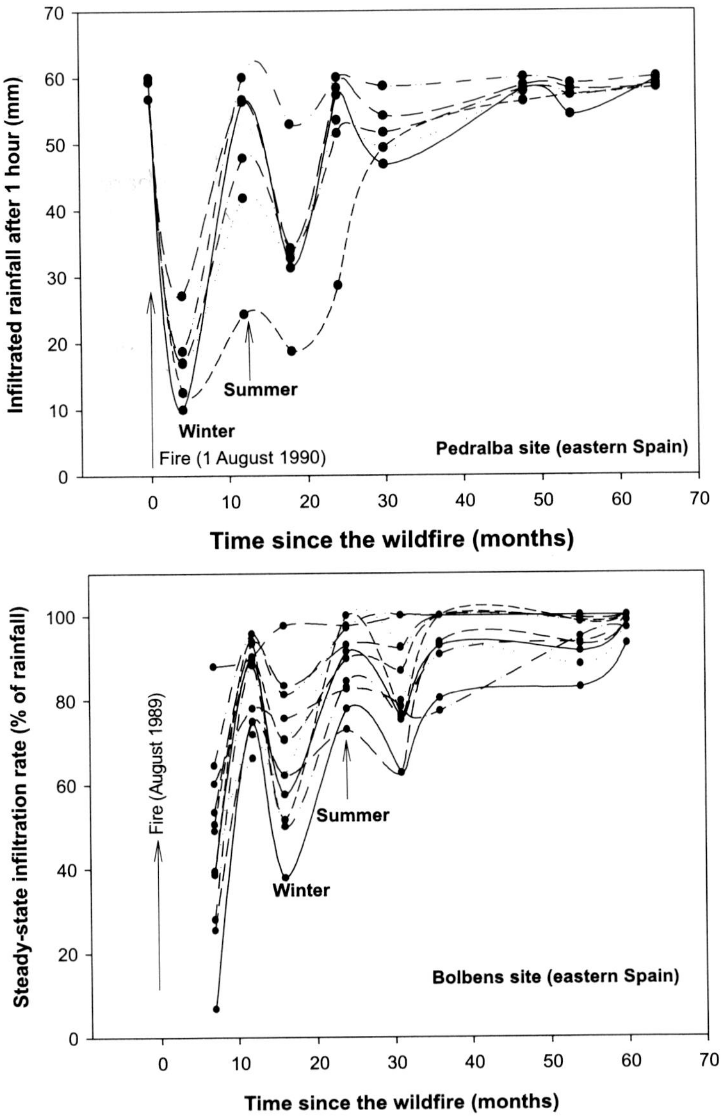

Fig. 6 Changes in the mean infiltration rate and variation coefficient for Pedralba and Bolbens sites. Soil moistures ranged from 20 to 30 percent in winter to 2 to 6 percent in summer on 5 to 50 percent slope angles.

later. The infiltration rates during the summer season, when the soils were dry, were greater than in winter with values of 84 to 99 percent of rainfall in the summer of post-fire year five (Fig. 6; Cerdà et al. 1995).

Soil water repellency, whether naturally-occurring or fire-induced, is not a stable phenomenon. While the soil is wet, hydrologic behavior is normal;

however, water repellency often returns as the soil dries (Dekker and Ritsema 2000). Consequently, soil water repellency is generally strongest in the drier summer months. Fire-induced soil water repellency slowly declines as the hydrophobic substances responsible for soil water repellency dissolve and soil wettability and infiltration increase (DeBano 1981, Letey 2001). Although some researchers have measured persistent soil water repellency from weeks to years (Holzhey 1969, DeBano et al. 1970), fire-induced water repellency is generally broken up after one to two years.

CONCLUSIONS AND FUTURE RESEARCH CHALLENGES

After a fire, the ash cover on the soil may increase the acceptance of water for a few days or weeks, after which infiltration rates drop to their lowest values due to ash removal via raindrop impact, overland flow, and/or wind. The ash removal process generally causes the ash (and in some cases, sediment) to clog soil pores and cause soil sealing. Fire also can create or enhance soil water repellency and often destroys existing surface water repellency and drives it deeper into the soil. As a result, fire can reduce infiltration rates and increase runoff for several years. During post-fire recovery of vegetation the spatial variability of infiltration and runoff increase, while the rates of infiltration tend to increase and runoff tends to decrease over time. Soil water repellency and infiltration rates also vary by soil moisture causing distinct seasonal fluctuations in many regions. Future research should focus on: a) the effects of recurrent fire intervals on infiltration changes; b) the relative effects of soil water repellency, reduced ground cover, soil sealing, and soil disaggregation on post-fire infiltration rates; c) long-term recovery of infiltration rates; d) influence of various plant species on post-fire infiltration changes; e) scaling effects on post-fire infiltration changes; and f) the effects of various rehabilitation treatments on infiltration and runoff.

Acknowledgements

Some of the research referenced in this chapter was financed by projects CGL2005-04863/CLI, GV-2005-02, BEST-2007-199; AINV/2007/076 and AE/07/125. Artemi Cerdà wishes to thank Professor Anton Charles Imeson for his comments and teaching the importance of fire on soil hydrology. The authors are grateful to the Organisation for Economic Co-operation and Development (OECD), Co-operative Research Programme, Biological Resource Management for Sustainable Agricultural Systems for providing financial support to Dr. Cerdà to participate in this study. In addition, Peter Robichaud gratefully acknowledges the U.S. Department of Interior and U.S. Department of Agriculture Forest Service Joint Fire Science Program (JFSP) for their generous support fire-related studies in the United States. Much of the work cited here and throughout this text are the results of JFSP supported projects.

References

Abrahams, A.D. and A.J. Parsons. 1991. Relation between infiltration and stone cover on a semiarid hillslope, southern Arizona. *Journal of Hydrology*, 122: 49-59.

Abrahams, A.D., A.J. Parsons, and S.H. Luk. 1988. Hydrologic and sediment response to simulated rainfall on desert hillslopes in southern Arizona. *Catena*, 15: 103-117.

Agassi, M. and G.J. Levy. 1991. Stone cover and rain intensity: effects on infiltration, erosion and water splash. *Australian Journal of Soil Research*, 29: 565-575.

Amerman, C.R. 1983. Infiltration measurement. pp. 201-214. *In* Proceedings of the National Conference on Advances in Infiltration, American Society of Agricultural Engineers, St. Joseph, Michigan, USA.

Amerman, C.R., D.I. Hillel, and A.E. Peterson. 1970. A variable-intensity sprinkling infiltrometer. *Soil Science Society of America Proceedings*, 34: 830-832.

Anderson, H.W. 1949. Does burning increase surface runoff? *Journal of Forestry*, 47: 54-57.

Andreu, V., A.C. Imeson, and J.L. Rubio. 2001. Temporal changes in soil aggregates and water erosion after a wildfire in a Mediterranean pine forest. *Catena*, 44: 69-44.

Arend, J.L. 1941. Infiltration rates of forest soils in the Missouri Ozarks as affected by burning and litter removal. *Journal of Forestry*, 39: 726-728.

Bautista, S. 1999. Regeneración post-incendio de un pinar (*Pinus halepensis*, Miller) en ambiente semiárido. Erosión del suelo y medidas de conservación a corto plazo. Tesis Doctoral, Facultad de Ciencias, Universidad de Alicante, Alicante, Spain.

Benavides Solorio, J. and L.H. MacDonald 2001. Post-fire runoff and erosion from simulated rainfall on small plots, Colorado Front Range. *Hydrological Processes*, 15: 2931-2952.

Berndtsson, R. and M. Larson. 1987. Spatial variability of infiltration in a semiarid environment. *Journal of Hydrology*, 90: 117-133.

Bernier, P.Y. 1985. Variable source area and storm-flow generation: an update of concept and a simulation effort. *Journal of Hydrology*, 79: 195-213.

Beutner, E.L., R.R. Gaebe, and R.E. Horton. 1940. Sprinkled-plot runoff and infiltration experiments on Arizona desert soils. *Transactions American Geophysical Union*, 21: 550-558.

Blackburn, W.H., F.B. Pierson, and M.S. Seyfred. 1990. Spatial and temporal influence of soil frost on infiltration an erosion of sagebrush rangeland. *Water Resources Research*, 17(2): 370-376.

Brady, J., P.R. Robichaud, and F.B. Pierson. 2001. Infiltration rates after wildfires in the Bitterroot Valley. ASAE paper number 01-8003, *Presented at* the 2001 ASAE Annual International Meeting, Technical Session 21: Forest soil erosion and water quality. American Society of Agricultural Engineers, St. Joseph, Michigan, USA.

Brown, J.H.A. 1972. Hydrological effects of a bushfire in a catchment in southeastern New South Wales. *Journal of Hydrology*, 15: 77-96.

Burch, G.J., I.D. Moore, and J. Burns. 1989. Soil hydrophic effects on infiltration and catchment runoff. *Hydrological Processes*, 3: 211-222.

Burgy, R.H. and V.H. Scott. 1952. Some effects of fire and ash on the infiltration capacity of soils. *Transactions American Geophysical Union*, 33: 405-416.

Burt, T.P. 1987. Slope and slope processes. *Progress in Physical Geography*, 11: 590-611.

Calvo, A. and A. Cerdà. 1994. An example of the changes in the hydrological and erosional response of soil after a forest fire, Pedralba, Valencia, Spain. pp. 99-110. *In* M. Sala and J.L. Rubio [eds.]. *Soil Erosion and Degradation as a Consequence of Forest Fire*. Geoforma Ediciones, Logroño, Spain.

Campbell, R.E., M.B. Baker, P.F. Ffolliott, F.R. Larson, and C.C. Avery. 1977. Wildfire effects on a ponderosa pine ecosystem: an Arizona case study, Research Paper RM-191. U.S. Department of Agriculture, Forest Service, Rocky Mountain Forest and Range Experimental Station, Fort Collins, Colorado, USA.

Cerdà, A. 1995. Factores y variaciones espacio-temporales de la infiltración en los ecosistemas mediterráneos. Monografías Científicas 5. Geoforma Ediciones, Logroño, Spain.

Cerdà, A. 1996. Seasonal variability of infiltration rates under contrasting slope conditions in Southeast Spain. *Geoderma*, 69: 217-232.

Cerdà, A. 1997a. Seasonal changes of the infiltration rates in a Mediterranean scrubland on limestones. *Journal of Hydrology*, 198: 209-225.

Cerdà, A. 1997b. The effect of patchy distribution of *Stipa tenacissima* L. on runoff and erosion. *Journal of Arid Environments*, 36: 37-51.

Cerdà, A. 1998a. Post-fire dynamics of erosional processes under Mediterranean climatic conditions. *Zeitschrift fur Geomorphologie, Neue Folge, Supplementbände*, 42(3): 373-398.

Cerdà, A. 1998b. Changes in overland flow and infiltration after a rangeland fire in a Mediterranean scrubland. *Hydrological Processes*, 12: 1031-1042.

Cerdà, A. 1999a. Parent material and vegetation affect soil erosion in eastern Spain. *Soil Science Society of America Journal*, 63: 362-368

Cerdà, A. 1999b. Seasonal and spatial variations in infiltration rates in badland surfaces under Mediterranean climatic conditions. *Water Resources Research*, 35: 319-328.

Cerdà, A. 2007. Soil water erosion on road embankments in eastern Spain. *Science of the Total Environment*, 378: 151-155.

Cerdà, A. and A. Calvo. 1991. Influencia del fuego en la escorrentía y erosión Serra Grossa, Valencia. pp. 23-27. *In* Territorio y Sociedad. AGE-Universitat de València, València, Spain.

Cerdà, A. and S.H. Doerr. 2005. Influence of vegetation recovery on soil hydrology and erodibility following fire: an eleven-year investigation. *International Journal of Wildland Fire*, 14(4): 423-437.

Cerdà, A. and S.H. Doerr. 2007. Soil wettability, runoff and erosion responses for major dry-Mediterranean land use types on calcareous soils. *Hydrological Processes*, 21: 2325-2336.

Cerdà, A. and S.H. Doerr. 2008. The effect of ash and needle cover on surface runoff and erosion in the immediate post-fire period. *Catena*, 74: 256-263. doi: 10.1016/j.catena.2008.03.010

Cerdà, A. and A. Lasanta. 2005. Long-term erosional responses after fire in the Central Spanish Pyrenees: 1. Water and sediment yield. *Catena*, 60: 59-80.

Cerdà, A., A.C. Imeson, and A. Calvo. 1995. Fire and aspect induced differences on the erodibility and hydrology of soils at la Costera, Valencia, Southeast Spain. *Catena*, 24: 289-304.

Certini, G. 2005. Effects of fire on properties of forest soils: a review. *Oecologia*, 143: 1-10. DOI 10.1007/s00442-004-1788-8.

DeBano, L.F. 1971. The effect of hydrophobic substances on water movement in soil during infiltration. *Soil Science Society of America Proceedings*, 35: 340-343.

DeBano, L.F. 1981. Water repellent soils: a state-of-the-art. General Technical Report PSW-46. US Department of Agriculture, Forest Service, Pacific Southwest Forest and Range Experiment Station, Berkeley, California, USA.

DeBano L.F. 2000. The role of fire and soil heating on water repellency in wildland environments: a review. *Journal of Hydrology*, 231-232: 195-206.

DeBano, L.F., L.D. Mann, and D.A. Hamilton. 1970. Translocation of hydrophobic substances into soil by burning organic litter. *Soil Science Society of America Journal*, 34: 130-133.

DeBano, L.F., D.G. Neary, and P.F. Ffolliott. 1998. *Fire's Effects on Ecosystems*. John Wiley, New York, New York, USA.

Dekker, L.W. and C.J. Ritsema. 2000. Wetting patterns and moisture variability in water repellent Dutch soils. *Journal of Hydrology*, 231-232: 148-164.

Doerr, S.H., W.H. Blake, R.A. Shakesby, F. Stagnitti, S.H. Vuurens, G.S. Humphreys, and P. Wallbrink. 2004. Heating effects on water repellency in Australian eucalypt forest soils and their value in estimating wildfire soil temperatures. *International Journal of Wildland Fire*, 13(2): 157-163.

Doerr, S.H., R.A. Shakesby, W.H. Blake, C.J. Chafer, G.S. Humphreys, and P. Wallbrink. 2006. Effects of differing wildfire severities on soil wettability and implications for hydrological response. *Journal of Hydrology*, 319: 295-311.

Douglas, I. 1977. *Humid Landforms*. Australian National University Press, Canberra, New South Wales, Australia.

Dunin, F.X. 1976. Infiltration: its simulation for field conditions. pp. 199-227. *In* J.C. Rodda [ed.]. *Facets of Hydrology*. John Wiley and Sons, Bristol, UK.

Dunkerly, D.L. and K.L. Brown. 1995. Runoff and runon areas in a patterned chenopod shrubland, arid Western New South Wales, Australia: characteristic and origin. *Journal of Arid Environments*, 20: 41-55.

Dunne, T. and L.B. Leopold. 1978. *Water in Environmental Planning*. W.H Freeman, San Francisco, California, USA.

Dunne, T. and W. E. Dietrich. 1980. Experimental study of Horton overland flow on *Zeitschrift fur Geomorphologie, Neue Folge, Supplementbände*, 35: 40-59.

Dunne, T., W. Zhang, and B.F. Aubry. 1991. Effects of rainfall, vegetation and microtopography on infiltration and runoff. *Water Resources Research*, 27(9): 2271-2285.

Emblenton, C. and J.B. Thornes. 1979. *Process in Geomorphology*. Arnold, London, UK.

Emmerich, W.E. and J.R. Cox. 1992. Hydrologic characteristics immediately after seasonal burning on introduced and native grasslands. *Journal of Range Management*, 45: 476-479.

Faulkner, H. 1990. Vegetation cover density variations and infiltration patterns on piped alkali sodic soils: implications for the modeling of overland flow in semi-arid areas. pp. 317-346. *In* J.B.Thornes [ed.]. *Vegetation and Erosion: Process and Environments*, John Wiley and Sons, Chichester, UK.

Ferreira, A.J.D., C.O.A. Coelho, R.P.D Walsh, R.A. Shakesby, A. Ceballos, and S.H. Doerr. 2000. Hydrological implications of soil water repellency in *Eucalyptus globulus* and *Pinus pinaster* forests, north-central Portugal. *Journal of Hydrology*, 213-232, 165-177.

Ferreira, A.J.D., C.O.A. Coelho, A.K. Boulet, and F.P. Lopes. 2005. Temporal patterns of solute loss following wildfires in Central Portugal. *International Journal of Wildland Fire*, 14(4): 401-412.

Garza, N.E. Jr. and W.H. Blackburn. 1985. The effect of early winter or spring burning on runoff, sediment, and vegetation in the Post Oak Savannah of Texas. *Journal of Range Management*, 38(3): 280-282.

Gimeno, E., V. Andreu, and J.C. Rubio. 2000. Changes in organic matter, nitrogen, phosphorus and cations in soil as a result of fire and water erosion in a Mediterranean landscape. *European Journal of Soil Science*, 51: 201-210.

Green, R.E and G. Ampt. 1911. Studies of soil physics 1: The flow of air and water through soils. *Journal of Arid Science*, 4: 1-24.

Heede, B.H., M.D. Harvey, and J.R. Laird. 1988. Sediment delivery linkages in a chaparral watershed following a wildfire. *Environmental Management*, 12: 349-358.

Helvey, J.D. 1980. Effects of a north central Washington wildfire on runoff and sediment production. *Water Resources Bulletin*, 16: 627-634.

Hendricks, B.A. and J.M. Johnson. 1944. Effects of fire on steep mountain slopes in central Arizona. *Journal of Forestry*, 42: 568-571.

Hendricks, J.M.H. 1990. Determination of hydraulic soil properties. pp. 43-92. *In* M.G. Anderson and T.P Burt [eds.]. *Process Studies in Hillslope Hydrology*. John Wiley and Sons, Chichester, UK.

Hester, J.W., T.L. Thurow, and C.A. Taylor, Jr. 1997. Hydrologic characteristics of vegetation types as affected by prescribed burning. *Journal of Range Management*, 50: 199-204.

Hewlett, J.H. and A.R. Hibbert. 1967. Factors affecting the response of small watersheds to precipitation in humid areas. pp. 275-290. *In* W.E. Sopper and H.W. Lull [eds.]. Proceedings of International Symposium of Forest Hydrology, National Science Foundation, Pennsylvania State University, Pennsylvania, USA.

Hibbert, A.R. 1985. Storm runoff and sediment production after wildfire in chaparral. *Hydrology and Water Resources in Arizona and the Southwest*, 15: 31-42.

Hillel, D. 1971. *Soil and Water. Physical Principles and Processes*. Academic Press, New York, New York, USA.

Hillel, D. 1982. *Introduction to Soil Physics*. Academic Press, New York, New York, USA.

Hills, R.C. 1970. The determination of the infiltration capacity on field soils using the cylinder infiltrometer. B.G.R.G., Technical bulletin 3, Goebooks Ltd., Norwich, UK.

Holzhey, C.S. 1969. Water-repellent soils in southern California. pp. 31-41. *In* L.F. DeBano and J. Letey [eds.]. Proceedings of Water-repellent Soils, University of California, Riverside, California, USA.

Horton, E. 1933. The role of infiltration in the hydrologic cycle. *Eos, Transactions, American Geophysical Union*, 14: 446-460.

Horton, R.E. 1940. An approach toward a physical interpretation of infiltration capacity. *Soil Science Society of America*, 5: 399-417.

Hubbert, K.R., H.K. Preisler, P.M. Wohlgemuth, R.C. Graham, and M.G. Narog. 2006. Prescribed burning effects on soil physical properties and soil water repellency in a steep chaparral watershed, southern California, USA. *Geoderma*, 130: 284-298.

Huffman E.L., L.H. MacDonald, and J.D. Stednick 2001. Strength and persistence of fire induced soil hydrophobicity under ponderosa and lodgepole pine, Colorado Front Range. *Hydrological Processes,* 15: 2877-2892.

Hungerford, R.D. 1996. Soils—fire in ecosystem management notes: unit II-I. U.S. Department of Agriculture, Forest Service, National Advanced Resource Technology Center, Marana, Arizona, USA.

Imeson, A.C. 1971. Heather burning and soil erosion on the North Yorkshire Moors. *Journal of Applied Ecology,* 8: 537-542.

Imeson, A.C., J.M. Verstraten, E.J. van Mulligen, and J. Sevink. 1992. The effect of fire and water repellency on infiltration and runoff under Mediterranean type forest. *Catena,* 19: 345-361.

Inbar, M., L. Wittenberg, and M. Tamir. 1997. Soil erosion and forestry management after wildfire in a Mediterranean woodland, Mt. Carmel, Israel. *International Journal of Wildland Fire,* 7: 285-294.

Inbar, M., M. Tamir, and L. Wittenberg. 1998. Runoff and erosion processes after a forest fire in Mount Carmel, a Mediterranean area. *Geomorphology,* 24: 17-33.

Kirkby, M.J. [ed.]. 1978. *Hillslope Hydrology.* Wiley-Interscience, Norwich, UK.

Knapp, B.J. 1978. Infiltration and storage of soil water. pp. 43-72. *In* M.J. Kirkby [ed.]. *Hillslope Hydrology.* Wiley-Interscience, Norwich, UK.

Knight, R.W., W.H. Blackburn, and C.J. Scifres. 1983. Infiltration rates and sediment production following herbicide/fire brush treatments. *Journal of Range Management,* 36: 154-157.

Kostiakov, A.N. 1932. On the dynamics of the coefficient of water percolation in soils and the necessity for studying it from a dynamic point of view for purposes of amelioration. pp. 17-21. *In Transactions of the 6th Commission of the International Society of Soil Science,* Russian Part A, International Soil Science Society, Groningen, Netherlands.

Lasanta, A. and A. Cerdà. 2005. Long-term erosional responses after fire in the Central Spanish Pyrenees: 2. Solute release. *Catena,* 60: 81-100.

Lavee, H., P. Kutiel, M. Segev, and Y. Benyamini. 1995. Effect of surface roughness on runoff and erosion in a Mediterranean ecosystem: the role of fire. *Geomorphology,* 11: 227-234.

Letey, J. 2001. Causes and consequences of fire-induced soil water repellency. *Hydrological Processes,* 15: 2867-2875.

Link, S.O., G.W. Gee, M.E. Thiede, and P.A. Beedlow. 1990. Response of a shrub-steppe ecosystem to fire: soil water and vegetational change. *Arid Soil Research and Rehabilitation,* 4: 163-172.

Marcos, E., R. Tarrega, and E. Luis-Calabuig. 2000. Comparative analysis of runoff and sediment yield with a rainfall simulator after experimental fire. *Arid Soil Research and Rehabilitation,* 14: 293-307.

Marqués, M.A. and E. Mora. 1992. The influence of aspect on runoff and soil loss in a Mediterranean burnt forest (Spain). *Catena,* 19: 333-344.

Martin, D.A. and J.A. Moody. 2001. Comparison of soil infiltration rates in burned and unburned mountainous watersheds. *Hydrological Processes,* 15: 2893-2903.

Meewing, R.O. 1969. Infiltration and soil erosion as influences by vegetation and soil in northern Utah. *Journal of Range Management,* 23(3): 185-188.

Meyer, G.A., J.L. Pierce, S.H. Wood, and A.J.T. Jull. 2001. Fire, storms, and erosional events in the Idaho batholith. *Hydrological Processes,* 15: 3025-3038.

Moody, J.A. and D.A. Martin. 2001a. Initial hydrologic and geomorphic response following a wildfire in the Colorado Front Range. *Earth Surface Processes and Landforms,* 26: 1049-1070.

Moody, J.A. and D.A. Martin. 2001b. Post-fire, rainfall intensity-peak discharge relation for three mountainous watersheds in the western USA. *Hydrological Processes,* 15: 2981-2993.

Munn, D.A., E.O. McLean, A. Ramirez, and T.J. Logan. 1973. Effect of soil, cover, slope, and rainfall factors on soil and phosphorus movement under simulated rainfall conditions. *Soil Science Society of America Proceedings,* 37: 428-431.

Musgrave, G.W. and H.N. Holtan. 1964. Infiltration. pp. 244-258. *In* V.T. Chow [ed.]. *Handbook of Applied Hydrology.* McGraw-Hill, New York, New York, USA.

Nassif, S.H. and E.M. Wilson. 1975. The influence of slope and rain intensity on runoff and infiltration. *Hydrologic Science Bulletin,* 20: 539-553.

Naveh, Z. 1990. Fire in the mediterranean – a landscape ecological perspective. pp. 1-20. *In* J.M. Goldammer and M.J. Jenkins [eds.]. *Fire in Ecosystems Dynamics. Mediterranean and Northern Perspective.* SPB Academic Publishing, The Hague, Netherlands.

Neal, J.H. 1938. The effect of the degree of slope and rainfall characteristics on runoff and soil erosion. Research Bulletin 280. U.S. Department of Agriculture, University of Missouri Agricultural Experiment Station, Columbia, Missouri, USA.

Nichols, M.L. and H.D. Sexton. 1932. A method of studying soil erosion. *Agricultural Engineering,* 13: 101-103.

Parr, J.F. and A.R. Bertrand. 1960. Water infiltration in soils. *Advanced Agriculture,* 12: 311-363.

Philip, J.R. 1957. Theory of infiltration: 1. The infiltration equation and its solution. *Soil Science,* 83(5): 345-357.

Pierson, F.B., P.R. Robichaud, C.A. Moffet, and K.E. Spaeth. 2006. Erosion following fire in a sagebrush ecosystem of the northern Great Basin, USA. *In* Proceedings of the 14th Conference of International Soil Conservation Organization, Marrakech, Morocco. *Online at* http://www.tucson. ars.ag.gov/isco/index_files/page416.htm [accessed 4 Sep. 08]

Pierson, F.B., P.R. Robichaud, and K.E. Spaeth. 2001. Spatial and temporal effects of wildfire on the hydrology of step rangeland watershed. *Hydrological Processes,* 15: 2905-2916.

Poesen, J. 1984. The influence of slope angle on infiltration rate and Hortonian overland flow volume. *Zeitschrift fur Geomorphologie, Neue Folge, Supplementbände,* 49: 117-131.

Ponce, V.M. 1989. *Engineering Hydrology: Principles and Practices.* Prentice-Hall, Englewood Cliffs, New Jersey, USA.

Prosser, I.P. and L. Williams. 1998. The effect of wildfire on runoff end erosion in native Eucalypt forest. *Hydrological Processes,* 12: 251-265.

Reid, K.D., B.P. Wilcox, D.D. Breshears, and L.H. MacDonald. 1999. Runoff and erosion in a piñon-juniper woodland: influence of vegetation patches. *Soil Science Society of America Journal,* 50: 1031-1036.

Robichaud, P.R. 2000. Fire effects on infiltration rates after prescribed fire in northern Rocky Mountain forests, USA. *Journal of Hydrology,* 231-232: 220-229.

Robichaud, P.R. 2005. Measurement of post-fire hillslope erosion to evaluate and model rehabilitation treatment effectiveness and recovery. *International Journal of Wildland Fire,* 14(4): 475-485.

Robichaud, P.R. and T.A. Waldrop. 1994. A comparison of surface runoff and sediment yields from low-and-high severity site preparation burns. *Water Research Bulletin*, 30: 27-34.

Robichaud, P.R. and S.M. Miller. 1999. Spatial interpolation and simulation of post-burn duff thickness after prescribed fire. *International Journal of Wildland Fire*, 9(2): 137-143.

Robichaud, P. R. and R.D. Hungerford. 2000. Water repellency by laboratory burning of four northern Rocky Mountain forest soils. *Journal of Hydrology*, 231 232(1-4): 207-219.

Robichaud, P.R., F.B. Pierson, and R.E. Brown. 2007. Runoff and erosion effects after prescribed fire and wildfire on volcanic ash-cap soils. pp. 83-94. *In* D.Page-Dumroese, R. Miller, J. Mital, P. Mc Daniel, and D. Miller [tech. eds.]. Proceedings of Volcanic-Ash-Derived Forest Soils of the Inland Northwest: Properties and Implications for Management and Restoration. Proceedings RMRS-P-44, U.S. Department of Agriculture, Forest Service, Rocky Mountain Research Station, Fort Collins, Colorado, USA.

Robichaud, P.R., F.B. Pierson, R.E. Brown, and J.W. Wagenbrenner. 2008a. Measuring effectiveness of three post-fire hillslope erosion barrier treatments, western Montana, USA. *Hydrological Processes*, 22: 159-170.

Robichaud, P.R., S.A. Lewis, and L.E. Ashmun. 2008b. New procedure for sampling infiltration to assess post-fire soil water repellency. Research Note RMRS-RN-33. U.S. Department of Agriculture, Forest Service, Rocky Mountain Research Station, Fort Collins, Colorado, USA.

Robichaud, P.R., J.R. Wagenbrenner, R.E. Brown, P.M. Wohlgemuth, and J.L. Beyers. 2008c. Evaluating the effectiveness of contour-felled log erosion barriers as a post-fire runoff and erosion mitigation treatment in the western United States. *International Journal of Wildland Fire*, 17: 255-273.

Rubio, J.L., J. Forteza, V. Andreu, and R. Cerni. 1997. Soil profile characteristics influencing runoff and soil erosion after forest fire: A case study (Valencia, Spain). *Soil Technology*, 11: 67-78.

Sànchez, J.R., V.J. Mangas, C. Ortiz, and J. Bellot. 1994. Forest fire effect on soil chemical properties and runoff. pp. 53-65. *In* M. Sala and J.L. Rubio [eds.]. *Soil Erosion and Degradation as a Consequence of Forest Fires*. Geoforma Ediciones, Logroño, Spain.

Scott, V.H. 1956. Relative infiltration rates of burned and unburned upland soils. *Transactions of the American Geophysical Union*, 37: 67-69.

Scott, D.F. 1993. The hydrological effects of fire in South African mountain catchments. *Journal of Hydrology*, 150: 409-432.

Scott, D.F. and D.B. van Wyk. 1990. The effects of wildfire on soil wettability and hydrological behaviour of an afforested catchment. *Journal of Hydrology*, 121: 239-256.

Seyfred, M.S. 1991. Infiltration patterns from simulated rainfall on semiarid rangeland soil. *Soil Science Society of America Journal*, 55: 1726-1734.

Shahlaee, A.K., W.L. Nutter, E.R. Burroughs, Jr., and L.A. Morris. 1991. Runoff and sediment production from burned forest sites in the Georgia Piedmont. *Water Resources Bulletin*, 27: 485-493.

Shakesby, R.A. and S.H. Doerr. 2006. Wildfire as a hydrological and geomorphological agent. *Earth-Science Reviews*, 74: 269-307.

Shakesby, R.A., CDA Coelho, A.D. Ferreira, J.P. Terry, and R.P.D. Walsh. 1993. Wildfire impacts on soil erosion and hydrology in wet Mediterranean forest, Portugal. *International Journal of Wildland Fire,* 3(2): 95-110.

Shakesby, R.A., S.H. Doerr, and R.P.D. Walsh. 2000. The erosional impact of soil hydrophobicity: current problems and future research directions. *Journal of Hydrology,* 231-232: 178-191.

Simmanton, J.R. and K.G. Renard. 1982. Seasonal change in infiltration and erosion from USLE plots in Southeastern Arizona. *Hydrology and Water Resources in Arizona and the Southwest,* 12: 37-46.

Soler, M. and M. Sala. 1992. Effects of fire and of clearing in a Mediterranean *Quercus ilex* woodland: an experimental approach. *Catena,* 19: 321-332.

Soni, P., S. Naithani, and H.N. Mathur. 1985. Infiltration studies under different vegetation cover. *Indian Journal of Forestry,* 8(3): 170-173.

Soto, B., R. Basanta, E. Benito, R. Pérez, and F. Díaz Fierros. 1994. Runoff and erosion from burnt soils in Northwest Spain. pp. 91-98. In M. Sala and J.L. Rubio [eds.]. *Soil Erosion and Degradation as a Consequence of Forest Fires.* Geoforma Ediciones, Logroño, Spain.

Springer, E.P. and G.F. Gifford. 1980. Spatial variability of rangeland infiltration rates. *Water Resources Bulletin,* 16(3): 550-552.

Starr, J.L. 1990. Spatial and temporal variation of ponded infiltration. *Soil Science Society of America Journal,* 54(3): 629-636.

Steichen J., M. Hooker, and D.M. Powell. 1987. Straw burning reduces infiltration in winter wheat. *Journal of Soil and Water Conservation,* 42(5): 364-366.

Summerfield, M.A. 1991. *Global Geomorphology.* Logman Scientifc and Technical, New York, New York, USA.

Trabaud, L. 1981. Man and fire: impacts on Mediterranean vegetation. pp. 523-537. *In* F. di Castri, D.W. Goodall and R.L. Specht [eds.]. *Mediterranean-type Shrublands.* Elsevier, Amsterdam, Netherlands.

Wondzell, S.M. and J.G. King. 2003. Post-fire erosional processes in the Pacific Northwest and Rocky Mountain regions. *Forest Ecology Management,* 178: 75-87.

Wood, M.K. and S.H. Luk. 1990. Vegetation effects on soil and water losses on weathered granitic hillslopes, south China. *Physical Geography,* 11(1): 1-16.

Wright, H.A., F.M. Churchill, and W.C. Clarck. 1978. Influence of prescribed burning on infiltration and sediment production in the pinyon-juniper woodland, Nevada. *Journal of Range Management,* 31: 250-253.

Wright, H.A., F.M. Churchill, and W.C. Clarck. 1982. Soil loss, runoff, and water quality of seeded and unseeded step watersheds following prescribed burning. *Journal of Range Management,* 35: 382-385.

Zwolinski, M.J. 1971. Effects of forest fire on water infiltration in a ponderosa pine stand. *Hydrology and Water Resources in Arizona and the Southwest,* 1: 107-109.

4

Physical and Chemical Effects of Fire on Soil

Xavier Úbeda[1]* and *Luís R. Outeiro*[1]

Abstract

Soils, like other natural systems, are composed of many highly interrelated components making it difficult to analyze individual characteristics and processes out of context. Physical properties (e.g., texture, structure, density, and porosity) and chemical properties (e.g., pH, reactivity, and mineral content) are only part of the complex soil system. Soil processes are the mechanisms that occur within the soil. Both soil properties and soil processes differ by soil type and by the environmental context being considered, which for this review, is fire. Fire, whether natural and intentional, has a large impact in most ecosystems, and in fire-dependent natural systems, the lack of fire can disrupt basic forest and soil dynamics. Research methodologies used to determine changes in soil physical and chemical properties have been adapted to study soil after wildland fires, prescribed fires, and laboratory fire simulations.

Direct fire effects on soil are caused by increased temperature, loss of matter through volatilization, combustion of vegetation and organic matter, and addition of substances by condensation. General post-fire changes in soil properties and processes include: a) decreased organic matter content, structure, stability, porosity, cation exchange capacity, water storage capacity, and infiltration; and b) increased bulk density, erodibility, water repellency, pH, and soil nutrient availability (short-term). The degree of change in soil properties and processes is directly affected by soil burn severity (the result of fire heat and duration), pre-fire soil water content, and post-fire rain and wind events. Fire effects on soil are important considerations as the use of fire as a management tool expands.

THE EFFECTS OF FIRE ON SOIL PROPERTIES

The earliest scientific literature about the effects of fire on soil properties dates from the 1930s, the same decade Herbert Stoddard and other advocates of fire

[1] *GRAM (Grup de Recerca Ambiental Mediterrània), Department of Physical Geography, Universitat de Barcelona, Barcelona, Spain.*

* *Corresponding author: Xavier Úbeda, GRAM (Grup de Recerca Ambiental Mediterrània), Department of Physical Geography, Universitat de Barcelona, Montalegre, 6. 08001 Barcelona, Spain. Tel: 934037892, e-mail: xubeda@ub.edu*

management were actively encouraging the use of prescribed fire to create healthy, productive environments. After several decades of fire suppression, land managers, scientists, and policy makers began to notice a gradual deterioration of forests and fields. Problematic levels of forest fuels were accumulating in some of the areas where prescribed burning had been discontinued, environmental integrity was declining, and the threat of catastrophic wildfires was increasing. Fire exclusion also led to an abundance of undesirable plant species (Baker and Hunter 2008). Several scientific publications appeared in the 1930s in favor of re-introducing prescribed burning, or the use of fire to manage the land. The benefits of 'Indian fires' for longleaf pine forests were reported by Greene (1931), Chapman (1932), and Stewart (2002). Thus, during the 1930s, prescribed fire was introduced across the southern United States as a forest management technique.

One of the first references on soil effects in literature was provided by Greene (1934). He analyzed the annual burning of grass in a virgin longleaf pine forest and the effects of the fires, particularly on the organic matter content, but also on other soil properties. Heyward (1937) studied the physical properties of soils in a similar longleaf pine forest and found that excluding fire for as little as 10 years resulted in more porous, penetrable soil. Heyward (1938) also found that single, low-intensity prescribed burns in the southern United States typically did not cause dramatic changes in soil structure and texture. He claimed that the higher soil temperatures during these prescribed fires usually lasted for a short period and, as a result, did not have dramatic consequences.

Wahlenberg (1935) also worked on these types of forest, although his results were more focused on the relationship between fire and soil and the reproduction of pines. He also found that infiltration rates were lower in soil from the southern United States that had been burned several times (Wahlenberg et al. 1939), a fact also reported by Meginnis (1935) and Arend (1941). Garren (1943) found that soil organic matter content is usually lower in soils that are repeatedly burned, and early researchers noted that burned soils were harder, denser and less permeable than unburned soils.

FIRES AND SOIL TEMPERATURE

The transfer of heat to soil is the principal mechanism by which fires affect the physical, chemical, and biological properties of soil (Neary et al. 1999). The effects of fire on the physical and chemical properties of soil depend on the intensity and severity of the fire, and fire intensity is directly proportional to the heat of combustion, amount of fuel consumed, and rate of fire spread. Therefore, fuel types, weather, and topography are important in determining the rate of heat released by a fire. The fuel properties that directly or indirectly affect fire intensity include fuel loading, moisture content, arrangement, chemical composition, and size. Wind speed and other weather conditions that affect fuel moisture also influence fire intensity (Baker and Hunter 2008).

The literature contains several definitions of *fire intensity*. For Byram (1959) the *fireline intensity* is the rate of heat energy released per unit time per unit length of fire front, regardless of the depth of the flame zone. According to Chandler (1983) and DeBano et al. (1998), the *reaction intensity* is calculated by estimating the amount of fuel burned per second, and assuming heat yields for the fuel, is usually expressed in kW m^{-2}. In relation to *fire intensity* and soil properties, Úbeda (1998) defined the fire intensity as the maximum temperature recorded at a certain point and the time that the temperature remains at a certain point, expressed in °C s^{-1}. According to Neary et al. (2005), *fire intensity* describes the rate at which a fire produces thermal energy.

Burn severity describes the response of ecosystems to fire and can be used to describe the effects of fire on the soil (soil burn severity), water system, ecosystem flora and fauna, atmosphere, and society. Burn severity, loosely, is a product of fire intensity and residence time and is generally considered to be light/low, moderate, or high. Given that burn severity is a response to the amount of energy (heat) released by a fire, it usually reflects fire intensity; however, the relationship between fire intensity and burn severity remains largely undefined because of difficulties encountered in relating resource responses to the burning process (Hungerford et al. 1990, Hartford and Frandsen 1992, Ryan 2002). Ryan and Noste (1985) and DeBano et al. (1998) developed the following criteria to solve this problem:

- Low burn severity—less than 2 percent of the area is severely burned, less than 15 percent is moderately burned and the remainder of the area is burned at a low severity or unburned.
- Moderate burn severity—less than 10 percent of the area is severely burned, but more than 15 percent is moderately burned and the remainder is burned at low severity or unburned.
- High burn severity—more than 10 percent of the area has spots that are burned at high severity, more than 80 percent is severely or moderately burned and the remainder is burned at a low severity.

According to Jain and Graham (2003), burn severity (sometimes referred to as fire severity) is not a narrowly defined entity but rather a general concept, and burn severity classification is a function of the measured units unique to the system of interest. These systems include flora and fauna, soil microbiology, hydrological processes, atmospheric inputs, fire management, and society, each with a set of measurable characteristics that can indicate level of response. For example, in fire management the indicator characteristics include the consumption of organic material, flame length, torching index, and other variables of fire behavior. In terms of the atmosphere, the characteristics reflective of burn severity include the types and amounts of particulates and toxic gases in the smoke. For society, the variables that indicate burn severity include the number of homes damaged, injuries and net value changes of valued resources. Measured attributes of flora and fauna, soil microbiology, and hydrological processes quantify the residual ecosystem structure after the fire and may indicate potential impacts on nutrient cycling, erosion, species diversity, and recovery rates. The difficulty in the classification of burn severity is to develop consistent and meaningful ecological information that can be easily related to secondary fire effects. Currently, burn severity classifications are either based on qualitative estimates or include detailed information about individual forest components that are difficult to condense into a single classification system. In addition, the measured forest structure may have no ecological relation to the secondary fire effects. The use of burn severity classification has limited ability to quantify and summarize the post-fire conditions and to predict secondary fire effects, such as erosion, tree mortality, nutrient cycling, and vegetation recovery.

The measure of fire intensity is vital to understanding changes in soil properties. Consequently, several authors have studied the changes in soil properties caused by fire (Hungerford et al. 1990, Dimitrakopoulos et al. 1994, Campbell et al. 1995, DeBano et al. 1998). However, these evaluations are sometimes made in laboratory assessments and can differ widely from prescribed fires and, to a lesser degree, from wildfires. It is difficult to take field measurements of fire intensity and it is generally only possible with prescribed fires. Although real temperature and duration data is essential to understanding the effects of wildfire, the safety risks inherent in obtaining these real-time measurements makes these research efforts very challenging.

Wildland Fires

One way to measure fire intensity in natural or wildfires is to reconstruct the fire. In a study of the Cadiretes Massif (northeast Spain), fire intensity categories were determined immediately after the fire. Three fire intensity zones were distinguished by examining the state and number of leaves and branches, the soil surface, and the quantity and color of ash: black-moderate, grey-medium and white-complete combustion (Moreno and Oechel 1989). This methodology resulted in three categories:

- *Low intensity fire:* The trees retained some leaves (although not all were green, and many would eventually fall) and a large number of branches, including smaller ones. A sizeable amount of litter deposited just after the fire on top of a two-centimeter burned layer of black humus (still remains two years after the fire). All *Quercus suber* and some *Pinus* survive. This area corresponds to the lower part of the burned area.
- *Medium intensity fire:* The trees had no leaves but retained a substantial number of their branches. There is a slight quantity of litter on the soil. This area is located between the low intensity burned area and the high intensity burned area.

- *High intensity fire:* The trees lost all their leaves and branches and only the trunks remained. The soil surface was completely exposed and showed a large area of grey and white ashes.

Prescribed Fires

Prescribed fires are used both for research and forest managements purposes (research: Soler et al. 1994; forest management: Grillo 2008; both: Outeiro and Úbeda et al. 2008). These fires provide the easiest means of recording fire temperatures and durations because the time and location of the fire is known in advance and the site can be instrumented. Some temperature recording methods are only able to register the maximum temperature while others take continuous measurements and can be used to classify fire intensity (Table 1). It is often valuable to use different methods to compare results (Odion and Davis 2000).

Paints

The use of temperature sensitive paints is an inexpensive method to record temperature maxima within a burned area. A change in paint color indicates the maximum temperature range reached (e.g., between 400° and 500°C). Several different paints may be used to obtain more refined data.

Evaporation cups

High heat-resistant porcelain evaporation cups containing the same quantity of water can be placed throughout the fire area. The difference in the volume of water remaining in the cups will indicate the temperatures reached at each location.

Metal

As the melting point differs for various metals, thin metal sticks can be placed throughout the fire. Based on the types of metals that were melted, the maximum temperature range in that area can be determined.

Laser thermometer

Although the laser thermometer was not designed to take field temperature measurements, it can be used in low fire intensity prescribed fires. This type of thermometer can detect differences in surface reflectometry that change with the temperature. Some laser thermometers have data loggers that record all measurements allowing temperature and duration to be determined. Although it is possible to take as many measurements as desired, only surface temperatures can be measured.

Thermocouples

Thermocouple thermometers can record the temperature at pre-determined intervals for periods of more than eight hours. These data allow fire intensity

Table 1 Soil temperatures measured during fires (Úbeda 1998; actualized by Mataix-Solera 2001)

Author (year)	Temp (°C)	Depth (cm)	Vegetation
Heyward (1938)	135	0.32-0.64	Pines
Cook (1939)	550	Surface	Grassland
Beadle (1940)	250	2.5	Dense Forest
	105	7.5	
	60	15	
Sampson (1944)	538	Surface	Scrubland
	149	3.8	
Masson (1949)	700	Surface	Savanna
Uggla (1958)	438	Surface	Conifers
	27	3	
	17	7	
Bentley and Fenner (1958)	590	Surface	Scrubland
	399	1	
Bentley and Fenner (1958)	177	Surface	Grassland
	93	1.3	
Uggla (1960)	1150	Surface	Pines
	500	3	
Humphreys and Lambert (1965)	900	Surface	Eucalyptus
	100	5	
Smith and Sparling (1966)	400-200	Surface	Shrubs
Floyd (1966)	510	Surface	Dense Forest
	44		
Cromer (1966)	666	Surface	Eucalyptus
	112		
Tothill and Shaw (1968)	245	Surface	Grassland
	68	1.3	
DeBano and Rice (1971)	716	Surface	Dense Forest
	166	2.5	(by afternoon)
	66	5	
	316	Surface	Dense Forest
	66	2.5	(by night)
	43	7.6	
Agee (1973)	93	Surface	Pines
Dunn and DeBano (1977)	800	Surface	Scrubland
	500	1	
Trabaud (1979)	250	Surface	Scrubland
	125	2.5	
	50	5	
DeBano et al. (1979)	700-250	Surface	Scrubland
	200-90	2.5	
Wells (1979)	250	Surface	Different kinds
	100	2	Black ashes

Contd.

Table 1 Contd.

Author (year)	Temp (°C)	Depth (cm)	Vegetation
Wells (1979)	500-750	Surface	Different kinds
	350-450	2	White ashes
	150-300	3	
	<100	5	
Biederbeek et al. (1980)	388-442	Surface	Masticated fuel
Rasmussen et al. (1986)	170-330	Surface	Wheat
Ventura et al. (1994)	700	Surface	Pines
	300	15	
Sánchez et al. (1994)	340	Surface	Pines
	740	Surface	*Cistus*
	280	Surface	Wheat
	51	Surface	No vegetation
Luchessi et al. (1994)	180	Surface	Scrubland
	50	2.5	
	475	Surface	Low Dense
	90	2.5	Scrubland
	40	5	
Úbeda (1998)	600	Surface	Grassland
	50	1	
Mataix-Solera (1999)	702	Surface	*Ulex parviflorus*
	22	5	

to be accurately determined. In addition, thermocouples can be placed on the surface and at different depths within the soil, which allows a temperature and heat duration soil profile to be determined.

Laboratory

Mufla

In the laboratory, fire intensity simulation can be performed using a Mufla oven (Guerrero 2003). Leaf consumption and ash generation can be measured under these laboratory conditions (Martin and Ubeda, unpublished data).

Near-infrared spectroscopy

Near-infrared spectroscopy (NIR) is a new laboratory technique developed by GEA to estimate the maxima temperatures reached on burned soils (Arcenegui et al. 2007). These data are used to determine fire effects on soil properties.

SAMPLING METHODS FOR MEASURING FIRE EFFECTS ON SOIL PHYSICAL AND CHEMICAL PROPERTIES

Soil Sampling After Wildland Fires

The main difficulty in soil sampling after wildland fires is the lack of information on the characteristics of the soil before the fire, particularly when considering the spatial variability of soil properties. Although sometimes criticized by researchers, the usual solution is to sample from a nearby area that was not affected by the fire to determine the pre-fire soil characteristics. The unburned area should be nearby and at least have comparable aspect, slope, and vegetation as the burned plot that will be sampled (Úbeda 1998). The burned areas to be sampled are often determined by burn severity. Sampling should be done immediately after the fire, before the first rain or a wind event. As ashes cause soil changes (see Chapter 3), sampling these ashes and the soil beneath them is important for determining fire effects.

Soil Sampling After Prescribed Fires

The advantage of studying fire effects in a forest area scheduled for prescribed burning is the ability to sample the same locations before and after the fire. The most common methodology is to reference the sampling points in a grid that may extend beyond the burned area. Grids may be square, circular, or triangular and the sample points within the grids can have a regular distribution or random pattern.

The number of sample points within the grid must provide for adequate statistical analysis. The relation between the area of the sampling plot and the number of sample points is usually expressed in terms of the separation distance between samples.

THE EFFECTS OF FIRE ON SOIL PROPERTIES

One of the main constraints when researching effects of fire on soil, or any variable in an ecosystem, is the lack of pre-fire data to compare to post-fire data. Researchers are often forced to simulate fire conditions in laboratories with heating techniques or use prescribed fire for research purposes. Increasingly, planning for prescribed fires includes a research monitoring plan to assess the effects of fire. Although this chapter focuses on the effects of wildfire on soil physics and chemistry, research results from laboratory heating, experimental fires, and research fires are included as these methods were used to determine the effects reported.

Since soil exists within a complex environment system, the reaction to fire is not fully explained by fire as the causal agent. Generally environmental complexity minimizes the primary effects of fire until the environmental system reaches a maximum threshold of resilience. Several smaller disturbances to the system may result in a more severe perturbation, but the system may remain resilient for a longer period. In mathematical notation:

$$f(R) = \frac{\sum_{i=1}^{N} W(t)}{r(t)}$$

where R = ecosystem resilience, $W(t)$ = low intensity wildfires and $r(t)$ = the given ecosystem response after these low intensity wildfires. Additionally, new terminology is needed to understand wildfire effects in this environmental context (Fig. 1).

Physical Soil Properties

In general, most fires do not cause enough soil heating to produce significant changes to soil physical properties (Hungerford et al. 1990). However, even small changes are important in a chain-related system, such as the soil system. Wells et al. (1979) suggest the magnitude of the change in physical properties depends largely on the severity of a fire, the proportions of over-story and under-story vegetation destroyed, forest floor consumed, heating of the soil, proportion of area burned, and length of fire intervals. Fire's effects on physical properties are significantly related to the amount of soil organic matter, which is essential for maintaining soil structure, consumed by the fire. According to Raison and Walker (1986), the physical fire-induced changes in soil and plant material become significant when the temperature reaches 400°C (Table 2). Consumption of soil organic material can also result in various indirect impacts including increased soil water repellency, which in turn is reflected in decreased infiltration, increased runoff, and subsequent increased erosion (Clark 2001).

Table 2 Physical changes in some soil properties as temperature increases (Walker et al. 1986)

Temperature (°C)	Property change
>1200	Loss of calcium as gas
950	Clay minerals converted to different phases
600	Maximum loss of potassium and phosphorus Fine ash produced
540	Little residual nitrogen or carbon left
420	Water lost from clay minerals causing change in type
400	Organic matter carbonizes

Soil bulk density

Fire-induced changes in soil bulk density are not immediately detrimental; however, numerous fires in a given period of time can produce a significant change in soil structure and other related properties, including bulk density.

Soil porosity

Pore space in soils allows water (soil solution) and air to move through the soil. Well-aggregated soils contain a balance of macropores (>0.6 mm in diameter) and micropores (<0.6 mm in diameter) (Singer and Munns 1996).

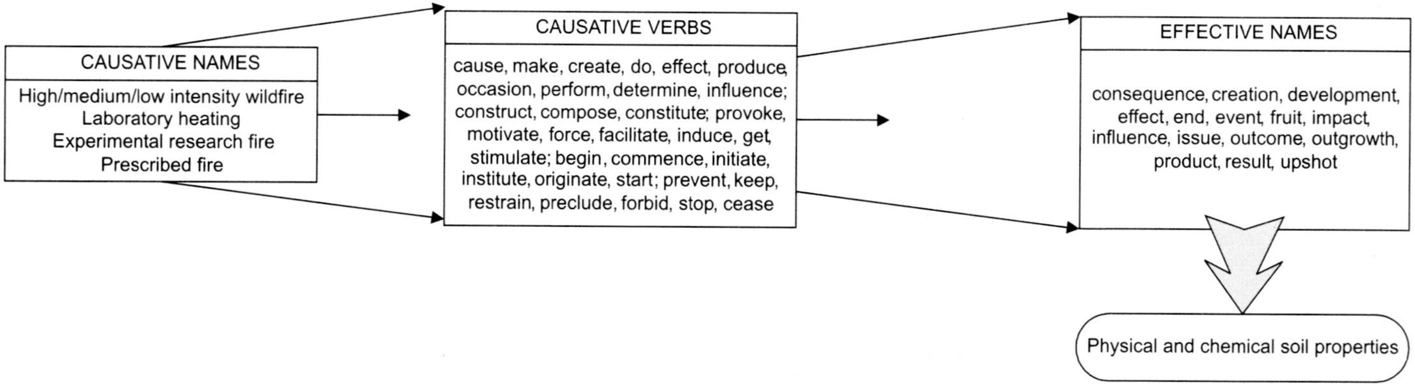

Fig. 1 Cause-effect diagram of physical and chemical changes on soil.

This balance allows a soil to rapidly transmit water and air through macropores and to retain water by capillarity in micropores. Macropores in the surface soil horizons are especially important pathways for infiltration of water into the soil. Fire and associated soil heating can destroy soil structure, resulting in a decrease in total porosity and pore size and affecting the micro- and macropore distribution in the surface horizons of soil. Loss of macropores in the soil surface reduces infiltration rates and produces overland flow. Combustion of organic matter above and within the soil can also lead to a water repellent soil condition, which further decreases infiltration rates (DeBano et al. 1998).

Soil temperature

Based on laboratory experiments, many fire-induced soil changes appear to have threshold temperatures for these changes. However, the lack of standardization in laboratory heating methods may lead to inconsistent results, and laboratory results do not necessarily translate directly into field scale fire effects. Researchers have used heating intensity methods with heating times that vary from 40 minutes to 48 hours. Busse et al. (2005) tested the importance of soil texture and soil moisture on soil lethal heating in 48 controlled burns of masticated fuels. They found that soil temperatures were strongly affected by soil moisture but not by soil texture; and that soil moisture as low as 20 percent by volume was sufficient to quench soil heating and inhibit lethal soil temperatures below the first 2.5 cm of soil.

Soil structure

Fire can affect the clay mineral and organic components in the upper portion of the soil profile. At the upper part of the A mineral soil horizon, organic matter plays an important role; whereas the clay mineral is more relevant at the B horizon. DeBano et al. (1998) suggests that clay minerals in soils are usually not significantly altered during a fire for two reasons: a) its capacity to tolerate relatively high temperatures before becoming irreversibly damaged (soil heating of 460°C or above is needed to remove OH groups from clays; Giovannini 1998); and b) clay content in the first centimeters of soil tends to be less than five percent, as clay minerals are formed deeper in the soil profile. Therefore, damage to clay minerals would only occur during severe fires on soils where the A horizon has been removed by some disturbance (such as erosion) or immature soils that have surface clays.

Soil structure, especially near the surface, mainly depends on soil organic matter content; thus, the amount of Soil Organic Carbon (SOC) consumed by fire is directly related to changes in soil structure. Unlike clay mineral components, soil organic matter is highly affected by fires because: a) destruction of organic matter occurs at low temperatures (begins at 200°C and there is complete destruction at 500°C); and b) soil organic matter is concentrated in the upper part of the soil profile where it is exposed more directly to heat radiated downward during combustion of surface fuels. In a laboratory heating experiment on Galician soils (schist and granidorites), Garcia-Corona et al. (2004) found that soil heating below 220°C had no adverse effect on aggregation-related soil properties and that water aggregate stability increased over the temperature range of 170° to 220°C. Combustion of organic matter favored dry disaggregation and reduced water aggregate stability (Garcia-Corona et al. 2004).

Soil water repellency

As stated by DeBano (1981), soil water repellency is a physical soil property that can be affected by fire. Soil water repellency occurs in several ways: a) drying of soil organic material (Jemison 1942, Gilmour 1968); b) organic substances leached from plant litter material coat mineral particles (Roberts and Carbon 1972, DeBano, 1981); c) hydrophobic by-products of microbial mycelium form in the upper soil (Bond and Harris, 1964, Savage et al. 1969); d) hydrophobic organic matter is intermixed with mineral soil particles (Das and Das 1972, McGhie and Posner 1980, DeBano 1981); and e) heated organic matter forms volatile substance that move into the soil column and condense on soil particles (DeBano and Krammes 1966, DeBano et al. 1976; see Chapter 7 for detailed discussion).

Water soil infiltration rate-Hydraulic conductivity

Garcia-Corona et al. (2004) state that the reduced saturated hydraulic conductivity (K_s) is the principal adverse effect of soil heating by fires, as the decreased infiltration capacity increases surface runoff and erosion (see Chapter 3 for detailed discussion). At two sites in the northern Rocky Mountains, a plot scale simulation rainfall experiment was used to determine K_s following prescribed fires of different burn severities (Robichaud 2000). In plots with water repellent surface conditions, the K_s ranged from 23 to 55 mm h^{-1} for both low and high severity burns (N=4). Plots burned at low severity, had K_s ranging from 60 to 89 mm h^{-1} in one site and 10 to 63 mm h^{-1} in the other (n=11).

Water storage capacity of soils and water retention

Water is retained in soil pores by capillarity. The smaller the pore size, the greater the water holding capacity. For example, small pore sizes in clay-sized inorganic materials have large water-holding capacities per unit volume and, as a consequence, permit retention of significant amounts of water for plant growth. Organic matter 'glues' mineral soil particles together to produce aggregates that create water-holding micropores and enhance soil structure. Therefore, a loss of soil organic matter and structure during a fire has an adverse effect on water retention by soil.

Soil erodibility

Soil texture is important in determining erodibility (see Chapter 6). Sandy soils are more easily detached than silt soils, but have lower runoff rates and are less likely to be transported. Clay soils are not easily detached but have lower infiltration rates that may lead to greater runoff and increased erosion. Silt soils tend to have the greatest erodibility values, since particles are easily detached and transported, and consolidation of subsoils, or subsoils with higher clay contents, can lead to greater runoff. As expected, Giovannini (1997) found the greatest post-fire erosion is often associated with high silt loess soils. In the same study, it was reported that erodibility of burned soils increases with increased soil heating; thus, more erodible soil is expected in high burn severity areas. In addition, recovery to pre-fire erodibility values was more rapid in less intensely burned soil (234 ± 72°C) than in more intensely burned soils (570 ± 122°C), where, after two post-fire years, the soil remained highly erodible. It was noted that erosion occurring after 15 and 24 months exposed a new erodible surface (Giovannini 1997). Using the Water Erosion Prediction Project (WEPP), Robichaud and Monroe (1997) predicted that on a uniform hill slope, the high-severity burn conditions produced about four times as much sediment as low-severity burn conditions for a 160 mm 6-hr storm.

Chemical Soil Properties

The effects of fire on soil chemistry involve complex biogeochemical interactions that impact biotic and abiotic cycling of soil components. The most studied fire-induced changes in the soil chemical processes are related losses and additions of soil nutrients and changes in nutrient availability for regenerating vegetation (Fig. 2).

Soil pH

Soil pH may be the variable most reported by authors working on fire-affected soil, and in most cases, wildfires increase soil pH. Many, although not all, of these authors report significant changes due to wildfire that could last for decades (Khanna and Raison 1986, Etiégni and Campbell 1991). Changes in soil pH after fire are related to the post-fire wind and rain events and longevity of the ashes on the site (Mataix-Solera 1999, Wells et al. 1979). Soil pH is an important factor affecting the availability of plant nutrients, with phosphorous, iron, and copper being highly affected by fire (Neary et al. 2005).

Soil electrical conductivity (EC)

Electrical conductivity (EC) is a measure of a material's ability to conduct an electric current. In soils the electrical conductivity gives a measure of total

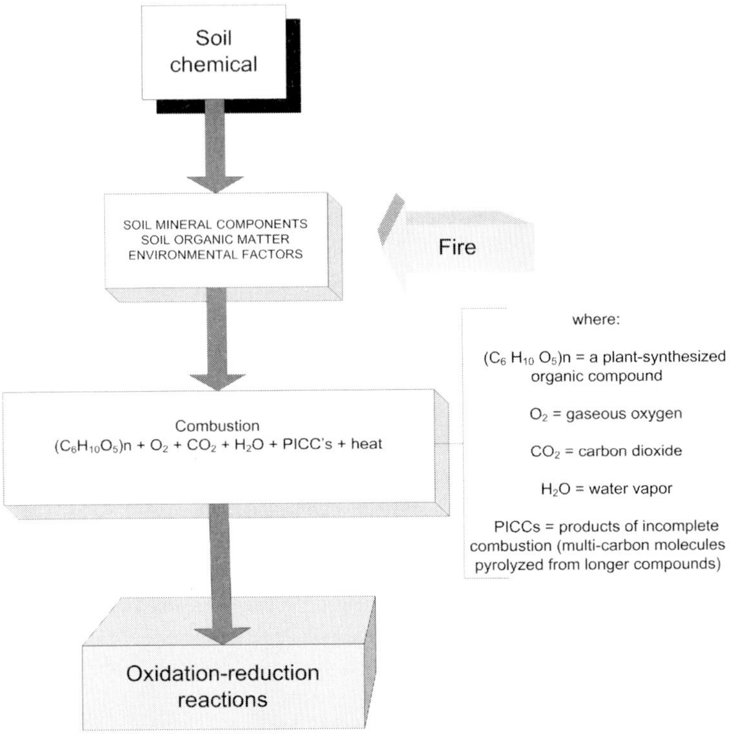

Fig. 2 Soil chemical cascade chain.

dissolved solids present. Although EC is not commonly reported in post-fire studies, it generally increases after fire. As in other soil variables, the significance of the EC response to fire is highly dependent on land use history of the area being studied. Outeiro and Úbeda (in review, *Catena*) found that EC had a significant change in forest plots, but, in the same study, an abandoned terrace did not have significant EC changes with prescribed fire. In a study on the northeastern Iberia Peninsula in Spain, Pardini et al. (2004) reported a post-fire EC increase (not statistically significant) due to the large amount of ashes incorporated into the soil.

EFFECTS OF FIRE ON SOIL COMPONENTS

Organic matter, arguably the most important soil constituent, is concentrated on the surface where it makes up most of the L-, F-, and H-layers and influences the properties of the underlying mineral soil horizons. Humus and soil organic matter (SOM) are both used to describe the total organic compounds in soils, excluding plant and animal tissues, their 'partial decomposition' products, and the soil biomass (Stevenson 1982). The quantity of SOM in a soil depends on the five soil-forming factors described by Jenny

(1941) including: time, climate, topography, vegetation, and parent material. The effects of fire on soil SOM are dependent on these five formation factors, and the 'sixth factor', the fire itself.

The fire-induced changes in SOM are related to fire effects on other soil physical properties, such as loss of structure, cation exchange capacity (CEC), and water retention capacity (Fig. 3). Fires generally increase SOM decomposition rates and the mineralization process, which increases nutrient availability. Burning, or combustion, is an instantaneous physical decomposition process that not only volatilizes nutrients, such as nitrogen (N), but also alters the remaining organic materials (St. John and Rundel 1976).

Soil Macro- and Micro-nutrients

Nitrogen

Soil nitrogen (N) is considered the most limiting nutrient in wildland ecosystems because it is not produced by decomposition of rocks and minerals (Macadam 1989). Consequently, N requires special consideration when managing fire, particularly in N deficient ecosystems (Neary et al. 2005, Maars et al. 1983). Most N found in the vegetation, water, and soil of wildland systems originates in the atmosphere. The rare exception is the addition of synthetic N-fertilizers produced industrially and used to fertilize forested areas. In any ecosystem N depletion can affect plant productivity, and even relatively small N losses can affect the long-term productivity in N-deficient ecosystems. Volatilization is the process most responsible for N losses during fire where temperatures frequently exceed the necessary 200° to 300°C. There is a gradual increase in N loss by volatilization as temperatures increase (Fig. 4; Knight 1966, White et al. 1973). Bustamante et al. (2006) report that N loss occurs through volatilization and via particulate matter formation during combustion and subsequent wind-borne transport. Pivello and Coutinho (1992) estimated that, during a cycle of six prescribed fires in a campo sujo area (grass-dominated vegetation type), about 95 percent of the N in plant biomass was released to the atmosphere.

The N that is not completely volatilized either remains as part of the unburned fuels or it is converted to highly available ammonium nitrogen (NH_4-N) that remains in the soil (DeBano et al. 1979, Covington and Sackett 1986, Kutiel and Naveh 1987, DeBano 1991). Spatial variability is an important factor to take into account when assessing the losses of N by fire. Neary et al. (2005) suggests estimates of the total N losses during prescribed fire must be based on both fire behavior and total fuel consumption because irregular burning patterns are common. As a result, combustion is not complete at all locations on the landscape (DeBano et al. 1998). For example, during a prescribed burn in southern California, total N loss only amounted to 10 percent of the total N contained in the plant, litter, and upper soil layers before burning (DeBano and Conrad 1978). The greatest loss of N occurred in

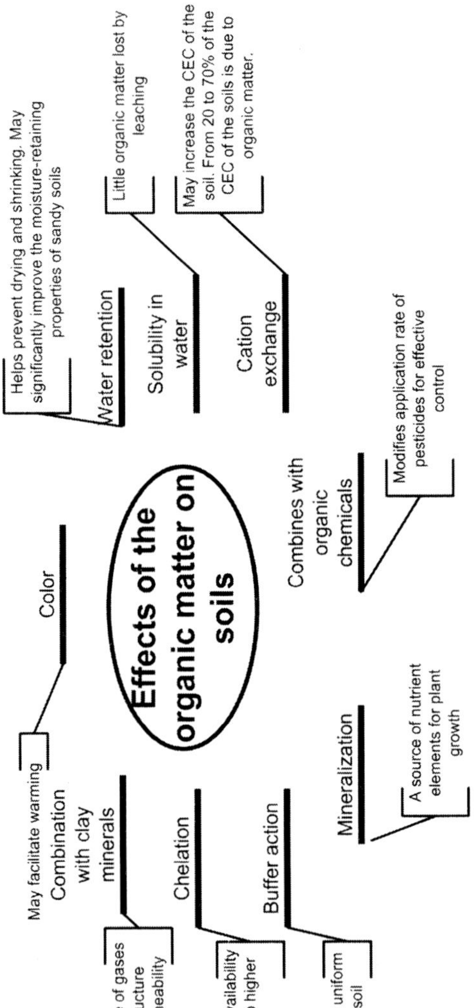

Fig. 3 Soil organic matter diagram. (Adapted from F.J. Stevenson 'Humus Chemistry' in Sparks, 2005.)

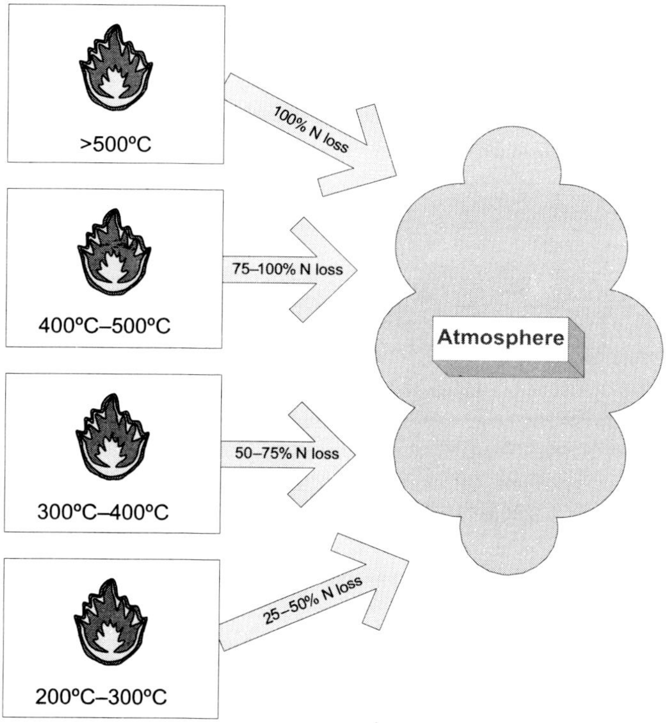

Fig. 4 Nitrogen loss-fire temperature diagram. (Adapted data from Knight, 1966.)

above-ground fuels and litter on the soil surface. Úbeda et al. (2005) reports an increase of total N just after a prescribed fire in western Mediterranean grassland. DeBano et al. (1979) reported N loss during a prescribed fire in which two-thirds of the total N was lost over dry soils compared to only 25 percent when the litter and soil were moist. Monleon et al. (1997) conducted under-story burns on ponderosa pine sites that had preciously burned 4 months, 5 years, and 12 years ago. The only significant response was in the surface soils (0 to 5 cm) where, at the 4-mo sites, an increase in total C, inorganic N, and C/N ratio were measured. After burning the 5-yr sites, a decrease in total soil C, N, and the C/N ratio were measured. Total soil C and N in the surface soils did not respond to burning on the 12-yr site.

The rate at which N becomes available depends on levels of microbial activity in the soil. This in turn depends on soil moisture, temperature, pH, and the presence and quality of organic materials. In general, the drier and poorer a site, the higher the proportion of total site N is contained in the forest floor and the more important it is to conserve it. Since natural rates of N accrual are extremely slow on most of the forest, it is important to prevent or minimize losses. Otherwise, chemical fertilizers or the introduction of nitrogen-fixing species may be required to offset losses (Macadam 1989).

Although details of the N cycle are beyond the scope of this chapter (see Chapter 8), the significance of fire in determining the form of N available is significant. For example, at 100°C the soil ammonium production begins while at 200°C loss of N from the atmosphere starts. Fire effects on the N cycle are dependent on the temperatures reached and have high spatial variability (Fig. 5).

Phosphorus

The combustion of vegetation and litter has significant impact on the biogeochemical cycle of phosphorus (P); however, P losses through volatilization and leaching are much smaller than N losses (Certini 2005). In northwest Melbourne (Australia), a study of fire effects on P soil levels from various fire treatments. There was a substantial decline in P across all fire treatments, including the unburned control (Hopmans 2003). Repeated low-intensity fires increased extractable P in the surface soils in some studies (Adams et al. 1994, McKee 1982), while some other works showed little if any change in soil P (Binkley et al. 1992, Boyer and Miller 1994). Pardini et al. (2004) reported a significant increase of available P in *Quercus suber* with *Erica arborea* shrub sites immediately after a wildfire, and after six months values remained one or two orders of magnitude higher than pre-fire values. In a

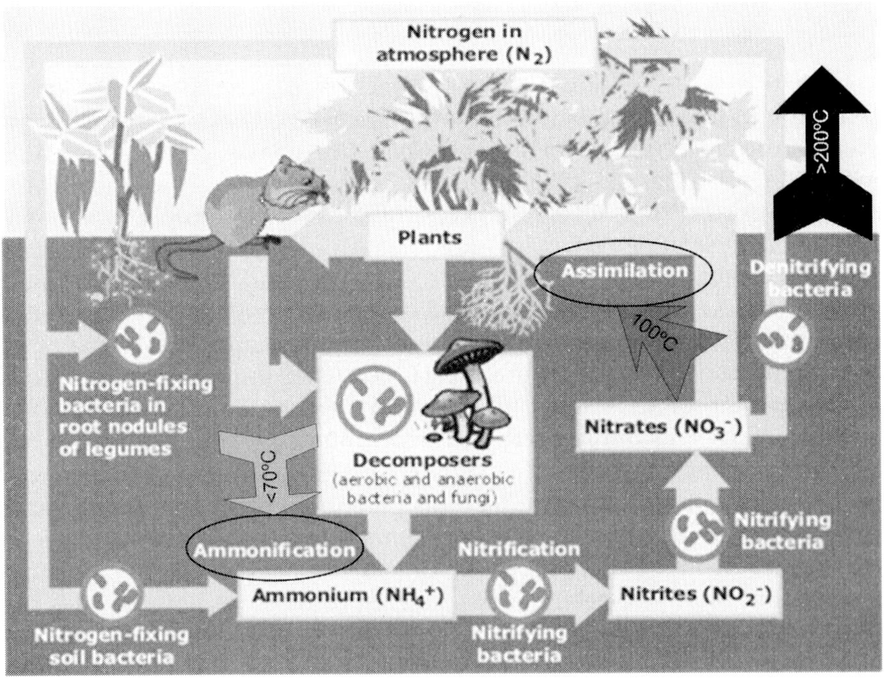

Fig. 5 Temperature dependent fire effects on the natural N cycle. (Adapted from U.S. Environmental Protection Agency, 2006.)

slash burning experiment, Romanyà (1993) reported an increase in post-fire labile organic P, which is the opposite of the conditions under the ashbed where decreases of total organic P and less labile organic P were observed. Despite all efforts to follow the effects of fire on the soil P pool, it is still poorly understood. The lack of explicit descriptions of chemical reactions and the wide variety of chemical analyses used to measure P have contributed to the range of measurements and contradictions among studies.

Phosphorous is particularly important because it is a macronutrient that is frequently limiting in soil systems, and it becomes insoluble at both high and low pHs. At low pH, P forms insoluble compounds with iron and at high pH, calcium compounds tend to immobilize it (Outeiro and Úbeda in review, *Catena*, Neary et al. 2005). Given that fire and ash formation can alter soil pH, this is significant to P availability. The long-term use of retardants (which contain P) and groundwater eutrophication (related to the amount of P in the water), are significant environmental problems directly related to fire effects on soil. An improved understanding of how fire modifies the soil P pool is needed to develop appropriate mitigations.

Inorganic cations: calcium, potassium, sodium, magnesium

Volatilization of cations are usually minor due to the high volatilization temperatures of these minerals, however, their loss from severely burned sites may be caused by surface erosion, leaching, or transport of ash (Wright and Bailey 1982). The proportions of cations produced by exposing fresh vegetation and litter from Mediterranean plots to high temperature heating (585°C) determined that potassium (K) was the predominant cation (1600 ppm) followed by sodium (Na; 860 ppm), calcium (Ca; 20 ppm), and magnesium (Mg; 2 ppm; Giovannini 1994). After fires, monovalent cations, such as Na and K, are present largely as chlorides and carbonates that are readily mobilized (DeBano et al. 1998, Soto and Diaz Fierros 1993). Divalent ions, including Ca and Mg, are also present as oxides and carbonates but are less mobile (Lewis 1974). The amount and composition of these inorganic cations determine base saturation which plays an important role in controlling pH regimes in soils.

Raison et al. (1990) noted that while K, Na, and Mg are relatively soluble and can leach into and possibly through the soil, Ca is most likely retained on the cation exchange sites. Surface soil Ca levels may remain elevated for many years following burning. Prevost (1994) found that after burning *Kalmia* spp. litter, only leaching of Mg increased. Although ash and forest floor cations were released in the burning, there was no detected change in surface soil (0 to 5 cm) cation concentrations.

Another issue to consider is the ability of cations to be leached into the soil and the effect of fire on this process. Soto and Diaz-Fierros (1993) noted changes in the pattern of cation leaching at differing temperatures for the six acid soils types (Fig. 6). Leaching of divalent cations, Ca, and Mg, increased as the temperature increased with a peak at 460°C. Monovalent cations, K and Na, decreased as temperature increased, reaching a minimum at 380°C, and then increased up to 700°C. The nutrients leached from the forest floor and the ashes were adsorbed in the mineral soil where surface soils retained 89 to 98 percent of the leached nutrients leached (Soto and Diaz-Fierros 1993). As the leachates moved through the mineral soil, the pH of the solution decreased; however, the texture and pH of the parent material, rather than the heating, may have significantly influence this behavior.

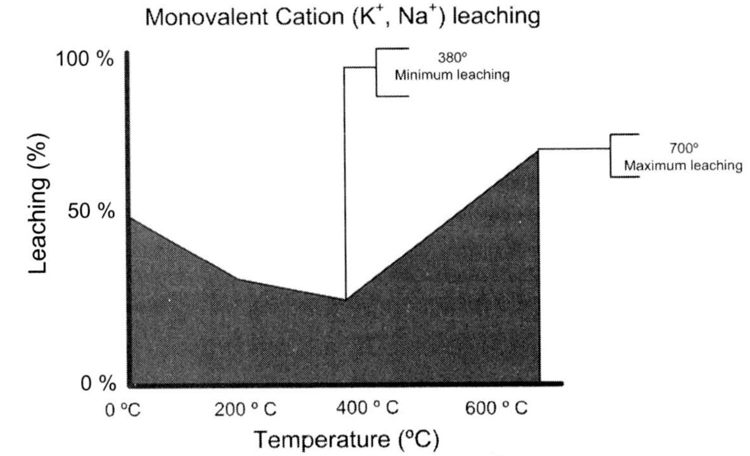

Fig. 6 Monovalent cation leaching versus temperature. (Adapted from Soto and Diaz-Fierros, 1993.)

Cation exchange capacity

The capacity of a soil for the exchange of positively charged ions between the soil and the soil solution is known as cation exchange capacity (CEC). CEC is used as a measure of fertility, nutrient retention capacity, and the capacity to protect groundwater from cation contamination. The CEC of a soil is dependent on the secondary clay minerals and organic matter fraction (Sparks 2005). Given that these factions are susceptible to alteration by fire, it is likely that the CEC will be reduced in proportion to the fire effects on organic matter and, to a lesser extent, clay mineral. In a study of Galician acidic soil, Fernández at al. (1997) reported a reduction of CEC as a consequence of consumption of organic matter due to fire. In a study of coarse wood dynamics in Quebec, Canada, Brais et al. (2005) showed that CEC was related to forest floor organic matter and buried wood content. Hatten et al. (2005), in a comparative study between soil affected by wildfire and unburned soils, found that the exposure of younger mineral materials (weathering) would tend to lower CEC while the deposition of organic rich substrates (originated by fire) would tend to increase CEC. Arocena and Opio (2003) in sub-boreal forest soils of British Columbia, Canada noted the importance of clay as the major sources of negative charges in soils. They observed a reduction in CEC and the generation of thermal cracks in feldspars and other soil aggregates could accelerate weathering due to increased surface area and may result in loss of K, Ca, and Mg through leaching.

Carbon

The role of the fire in carbon sequestration

Soil carbon (C) sequestration is the process of transferring carbon dioxide from the atmosphere into the soil through crop residues and other organic solids, and in a form that is not immediately re-emitted. This transfer or 'sequestering' of C helps off-set emissions from fossil fuel combustion and other C-emitting activities while enhancing soil quality and long-term agronomic productivity. Soil C sequestration can be accomplished by management systems that add high amounts of biomass to the soil, cause minimal soil disturbance, conserve soil and water, improve soil structure, and enhance soil fauna activity. Continuous no-till crop production is a prime example of a soil C sequestration technique (Sundermeier et al. 2006). The land use history of a soil determines the quantity and quality of the C stored, and a mature forest is able to store large and stable amounts of C (sink) whereas a new forest plantation is actually a source of CO_2. Increasingly soil scientists are concerned about fire management as a potential mechanism to sequester C in soils in a recalcitrant form (charcoals) which is not immediately re-emitted. Lal (2004) reports that 3.7 billion hectares of rangelands and grasslands in the sub-humid and sub-arid regions of the world could store between 50 to 100 kg $ha^{-1}yr^{-1}$, and terra preta soil, an anthropogenic, high-carbon soil, is also being investigated as a sequestration mechanism.

SOC and SIC pools

Lal (2001) reported that the soil organic carbon (SOC) pool in the global carbon cycle is between 1500 Pg (to 1 m depth) and 2400 Pg (to 2 m depth) [Note: 1 Pg = 10^{15} g], and soils contain about 75 percent of the C pool on land – three times more than is found living plants and animals. The soil inorganic carbon (SIC) pool ranges between 695 and 748 Pg of CO_3^{2-}, which include lithogenic and pedogenic carbonates, with the latter more important in C sequestration (Sparks 2005). Declines in the SOC pool are due to mineralization of SOC, transport by soil erosion processes, and leaching into subsoil or groundwater. The stability of the SOC is especially important when C sequestration in soils becomes the objective. Consequently, many researchers are concerned about the role of fire as a factor in soil storage of stable C.

Charcoal, soot, and black carbon

These are three different forms of charcoal as a dark-colored, solid carbonaceous material that results from the incomplete combustion of organic material. Kuhlbusch (1998) has proposed some useful definitions of charcoal and related materials:

- Charcoal is the carbonaceous residue formed in-place during combustion. It retains the gross anatomical features of the original material.
- Soot is the solid-phase carbonaceous particulate material formed above a fire by gas-to-solid conversion.
- Black carbon (BC) is a component of charcoal and soot. BC is highly aromatic, resistant to oxidation at 340°C under pure oxygen, and has a molar H/C ratio of 0.2 or less.

As BC has a highly aromatic structure with few functional groups, it is resistant to decay and therefore contributes to the recalcitrant fraction of soil carbon. At the global scale, formation of BC transfers fast-cycling C from the atmosphere-biosphere system to much slower-cycling geological forms that may persist for millennia, and therefore represents a sink for atmospheric CO_2 (Dai et al. 2005, Kuhlbusch 1998, Masiello and Druffel 1998). Despite this, Dai et al. (2005), in a study of four different fire treatments on temperate mixed grass-savanna, reported no statistical significant difference in BC found on the treated fields compared to the control field.

There is a large knowledge gap concerning the reactivity and stability of pyrogenic carbon in the natural environment. Pyrogenic carbon has considerable capacity to absorb a range of organic and inorganic compounds, including a range of pollutants, and does chemically interact with its environment after formation. In situ degradation, erosion, and translocation within the soil profile is likely to be responsible for any obvious BC loss (Knicker et al. 2005).

In terms of vegetation synecology, Wardle et al. (1998) provides clear evidence that immediately after wildfire in boreal forest ecosystems, fresh charcoal contributed to the rejuvenating effects of wildfire on the ecosystems. However, peatlands, a soil system dominated by partially decayed plant matter, maintain a highly stable C pool. When peat forests burn, this sequestered C is released into the atmosphere.

References

Adams, M.A., J. Iser, A.D. Keleher, and D.C. Cheal. 1994. Nitrogen and phosphorus availability and the role of fire in heathlands at Wilsons Promontory. *Australian Journal of Botany*, 42: 269-281.

Arcenegui, V.C., C. Guerrero, J. Mataix-Solera, R. Zornoza, J. Mataix-Beneyto, F. García-Orenes, J. Morales, and A. Martínez-Mayoral. 2007. Near-infrared spectroscopy (NIR) to estimate the maximum temperatures reached on burned soils: the presence of ashes as interference on the estimation. *In* X. Úbeda and L.R. Outeiro [eds.]. The Proceedings of the International Meeting of Fire Effects on Soil Properties, University of Barcelona, Barcelona, Spain.

Arend, J.L. 1941. Infiltration rates of forest soils in the Missouri Ozarks as affected by woods burning and litter removal. *Journal of Forestry*, 39: 726-728.

Arocena J.M. and C. Opio. 2003. Prescribed fire-induced changes in properties of sub-boreal forest soils. *Geoderma*, 113(1-2): 1-16.

Baker, J. and C. Hunter. 2008. Effects of Forest Management On Terrestrial Ecosystems. Encyclopedia ID p2411. In Web-Based Encyclopedia Compiling the Effects of Fire on Southeastern Ecosystems, U.S. Department of Agriculture, Forest Service, <http://fire.forestencyclopedia.net/p/p2411/view>. [accessed 4 Aug 2008]

Boyer, W.D. and J.H. Miller. 1994. Effect of burning and brush treatments on nutrient and soil physical properties in young longleaf pine stands. *Forest Ecology and Management*, 70: 311-318.

Bond, R.D. and J.R. Harris. 1964. The influence of the microflora on physical properties of soils: I. Effects associated with filamentous algae and fungi. *Australian Journal of Soil Research*, 2: 111-122.

Binkley, D., D. Richter, M.B. David, and B. Caldwell. 1992. Soil chemistry in a Loblolly longleaf pine forest with interval burning. *Ecological Applications*, 2: 157-164.

Brais, S., F. Sadi, Y. Bergeron, and Y. Grenier. 2005. Coarse woody debris dynamics in a post-fire jack pine chronosequence and its relation with site productivity. *Forest Ecology and Management*, 220: 216-226.

Busse, M.D., K.R. Hubbert, G.O. Fiddler, C.J. Shestak, and R.F. Powers. 2005. Lethal soil temperatures during burning of masticated forest residues. *International Journal of Wildland Fire*, 14: 267-276.

Bustamante, M., M.C.E. Medina, G.P. Asner, G.B. Nardoto, and D.C. Garcia-Montiel. 2006. Nitrogen cycling in tropical and temperate savannas. *Biogeochemistry*, 79: 209-237.

Byram, G.M. 1959. Combustion of forest fuels. *In* K.P. Davis [eds.]. *Forest Fire: Control and Use*. McGraw Hill, New York, USA.

Campbell, G.S., J.D. Jungbauer, K.L. Bristow, and R.D. Hungerford. 1995. Soil temperature and water beneath a surface fire. *Soil Science*, 159(6): 363-374.

Certini, G. 2005. Effects of fire on properties of forest soils: a review. *Oecologia*, 143: 1-10.

Chandler, C. 1983. *Fire in Forestry*. Volume 1. John Wiley, New York, New York, USA.

Chapman, H.H. 1932. Is the longleaf type a climax? *Ecology*, 13(4): 328-334.

Clark, B. 2001. Fire effects guide. National Wildfire Coordinating Group. National Interagency Fire Center, Boise, Idaho, USA. http://www.nwcg.gov/pms/RxFire/FEG.pdf [accessed 4 August 2008]

Covington, W.W. and S.S. Sackett. 1986. Effect of periodic burning on soil nitrogen concentrations in ponderosa pine. *Soil Science Society of America Journal*, 50: 452–457.

Dai, X., T.W. Boutton, B. Glaser, R.J. Ansley, and W. Zech. 2005. Black carbon in a temperate mixed-grass savanna. *Soil Biology and Biochemistry*, 37: 1879-1881.

Das, D.K. and B. Das. 1972. Characterization of water repellency in Indian soils. *Indian Journal of Agricultural Science*, 42: 1099-1102.

DeBano, L.F. 1981. Water repellent soils: a state-of-the-art. General Technical Report PSW-46. U.S. Department of Agriculture, Forest Service, Pacific Southwest Forest and Range Experiment Station, Berkeley, California, USA.

DeBano, L.F. 1991. The effect of fire on soil. pp. 32-50. *In* A.E. Harvey and L.F. Neuenschwander [eds.]. Management and Productivity of Western-montane forest soils. General Technical Report INT-280, U.S. Department of Agriculture, Forest Service, Pacific Southwest Forest and Range Experiment Station, Berkeley, California, USA.

DeBano L.F. and J.S. Krammes. 1966. Water repellent soils and their relation to wildfire temperatures. *International Association of Hydrological Sciences Bulletin,* 2: 14-19.

DeBano, L.F. and C.E. Conrad. 1978. Effects of fire on nutrients in a chaparral ecosystem. *Ecology,* 59: 489-497.

DeBano, L.F., S.M. Savage, and D.A. Hamilton. 1976. The transfer of heat and hydrophobic substances during burning. *Soil Science Society American Journal,* 40: 779-782

DeBano, L.F., G.E. Eberlein, and P.H. Dunn. 1979. Effects of burning on chaparral soils: I. Soil nitrogen. *Soil Science Society of American Journal,* 43: 504-509.

DeBano, L.F., R.M. Rice, and C.E. Conrad. 1979. Soil heating in chaparral fires: effects on soil properties, plant nutrients, erosion and runoff. Research Paper PSW-145. U.S. Department of Agriculture, Forest Service, Pacific Southwest Research Station, Berkeley, California, USA.

DeBano, L.F., D.G. Neary, and P.F. Ffolliott. 1998. *Fire's Effects on Ecosystems.* John Wiley & Sons, New York, New York, USA.

Dimitrakopoulos, A.P., R.E. Martin, and N.T. Papamichos. 1994. A simulation model of soil heating during wildland fires. pp. 199-204. *In* M. Sala and J.L. Rubio [eds.]. *Soil Erosion and Degradation as a Consequence of Forest Fires.* Geoforma Ediciones, Logroño, Spain.

Etiégni, L. and A.G. Campbell. 1991. Physical and chemical characteristics of wood ash. *Bioresource Technology,* 37: 173-178.

Fernández, I,A., C. Cabaneiro, and T. Carballas. 1997. Organic matter changes immediately after a wildfire in an Atlantic forest soil and comparison with laboratory soil heating. *Soil Biology and Biochemistry,* 29(I): 1-11.

Garcia-Corona R., E. Benito, E. de Blas, and M.E. Varela. 2004. Effects of heating on some soil physical properties related to its hydrological behavior in two north-western Spanish soils. *International Journal of Wildland Fire,* 13: 195-199.

Garren, K.H. 1943. Effects of Fire on Vegetation of the Southeastern United States. *Botany Research,* 9(9): 617-654.

Gilmour, D.A. 1968. Water repellence of soils related to surface dryness. *Australian Forestry,* 32: 145-148.

Giovannini, G. 1994. Effect of fire on soil quality. pp. 15-27. *In* M. Sala and J.L. Rubio [eds.]. *Soil Erosion as a Consequence of Forest Fires.* Geoforma ediciones, Logroño, Spain.

Giovannini, G. 1997. Post fire soil erosion risk: how to predict and how to prevent. pp. 305-321. *In* G. Eftichidis, P. Balabanis, and A. Ghazi [eds.]. Proceedings of the Advanced Study Course on Wildfire Management. European Commission, Brussels, Belgium.

Greene, S.W. 1931. The forest that fire made. *American Forests,* October: 53-54.

Greene, S.W. 1934. Effect of annual grass fires on the organic matter and other constituents of virgin longleaf pine soils. *Journal of Agricultural Research,* 50: 809-822.

Grillo, F., M. Castellnou, D. Molina, E. Martínez, and D. Díaz. 2008. Análisis del incerdio forestal: Planificacíon de la extinción. Aifema ediciones, El Chaparral, Granada, Spain.

Guerrero, C. 2003. Uso de diferentes residuos orgánicos en la restauración de suelos foretales quemados. Ph.D. thesis, University Miguel Hernández. Elche, Alicante, Spain.

Hartford, R.A. and W.H. Frandsen. 1992. When it's hot, it's hot ... or maybe it's not (surface flaming may not portend extensive soil heating). *International Journal of Wildland Fire,* 2: 139-144.

Hatten, J., D. Zabowski, G. Scherer, and E. Dolan. 2005. A comparison of soil properties after contemporary wildfire and fire suppression. *Forest Ecology and Management,* 220: 227-241.

Heyward, F. 1937. The effect of frequent fires on profile development of longleaf forest soils. *Journal of Forestry,* 35: 23-27.

Heyward, F. 1938. Soil temperatures during forest fires in the longleaf pine region. *Journal of Forestry,* 36: 478-491.

Hopmans, P. 2003. Effects of repeated low-intensity fire on carbon, nitrogen and phosphorus in the soils of a mixed eucalypt foothill forest in south-eastern Australia. Research Report No. 60. Forest Science Centre, Department of Sustainability and Environment, Heidelberg, Germany.

Hungerford, R.D., M.G. Harrington, W.H. Frandsen, R.C. Ryan, and J.G. Niehoff. 1990. Influence of fire on factors that affect site productivity. pp. 32-51. *In* A.E. Harvey and L.F. Neuenschwander [compilers]. Symposium on Management and Productivity of Western-Montane Forest Soils, General Technical Report INT-280, U.S. Department of Agriculture, Forest Service, Intermountain Research Station, Ogden, Utah, USA.

Hungerford, R.D. and G.S. Campbell. 1991. Evaluation of models predicting soil heating under fires. pp. 186-187. *In* P.L. Andrews and D.F. Potts [eds.]. Conference Proceeding, SAF Publication 4-16-1991. Society of American Foresters, Bethesda, Maryland, USA.

Jain, T.H. and R.T. Graham. 2003. Fire severity classification: uses and abuses. Abstract. The 5th Symposium on Fire and Forest Meteorology and 2nd International Wildland Fire Ecology and Fire Management Congress. American Metheorological Society, Boston, Massachusetts, USA.

Jemison, V.C. 1942. The slow reversible drying of sandy surface soils beneath citrus trees in central Florida. *Soil Science Society American Proceedings,* 7: 36-41.

Jenny, H. 1941. *Factors of Soil Formation.* McGraw-Hill, New York, New York, USA.

Khanna, P.K. and R.J. Raison. 1986. Effect of fire intensity on solution chemistry of surface soil under a *Eucalyptus pauciflora* forest. *Australian Journal of Soil Research,* 24: 423-434.

Knicker, H., A. Hilscher, F.J. González-Vila, J.A. González, and G. Almendros. 2005. Transport of black carbon in fire-affected soils. *Geophysical Research Abstracts,* 7: 01529.

Knight, H. 1966. Loss of nitrogen from the forest floor by burning. *Forestry Chronicle*, 42: 149-152.

Kutiel, P. and Z. Naveh. 1987. Soil properties beneath *Pinus halepensis* and *Quercus calliprinos* trees on burned and unburned mixed forest on Mt. Carmel, Israel. *Forest Ecology and Management*, 20: 11-24.

Kuhlbusch, T.A.J. 1998. Black carbon in soils, sediments, and ice cores. *In* R.A. Meyers [eds.]. *Encyclopedia of Environmental Analysis and Remediation*. Wiley, New York, New York, USA.

Lal, R. 2001. World cropland soils as a source or sink for atmospheric carbon. *Advances in Agronomy*, 22: 75-117.

Lal, R. 2004. Soil carbon sequestration impacts on global climate change and food security. *Science*, 304: 16-23.

Lewis, W.M. 1974. Effects of fire on nutrient movement in a South Carolina pine forest. *Ecology*, 55: 1120-1127.

Maars, R.H., R.D. Roberts, R.A. Skeffinton, and A.D. Bradshaw. 1983. Nitrogen in the development of ecosystems. pp. 131-137. *In* J.A. Lee, S. McNeill, and I.H. Rorison [eds.]. *Nitrogen as an Ecological Factor*. Blackwell Science Publishing, Oxford, UK.

Macadam, A. 1989. Effects of prescribed fire on forest soils. Research Report, 89001-PR., B.C. Ministry of Forests, Smithers, British Columbia, Canada.

Masiello, C.A. and E.R.M. Druffel. 1998. Black carbon in deep-sea sediments. *Science*, 280: 1911-1913.

Mataix-Solera, J. 1999. Alteraciones físicas, químicas y biológicas en suelos afectados por incendios forestales. Contribución a su conservación y regeneración. Ph.D. thesis, University of Alicante, Alicante, Spain.

Mataix-Solera, J. 2001. Forest fires in Mediterranean environments and their effects on soils. Seminar Presentation at Department of Geography, University of Wales Swansea, Swansea, UK.

McGhie, D.A. and A.M. Posner. 1980. Water repellence of a heavy textured western Australian surface soil. *Australian Journal of Soil Research*, 18(3): 309-323.

McKee, W.H. 1982. Changes in soil fertility following prescribed burning on coastal plains pine sites, Research Paper SE 234. U.S. Department of Agriculture, Forest Service, Southeastern Forest Experiment Station, Ashville, North Carolina, USA.

Meginnis, H.G. 1935. Effect of cover on surface run-off and erosion in the loessial uplands of Mississippi, Circulation 347. U.S. Department of Agriculture, Washington, D.C., USA.

Monleon, V.J., K. Cromack Jr., and J.D. Landsberg. 1997. Short and long-term effects of prescribed underburning on nitrogen availability in ponderosa pine stands in central Oregon. *Canadian Journal of Forest Research*, 27: 369-378.

Moreno, J.M. and W.C. Oechel. 1989. A simple method for estimating fire intensity after a burn in California Chaparral. *Acta Oecologica*, 10/1: 57-68.

Neary, D.G., C.C. Klopatek, L.F. Debano, and P.F. Folliott. 1999. Fire effects on belowground sustainability: a review and synthesis. *Forest Ecology and Management*, 122: 51–71.

Neary, D.G., K.C. Ryan, L.F. DeBano, J.D. Lansberg, and J.K. Brown. 2005. Chapter 1: Introduction. pp. 1-17. *In* D.G. Neary, K.C. Ryan, and L.F. DeBano [eds.]. Wildland Fire in Ecosystems: Effects of Fire on Soils and Water. General

Technical Report RMRS-GTR-42-vol 4. U.S. Department of Agriculture, Forest Service, Rocky Mountain Research Station, Ogden, Utah, USA.

Neary, D.G., K.C. Ryan, L.F. DeBano, and F. Leonard. 2005. Wildland fire in ecosystems: effects of fire on soils and water. General Technical Report RMRS-GTR-42-Vol. 4. U.S. Department of Agriculture, Forest Service, Rocky Mountain Research Station, Ogden, Utah, USA.

Odion, C.D. and F.W. Davis. 2000. Fire, soil heating, and the formation of vegetation patterns in chaparral. *Ecological Monographs*, 70(1): 149-169.

Outeiro, L., F. Asperó, and X. Úbeda. 2008. Geostatical methods to study spatial variability of soil cations after a prescribed fire and rainfall. *Catena*, 74(3): 310-320.

Pardini, G., M. Gispert, and G. Dunjó. 2004. Relative influence of wildfire on soil properties and erosion processes in different Mediterranean environments in NE Spain. *Science of Total Environment*, 328: 237-246.

Prevost, M. 1994. Scalping and burning of *Kalmia angustifolia* (Ericaceae) litter: Effects on *Picea mariana* establishment and ion leaching in a greenhouse experiment. *Forest Ecology and Management*, 63: 199-211.

Pivello, V.R. and L.M. Coutinho. 1992. Transfer of macro-nutrients to the atmosphere during experimental burnings in an open cerrado (Brazilian savanna). *Journal of Tropical Ecology*, 8: 487-497.

Raison, R.J., H. Keith, and P.K. Khanna. 1990. Effects of fire on the nutrient supplying capacity of forest soils. pp. 39-54. *In* W.J. Dyck and C.A. Meeg [eds.]. Impact of Intensive Harvesting on Forest Site Productivity. Bulletin Number 159. Forest Research Institute, Rotorua, New Zealand.

Roberts, F.J. and B.A. Carbon. 1972. Water repellence in sandy soils of southwestern Australia. *Australian Journal of Soil Research*, 10: 35-42.

Robichaud, P.R. 2000. Fire effects on infiltration rates after prescribed fire in Northern Rocky Mountain forests, USA. *Journal of Hydrology*, 231-232: 220-229.

Robichaud, P.R. and T.M. Monroe. 1997. Spatially-varied erosion modeling using WEPP for timber harvested and burned hill slopes. Paper Number 97-5015. *In* ASAE Annual International Meeting. American Society of Agricultural Engineers, St. Joseph, Michigan, USA.

Romanyà, J. 1993. Phosphorus cycling in fast growing forest plantations: availability, plant uptake and the role of forest floor. Ph.D. thesis, University of Barcelona, Barcelona, Spain.

Ryan, K.C. 2002. Dynamic interactions between forest structure and fire behavior in boreal ecosystems. *Silva Fennica*, 36(1): 13-39.

Ryan, K.C. and N.V. Noste, 1985. Evaluating prescribed fires. pp. 230-238. *In* J.E. Lotan, B.M. Kilgore, W.C. Fischer, and R.W. Mutch [eds.]. Proceedings—symposium and workshop on wilderness fire. General Technical Report INT-182. U.S. Department of Agriculture, Forest Service, Intermountain Forest and Range Experiment Station, Ogden, Utah, USA.

Savage, S.M., J.P. Martin, and J. Letey. 1969. Contribution of humic acid and a polysaccharide to water repellency in sand and soil. *Soil Science Society of America Proceedings*, 33: 149-151.

Singer, M.J. and D.N. Munns. 1996. *Soils: An Introduction. 3rd edition*. Prentice Hall, Upper Saddle River, New Jersey, USA.

Soler, M., M. Sala, and F. Gallart. 1994. Post-fire evolution of runoff and erosion during an eighteen-month period. pp. 149-161. *In* M. Sala and J.L. Rubio [eds.]. *Soil Erosion as a Consequence of Forest Fires.* Geoforma ediciones Logroño, Spain.

Soto, B. and F. Diaz-Fierros, 1993. Interactions between plant ash leachates and soil. *International Journal of Wildland Fire*, 3(4): 207-216.

Sparks, D.L. 2005. *Environmental Soil Chemistry.* Elsevier Academic Press, Amsterdam, The Netherlands.

Stevenson, F.J. 1982. *Humus Chemistry: Genesis, Composition, Reactions.* John Wiley and Sons, New York, New York, USA.

Stewart, O.C. 2002. The effects of burning of grasslands and forests by aborigines the world over. pp. 67-354. *In* O.C. Stewart [ed.]. *Forgotten Fires: Native Americans and the Transient Wilderness.* University of Oklahoma Press, Norman, Oklahoma, USA.

St. John, T.V. and P.W. Rundel. 1976. The role of fire as a mineralizing agent in the Sierran coniferous forest. *Oecologia*, 25: 35-45.

Sundermeier, A., R. Reeder, and R. Lal. 2006. Soil carbon sequestration-Fundamentals. Ohio State University Extension Fact Sheet. Online at: http://ohioline.osu.edu/aex-fact/0510.html [accessed 5 Sep. 2008]

Úbeda, X. 1998. Efectes de les diferents intensitats de foc, Durant els incendis forestals, en els paràmetres físics i químics del sòl i en l'increment de l'escolament i l'erosió. Ph.D. thesis, University of Barcelona, Barcelona, Spain.

Úbeda, X., M. Lorca, L.R. Outeiro, S. Bernia, and M. Castellnou. 2005. Effects of prescribed fire on soil quality in Mediterranean grassland (Prades Mountains, Tarragona Province). *International Journal of Wildland Fire*, 14: 379-384.

Wahlenberg, W.G. 1935. Effects of fire and grazing on soil properties and the natural reproduction of longleaf pine. *Journal of Forestry*, 33: 331-337.

Wahlenberg, W.G., S.W. Greens, and H.R. Reed. 1939. Effects of fire and grazing on soil properties and the natural reproduction of longleaf pine. *Journal of Forestry*, 33: 331-338.

Wardle, D.A., O. Zackrisson, and M.C. Nilsson. 1998. The charcoal effect in Boreal forests: mechanisms and ecological consequences. *Oecologia*, 115(3): 419-426.

Wells, C.G., R.E. Campbell, L.F. DeBano, C.E. Lewis, R.L. Fredricksen, E.C. Franklin, R.C. Frolich, and P.H. Dunn. 1979. Effects of fire on Soil: A state-of-knowledge review. General Technical Report WO-7. U.S. Department of Agriculture, Forest Service, Washington D.C., USA.

White, E.M., W.W. Thompson, and F.R. Gartner. 1973. Heat effects on nutrient release from soils under ponderosa pine. *Journal of Range Management*, 26: 22-24.

Wright, H.A. and A.W. Bailey. 1982. *Fire Ecology, United States and southern Canada.* John Wiley, New York, New York, USA.

5

Forest Fire Effects on Soil Microbiology

Jorge Mataix-Solera[1*], *César Guerrero*[1], *Fuensanta García-Orenes*[1], *Gema M. Bárcenas*[1] and *M. Pilar Torres*[2]

Abstract

Soil microbiology is crucial for soil system functioning. Fire can affect soil microbes directly through heating and indirectly by modifying soil properties. Microbes will also be affected by post-fire environmental factors and the reestablishment of vegetation. The most important factor affecting soil microbes seems to be the burn severity, which is controlled by such factors as fire intensity, duration, and soil properties which normally causes a decrease in the numbers of microbes. The temperatures reached in the topsoil are often sufficient to affect soil microorganisms and other soil properties related to the post-fire microbial recolonization. In extreme cases, the topsoil can undergo complete sterilization. Fungi seem to be more sensitive to heating than bacteria and actinomycetes, and a higher impact under wet soil conditions has been reported. In the case of fungi that form arbuscular mycorrhizas, almost all the studies show a negative influence resulting in a reduced number of propagules. An important factor is the presence of fungal resistant structures, such as sclerotia, from which new mycelia originate to colonize new plants. The activity of soil microorganisms also decreases due to changes in the quality of organic matter. In the short-term, mainly due to the increase in soluble carbon and nutrients in affected soils, an increase in heterotrophic bacteria population basal respiration is commonly observed. After depletion of the easily mineralized organic compounds, this initial increase in microbial basal respiration is generally followed by a decrease as the remaining carbon and nitrogen forms are more recalcitrant to microbial attack. Some other changes in soil properties such as

[1] *GEA (Grupo de Edafología Ambiental), Departamento de Agroquímica y Medio Ambiente, Universidad Miguel Hernández, Avenida de la Universidad, Alicante, Spain.*

[2] *Departamento de Biología Aplicada, Universidad Miguel Hernández, Avenida de la Universidad, Alicante, Spain.*

* *Corresponding author: Jorge Mataix-Solera, GEA (Grupo de Edafología Ambiental), Departamento de Agroquímica y Medio Ambiente, Universidad Miguel Hernández, Avenida de la Universidad s/n, 03202-Elche, Alicante, Spain. Tel.: +34 966658334, Fax: +34 966658340, e-mail: jorge.mataix@umh.es*

increase in pH (due to ash deposition) have been argued as the cause of the post-fire increase in the bacteria/fungi ratio. As pioneer species, a rapid recolonization of soil by photoautotroph microorganism (such as algae) has been reported after moderate and high intensity fires. The diversity of microorganisms can be modified by fire in several ways as a consequence of differences in heat sensitivity, survival strategies, colonization mechanisms, and sensitivity to soil and microclimate changes. Destruction and creation of new ecological niches and changes in biomass and composition of above-ground vegetal species can also impact microogranism diversity.

INTRODUCTION

Soil microorganisms are fundamental for the regulation of biogeochemical cycles in terrestrial ecosystems because they are responsible (together with soil fauna) for the decomposition, sooner or later, of virtually all organic compounds (Nannipieri et al. 2003). The relevance of soil microbes and invertebrates for the conservation and management of biological diversity after fire is of key importance. Soil organisms play important roles in modifying the structure of the soil by creating habitats with widely varying aeration, moisture relations, nutrient status, and penetrability. These factors directly influence the growth and health of plants. It is known that biodiversity depletion not only decreases genetic resources, but also reduces the productivity of ecosystems and alters their buffering capacity against disturbance (Naeem et al. 1994). As microorganisms have key functions, they are useful indicators of soil quality and ecosystem health. Although soil organisms include macro, meso and microfauna, and microflora, we have focused this review on fire effects on microbiota because 80 to 90 percent of the soil processes are reactions mediated by microbes (Coleman and Crossley 1996, Nannipieri and Badalucco 2003).

Fire effects on soil microbes can vary widely and are dependent on fire severity, changes in some soil properties, and post-fire environmental conditions (Fig. 1). Each of these factors has a range of affects depending on the type of soil organism being considered (Widden and Parkinson 1975, Bisset and Parkinson 1980, Dunn et al. 1985, Bartoli et al. 1991, Vázquez et al. 1993, Acea and Carballas 1996, Pietikäinen et al. 2000).

Several authors have reviewed the effects of fire on soils: e.g., Raison (1979) with particular focus on nitrogen transformations, DeBano (1981, 2000) and Doerr et al. (2000) examined the effect of fire on soil water repellency. Neary et al. (1999) and more recently Certini (2005) have provided general reviews of fire effects on the physical, chemical, and biological soil properties. Shakesby and Doerr (2006) reviewed the hydrological and geomorphological effects of fire on soil. Soil microbiology after fire has received less attention than physical and chemical soil properties, and no previous reviews have specifically focused on the effects of fire on soil microbiology. Although there

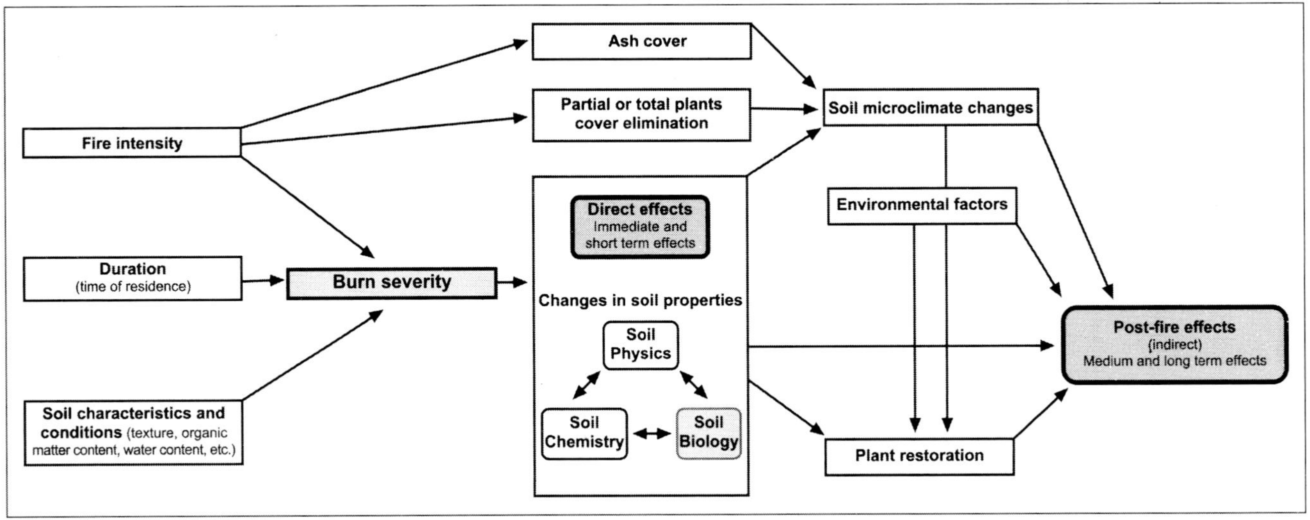

Fig. 1 Factors controlling direct and indirect effects on soil properties with particular focus on soil microbiology.

are a number of ways to examine change in soil microbiology, this review will generally describe the effects of fire on microorganism numbers, biomass, activity, and diversity.

Soils are complex and very heterogeneous environments, which may contain 10^8 to 10^{12} bacterial cells per gram, in addition to large numbers of other microorganisms such as fungi and protozoa. The number of different species of microorganisms in one gram of soil may be in the range of several thousands. The enumeration of soil microorganisms could be made directly by counting cells (or the length of fungal mycelium) after staining, but this method is laborious. Enumeration of microorganisms using liquid or solid (agar) cultural methods must take into account that at least 95 percent of soil bacteria cannot be cultured (and thus enumerated by plate count); it is estimated that there are between 500 and 5000 types. Some culture media are selective for growth of specific groups, such as fungi. Other culture media can be used to enumerate physiological groups based on activity, for example cellulolytics or ammonifiers, but without discrimination of the type of microorganisms (both fungi and bacteria are capable of being ammonifiers and/or cellulolytic). Consequently, several selective cultures must be performed to get an approximation of the overall microbial population in a soil. Thus, the estimation of the total number of microorganisms is complicated. For this reason, most studies have estimated microbial biomass rather than enumerating individual organisms.

Microbial biomass has been defined as the portion of organic matter that consists of living microorganisms smaller than 5 to 10 μm^3. This pool comprises the overall microorganisms in the soil (fungi, bacteria, protozoa, etc.). The carbon contained in this living fraction usually ranges from 0.5 to 5 percent of the total organic carbon present in a soil. The laboratory methods for quantification of the microbial biomass are mainly based on the measurement of compounds present in microbes (i.e., carbon by fumigation-extraction: C_{mic}), or based on physiological measurements (ATP, SIR: substrate induced respiration). Depending on the method used, the estimation corresponds to total or active (non-latent) microbial biomass.

The activity of microorganisms is another important measurement. As previously stated, physiological activity can be used to enumerate microbes, but this approach assumes similar activity for different groups. The quantification of soil respiration is a measure of the overall microbial heterotrophic activity and organic carbon decomposition. With respect to biological activity, the measurements of enzymatic activities have interesting characteristics. As enzymatic activities have extraordinary specificity in catalyzing biological reactions, they offer crucial information of potential reaction rates related to production of essential elements in biogeochemical cycles. However, it must be indicated that some of the enzymes are not exclusive to microorganisms.

The soil diversity and community composition in soils can be evaluated with three main approaches:

1) *Direct counts:*
    - Macroscopic organisms
    - Microscopic staining
2) *Biochemical-based techniques:*
    - Counts for specific groups in selective culture media (i.e., plate counts on agar for fungi).
    - Physiological methods based on the metabolic diversity, with the ability to degrade different substrates (CLPP: community-level physiological profiles), using for example Biolog^{TM} ECO plates.
3) *Biomarkers and fingerprints:*
    - PLFA (phospholipids fatty acids): certain PLFAs are constituents of cell membranes that could be used as a unique signature for specific functional groups of microorganisms (i.e., fungi, bacteria, protozoa), not for species.
    - Based on total community DNA- or RNA-based profiling methods.

HEATING EFFECTS (DIRECT EFFECTS)

Factors Controlling Direct Effects

The direct effects of fire on soil microbiology will depend mainly on burn severity, which is related to the fire behavior, type of soil, and particular soil conditions during fire spread. Hartford and Frandsen (1992) describe burn severity as a qualitative measure of the effects of fire on soil, and which consists of two components: intensity and duration. Intensity refers to the rate at which a fire is producing thermal energy in the fuel-climate environment where it occurs (Byram 1959, DeBano et al. 1998). The duration (i.e., the length of time burning occurs at a particular point) is the component of burn severity that results in the greatest below ground damage to the soil system. The heat is transferred from burning fuels to the organic and mineral layers of soils in different ways: radiation, convection, conduction, and vaporization/condensation (Chandler et al. 1983). Radiation and convection are responsible for most heat transfer from light fuels like grass, foliage, and branches to the soil. Conduction is the major heat transfer mechanism in heavy fuels like duff, organic soils, and slash piles (Neary et al. 1999). Vaporization and condensation play an important role in rapid heat movement into soil. These processes involve phase changes of water and organic compounds distilled by the combustion process.

Related to fire behavior

Burn severity depends on interrelated factors of fire behavior (i.e., rate of spread, flame length, intensity duration, etc.), weather (i.e., temperature, relative humidity, antecedent rainfall and wind), topography, and fuel characteristics (amount, size, and moisture of live and dead fuel, fuel structure and chemical composition) (Ahlgren and Ahlgren 1960, Raison 1979, Chandler et al. 1983, Pritchett and Fisher 1987). The spatial variations of these factors implies a heterogeneous spatial distribution of fire behavior and thus in burn severity. Burned soils often appear as chaotic mosaics of areas with minor fire effects intermingled with other areas that have been severely impacted (Rab 1996).

Fast-moving fires with fine fuels may be intense in terms of energy release per unit area, but the transfer of heat to a forest floor and mineral soil is much lower than slow moving fires with moderate to heavy fuels (Neary et al. 1999). The quantity and duration of the heat transfer to soil determines the severity of the impact to the physical soil system, its chemical constituents, and biological components. Threshold temperatures for biological disruptions in soils are relatively low (40 to 121°C; Neary et al. 1999) and these temperatures are commonly reached in the first few centimeters of mineral soil. Temperatures reached in soils during fires can vary widely. Maximum soil surface temperatures are typically in the range of 200 to 300°C for forest fires (Rundel 1983), but greater where high loads of heavy fuels are present (500 to 700°C), and more than 1500°C has been instantaneously registered (Dunn and DeBano 1977). The range of maximum soil temperatures in shrublands is higher (300 to 700°C), and lower maximum soil surface temperatures are usually registered in grasslands (200 to 300°C; Neary et al. 1999). Clearly, the maximum temperatures attained in the topsoil during a fire are sufficient to affect soil microorganisms as well as other soil properties related to the post-fire microbial recolonization. Further data about soil temperatures during fires can be found in Chapter 4 of this book.

Related to soil characteristics and conditions

Some soil characteristics, such as organic matter content and soil texture, can play an important role in the heat transfer into soil profile. Soil heating generally decreases rapidly with soil depth in a dry soil because dry soil is a poor conductor of heat (Bradstock et al. 1992, Valette et al. 1994, Bradstock and Auld 1995). In the case of soil microorganisms the impact of fire depends strictly on soil moisture. Latent heat of vaporization prevents soil temperature exceeding 95°C until water completely vaporizes (Jury et al. 1991, Campbell et al. 1994). While convection and vaporization-condensation are the most important mechanisms for heat transfer in a dry soil, in a wet or moist soil, conduction can contribute significantly to heat transfer (Chandler et al. 1983). Soil moisture prevents sudden temperature rises at the surface; however, it increases the thermal conductivity. Moist heat is more effective for killing soil microorganisms than dry heat; lethal temperatures (50 to 210°C) for specific microbial groups maybe reduced by as much as one-half in moist compared to dry soils (Wells et al. 1979). A higher impact under wet conditions has been reported (Klopatek and Klopatek 1987, Choromanska and DeLuca 2002), as water is a better heat conductor than air and microorganisms are more sensitive to moist heat than dry heat (Powlson and Jenkinson 1976, Wolf and Skipper 1994). In addition, the survival of heterotrophic bacteria is higher in dry than moist soil, in part due to a higher presence of dormant forms when the soil is dry (Dunn and DeBano 1977, Dunn et al. 1985, Choromanska and DeLuca 2002).

Direct Effects of Heating

The immediate direct effect of fire on soil microbiology is normally a reduction of microbial biomass. The maximum temperatures reached in topsoil often considerably exceed those required for killing most organisms (DeBano et al. 1998). In extreme cases, the topsoil can undergo complete sterilization. Several authors have found a dramatic negative effect on microbes immediately after fire (Meiklejohn 1955, Ahlgren and Ahlgren 1965, Dunn et al. 1979, Theodorou and Bowen 1982, Deka and Mishra 1983), and especially on fungi (Wright and Bollen 1961, Ahlgren and Ahlgren 1965, Widden and Parkinson 1975, Tiwari and Rai 1977).

In temperatures ranging from several degrees below 0°C and up to 40°C, the rates of biochemical reactions are related to temperature, generally following a pattern described by Q_{10}. This parameter describes the change in the reaction rate when temperature changes 10°C (the ratio of activity at two temperatures differing by 10°C). For example, soil respiration has a Q_{10} of 8 at 0°C and 2 at 30°C (Kirschbaum 1995). The disruption and denaturation of biological molecules start at 40 to 70°C, and cellular components, such as proteins, membrane lipids, and nucleic acids, are negatively affected at elevated temperatures (higher than 100°C).

Several soil-heating experiments have been done in laboratory conditions to isolate the effect of temperature on soil microbes. In general, temperatures greater than 70 to 80°C kill many soil microbes. Some groups of microorganisms are more sensitive to heating than others. Some nitrifying bacteria, protozoa, and non-spore forming fungi will be killed at 70°C. Arbuscular mycorrhizae are killed at temperatures of 80 to 90°C (Klopatek et al. 1988, Pattison et al. 1999). Ciardi (1998) found a strong decrease in ATP when the soil was heated to 60°C and a small decrease at 80°C. However, at 80°C some bacteria were cultured by Ciardi (1998), but microbes totally disappeared between 115 to 150°C. In the case of some microorganisms, a low short-time heating (50 to 60°C) can stimulate spore germination (Bollen 1969). Several microorganisms are able to survive as thermotolerant spores (Alexander 1967, Widden and Parkinson 1975). For example, the fungi *Neosartorya fischeri* survives fire in the form of spores and becomes dominant in the post-fire environment (Bartoli et al. 1991) because its ascospores are thermotolerant and their germination is stimulated by thermic stress (Warcup and Baker 1963). The sterilizing effect of heat is highly related to temperatures, but the duration of heating is also important. Raison (1979) found that temperatures above 127°C could sterilize the soil.

The results from field studies on fire effects on microbes vary substantially. This variability is the result of several factors operating at the same time: soil moisture, soil sampling depth, soil burn severity, and timing of post-fire soil sampling. When burned soils have been analyzed after incubation in the laboratory, or sampled several days after a fire, the data obtained correspond directly with the effect of heating and also to the indirect short-term effects (changes in soil properties). The first post-fire rainfall determines when data are related with only the direct effects, or with both the direct and indirect effects (Ahlgren and Ahlgren 1965). Microbes rapidly react and change with new conditions (soil and climate). Those studies will be described in the post-fire effects section.

Bacterial populations

The sterilizing effect of heat is greater for fungi than for bacteria (Bollen 1969, 1985). After heating soil (100 to 700°C) for 15 min, Guerrero et al. (2005) found a large reduction in bacterial counts, with reduction directly related to temperature and bacteria below detectable limits at 500°C.

Some authors report 200°C for 2 h as the sterilization temperature in the first centimeters of soil (Acea and Carballas 1999), and a reduction in bacterial population with temperatures above 100°C for 1 h (Pritchett and Fisher 1987). However, other authors such as Labeda et al. (1975) have shown that even exposure to 160°C for 3 h is not enough for complete sterilization. Grasso et al. (1996) found a small decrease in bacteria immediately after heating at 450°C for 3 min. Soil bacteria decreases are usually reported for the soil surface layer (0 to 2 cm) only (Theodorou and Bowen 1982, Deka and Mishra 1983).

In contrast, no significant differences were found by Picone et al. (2003) between heated and control soils. Mataix-Solera et al. (2002) did not find differences in bacterial populations after an experimental fire under field conditions. The elevated soil moisture prevented killing temperatures and the high spread rate of the fire (i.e., short fire duration) contributed to these results. Dunn et al. (1985) found that heat and soil moisture act together to kill soil microbes. They found that heterotrophic bacteria were more sensitive to heat in wet soil (20 percent soil moisture) than in dry soil (3 percent soil moisture); however, after heating the soil to 120°C for 30 min, a low survival rate (<1 percent) of heterotrophic bacteria was observed. As previously mentioned, soil moisture prevents high temperatures from occurring, but favors heat transfer into soil profile, which can influence bacterial physiological activity and heat resistance (Dunn et al. 1985, Ciardi 1998).

When we analyze the behavior of different bacterial groups, we find that gram-negative bacteria are more sensitive to heating than gram-positive bacteria (Mabuhay et al. 2003, 2006a). Theodorou and Bowen (1982) found that the fluorescent *Pseudomonas* (gram-negative bacillus) were the most heat sensitive bacterial group in their analysis. Bacteria groups, such as *Bacillus* spp. or *Clostridium* spp., can create heat resistant forms and survive temperatures between 100 and 120°C, which has been observed in different ecosystems after fire (Yeager et al. 2005).

Actinomycetes

Similar behaviors to bacterial populations have been observed for actinomycetes. In the first centimeters of the soil surface, this group can significantly decrease immediately after a fire (Meiklejohn 1955, Deka and Mishra 1983). But other authors have found the opposite trend, where burning does not effect the number of actinomycetes in the topsoil (Picone et al. 2003, Gundale et al. 2005). Soil moisture is one of the factors that may explain this behavior according to Picone et al. (2003). However, various heat treatments are used to isolate specific actinomycetes. Air-dried soil heated to 120°C for 1 h is selective for isolating *Actinomadura, Micromonospora, Microbispora, Microtetraspora, Streptosporangium* and *Thermomonspora* in soil. Selective isolation of genus *Streptomyces* (considered the most numerous actinomycetes genera in soil) uses a 60°C heating process (Wellington and Toth 1994). This indicates that heat resistance is likely related to the ability of these microorganisms to form spores.

Fungi

The taxonomic and functional fungi groups that are a part of soil are numerous and varied, ranging from saprophytic fungi, parasites, and symbionts to the 'predatory' types (nematophagous fungi); however, mycorrhizal fungi (those which establish symbiosis with plants) form the most important group from both a quantitative and qualitative point of view. In general, most of the published information refers to filamentous fungi, or more specifically, to mycorrhizal fungi. The existing literature is somewhat contradictory, which is probably a consequence of differences in the fires studied (intensity, severity, the vegetation affected, etc.), as well as in the experimental designs. Filamentous fungi appear to be the microorganisms most affected by fire (Ahlgren and Ahlgren 1965, Bollen 1969, Jalaluddin 1969, Vázquez et al. 1993, Acea and Carballas 1996), and they are more susceptible than bacteria (Dunn et al. 1985, Fonturbel et al. 1993, 1995, Bååth et al. 1995). This is likely due to their greater sensitivity to temperature, as Dunn et al. (1985) point out. In studies where burn severity was low (prescribed fire), some authors have found little change (Jorgensen and Hodges 1970) in fungi. One day after an experimental fire, Mataix-Solera et al. (2002) found an increase in fungal propagules and no effects on culturable bacteria, as a consequence of a small increase in soil temperature. This mild heating did not kill microbes, but did stimulate the germination of fungal spores (Warcup and Baker 1963).

An important factor is the effect of soil heating on the presence of fungal resistance structures, such as sclerotia, from which new mycelia originate and colonize new plants. In the case of fungi that form arbuscular mycorrhizas (AM), nearly all studies on fire effects found a negative influence that resulted in a reduced number of propagules. The degree of this reduction and its persistence over time depended on the fire intensity and resulting burn severity as well as the effects on host plants. With low-intensity fires, no significant decreases in AM propagules were found (Bellgard et al. 1994, Dhillion et al. 1988). However, with high-intensity fires, significant reductions in AM fungi propagules (spores and mycelium) have been observed (Vilariño and Arines 1991, Amaranthus and Trappe 1993, Rashid et al. 1997). Pattinson et al. (1999) verified that major fire effects occur in the first centimeters of soil. In their experiment, a simulation forest fire, with soil temperatures reaching 200°C on the surface and 45 to 70°C at a depth of 5 cm, a significant decrease in the quantity of viable propagules was measured, but propagule mortality decreased with depth and was imperceptible at depths of 10 cm or more.

It is obvious that fire causes a significant loss of fungal biomass in the organic horizons of the soil (Jonsson et al. 1999), thus the abundance of mycorrhizas at this level is drastically decreased. However, the mycorrhizas that develop in the deepest soil substrata are more protected against high temperatures, and survive. After fires in ponderosa pine forests in the USA, Stendell et al. (1999) found insignificant reduction of mycorrhizas in the mineral substratum, irrespective of the burn severity, while the reduction of mycorrhizal biomass was up to eight times in the organic horizon. It should be noted that in dry or semiarid environments mycorrhizas are located in deeper mineral soil than in temperate or humid ecosystems (mycorrhizas are located as deep as roots). Therefore fire effects on mycorrhizas are lower in dry environments (Dahlberg 2002).

Meso and microfauna

One method of classifying living organisms is to divide them into two groups—flora and fauna. Soil fauna has been further divided into micro-, meso-, and macrofauna based on body lengths of less than 0.2 mm, 0.2 to 10.4 mm, and greater than 10.4 mm, respectively (Wallwork 1970). As many living organisms and the organic matter in soils are located on, or near, the soil surface, they are exposed to heat radiated by flaming surface fuels and smoldering forest floor fuels. Consequently, soil micro- and meso-organisms are often directly killed or injured by fire.

The effect of burn severity and frequency on invertebrates has been studied. On a *P. sylvestris* forest in Sweden, Wikars and Schimmel (2001) showed that the overall mortality of invertebrates ranged from 59 to 100 percent and depended on the proportion of organic soil consumed by fire. Other studies on the effect of fire on soil invertebrates indicate that the effects of fire can occur by several mechanisms including direct mortality, forced migration, or pyrophilous species (DeBano et al. 1998, Andersen and Müller 2000, Lyon et al. 2000a, 2000b).

Generally the direct effects of fire on soil-dwelling invertebrates are less marked than those on microorganisms, due to a greater mobility which increases the potential for invertebrates to escape heating by burrowing deep into the soil (Certini 2005). Several studies have reported decreased microarthropod abundance immediately following fire (e.g., Sgardelis and Magaris 1993); however, the majority of these results correspond to studies where burn severity was high. In contrast, others studies have found no effect of burning on microarthropod abundance. Lussenhop (1976) reported greater microarthopod abundance in a biennially burned prairie compared to an unburned prairie. Individual studies of the effects of repeated burns on microarthropod populations have produced inconsistent results. In a Eucalyptus forest in Australia, *Collembola* abundance was reduced after a single prescribed burn in 1987; but, a second fire in 1992, of the same intensity and at the same time of the year, had no effect on *Collembola* abundance (Collett 1998). This indicates that prescribed burns may have unpredictable effects on microarthropod populations and that such effects may not be readily predicted by fire alone. Dress and Boerner (2004) studied the effects of prescribed burns on microarthropods in oak-hickory forests ecosystems in southern Ohio (USA) and showed that frequency of fire and landscape position have significant effects on abundance. The combined effect of fire frequency and time of burn on arthropod taxa was reported for tropical savannas found in Australia (Andresen and Müller 2000). A substantial resilience to fire in arthorpods has also been documented with only four of the eleven arthropod taxa being significantly affected by fire. Coleman and Rieske (2006) examined the effect of early spring prescribed fires on forest floor arthropod abundance and diversity in mixed hardwood-pine of southeastern Kentucky (USA), and found that leaf-litter arthropod abundance, diversity, and richness did not differ among the pre-burned, unburned and single burned areas. Swengel (2001) showed that leaf-litter and soil-dwelling arthropods might be directly affected by increases in temperature and exposure or indirectly affected through changes in habitat availability and quality. In this sense, some studies have shown immediate reductions in arthropod abundance (Paquin and Coderre 1997, Siemann et al. 1997), whereas others have shown some resilience in the arthropod populations (Panzer 1988, Moretti et al. 2004). Taken altogether, there is no pattern of micro and mesofauna response to fire, instead several factors are implicated in the responses of these organisms to fire.

Biomass and diversity

Some researchers have studied the effects of fire on microorganisms through measurement of microbial biomass, a pool comprising all microbes present in soil. Generally, the total microbial biomass is composed of bacterial and fungal biomass. Depending on several factors, such as soil pH, the fungal biomass contributes from 30 to 80 percent of the total soil microbial biomass, and is generally greater than the bacterial soil biomass. Given that fungi have higher heat sensitivity as compared to bacteria, a decrease in microbial biomass due to fire is expected and has been reported in most studies.

Diaz-Raviña et al. (1992) observed that heating soil to 160°C for 30 min decreased the microbial biomass carbon (C_{mic}) from 1011 mg kg^{-1} to undetectable values. Basanta et al. (2002) found the same pattern with soil heated to 350°C for 25 min. A gradual decrease in the microbial biomass carbon was observed by Choromanska and DeLuca (2002) after heating soil to 160 and 380°C for 30 min. Similarly, a negative trend of microbial biomass with temperature was observed by Guerrero et al. (2005), where the C_{mic} decreased below detectable limits when samples were heated to 400°C for 15 min. In most cases, data collected after field burning showed a similar negative effect when soils were sampled within a few days of the fire and before any rainfall. One day after a wildfire, Prieto-Fernández et al. (1998) observed significantly less microbial biomass in burned soil (54 mg C kg^{-1}) compared to unburned soil (1242 mg C kg^{-1}). However, due to the low thermal conductivity of the soils, the decrease in biomass was about 50 percent in the subsurface soil layer (5 to 10 cm depth). Mabuhay et al. (2006a) reported a very strong decrease in microbial biomass in topsoil (0 to 5 cm) 6 days after a wildfire. Pietikäinen and Fritze (1993) also found an immediate decrease in microbial biomass after a wildfire and after a prescribed fire (Fritze et al. 1993). Higher temperatures are measured when greater fuel quantities are available, which generally results in greater decreases in microbial biomass (Wüthrich et al. 2002). Table 1 summarizes results from several studies of fire effects on biomass.

Enzymes

The activity of many soil enzymes increases up to 60 to 70°C, and then decreases due to thermal denaturation. The complete inactivation of enzymes occurs at 180°C (Skujins 1967, Tabatabai and Bremner 1970). Ciardi (1998) suggests that activities at temperatures higher than 100°C could be due to abiotic reactions. Raison (1979), in an exhaustive review, reports references of abiotic reactions in heated soils such as chemical oxidation and decarboxylation of soil organic matter as a consequence of heat stable decarboxylases.

POST-FIRE SOIL MICROBIOLOGY (INDIRECT EFFECTS)

Factors Controlling Indirect Effects

Direct fire effects are the first factors controlling the soil microbial population response. But fire can also affect survival and recolonization of soil organisms

Table 1 Fire effects on soil microbial biomass (a summary of results from various studies)

Reference	Values[1]	Fire[2]	Time after fire	Additional information
Andersson et al. 2004	H	EF	12 d	Short-lived increase
Bååth et al. 1995	L	EF	1 yr	Fungal PLFA decrease
Carballas et al. 1993	H	WF	7 mo	Increase in labile compounds
Choromanska and DeLuca 2001	L	PB	1 mo-2 yr	Short-term increase in sugars
Choromanska and DeLuca 2001	L	WF	1 mo-2 yr	Increase in total C and total N
D'Ascoli et al. 2005	H	EF	7 d	Decrease in fungi
De la Torre et al. 2002	L	WF	15 mo	Calcareous soil (Typic Xerorthent)
De la Torre et al. 2002	S-H	WF	15 mo	"Terra rossae" (Lithic Rhodoxeralf)
De Marco et al. 2005	H	EF	7-84 d	Increase in organic matter
Dumontet et al. 1996	H	WF	1 yr	High content in organic C
Dumontet et al. 1996	L	WF	6-12 yr	Dunes (sandy soils)
Fenn et al. 1993	S	WF	50 yr	No clear trend during studied period
Fritze et al. 1993	L	PB	12 yr	Recovery after 12yr (C_{mic}/C_{org})
Guerrero 2003	L	WF	6 yr	Mediterranean (semiarid)
Guerrero et al. 2002	L	WF	125-1,371 d	Influence of climate
Hernández et al. 1997	L	WF	9 mo	Decrease in enzymes
Mabuhay et al. 2006a	L	WF	1 yr	Decrease in diversity
Mabuhay et al. 2006b	L	WF	25 yr	Topographic effects
Mataix-Solera 1999	S	EF	1 d-1 yr	Increase in bacteria fungi ratio
Mataix-Solera et al. 2006	L	WF	11 yr	Lower basal respiration and qCO_2
Ojima et al. 1994	H	EF	2 yr	High aboveground biomass
Ojima et al. 1994	L	EF	2 yr	Annually burned
Palese et al. 2004	L	EF	12-18 mo	Increase in bacteria/ fungi ratio
Picone et al. 2003	S	EF	1-360 d	Influence of climate
Pietikäinen and Fritze 1993	L	EF	800 d	Humus layer
Pietikäinen and Fritze 1993	L	PB	800 d	Humus layer
Pietikäinen and Fritze 1995	L	PB	1 yr	Decrease in ergosterol
Prieto-Fernández et al. 1998	L	WF	1 d-13 yr	Humid Atlantic forests
Tateishi and Horikoshi 1995	H-S	WF	54-79 mo	Higher C_{mic}/C_{org} in burned soil
Tateishi et al. 1989a	S	WF	3.5 yr	Topsoil mineral layer (0-2 cm)
Zornoza 2006	L	WF	1-6 yr	Decrease in enzymes

[1] Lower (L), similar (S) or higher (H) microbial biomass in burned with respect unburned soils.
[2] Fire type: WF (wildfire); PB (prescribed burning); EF (experimental fire)

indirectly through modification of soil properties, especially those properties related to organic substrates (Bissett and Parkinson 1980, Monleon and Cromack 1996). Microclimate modification and the presence or absense of host plants can also influence soil organisms.

Related to soil status

Post-fire changes in some soil properties are directly attributed to heating, and the magnitude of change is usually related to burn severity. These soils changes can include: quantitative and qualitative modifications of the soil organic matter (González-Pérez et al. 2004); increase in soil pH due to the production of K and Na oxides, hydroxides, and carbonates through ash deposition (Ulery et al. 1993, Mataix-Solera et al. 2002); the deterioration of soil structure and stability; and the formation of hydrophobic films on soil aggregates (DeBano 2000). The post-fire response of soil microbes is highly dependent on soil status and differs in the short- and long-term (Guerrero et al. 2005). Immediately after heating, the soluble organic carbon increases with temperature, reaching maxima of around 400°C (Guerrero et al. 2005). These increases are likely due to lysis of microbial cells, which releases organic cytoplasmatic compounds (Serrasolsas and Khanna 1995a, Prieto-Fernández et al. 1998, Choromanska and DeLuca 2002). Plant tissue debris and the increased organic carbon extractability due to desiccation are likely to contribute to the increased soluble organic carbon (Powlson and Jenkinson 1976, Birch 1958). The increase in soluble carbon and nutrients in burned soils encourages rapid recolonization of some types of microorganisms, such as heterothrophic bacteria, and increases basal respiration rates. In the medium-term these easily mineralized compounds are consumed, and the remaining organic fractions of carbon and nitrogen become more recalcitrant to microbial attack (González-Pérez et al. 2004), at which point a decrease in microbial populations is usually observed. More information on soil physical and chemical fire effects and nutrient dynamics can be found in Chapters 4 and 8 of this book.

Post-fire environmental factors

Plants exert a strong influence on the composition of soil microbial communities through rhizodeposition, decay of litter and roots, and the influence on water dynamics. The co-evolution of plant species and microbial communities in the rhizosphere soil has made a tight link between these organisms (Brimecombe et al. 2001). Consequently the temporary loss of plant cover after a fire has significant impact on the soil microbial communities, and the recolonization of burned areas by microorganisms will depend on the plant restoration status. Hart et al. (2005) hypothesize that the strong links between some plant species and soil microbial communities suggest that changes in vegetation community structure in the years following fire have the potential to be the more dominant driver and shaper of the soil microflora than the direct impact of the fire disturbance itself.

Post-fire soil color changes (usually darkened as a consequence of the incorporation of partially pyrolized material from burning vegetation to soil), are evidence of microclimatic modifications of the soil temperature and water content. Not only is there a direct loss of water from the soil during heating, but the evaporation rate from burned soil is usually higher as a consequence of the greater radiation received due to the loss of plant cover and blackened soil surface (Campbell et al. 1977, Wells et al. 1979, Pietikäinen and Fritze 1995). This implies warmer soil temperatures during the day, and more rapid heat loss and cooler soil temperatures at night. Differences in maxima daily temperatures measured in the upper centimeters of burned and nearby unburned soil, ranged between 6 and 25°C during the warm season (Chandler et al. 1983, Díaz-Fierros et al. 1990, Mataix-Solera 1999). Although more water can be available in burned soil after some light rains, soil water availability is generally lower in burned soils than in unburned soils. The presence of hydrophobic substances after a fire can also contribute to reduction in soil moisture (Doerr et al. 2000, Úbeda et al. 2002) as does wind desiccation and modifications in some soil properties related to water holding capacity, such as organic matter content, aggregate stability, porosity, bulk density, etc. (see Chapter 4).

The susceptibility of soil to erosion after fire is normally increased because of changes in soil properties and decrease in plant cover (Cerdà et al. 1995). The environmental conditions during the first post-fire months are of crucial importance in natural revegetation and recovery. Post-fire months with high intensity rainfalls can produce significant soil losses through erosion (Atkinson 1984, Marqués and Mora 1992; see Chapters 2 and 6), and consequent soil degradation. A drought during the year after a fire can also affect the soil system and plant recovery. Soil microbiology is highly dependent on both scenarios.

Post-fire Effects on Soil Organisms

Bacterial populations

Indirect fire effects on bacterial population vary based on the post-fire soil parameters. In the short-term, a marked increase in the heterotrophic bacterial population has been reported in numerous studies where higher abundance values were measured in burned than unburned soils (Meiklejohn 1955, Theodorou and Bowen 1982, Adedeji 1983, Vazquez et al. 1993, Grasso et al. 1996, Mataix-Solera et al. 2002, Badía and Martí 2003). For example, soil was heated to 400°C for 15 min (practically sterilized) and during the subsequent 100 days of recovery, the abundance of bacteria was about two orders of magnitude higher in the heated soil than in the unheated soil (Guerrero et al. 2005). Such surviving colonizing bacteria find post-fire conditions are beneficial for proliferation. Researchers often suggest that the post-fire increase in bacteria is due to ash deposition (causes an increase in soil pH and nutrient availability) and the fact that the bacteria have a greater ability to use the soluble organic compounds released by the heat. Results from Jokinen et al. (2006) suggest that direct and indirect pH effects are mainly responsible for the increase in microbial activities following ash fertilization. Other researchers added rainfall as a critical factor that initiates this increase (Deka and Mishra 1983) as mineral nutrients released from combustion of organic matter are not available until dissolved by rain (Ahlgren and Ahlgren 1965). Other factors, such as an increase in temperature (Adedeji 1983) or decrease in competition for different nutrients (Bauhus et al. 1993), can also influence short-term bacterial proliferation.

In the long-term, bacteria populations tend to return to pre-fire levels (Theodorou and Bowen 1982, Vazquez et al. 1993, Grasso et al. 1996, Mataix-Solera et al. 2002) and are related to vegetation succession. However, three years after a fire, negative fire effects on bacteria were reported by Mabuhay et al. (2003) who found bacterial abundance in burned areas was still lower than in unburned areas. Those microorganisms that intimately interact with vegetation, through mutualism, synergism, antagonism, or competition, will respond to changes in vegetal composition. An example of mutualism is symbiotic N-fixer bacteria associated with the rhizosphere of pioneer woody legumes in tropical forests. These microorganisms are more abundant in early succession when high rates of N fixation are produced and decrease in later succession when pioneer species disappear from the forest (Pérez et al. 2004).

Actinomycetes

Similar to bacterial populations, actinomycetes populations increase during the first few weeks after a fire (Meiklejohn 1955, Ahlgren and Ahlgren 1965, Deka and Mishra 1983). Nutrient release by ash is the most probable cause of the increases observed. Soil pH and temperature increases with subsequent loss of moisture could favor actinomycetes proliferation, as these microorganisms are abundant in alkaline soils and their spores provide a survival advantage in dry high-temperature sites (Coyne 2000).

Photoautotroph microorganisms

In the short-term after fires, photoautotroph microorganisms decrease in abundance (Vazquez et al. 1993, Myers and Davis 2003) with cyanobacteria decreasing more than other crust components (Bowker et al. 2004). However, cyanobacteria and algae are commonly pioneering species during the revegetation of degraded ecosystem (Acea et al. 2001). Rapid colonization by algae has been reported after moderate and intense fire (Myers and Davis 2003). Vazquez et al. (1993) found that, in the long-term, photoautotroph microorganisms are clearly favored in the post-fire environment. Given that photoautotroph microorganism habitats are characterized by wide fluctuations in environmental factors, such as pH, nutrient content and availability, light, aeration, texture, temperature, and moisture (Acea et al. 2001), most post-fire fire parameters favor their growth. One major factor is the absence of light-competitors, which facilitate photoautotroph colonization through autotrophic succession (Atlas and Bartha 2001). Soil moisture is another decisive factor for photoautotroph microorganisms. With adequate soil surface moisture, biological soil crusts are metabolically active; and as a result, abundant precipitation following fire promotes a faster recovery (Ford and Johnson 2006).

Cyanobacteria and algae establishment promotes the colonization of other microorganisms and improves soil stability through biological crust formation (Acea et al. 2001, 2003). When plant cover is severely damaged by wildfire or prescribed burning, the biochemical activity of photoautotrophs provides important soil ecosystem inputs, such as organic carbon and nitrogen (López-Hernández et al. 2006). In the longer-term, as vegetation recovers, cyanobacteria and algae will decrease and vary their composition and other microorganisms will be dominated in the soil ecosystem.

Fungi

With respect to the post-fire saprophytic fungi, results show several fire effects: a) they are killed at lower temperatures than bacteria, and strong decreases in the length of the mycelium has been observed (Vázquez et al. 1993); b) they are favored by the rise in soil pH (consequence of ash deposition) favoring bacteria (Ahlgren 1974, Vázquez et al. 1993, Adedeji 1983, Bååth et al. 1995); c) they are sensitive to some toxic compounds present in burned soils (Widden and Parkinson 1975, Fritze et al.1998); and d) they are negatively affected by some post-fire changes in specific organic fractions, such cellulose. This explains the positive effect on fungal mycelia observed after organic amendments (rich in cellulose) were added to heated and burned soils (Prieto-Fernández et al. 1998, Acea and Carballas 1999).

The colonization of burned areas by mycorrhizal fungi may take place from different inoculum sources (propagules), such as surviving mycelium, heat-resistant or disseminated spores, and heat-resistant structures (sclerotium and mycelium) originating from the mycorrhized roots of living or dead plants. When a disturbance like a fire occurs in an ecosystem, the mycorrhizal fungi associated with organic horizons (particularly in the case of ectomycorrhizal fungi) would theoretically be more susceptible than those found in mineral substrata (like those found in deeper layers, in the rhizosphere, as is the case of arbuscular fungi). Nonetheless, the main factors which determine the post-fire response of mycorrhizal fungi populations are the intensity of the fire, the extent of the affected area, and the type of strategy used to regenerate the dominant vegetation.

High intensity fires which destroy host plants and significantly alter soil properties clearly reduce mycorrhizal fungi diversity by temporarily modifying the fungal community (Rashid et al. 1997, Smith et al. 2005). However, low intensity fires, where many host species survive, the effect on diversity of the mycorrhizal fungi populations are less significant and recovery takes place in the medium- and short-term (between 6 mo to 1 yr in most cases; Bellgard et al. 1994, De Román and De Miguel 2005).

In fires that affect small areas or have uneven effects (i.e., form 'islands' of unburned vegetation), recolonization by mycorrhizal fungi is much faster since the unaffected areas act as an inoculum source (Torres and Honrubia 1997). Many plant species are adapted to surviving after a fire with regeneration strategies that enable them to be a resprouter or germinator species. The former are those species which, due to the survival of subterranean organs (and of the associated fungi), resprout after the fire and act as a reserve of inoculum while the environmental conditions are established once more. Species whose germination and expansion are favored by the temperatures reached in the fire are germinator species. During the first stages of their development, they are able to mycorrhize with certain fungi species whose fructification increases after a fire (pyrophilous species). These fungi are 'bridge' mycorrhizal fungi that will be replaced by other typical mycorrhizal fungi species in the short- or medium-term (Torres and Honrubia 1997). In other cases, seedlings of a germinator species and a typical ectomycorrhizal species, such as pines, are colonized by both ectomycorrhizal fungi and arbuscular fungi, in such a way that the latter could survive until their host plant appears (Horton et al. 1998). In this sense, it seems that arbuscular fungi are less affected by the destruction of certain plants by fire than ectomycorrhizal fungi since they are capable of establishing mycorrhizas with a greater number of plant species (there is lower fungi-plant specificity). In contrast however, arbuscular fungi are less likely to survive in the absence of their host plants than ectomycorrhizal fungi.

In recent attempts to determine the effect of a low intensity fire on arbuscular fungi activity by analyzing the glomalin protein content in soil, no differences were observed between the burned areas and unburned areas (Knorr et al. 2003). This may be due to the limited plant mortality from the fire.

Fire effects on ectomycorrhizal fungi have been studied more than on the arbuscular species, which is mainly due to quantitative and economic importance in forest ecosystems and their easily detectable fructifications that enable post-fire changes to be readily observed. Fire effects on ectomycorrhizal fungi may be grouped as follows: a) fructification produced; b) abundance of mycorrhizas or mycorrhizal roots; and c) dynamics and composition of species within the ecosystem. High burn severity with high tree mortality often results in near total disappearance of ectomycorrhizal fungi fructifications (Visser 1995, Torres and Honrubia 1997). When burn severity is low, only the superficial part of the organic horizon is burned, trees survive, and the number and biomass of ectomycorrhizal fungi fructifications remain virtually unaltered (Dahlberg et al. 2001). In the case of recurrent fires, significant reductions have been observed in the number, biomass, and diversity of these fructifications (Bastias et al. 2006).

No model of post-fire succession has been observed to describe fire effects on the dynamics and composition of species in the ectomycorrhizal fungi community. Fire may modify the mycorrhizal fungi communities such that new species appear and others disappear or some families may become more abundant. For example, Visser (1995) observed significant post-fire changes in the composition of ectomycorrhizal species in forests of *Pinus banksiana* in Canada. Bastias et al. (2006) confirmed that some families of fungi were substituted by others after fire in forests of *Eucalyptus pilularis* in Australia. It is well known that pyrophilous species appear after a fire (especially the ascomycetes) and some of these are ectomycorrhizal species, which will disappear once the fire effects are alleviated (Torres and Honrubia 1997). In general, the potential changes in the ectomycorrhizal fungi community will depend on both burn severity and the pre-fire community composition.

In general, a post-fire decrease in ectomycorrhizal fungi species richness and diversity has been observed (Jonsson et al. 1999, Grogan et al. 2000), although no studies indicate if these changes persist for the long-term. Fire effects after low severity fires are almost nil; in fact, De Román and De Miguel (2005) did not detect any changes in the diversity of ectomycorrhizal species in *Quercus ilex* woods after a moderate severity fire. In the same study, they confirmed that, despite a reduction in the colonization percentages of ectomycorrhizal species during the first year after the fire, these percentages were recovered by the second year. On the other hand, Torres and Honrubia (1997) discovered significant reductions in the colonization percentages of ectomycorrhizal fungi in *Pinus halepensis* forests in southeastern Spain, as did Miller et al. (1998) in *Pinus contorta* stands in Wyoming, USA; in both cases the fires were high burn severity with elevated tree mortality and a total loss of organic soil horizons.

Depending on the species, the propagules of ectomycorrhizal fungi are spores or fragments of mycelium, sclerotia and mycorrhized roots (ectomycorrhizas) which have survived the fire given their location in the soil profile. Given their greater persistence, it is likely that spores and sclerotia are responsible for formation of new fungi following high severity fires where vegetation is almost completely destroyed; however, after moderate or low severity fires, mycorrhiza formation would fundamentally take place from the mycelium (Fig. 2).

Sclerotia are persistent propagules which can withstand unfavorable conditions for years (fires, drought, heavy metals in the soil, etc.), longer than any other resistance structure formed by fungi. These resistance structures, whose functions include recolonizing after natural disasters, play an important role in reestablishing vegetation after a fire (Miller et al. 1994). After a disturbance such as fire, the number of sclerotia present in the soil increases to a significantly greater or lesser extent (Torres and Honrubia 1997). The

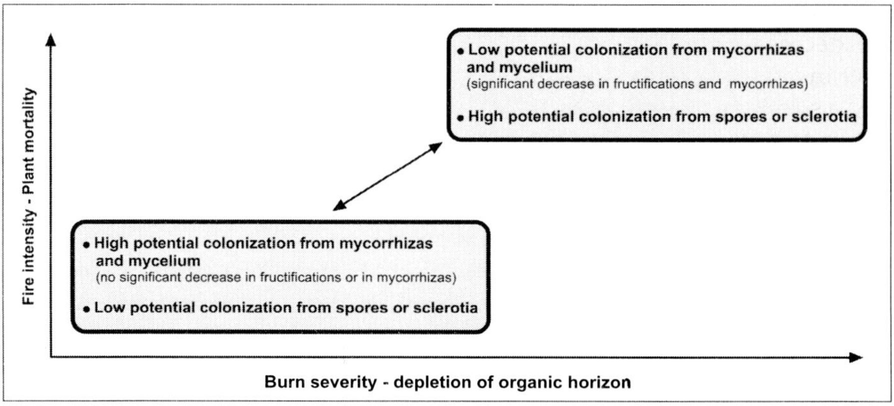

Fig. 2 The effect of fire on ectomycorrhizal fungi propagules (based on Dahlberg 2002).

tendency of sclerotia to appear in greater numbers in fire-affected areas may be provoked by the mortality of root systems with which ectomycorrhizal fungi are associated. The slow death of roots after a fire would favor the formation of fungal resistant structures. The role that sclerotia plays in the recovery of fungal flora, and therefore in the associated vascular flora, may be an important one, particularly after high severity fires and/or frequent fires where vegetation disappears completely along with the fungal component associated with its roots.

Meso and microfauna

Fire may indirectly affect invertebrate communities by changing plant species composition and vegetation structure, reducing the litter layer, and modifying soil moisture and temperature (Mitchell 1990). Burning can lead to increase soil pH and greater fluctuation in temperature and moisture, which in turn, influence vegetation composition (Springett 1976, Haimi et al. 2000). These modifications in soil habitat expose arthropods to greater extremes of temperature, light, and moisture (Buffington 1967). Various studies indicate that post-fire effects can modify these populations in different ways. New and Hanula (1998) did not find differences in arthropods abundance 2 and 3 yr following a fire, possibly due to the buildup of litter habitat. Studies developed in dry eucalyptus forest of Australia show that arthropod abundance was not affected by burning (Abbott 1984, Majer 1984). However, in mixed hardwood-pine of southeastern Kentucky, Coleman and Rieske (2006) found that prescribed fires that included reduction of litter depth resulted in reduction in arthropod abundance for at least 1.5 yr following a fire. These authors suggested that these indirect effects of fire might be due to changes in leaf-litter habitat complexity or stratification, accumulation of leaf-litter duff, and freshly fallen leaves, rather than simply quantity or availability. Fifteen months after a fire in a Mediterranean forest, Serra et al. (1992) observed a large decrease in the density of soil arthropods, from 123,000 ind m^{-2} to 33,000 ind m^{-2}, on average from 30 different taxa studied. Moreover, they found a strong disequilibrium in the community structure. They attributed the changes to the destruction of vegetal cover and the subsequent decrease in litter and upper organic horizons of soil. They also attributed some of the observed change to the decrease in soil water content (consequence of decrease in organic matter and darkening of soil surface) and increasing soil temperature. This may explain why *Collembola* was the most affected group and *Acari* the least affected, as *Collembola* are more sensitive to desiccation stress.

Certini (2005), in his review of fire effects on soil properties, indicates that the indirect effect of fire, particularly litter mass reduction, drastically decreases both total mass and number of species of soil-dwelling invertebrates. This work cites several studies about the indirect effect on micro and mesofauna. McSorley (1993) noticed that within 6 wk after controlled burning, total numbers of omnivores and predators were increased, while herbivore numbers stayed the same. After taking inventory of soil nematodes at 99 burned and unburned forested sites, Matlack (2001) concluded that, in the long run, fire does not significantly affect the nematode community either in number of individuals or diversity. Wanner and Xylander (2003) found that total biomass and species inventory of testate amoebae in pine forest are considerably reduced by a prescribed fire, but within 1 yr they return to the original level.

Biomass and diversity

The recolonization pattern of burned soils is not the same for fungi, heterotrophic bacteria, and non-heterotrophic bacteria. Moreover, this pattern is dependent on burn severity, and thus, dependent on some changes in soil. Some of these changes are short-lived or ephemeral, but others continue for years. In laboratory experiments, isolating the fire severity as a unique source of variation, common patterns have been found in the evolution of microbial biomass after fire: high temperature kills microbes and the recovery of the microbial biomass is dependent on the conditions of the burned soils (Díaz-Raviña et al. 1992, Acea and Carballas 1996, Pietikäinen et al. 2000, Badía and Martí 2003, Guerrero et al. 2005). In those soils where the fungal biomass is greater than the bacterial biomass, the negative effects of fire will be more evident. Similarly, the post-fire recovery of the microbial biomass will be the sum of the recovery of bacterial biomass and fungal biomass. It has been observed that the recovery pattern of both groups is different.

Heating decreases the total organic carbon (mineralized by combustion), but increases the pool of soluble organic compounds. The released soluble organic carbon acts as readily metabolizable compounds for recolonizating microbes, allowing a rapid increase in populations of microorganisms, mainly heterotrophic bacteria (Grasso et al. 1996, Badía and Martí 2003, Guerrero et al. 2005). Some studies observed partial microbial biomass recovery in the first days after soil heating. Guerrero et al. (2005) after 8 d of incubation, found that soil samples exposed at 400°C had more microbial biomass (bacteria mainly) than those exposed to 200 and 300°C, due to greater abundance of soluble organic carbon. Choromanska and DeLuca (2002) indicate that the initial carbon availability had important effects on the recovery (first 14 d) of microbial biomass in heated soils. After 1 mo of incubation, Pietikäinen et al. (2000) found that samples of humus heated at 160°C presented higher values of microbial biomass than samples heated at 100°C. Similarly, Badía and Martí (2003) found that after 1 mo of incubation, the microbial biomass in a calcareous soil burned at 250°C was higher than the unburned soil. However, these increases in biomass are short-lived and ephemeral, especially in severely heated soils, because the pool of soluble organic compounds released are rapidly mineralized (and exhausted) by the recolonizing microbes (Diaz-Raviña et al. 1992, Guerrero et al. 2005). Heating causes decreases in soil organic matter and large modifications in its quality. Most of these modifications lead to higher recalcitrance (Almendros et al. 1990, Fernández et al. 2001, González-Pérez et al. 2004), and then decreases the pool and replenishment rate of the easily mineralized compounds. Thus, the remaining organic matter is not able to maintain high populations of heterotrophic microbes (main contributors to microbial biomass). This could be the reason for the decrease in biomass (after the ephemeral increase) observed by Pietikäinen et al. (2000), Badía and Martí (2003) and Guerrero et al. (2005) in laboratory experiments.

In soil heated at 160° and 350°C for 30 min, Díaz-Raviña et al. (1992) observed a small increase of the microbial biomass during the second week after heating, with values 2-fold and 14-fold lower than the unburned soil. The low recovery of the mycelium length may be the reason for the low recovery of biomass. In soils heated at 600°C, a small recovery of microbes was observed by Díaz-Raviña et al. (1992) and Guerrero et al. (2005), as a consequence of a net decrease in total and soluble organic carbon. Given the large contribution of fungal biomass to the total microbial biomass (from 30 to 80 percent; Anderson and Domsch 1975), the poor recovery of microbial biomass could be explained by the low recovery of fungi (Vázquez et al. 1993, Bååth et al. 1995, Pietikäinen and Fritze 1995, Prieto-Fernández et al. 1998) as was previously discussed.

With respect to the post-fire recovery of microbial biomass under field conditions, a wide range of results can be found in the literature. This variability is the result of several different factors (see section about factors controlling indirect effects) as well as time of sampling after fire and soil sampling depth. In those experiments where the soil is heated in a laboratory and then incubated (rewetted), the post-fire short-term evolution of microbes is easy to follow. But under field conditions the evolution is more complicated, because of differences of many significant variables, such as climate, soil type, etc., found in field sites. In general, field observations follow the same sequence as laboratory experiments—the fire kills microbes (more or less depending on burn severity and microbial group being considered) and after the first rainfalls, the burned soil will be recolonized to a greater or lesser extent depending on soil changes and types of microorganisms. In this sense, field results mimic laboratory experiments where ephemeral increases of heterotrophic bacteria have been observed after the first rainfalls and a slow recovery of fungal mycelium length occurs over time (Ahlgren and Ahlgren 1965, Vázquez et al. 1993, Acea and Carballas 1996). Most field studies report a decrease in microbial biomass after fire (Fritze et al. 1993, Pietikäinen and Fritze 1993, 1995, Bååth et al. 1995, Hernández et al. 1997, Prieto-Fernández et al. 1998, Choromanska and DeLuca 2001, Guerrero et al. 2002, Palese et al. 2004, Mabuhay et al. 2006a,b), but no effect or slight increases (short-lived in some cases) have also been observed (Tateishi and Horikoshi 1995, Dumontet et al. 1996, Andersson et al. 2004, De Marco et al. 2005, D'Ascoli et al. 2005; see Table 1).

The ephemeral increase in the content of some nutrients (soluble organic C, ammonium, phosphorous) generally leads to explosive growth of heterotrophic bacteria, and it could increase the microbial biomass to values higher than for unburned. Given that fungal biomass is typically higher than bacterial biomass, the increase in bacterial biomass may not be always reflected as an increase in microbial biomass. Also, the greater effect of heat on fungal mycelia (Vázquez et al. 1993, Bååth et al. 1995, Acea and Carballas 1999) results in greater post-fire loss of microbial biomass in soils where fungal biomass is considerably greater than bacterial biomass. Conversely, in soils where the bacterial biomass is higher than the fungal biomass, the short-term increase in bacteria may be the result of an increase in microbial biomass (Rutigliano et al. 2002, D'Ascoli et al. 2005).

The reestablishment of microbial biomass depends on fire severity, changes in soil, degree of vegetation recovery, and post-fire climate. Many authors have reported the recovery of microbial biomass with the reestablishment in fungi (fungal biomass). Fritze et al. (1993) observed that the recovery of the microbial biomass around 12 to 15 yr after a prescribed fire was strongly correlated with the reestablishment of fungal biomass (measured through ergosterol). The recovery of fungi is often slow and dependent on changes of organic fractions such as celluloses (Vázquez et al. 1993, Prieto-Fernández et al. 1998). Sometimes post-fire soil organic matter increases due to deposition of partially combusted vegetation. If this organic matter is not recalcitrant, it may enhance the recovery of microbial biomass above pre-fire levels.

The diversity of soil microorganisms may be modified by fire in several ways: a) by differential sensitivity to heat; b) by different survival or colonization strategies; c) by differential sensitivity to modifications in soil properties (mainly organic matter and pH); d) by changes in soil microclimate (moisture and temperature); e) by the destruction and creation of new ecological niches; and f) through the changes in the biomass and composition of the above-ground vegetal species. As a result, changes in soil microorganism diversity may be measured in different ways.

Some authors have found differential changes by fire through the study of physiological groups, which are often dependent on the time since the fire (Vázquez et al. 1993, Acea and Carballas 1996). It is obvious that the changes caused by fire in soil and the environment will lead to changes in microbes. For example, changes in soil pH due to ash deposition results in changes in proportions of fungi and bacteria (Entry et al. 1986, Mataix-Solera et al. 2002), because they have different pH optima (Bååth et al. 1995). Bååth et al. (1995) observed different community structure (using PLFAs, or phospholipid fatty acids) as a consequence of fire (modifying the quality of the organic matter) and ash deposition (modifying the pH). Pietikäinen et al. (2000) observed gradual changes in the microbial community structure (also using PLFAs) in response to the soil temperature as heating went from 45 to 160°C, but greater changes were observed when samples were heated to 230°C. Changes in diversity of fungi, one of the most important microorganisms, are related to time after fire and burn severity (Jalaluddin 1969, Cooke 1970, Widden and Parkinson 1975, Bisset and Parkinson 1980, Bartoli et al. 1991, Persiani et al. 2002). Bartoli et al. (1991) observed an increase in numbers of fungi species during the first 4 yr after fire, some of these species being different than those in unburned soils (e.g., some carbonicolous fungi). They found that the number of fungal species found in both burned and unburned soils tended to increase during the first 8 yr after fire. Bisset and Parkinson (1980) report changes in fungal composition in organic horizons 6 yr after the fire. Widden and Parkinson (1975) found that aqueous extracts of burned soils inhibited the growth of most fungi, except some species typically found on burned soils. In contrast, Deka and Mishra (1983) found that slash burning caused quantitative decrease in fungi, but scant changes in species composition, similar to the study by Meiklejohn (1955).

Burned soils are more exposed to light, and the amplitude of variation of temperatures (daily and seasonal) is higher than in unburned soils. Moreover, the spatial and temporal distribution of water is different between the burned and unburned soils. This induces changes in diversity, such as increases in photoautrotophic microorganisms, for example algae and cyanobacteria (Vázquez et al. 1993). Another way to view microbial diversity is to observe functional diversity by examining the substrate utilization pattern. Microbial communities are thought to be functionally redundant where many taxonomically distinct members of the community can utilize the same carbon source. Consequently, differences in functional diversity may underestimate differences in taxonomic diversity. Pietikäinen et al. (2000) found changes in substrate utilization pattern according to the temperature of heating. Staddon et al. (1998a) found decreases in functional diversity 5 yr after prescribed burning. Four years after fire, Liu et al. (2000) found that burned soils had lower functional diversity, based on substrate utilization capabilities using Biolog^{TM} ECO plates. Mabuhay et al. (2006b) reported a negative effect on microbial biomass and diversity even 25 yr after a fire. Other studies of fire effects on microbial diversity are based on catabolic evenness; D'Ascoli et al. (2005) found a short-term decrease in diversity, which was influenced by burn severity.

Indicators of Microbial Activity

Basal respiration, carbon mineralization and eco-physiological indexes

With respect to the effects of heating on microbial activity, the division between direct and the indirect effects is clear. As previously discussed, at elevated temperatures (direct effects), most microbes are killed and most enzymes are denaturalized. But some types of microbes, such as heterotrophic bacteria, can quickly recover. The changes in the substrate (the soil) caused by the fire explain a considerable portion of the post-fire microbial evolution.

The measurement of CO_2 produced (or oxygen consumed) from the soil has been widely used as a measure of the total heterotrophic activity in the soil. Soil heating between 100 to 500°C causes a net increase of easy mineralizable compounds (such as carbohydrates) released from killed microbes (Serrasolsas and Khanna 1995a) and plant debris. Many authors have found a net increase in soluble organic carbon in heated soils as compared to unheated soils: 3-fold increase (Guerrero et al. 2005); 4-fold and 5.5-fold increase (Guerrero, personal communication); 5.8-fold increase (Díaz-Raviña et al. 1996); 7.5-fold increase (Díaz-Raviña et al. 1992); and 12-fold increase (in soluble sugars; Choromanska and DeLuca 2002). Moreover, the ammonium and the available P also increase in the burned soils. This increase in nutrients enhances the heterotrophic bacteria recovery and results in a sharp increase in the basal respiration (measured as CO_2 produced or oxygen consumed, Fig. 3; Almendros et al. 1990, Bauhus et al. 1993, Serrasolsas and Khanna 1995a, Choromanska and DeLuca 2001, Badía and Martí 2003, Guerrero et al. 2005). This increase in basal respiration is highly correlated with the numbers of bacteria and soluble organic carbon. Prior to this sharp increase, a lag phase without activity was observed. The magnitude of the lag phase is directly related to the temperature, which indicates the degree of sterilization. This increase in basal metabolism was observed during the first 48 h after the rewetting of burned soils (100 to 500°C). During this stage, a large increase in bacteria occurs. But Díaz-Raviña et al. (1996) found a very long lag phase (13 wk) in the bacterial activity (measured by the thymidine and leucine incorporation techniques), probably due to the strong changes in soil caused by the relatively long heating time (1 h at 200°C). The higher rates of basal respiration observed in the burned soils (100 to

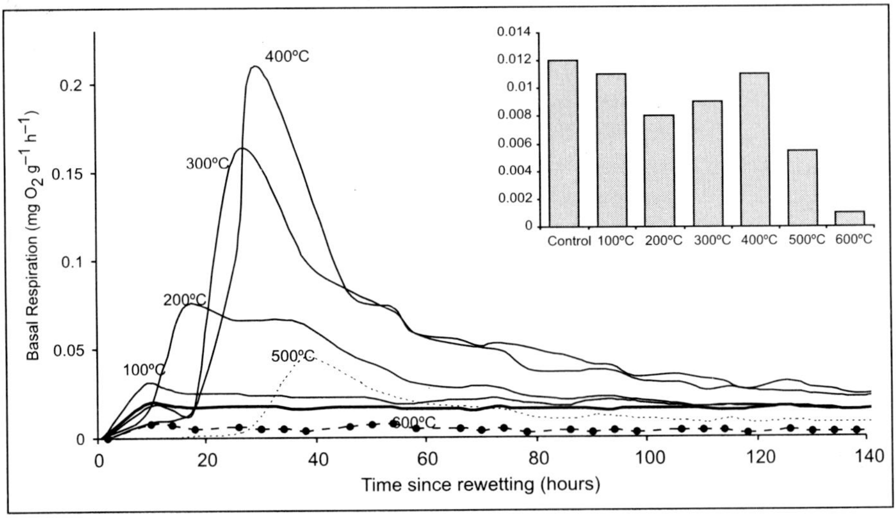

Fig. 3 Lines represent rates of basal respiration (measured as oxygen uptake) during the first 140 h after rewetting heated soils (100 to 600°C). Double-sized line is used for unheated soil. Bars are representing the rates after 1 mo (C. Guerrero, personal communication).

500°C; Fig. 3) for several weeks in laboratory incubations (Choromanska and DeLuca 2001, Guerrero et al. 2005), and, in some cases, for up to 9 mo (Badía and Martí 2003). The time was dependent on the temperature of heating and the exhaustion of easily absorbed compounds (Almendros et al. 1990, Guerrero et al. 2005).

In severely heated soils (>500°C), a net decrease (or complete disappearance) of soluble organic compounds occurs because they are mineralized by combustion. When this occurs, the basal respiration rates of the burned soils are always lower than those observed for unburned soils (Fig. 3). In some cases, negative rates of CO_2 (sequestration) have been observed during the first days after severe heating. This sequestration could be due to the carbonation of hydroxides formed during heating.

Under field conditions, apparently contradictory data are available. Prior to commenting on some available literature, it is important to note that the respiration measured on soil samples in the laboratory is generally called *basal respiration*, which differs from *soil respiration* measured in the field. Although both measurements provide interesting data, basal respiration reflects microbial respiration while soil respiration includes microbial, roots, and soil fauna respiration. Moreover, in basal respiration the environmental conditions (temperature and soil moisture) are controlled, allowing for better comparability of results as well as possible inferences about organic matter quality (carbon mineralization) and microbial status. In soil respiration, the measure is integrating site-specific environmental factors. For example, field sites with differential moisture and/or temperatures between the burned and unburned soil will lead to soil respiration values not directly related to soil organic matter content and quality or quantities and types of microbes. Bissett and Parkinson (1980) incubated burned and unburned field soil at uniform conditions (in laboratory, at controlled moisture and temperature) and observed that the burned samples had lower rates of basal respiration and cellulose decomposition than unburned samples. However, the rate of cellulose decomposition in the field was much higher at the burned sites. The authors suggested that it was due to higher temperatures in the burned soils as a consequence of the increase in solar radiation. Similar results were found by Pietikäinen and Fritze (1993) on the decomposition of coniferous needle litter in a soil affected by prescribed fire.

The sequence of post-fire microbial activity recovery observed in laboratory experiments will occur in the field after the first post-fire rainfall (Ahlgren and Ahlgren 1965), because very little respiration (activity) occurs when the soil is dry. Given the large influence of soil moisture and/or temperature (depending on climate type) on soil respiration, the magnitude of the described stages will vary. After the initial sharp increase in respiration, the subsequent respiration rate is dependent on the quantity and quality of the remaining organic matter and microbial status. In field conditions, the rates of soil respiration (in situ) will also be dependent on environmental conditions.

Most studies reported lower respiration values in burned soils as compared to unburned, but some of them report increases (Bisset and Parkinson 1980, Almendros et al. 1984b, White 1986, Pietikäinen and Fritze 1995, Hernández et al. 1997, Fernández et al. 1999, Wüthrich et al. 2002, Mataix-Solera et al. 2006). In a Mediterranean pine forest, Hernández et al. (1997) observed lower rates of basal respiration 9 mo after fire. Pietikäinen and Fritze (1993) observed lower values of basal respiration in soils affected by prescribed fire and experimental fire for 3 yr after the fire, with the exception of the first 3 d. Ahlgren and Ahlgren (1965) observed an initial decrease in basal respiration, followed by an increase after the first rainfall, and decreasing again for 3 yr. Fernández et al. (1997) incubated (11 wk) samples collected 1 d after a high severity wildfire. They observed higher rates of mineralization on burned samples during the first two weeks, which tended to decline for the remainder of the study. Almendros et al. (1990) collected soil samples 15 mo after a wildfire; when they are incubated for 2 mo, burned samples showed lower rates with respect to unburned samples only during the second month. In another work, Almendros et al. (1984a) found lower rates of basal respiration in samples burned 2 yr earlier.

Some authors have found changes in the organic carbon mineralization kinetics (Almendros et al. 1990, Guerrero et al. 2005). In agreement with other authors, Bauhus et al. (1993) and Guerrero et al. (2005) observed an ephemeral increase in the organic matter lability and a subsequent and more intense increase in the organic matter stabilization. Fernández et al. (1997, 1999), using several burned samples from 1 d to 2 yr after wildfires, found that changes in carbon mineralization kinetics did not completely recovered after 2 yr. In other words, after the initial consumption of easy biodegradable compounds, the organic fraction remains more resistant to degradation (Fernández et al. 1997, Hernández et al. 1997).

High fuel loads can enhance higher fire intensities, but some climatic conditions (such as high moisture) lead to a net increase in organic debris in soil from deposition of incompletely or partially burned vegetation. In this situation, an increase in respiration will be expected for several months after the fire. After an experimental fire, Wüthrich et al. (2002) found higher rates of soil respiration (in situ) in plots where fuel loads were higher. Lower values of basal respiration were measured 11 yr after a wildfire in Mediterranean conditions (Mataix-Solera et al. 2006), where slow recovery of vegetation leads to poor recovery of soil organic matter.

The metabolic quotient qCO_2, is the ratio of the basal respiration divided by the microbial biomass (carbon respired per unit of carbon in microbial biomass). In the last decades, qCO_2 has been widely used as an index to detect stress, impacts (including forest fire), efficiency using carbon, evolution of ecological succession, and changes in microbial activity and composition (Insam and Domsch 1988, Insam and Haselwandter 1989, Anderson and Domsch 1990, 1993, Fritze et al. 1993, Pietikäinen and Fritze 1993, Sakamoto and Oba 1994, Wardle and Ghani 1995). Generally, this index is high in the first stages of chrono-sequences and then decreases as ecological succession proceeds (Insam and Domsch 1988, Insam and Haselwandter 1989). Due to a combination of several factors acting together, this quotient increases immediately after disturbances such as fire (Fritze et al. 1993, Pietikäinen and Fritze 1993, Choromanska and DeLuca 2001, Rutigliano et al. 2002, Badía and Martí 2003, Guerrero et al. 2005). Generally, in unburned soils, fungal biomass is greater than bacterial biomass and fire has a greater negative effect on fungal biomass than on bacterial biomass. In some burned soils, the bacterial biomass could rise to become 80 percent of the total microbial biomass (Entry et al. 1986, Pietikäinen et al. 2000). The fire modifies soil organic matter quality and increases quantities of easily mineralized compounds, which can rapidly be used by bacteria with higher metabolic capabilities. Some authors (Alexander 1967, Miller 1993) observed that bacteria are less efficient than fungi in using carbon, resulting in higher qCO_2 when the fungal-to-bacterial microbial biomass ratio decreases (Sakamoto and Oba 1994). The ratio C_{mic}:C_{org} could also be used as eco-physiological index as an indication of carbon available for growth, whereas qCO_2 could be associated with maintenance of energy requirements and efficiency using carbon (Anderson 2003).

Immediately after a fire, the soil is enriched in available nutrients released from combusted organic material and killed microbes. Thus, a change in the type of microorganisms occurs, with a predominance of *r*-type (zymogenous, faster growing microbes) versus *k*-type strategist (slower growing specialists, more efficient) (Fritze et al. 1993, Odum 1969, Pianka 1970). Consequently, during the first stages of recolonization the qCO_2 increases. Several months after the immediate increase, the qCO_2 trends to decrease, which may explain the progressive displacement of the zymogenous (*r*-type strategist) microorganisms by slower growing specialist (*k*-type strategist) (Fritze et al. 1993). In addition, the decrease in easily mineralized compounds (Hernández et al. 1997) and the decrease in nutrients (immobilized by other microbes or vegetation, or lixiviated by erosion) also affect the shift from zymogenous to growing specialist microorganisms. Moreover, under stress conditions, such as changes in pH, microbes have to divert energy from growth and production to maintenance (Odum 1985), resulting in higher qCO_2 (Anderson and Domsch 1993). But microorganisms tend to adapt to new stressing conditions, and after the initial increase, the qCO_2 will tend to diminish. However, some authors disagree with the use of qCO_2 as an indicator of stress. For example, Wardle and Ghani (1995) mentioned "the metabolic quotient fails to distinguish between effects of disturbance and stress." The qCO_2 could also indicate the change in organic matter recalcitrance. Eleven years after a wildfire, Mataix-Solera et al. (2006) found lower values of qCO_2 (with respect to unburned) where the carbon mineralization coefficients were also low.

With respect to the ratio C_{mic}:C_{org}, the organic carbon (C_{org}) content in soils has a strong influence on levels of soil microbial biomass carbon (C_{mic}). Moreover, the spatial distribution of vegetation has influence on the spatial distribution of C_{org}. Consquently, there are times where it is more appropriate to relate the soil C_{mic}:C_{org} to detect fire effects on C_{mic}. Moreover, this ratio can provide information about the carbon available for microbial growth. This ratio generally decreases after fire (Fritze et al. 1993, Prieto-Fernández et al. 1998, Guerrero et al. 2002). As C_{mic}:C_{org} reaches an equilibrium in non-perturbed ecosystems, this ratio could be used to follow the trend of microbial reestablishment. It must be noted that the C_{mic}:C_{org} equilibrium is influenced by macroclimate (Insam et al. 1989).

Enzymatic activities

The measurement of enzymatic activities has interesting characteristics because they show extraordinary specificity in catalyzing biological reactions, and thus offers crucial information of potential rates in specific steps of essential elements in biogeochemical cycles. It should be noted that some enzymes, which can and are produced by plants, are mainly obtained from microorganisms in the soil.

The chemical activity of many soil enzymes increases with temperature up to 60 to 70°C, and then decreases with further temperature rise due to thermal denaturation. The complete inactivation of enzymes occurs at 180°C (Skujins 1967, Tabatabai and Bremner 1970). Immediately after forest fires, many authors found lower values of enzymatic activities in the burned soils (Saá et al. 1993, 1998, Carballas et al. 1993, Eivazi and Bayan 1996, Hernández et al. 1997), due to the disappearance of vegetation, decrease in microbial biomass, changes in soil properties (organic matter mainly), and direct thermal denaturation.

By comparing controlled fires with wildfires, Saá et al. (1933) reported that the phosphatase enzyme is negatively affected by burn severity. Serrasolsas and Khanna (1995b) reported decreases in phosphatase activity after heating at 60° and 120°C, with the enzyme being below detectable limits in soil heated at 250°C. But some enzymes can be also negatively affected by changes not directly related with the heating. Lower values of phosphatase activity were observed by Staddon et al. (1998b) 4 yr after a prescribed fire, being related to an increase in soil pH. Moreover, fire typically increases the available phosphorous, which can act as an enzyme (phosphatase) inhibitor (Hernández et al. 1997, Saá et al. 1998). However, decreases of phosphatase in burned soils have been observed without an increase of available phosphorous (Eivazi and Bayan 1996), which, in this case, reflects a direct effect of heat on the enzyme. In contrast, 1 yr after a fire, Palese et al. (2004) found that enzymatic activities were not related to burn severity.

Fungi are an important cellulose-degrading microorganisms in forest soils (Acea and Carballas 1996). Due to the high heat sensitivity of fungi, cellulase activity is generally lower in recently burned soils (Fioretto et al. 2002) than in non burned soils. Moreover, celluloses are one of the most heat-affected organic fractions (Fernández et al. 2001). The enzyme chitinase, after the initial decrease as a consequence of heat-denaturation, could increase above pre-fire levels, because it is mainly excreted by bacteria that rapidly colonize the burned soil (Carballas et al. 1993, Boerner et al. 2000). Some enzymes such as dehydrogenase are intracellular, and will be related with active microbes. Hernández et al. (1997) found significant lower values of dehydrogenase 9 mon after fire, coinciding with low recovery of microbial biomass.

Eivazi and Bayan (1996) found that frequent prescribed burning (annual or every 4 yr) diminishes in the long-term (40 yr) microbial biomass and activity of several enzymes (phosphatase, α and β-glucosidase, arylsulfatase and urease). Similarly, in a study of soil enzyme activity after annual and periodic fires over 4 yr, multiple fires resulted in a decrease in acid phosphatase and β-glucosidase activity and an increase in phenol oxidase activity (Boerner and Brinkman, 2003). Under Mediterranean conditions, lower values in urease, phosphatase and β-glucosidase were observed 6 yr after a wildfire (Zornoza 2006).

Changes in organic matter quality occur frequently because it is dependent on the presence, absence or accumulation of compounds where some enzymes could be excreted or inhibited. Eight years after burning, Yong-Mei et al. (2005) observed increases and decreases in organic matter quality depending on the enzyme considered. Similarly, Ajwa et al. (1999) found that some enzymes were increased and others were decreased by the fire. According to Boerner et al. (2005), some enzymes such as chitinase and acid phosphatase are regulated primarily by microclimate and soil chemical factors, whereas others such as glucosidase and phenol oxidase are more regulated by substrate availability. During post-fire succession, changes in soil organic matter quality may be explained by the pattern of availability of these enzymes related to substrate quality. When the vegetation recovers, and the rates of organic inputs to soil are reestablished, a trend toward pre-fire values is expected. Some results illustrate that burning could influence the enzyme activities of soil not only in the upper layer but also in lower soil layers (Saá et al. 1998, Yong-Mei et al. 2005).

Acknowledgments

The authors thank the Ministry of Education and Science of the Government of Spain for the support given in the co-financed FEDER projects REN2003-08424-C02-01, and CGL2006-11107-C02-01/BOS and the grant received by Gema Bárcenas for Ph.D. studies. The authors also wish to thank the 'Generalitat Valenciana' for funding project GV05/018, and Frances Young for editing the English.

References

Abbott, I. 1984. Changes in the abundance and activity of certain soil and litter fauna in the Jarrah forest of western Australia after a moderate intensity fire. *Australian Journal of Soil Research*, 22: 463-469.

Acea, M.J. and T. Carballas. 1996. Changes in physiological groups of microorganisms in soil following wildfire. *FEMS Microbiology Ecology*, 20: 33-39.

Acea, M.J. and T. Carballas. 1999. Microbial fluctuations after soil heating and organic amendment. *Bioresource Technology*, 67: 65-71.

Acea, M.J., A. Prieto-Fernández, and N. Diz-Cid. 2003. Cyanobacterial inoculation of heated soils: effect on microorganisms of C and N cycles and on chemical composition in soil surface. *Soil Biology & Biochemistry*, 35: 513-524.

Acea, M.J., N. Diz, and A. Prieto-Fernández. 2001. Microbial populations in heated soils inoculated with cyanobacteria. *Biology and Fertility of Soils*, 33: 118-125.

Adedeji, F.O. 1983. Effect of fire on soil microbial activity in Nigerian Southern Guinea savanna. *Revue d'Écologie et de Biologie du Sol*, 20: 483-492.

Ahlgren, I.F. 1974. The effect of fire on soil organisms. pp. 47-72. *In* T.T. Kozlowski and C.E. Ahlgren [eds.]. *Fire and Ecosystems*. Academic Press, New York, New York, USA.

Ahlgren, I.F. and C.E. Ahlgren. 1960. Ecological effects of forest fires. *Botanical Review*, 26: 483-533.

Ahlgren, I.F. and C.E. Ahlgren. 1965. Effects of prescribed burning on soil microorganisms in a Minnesota jack pine forest. *Ecology*, 46: 304-310.

Ajwa, H.A., C.J. Dell, and C.W. Rice. 1999. Changes in enzyme activities and microbial biomass of tallgrass prairie soil as related to burning and nitrogen fertilization. *Soil Biology & Biochemistry*, 31: 769-777.

Alexander, M. 1967. *Introduction to Soil Microbiology*. John Wiley & Sons, New York, New York, USA.

Almendros, G., A. Polo, J.J. Ibañez, and M.C. Lobo. 1984a. Contribución al estudio de los incendios forestales en las características de la materia orgánica del suelo. I. Transformaciones del humus en un bosque de *Pinus pinea* del centro de España. *Revue d'Écologie et de Biologie du Sol*, 21: 7-20.

Almendros, G., A. Polo, M.C. Lobo, and J.J. Ibañez. 1984b. Contribución al estudio de los incendios forestales en las características de la materia orgánica del suelo. II. Transformaciones controladas del humus por ignición en condiciones de laboratorio. *Revue d'Écologie et de Biologie du Sol*, 21: 145-160.

Almendros, G., F.J. González-Vila, and F. Martín. 1990. Fire-induced transformation of soil organic matter from an oak forest: an experimental approach to the effects of fire on humic substances. *Soil Science*, 149: 158-168.

Amaranthus, M. and J.M. Trappe. 1993. Effects of erosion on ecto- and VA-mycorrhizal inoculum potential of soil following forest fire in southwest Oregon. *Plant and Soil*, 150: 41-49.

Andersen, A.N. and W.J. Müller. 2000. Arthropods response to experimental fire regimes in an Australian tropical savannah: ordinal-level analysis. *Australian Ecology*, 25: 199-209.

Anderson, T.H. 2003. Microbial eco-physiological indicators to asses soil quality. *Agriculture Ecosystems & Environment*, 98: 285-293.

Anderson, J.P.E. and K.H. Domsch. 1975. Measurement of bacterial and fungal contributions to respiration of selected agricultural and forest soils. *Canadian Journal of Microbiology*, 21: 314-322.

Anderson, T.H. and K.H. Domsch. 1990. Application of eco-physiological quotients qCO_2 and qD on microbial biomasses from soils of different cropping histories. *Soil Biology & Biochemistry*, 22: 251-255.

Anderson, T.H. and K.H. Domsch. 1993. The metabolic quotient for $CO_2 \cdot qCO_2$ as a specific activity parameter to assess the effects of environmental conditions, such as pH, on the microbial biomass of forest soils. *Soil Biology & Biochemistry*, 25: 393-395.

Andersson, M., A. Michelsen, M. Jensen, and A. Kjøller. 2004. Tropical savannah woodland: effects of experimental fire on soil microorganisms and soil emissions of carbon dioxide. *Soil Biology & Biochemistry*, 36: 849-858.

Atkinson, G. 1984. Erosion damage following bushfires. *Journal of Soil Conservation Service NSW*, 40: 4-9.

Atlas, R.M. and R. Bartha. 2001. *Ecología Microbiana y Microbiología Ambiental*. Pearson Educación, Madrid, Spain.

Bååth, E., Å. Frostegård, T. Pennanen, and H. Fritze. 1995. Microbial community structure and pH response in relation to soil organic matter quality in wood-ash fertilized, clear-cut or burned coniferous forest soils. *Soil Biology & Biochemistry*, 27: 229-240.

Badía, D. and C. Martí. 2003. Effect of simulated fire on organic matter and selected microbiological properties of two contrasting soils. *Arid Land Research and Management*, 17: 55-69.

Bartoli A., R. Gerdol, and G. Massari. 1991. Soil fungi succesion in a Mediterranean 'macchia' after fire. *Revue d'Écologie et de Biologie du Sol,* 28: 387-402.

Basanta, M.R., M. Díaz-Raviña, and S.J. González-Prieto. 2002. Biochemical properties of forest soils as affected by a fire retardant. *Biology and Fertility of Soils,* 5: 377-383.

Bastias, B.A., X. Zhihong, and J.W.G. Cairney. 2006. Influence of long-term repeated prescribed burning on mycelial communities of ectomycorrhizal fungi. *New Phytologist,* 172: 149-158.

Bauhus, J., P.K. Khanna, and R.J. Raison. 1993. The effect of fire on carbon and nitrogen mineralization and nitrification in an Australian forest soil. *Australian Journal of Soil Research,* 31: 621-639.

Bellgard, S.E., R.J. Whelan, and R.M. Muston. 1994. The impact of wildfire on vesicular-arbuscular mycorrhizal fungi and their potencial to influence the re-establishment of post-fire plant communities. *Mycorrhiza,* 4: 139-146.

Birch, H.F. 1958. The effect of soil drying on humus decomposition and nitrogen availability. *Plant and Soil,* 10: 9-31.

Bissett, J. and D. Parkinson. 1980. Long-term effects of fire on the composition and activity of the soil microflora of a subalpine, coniferous forest. *Canadian Journal of Botany,* 58: 1704-1721.

Boerner, R.E.J. and J.A. Brinkman. 2003. Fire frequency and soil enzyme activity in southern Ohio oak–hickory forests. *Applied Soil Ecology,* 23: 137-146

Boerner, R.E.J., K.L.M. Decker, and E.K. Sutherland. 2000. Prescribed burning effects on soil enzyme activity in a southern Ohio hardwood forest: a landscape-scale analysis. *Soil Biology & Biochemistry,* 32: 899-908.

Boerner, R.E.J., J.A. Brinkman, and A. Smith. 2005. Seasonal variations in enzyme activity and organic carbon in soil of a burned and unburned hardwood forest. *Soil Biology & Biochemistry,* 37: 1419-1426.

Bollen, G.J. 1969. The selective effect of heat treatment on the microflora of a greenhouse soil. *Netherlands Journal of Plant Patology,* 75: 157-163.

Bollen, G.J. 1985. Lethal temperatures of soil fungi. pp. 191-193. *In* C.A. Parker, A.D. Rovira, K.G. Moore, and P.T.W. Wong [eds.]. *Ecology and Management of Soil-borne Plant Pathogens.* American Phytopatogens Society Press, St. Paul, Minnesota, USA.

Bowker, M.A., J. Belnap, R. Rosentreter, and B. Graham. 2004. Wildfire-resistant biological soil crust and fire-induced loss of soil stability in a Palouse prairies, USA. *Applied Soil Ecology,* 26: 41-52.

Bradstock, R.A. and T.D. Auld. 1995. Soil temperatures during experimental bushfires in relation to fire intensity: consequences for legume germination and fire management in south-eastern Australia. *Australian Journal of Applied Ecology,* 32: 76-84.

Bradstock, R.A., T.D. Auld, M.E. Ellis, and J.S. Cohn. 1992. Soil temperatures during bushfires in semi-arid, mallee shrublands. *Australian Journal of Ecology,* 17: 433-440.

Brimecombe, M.J., F.A. De Leij, and J.M. Lynch. 2001. The effect of root exudates on rhizosphere microbial populations. pp. 95-140. *In* R. Pinton, Z. Varaninin, and P. Nannipieri [eds.]. *The Rhizosphere: Biochemistry and Organic Substances at the Soil–Plant Interface.* Marcel Dekker, New York, New York, USA.

Buffington, J.D. 1967. Soil arthropod populations of New Jersey pine barrens as affected by fire. *Annals of Entomological Society of America,* 60: 530-535.

Byram, G.M. 1959. Combustion of forest fuels. pp. 61-89. *In* K.P. Davis. [ed.]. *Forest Fire: Control and Use.* McGraw-Hill, New York, USA.

Campbell, G.S., J.D. Jungbauer Jr., W.R. Bidlake, and R.D. Hungerford. 1994. Predicting the effect of temperature on soil thermal conductivity. *Soil Science,* 158: 307-313.

Campbell, R.E., M.B. Baker Jr., P.F. Ffolliott, F.R. Larson, and C.C. Avery. 1977. Wildfire effects on a ponderosa pine ecosystem: An Arizona case study. Research Paper RM-191, U.S. Department of Agriculture, Forest Service, Rocky Mountain Forest and Range Experimental Station, Fort Collins, Colorado, USA.

Carballas, M., M.J. Acea, A. Cabaneiro, C. Trasar, M.C. Villar, M. Díaz-Raviña, I. Fernández, A. Prieto, A. Saa, F.J. Vázquez, R. Zëhner, and T. Carballas. 1993. Organic matter, nitrogen, phosphorus and microbial population evolution in forest humiferous acid soils after wildfires. pp. 379-385. *In* L. Trabaud and R. Prodon [eds.]. Fire in Mediterranean Ecosystems. Report Number 5, Ecosystems Research Series, European Economic Community, Brussels, Belgium.

Cerdà, A., A.C. Imeson, and A. Calvo. 1995. Fire and aspect induced differences on the erodibility and hydrology of soils at La Costera, Valencia, southeast Spain. *Catena,* 24: 289-304.

Certini, G. 2005. Effects of fire on properties of forest soils: a review. *Oecologia,* 143: 1-10.

Chandler, C., Ph. Cheney, Ph. Thomas, L. Trabaud, and D. Williams. 1983. Fire in forestry. Vol. I, pp. 171-202. *In* C. Chandler, Ph. Cheney, Ph. Thomas, L. Trabaud, and D. Williams [eds.]. *Forest Fire Behavior and Effects.* John Wiley and Sons, New York, New York, USA.

Choromanska, U. and T.H. DeLuca. 2001. Prescribed fire alters the impact of wildfire on soil biochemical properties in a ponderosa pine forest. *Soil Science Society of America Journal,* 65: 232-238.

Choromanska U. and T.H. DeLuca. 2002. Microbial activity and nitrogen mineralization in forest mineral soils following heating: evaluation of post-fire effects. *Soil Biology & Biochemistry,* 34: 263-271.

Ciardi, C. 1998. Soil heating to distinguish the contribution of abiotic and extracellular activities to the overall enzyme activity in soil. *Agrochimica,* XLII: 104-117.

Coleman, D.C. and D.A. Crossley. 1996. *Fundamentals of Soil Ecology.* Academic Press, London, UK.

Coleman, T.W. and L.K. Rieske. 2006. Arthropod response to prescription burning at the soil-litter interface in oak-pine forests. *Forest Ecology and Management,* 233: 52-60.

Collett, N.G. 1998. Effects of two short rotation prescribed fires in autumn on surface-active arthropods in sclerophyll eucalypt forest of west-central Victoria. *Forest Ecology and Management,* 107: 253-273.

Cooke, W.B. 1970. Fungi in burned and unburned chaparral soils. *Sydowia Annales Mycologicae Sereies II,* 24: 164-168.

Coyne, M. 2000. *Microbiología del suelo: un enfoque explorativo.* ITP-Paramomfo, Madrid, Spain.

D'Ascoli, R., F.A. Rutigliano, R.A. De Pascale, A. Gentile, and A. Virzo De Santo. 2005. Functional diversity of the microbial community in Mediterranean maquis soils as affected by fires. *International Journal of Wildland Fire,* 14: 355-363.

Dahlberg, A. 2002. Effects of fire on ectomycorrhizal fungi in Fennoscandian boreal forest. *Silva Fennica,* 36: 69-80.

Dahlberg, A., J. Schimmel, A. Taylor, and H. Johannesson. 2001. The legacy of ectomycorrhizal fungal communities as mycorrhizas in boreal, following cutting and prescribed burning ay different intensities. *Biological Conservation*, 100: 151-161.

De la Torre, L., J. Mataix-Solera, J. Navarro, C. Guerrero, R. Zornoza, and M. Herrero. 2002. Evaluación del nivel de degradación de suelos por efecto del fuego en el paraje de la Granadilla. pp. 99-103. *In* The Proceedings of the II Congreso Nacional de Estudiantes de Ciencias Ambientales. Elche, Universidad Miguel Hernández, Elche, Spain.

De Marco, A., A.E. Gentile, C. Arena, and A. Virzo De Santo. 2005. Organic matter, nurtrient content and biological activity in burned and unburned soils of a Mediterranean maquis area of southern Italy. *International Journal of Wildland Fire*, 14: 365-377.

De Román, M. and A.M. De Miguel. 2005. Post-fire, seasonal and annual dynamics of the ectomycorrhizal community in a *Quercus ilex* L. forest over a 3-year period. *Mycorrhiza*, 15: 471-482.

DeBano, L.F. 1981. Water repellent soils: a state-of-the-art. USDA General Technical Report PSW-46, U.S. Department of Agriculture, Forest Service, Pacific Southwest Forest and Range Experimental Station, Berkeley, California, USA.

DeBano, L.F. 2000. Water repellency in soils: a historical overview. *Journal of Hydrology*, 231: 4-32.

DeBano, L.F., G.E. Eberlein, and P.H. Dunn. 1979. Effects of burning on chaparral soils: I. Soil nitrogen. *Soil Science Society of America Journal*, 43: 504-509.

DeBano, L.F., D.G. Neary, and P.F. Ffolliott. 1998. *Fire's Effects on Ecosystems*. John Wiley and Sons Inc., New York, New York, USA.

Deka, H.K. and R.R. Mishra. 1983. The effect of slash burning on soil microflora. *Plant and Soil*, 73: 167-175.

Dhillion, S.S., R.C. Anderson, and A.E. Liberta. 1988. Effect of fire on the mycorrhizal ecology of little bluestern (*Schizachyrium scoparium*). *Canadian Journal Botany*, 66: 706-713.

Díaz-Fierros, F., E. Benito, J.A. Vega, A. Castelao, B. Soto, R. Pérez, and T. Taboada. 1990. Solute loss and soil erosion in burnt soil from Galicia (NW Spain), pp. 103-116. *In* J.G. Goldammer and M.J. Jenkins [eds.]. *Fire in Ecosystem Dynamics. Mediterranean and Northern Perspectives*. SPB Academic Publishers,The Hague, Netherlands.

Díaz-Raviña, M., A. Prieto, M.J. Acea, and T. Carballas. 1992. Fumigation–extraction method to estimate microbial biomass in heated soils. *Soil Biology & Biochemistry*, 24: 259-264.

Díaz-Raviña, M., A. Prieto, and E. Bååth. 1996. Bacterial activity in a forest soil after heating and organic amendments measured by thymidine and leucine incorporation techniques. *Soil Biology & Biochemistry*, 28: 419-426.

Doerr, S.H., R.A. Shakesby, and R.P.D. Walsh. 2000. Soil water repellency: its causes, characteristics and hydro-geomorphological significance. *Earth Sciences Reviews*, 51: 33-65.

Dress, W.J. and R.E.J. Boerner. 2004. Patters of microarthropod abundance in oak-history forest ecosystems in relation to prescribed fire and landscape position. *Pedobiologia*, 48: 1-8.

Dumontet, S., H. Dinel, A. Scopa, A. Mazzatura, and A. Saracino. 1996. Post-fire soil microbial biomass and nutrient content of a pine forest soil from a dunal Mediterranean environment. *Soil Biology & Biochemistry*, 28: 1467-1475.

Dunn, P.H. and L.F. DeBano. 1977. Fire's effect on biological and chemical properties of chaparral soils. pp. 75-84. *In* H.A. Mooney and C.E. Conrad [eds.]. The Proceedings of Symposium on Environmental Conservation: Fire and Fuel Management in Mediterranean Ecosystems. General Technical Report WO-3, U.S. Department of Agriculture, Forest Service, Washington D.C., USA.

Dunn, P.H., L.F. DeBano, and G.E. Eberlein. 1979. Effects of burning on chaparral soils: II Soil microbes and nitrogen mineralization. *Soil Science Society of America Journal,* 43: 509-514.

Dunn, P.H., S.C. Barro, and M. Poth. 1985. Soil moisture affects survival of microorganisms in heated chaparral soil. *Soil Biology & Biochemistry,* 17: 143-148.

Eivazi, F. and M.R. Bayan. 1996. Effects of long-term prescribed burning on the activity of select soil enzymes in an oak-hickory forest. *Canadian Journal of Forest Research,* 26: 1799-1804.

Entry, J.A., N.M. Stark, and H. Loewenstein. 1986. Effect of timber harvesting on microbial biomass fluxes in a northern Rocky Mountain forest soil. *Canadian Journal of Forest Research,* 16: 1076-1081.

Fenn, M.E., M.A. Poth, P.H. Dunn, and S.C. Barro. 1993. Microbial N and biomass, respiration and N mineralization in soils beneath two chaparral species along a fire-induced age gradient. *Soil Biology & Biochemistry,* 25: 457-466.

Fernández, I., A. Cabaneiro, and T. Carballas. 1997. Organic matter changes immediately after a wildfire in an Atlantic forest soil and comparison with laboratory soil heating. *Soil Biology & Biochemistry,* 29: 1-11.

Fernández, I., A. Cabaneiro, and T. Carballas. 1999. Carbon mineralization dynamics in soils after wildfires in two Galician forests. *Soil Biology & Biochemistry,* 31: 1853-1865.

Fernández, I., A. Cabaneiro, and T. Carballas. 2001. Thermal resistance to high temperatures of different organic fractions from soils under pine forest. *Geoderma,* 104: 281-298.

Fioretto, A., S. Papa, M. Aniello, R. Merola, and A. Pellegrino. 2002. Microbial activities in burned and unburned soils in a low shurbland ecosystem. pp. 151-162. *In* L. Trabaud and R. Prodon [eds.]. *Fire and Biological Processes.* Backhuys Publishers, Leiden, The Netherlands.

Fonturbel, M.T., J.A. Vega, P. Cuiñas, and J.M. Gras. 1993. Variaciones cuantitativas de las poblaciones microbianas edáficas en pinares de Galicia incendiados pp. 197-202. *In* Actas del Congreso Forestal Español, Tomo III. Lourizán, Pontevedra, Spain.

Fonturbel, M.T., J.A. Vega, S. Bara, and I. Bernardez. 1995. Influence of prescribed burning of pine stands in NW Spain on soil microorganisms. *European Journal of Soil Biology,* 31(1): 13-20.

Ford, P.L. and G.V. Johnson. 2006. Effects of dormant- vs. growing-season fire in shortgrass steppe: Biological soil crust and perennial grass responses. *Journal of Arid Environments,* 67: 1-14.

Fritze, H., T. Pennanen, and J. Pietikäinen. 1993. Recovery of soil microbial biomass and activity from prescribed burning. *Canadian Journal of Forest Research,* 23: 1286-1290.

Fritze, H., T. Pennanen, and V. Kitunen. 1998. Characterization of dissolved organic carbon from burned humus and its effects on microbial activity and community structure. *Soil Biology & Biochemistry,* 30: 687-693.

González-Pérez, J.A., F.J. González-Vila, G. Almendros, and H. Knicker. 2004. The effect of fire on soil organic matter – a review. *Environmental International,* 30: 855-870.

Grasso, G.M., G. Ripabelli, M.L. Sammarco, and S. Mazzzoleni. 1996. Effects of heating on the microbial populations of a grassland soil. *International Journal of Wildland Fire,* 6: 67-70.

Grogan, P., J. Baar, and T.D. Bruns, 2000. Below-ground ectomycorrhizal community structure in a recently burned bishop pine forest. *Journal of Ecology,* 88: 1051-1062.

Guerrero, C. 2003. Uso de diferentes residuos orgánicos en la restauración de suelos forestales quemados. Ph.D. thesis, Universidad Miguel Hernández, Elche, Spain.

Guerrero, C., J. Mataix-Solera, J. Navarro-Pedreño, J. Mataix-Beneyto, and I. Gómez. 2002. Evolution of indices of soil quality in a chronosequence of semiarid Mediterranean burned soils: influenced by climatic conditions or age of fire? pp. 1-11. *In* D.X. Viegas [ed.]. *Forest Fire Research & Wildland Fire Safety.* Millpress, Rotterdam, The Netherlands.

Guerrero, C., J. Mataix-Solera, I. Gómez, F. García-Orenes, and M.M. Jordán. 2005. Microbial recolonization and chemical changes in a soil heated at different temperatures. *International Journal of Wildland Fire,* 14: 385-400.

Gundale, M.J., T.H. DeLuca, C.E. Fiedler, P.W. Ramsey, M.G. Harrington, and J.E. Gannon. 2005. Restoration treatments in a Montana ponderosa pine forest: Effects on soil physical, chemical and biological properties. *Forest Ecology and Management,* 213: 25-38.

Haimi, J., H. Fritze, and P. Moilanen. 2000. Response of soil decomposer animals to wood-ash fertilization and burning in a coniferous forest stand. *Forest Ecology and Management,* 129: 53-61.

Hart, S.C., T.H. DeLuca, G.S. Newman, M. Derek Mackenzie, and S.I. Boyle. 2005. Post-fire vegetative dynamics as drivers of microbial community structure and function in forest soils. *Forest Ecology and Management,* 220: 166-184.

Hartford, R.A. and W.H. Frandsen. 1992. When it's hot, it's hot - or maybe it's not (surface flaming may not portend extensive soil heating). *International Journal of Wildland Fire,* 2: 139-144.

Hernández, T., C. García, and I. Reinhardt. 1997. Short-term effect of wildfire on the chemical, biochemical and microbiological properties of Mediterranean pine forest soils. *Biology and Fertility of Soils,* 25: 109-116.

Horton, T.R., E. Cázares, and T.D. Bruns. 1998. Ectomycorrhizal, vesicular-arbuscular and dark septate fungal colonization of bishop pine (*Pinus muricata*) seedlings in the first 5 months of growth after wildfire. *Mycorrhiza,* 8: 11-18.

Insam, H. and K.H. Domsch. 1988. Relationship between soil organic carbon and microbial biomass on chronosequences of reclamation sites. *Microbial Ecology,* 15: 177-188.

Insam, H. and K. Haselwandter. 1989. Metabolic quotient of the soil microflora in relation to plant succession. *Oecologia,* 79: 174-178.

Insam, H., D. Parkinson, and K.H. Domsch. 1989. Influence of macroclimate on soil microbial biomass. *Soil Biology & Biochemistry,* 21: 211-221.

Jalaluddin, M. 1969. Micro-organisms colonization of forest soil after burning. *Plant and Soil,* 30: 150-152.

Jokinen, H.K., O. Kiikkilä, and H. Fritze H. 2006. Exploring the mechanisms behind elevated microbial activity after wood ash application. *Soil Biology & Biochemistry,* 38: 2285-2291.

Jonsson, L., A. Dahlberg, M.C. Nilsson, O. Zarickrisson, and O. Karen. 1999. Ectomycorrhizal fungal communities in late-successional Swedish boreal forests and composition following wildfire. *Molecular Ecology*, 8: 205-217.

Jorgensen, J.R. and C.S. Hodges Jr. 1970. Microbial characteristics of a forest soil after twenty years of prescribed burning. *Mycologia*, 62: 721-726.

Jury, W.A., W.R. Gardner, and W.H. Gardner. 1991. *Soil Physics*. John Wiley and Sons, New York, New York, USA.

Kirschbaum, M.U.F. 1995. The temperature dependence of soil organic matter decomposition and the effect of global warming on soil organic C storage. *Soil Biology and Biochemistry*, 27(6): 753-760.

Klopatek, C.C. and J.M. Klopatek. 1987. Mycorrhizae, microbes, and nutrient cycling processes in pinyon-juniper systems. pp. 360-367. *In* R.L. Everett [ed.]. Proceedings of the Pinyon-Juniper Conference, General Technical Report INT-215, U.S. Department of Agriculture, Forest Service, Intermountain Research Station, Ogden, Utah, USA.

Klopatek, C.C., L.F. DeBano, and J.M. Klopaek. 1988. Effects of simulated fire on vesicular-arbuscular mycorrhizae in pinyon-juniper woodland soil. *Plant and Soil*, 109(2): 245-249.

Knorr, M.A., R.E.J. Boerner, and M.C. Rilling. 2003. Glomalin content of forest soils in relation to fire frequency and landscape position. *Mycorrhiza*, 13: 205-210.

Labeda, D.P., D.L. Balkwill, and L.E.J. Casida. 1975. Soil sterilization effects on in situ indigenous microbial cells in soil. *Canadian Journal of Microbiology*, 21: 263-269.

Liu, X., W.C. Lindemann, W.G. Whitford, and R.L. Steiner. 2000. Microbial diversity and activity of disturbed soil in the northen Chihuahuan desert. *Biology and Fertility of Soils*, 32: 243-249.

López-Hernández, D., S. Santaella, and P. Chacon. 2006. Contribution of nitrogen-fixing organisms to the N budget in *Trachypogon* savannas. *European Journal of Soil Biology*, 42: 43-50.

Lussenhop, J. 1976. Soil arthropod response to prairie burning. *Ecology*, 57: 88-98.

Lyon, L.J., E.S. Telfer, and D.S. Schreinder. 2000a. Direct effects of fire effects and animal responses. pp. 17-24. *In* J.K. Smith [ed.]. Wildland Fire in Ecosystems: Effects of Fire on Fauna. General Technical Report RMRS-GTR-42-vol. 1. U.S. Department of Agriculture, Forest Service, Rocky Mountain Research Station, Ogden, Utah, USA.

Lyon, L.J., M.H. Huff, E.S. Telfer D.S. Schreinder, and J.K. Smith. 2000b. Fire effects on animal populations. pp. 25-34. *In* J.K. Smith [ed.]. Wildland Fire in Ecosystems: Effects of Fire on Fauna. General Technical Report RMRS-GTR-42-vol. 1. U.S. Department of Agriculture Forest Service, Rocky Mountain Research Station, Ogden, Utah, USA.

Mabuhay, J.A., N. Nakagoshi, and T. Horikoshi. 2003. Microbial biomass and abundance after forest fire in pine forest in Japan. *Ecological Research*, 18: 431-441.

Mabuhay, J.A., N. Nakagoshi, and Y. Isagi. 2006a. Soil microbial biomass, abundance, and diversity in a Japanese red pine forest: first year after fire. *Journal of Forest Research*, 11: 165-173.

Mabuhay, J.A., Y. Isagi, and N. Nakagoshi. 2006b. Wildfire effects on microbial biomass and diversity in pine forest at three topographic positions. *Ecological Research*, 21: 54-63.

Majer, J.D. 1984. Short term response of soil and litter invertebrates to a cool autumn burn in Jarrah (*Eucalyptus marginata*) Forest of western Australia. *Pedobiologia*, 26: 229-247.

Marqués, M.A. and E. Mora. 1992. The influence of aspect on runoff and soil loss in a Mediterranean burnt forest (Spain). *Catena*, 19: 333-344.

Mataix-Solera, J. 1999. Alteraciones físicas, químicas y biológicas en suelos afectados por incendios forestales. Contribución a su conservación y regeneración. Ph.D. thesis, Universidad de Alicante, Alicante, Spain.

Mataix-Solera, J., J. Navarro-Pedreño, C. Guerrero, I. Gómez, B. Marco, and J. Mataix. 2002. Effects of an experimental fire on soil microbial populations in a Mediterranean environment. pp. 1607-1614. *In* J.L. Rubio, R.P.C. Morgan, S. Asins, and V. Andreu [eds.]. *Man and Soil at the Third Millennium*. Geoforma Ediciones, Logroño, Spain.

Mataix-Solera, J., E. Lloret, R. Zornoza, C. Guerrero, F. García-Orenes, J. Mataix-Beneyto, J. Navarro, and I. Gómez. 2006. Soil organic carbon, microbial biomass and basal respiration 11 years after a forest fire in a Mediterranean environment. *Geophysical Research Abstracts*, 8: 01249.

Matlack, G.R. 2001. Factors determining the distribution of soil nematodes in a commercial forest landscape. *Forest Ecology and Management*, 146: 129-143.

McSorley, R. 1993. Short-term effects of fire on the nematode community in a pine forest. *Pedobiologia*, 37: 39-48.

Meiklejohn, J. 1955. The effect of bush burning on the microflora of a Kenya upland soil. *Journal of Soil Science*, 6: 111-118.

Miller, C.M. 1993. Composting as a process based on the control of ecologically selective factors. pp. 515-544. *In* F.B. Metting Jr. [ed.]. *Soil Microbial Ecology*. Marcel Dekker, New York, New York, USA.

Miller, S.L., P. Torres, and T. McLean. 1994. Persistence of basidiospores and sclerotia of mycorrhizal fungi and *Morchella* in soil. *Mycologia*, 86: 89-95.

Miller, S.L., T.M. McLean, N.L. Stanton, and S.E. Williams. 1998. Mycorrhization, survivability of conifer seedlings following natural fire in Grand Teton National Park. *Canadian Journal of Forest Research*, 28: 115-122..

Mitchell, R.G. 1990. Effects of prescribed fire on insect pest. pp. 111-121. *In* J.D. Walstad, S.R. Radosevich, and D.V. Sandberg [eds.]. *Natural and Prescribed Fires in Pacific Northwest Forests*. Oregon State University Press, Corvallis, Oregon, USA.

Monleon, V.J. and K. Cromack Jr. 1996. Long-term effects of prescribed underburning on litter decomposition and nutrient release in ponderosa pine stands in central Oregon. *Forest Ecology and Management*, 81: 143-152.

Moretti, M., M.K. Obrist, and P. Duelli. 2004. Arthropods biodiversity after forest fires: Winners and losers in the winter fire regime of the southern Alps. *Ecography*, 27: 173-186.

Myers, P.E. and J.S. Davis. 2003. Recolonization of soils by algae in a north central Florida pine forest after controlled fire and soil sterilization. *Nova Hedwig*, 76: 207-219.

Naeem, S., L.J. Thompson, S.P. Lawler, J.H. Lawton, and R.M. Woodfin. 1994. Declining in biodiversity can alter performance of ecosystems. *Nature*, 268: 734-737.

Nannipieri, P. and L. Badalucco. 2003. Biological processes. *In* D.K. Bembi and R. Nieder [eds.]. *Processes in the Soil-Plant System: Modelling Concepts and Applications*. The Haworth Press, Binghamton, New York, New York, USA.

Nannipieri, P., J. Ascher, M.T. Ceccherini, L. Landi, G. Pietramellara, and G. Renella. 2003. Microbial diversity and soil functions. *European Journal of Soil Science*, 54: 655-670.

Neary, D.G., C.C. Klopatek, L.F. DeBano, and P.F. Ffolliott. 1999. Fire effects on belowground sustainability: a review and synthesis. *Forest Ecology and Management*, 122: 51-71.

New, K.C. and J.L. Hanula. 1998. Effects of time elapsed after prescribed burning in longleaf pine stands on potential prey of the red-cockaded woodpecker. *Southern Journal of Applied Forestry*, 22: 175-183.

Odum, E.P. 1985. Trends expected in stressed ecosystems. *BioScience*, 35: 419-422.

Odum, E.P. 1969. The strategy of ecosystem development. *Science*, 164: 262-270.

Ojima, D.S., D.S. Schimel, W.J. Parton, and C.E. Owensby. 1994. Long- and short-term effects of fire on nitrogen cycling in tallgrass praire. *Biogeochemistry*, 24: 67-84.

Palese, A.M., G. Giovannini, S. Lucchesi, S. Dumontet, and P. Perucci. 2004. Effect of fire on soil C, N and microbial biomass. *Agronomie*, 24: 47-53.

Panzer, R. 1988. Managing prairie remnants for insect conservation. *Natural Areas Journal*, 8: 83-90.

Paquin, P. and D. Coderre. 1997. Deforestation and fire impact on edaphic insect larvae and other macroarthropods. *Environmental Entomology*, 26: 21-30.

Pattinson, G.S., K.A. Hammill, B.G. Sutton, and P.A. McGee. 1999. Simulated fire reduces the density of arbuscular mycorrhizal fungi at the soil surface. *Mycological Research*, 103: 491-496.

Pèrez, C.A., M.R. Carmona, J.C. Aravena, and J.J. Armesta. 2004. Successional changes in soil nitrogen availability, non-symbiotic nitrogen fixation and carbon/nitrogen ratios in southern chilean forest ecosystems. *Oecologia*, 140: 617-625.

Persiani, A.M., O. Maggi, and G. Castelli. 2002. Biodiversity and composition of post-fire soil microfungal communities of a Mediterranean maquis (Southern Italy). pp. 163-171. *In* L. Trabaud and R. Prodon [eds.]. *Fire and Biological Processes*. Backhuys Publishers, Leiden, The Netherlands.

Pianka, E.R. 1970. On r- and k-selection. *American Naturalist*, 104: 592-597.

Picone, L.I., G. Quaglia, F.O. García, and P. Laterra. 2003. Biological and chemical response of a grassland soil to burning. *Journal of Range Management*, 56: 291-297.

Pietikäinen, J. and H. Fritze. 1993. Microbial biomass and activity in the humus layer following burning: short-term effects of two different fires. *Canadian Journal of Forest Research*, 23: 1275-1285.

Pietikäinen, J. and H. Fritze. 1995. Clear-cutting and prescribed burning in coniferous forest: comparison of effects on soil fungal and total microbial biomass, respiration activity and nitrification. *Soil Biology & Biochemistry*, 27: 101-109.

Pietikäinen, J., R. Hiukka, and H. Fritze. 2000. Does short-term heating of forest humus change its properties as a substrate for microbes? *Soil Biology & Biochemistry*, 32: 277-288.

Powlson, D.S. and D.S. Jenkinson. 1976. The effects of biocidal treatments on metabolism in soil–II. Gamma irradiation, autoclaving, air-drying and fumigation. *Soil Biology & Biochemistry*, 8: 179-188.

Prieto-Fernández, A., M.J. Acea and T. Carballas. 1998. Soil microbial and extractable C and N after wildfire. *Biology and Fertility of Soils*, 27: 132-142.

Pritchett, W.L. and R.F. Fisher. 1987. *Properties and Management of Forest Soils. 2nd Edition.* Wiley, New York, New York, USA.

Rab, M.A. 1996. Soil physical and hydrological properties following logging and slash burning in the *Eucalyptus regnans* forest of southeastern Australia. *Forest Ecology and Management*, 84: 159-175.

Raison, R.J. 1979. Modifications of the soil environment by vegetation fires, with particular reference to nitrogen transformations: a review. *Plant and Soil*, 51: 73-108.

Rashid, A., T. Ahmed, N. Ayub, and A.G. Khan. 1997. Effect of forest fire on number, viability and post-fire re-establishment of arbuscular mycorrhizae. *Mycorrhiza*, 7: 217-220.

Rundel, P.W. 1983. Impact of fire on nutrient cycles in Mediterranean-type ecosystems with reference to chaparral. pp. 192-2207. *In* F.J. Kruger, D.T. Mitchell, and J.U.M. Jarvis [eds.]. *Mediterranean-type Ecosystems: The Role of Nutrients.* Springer-Verlag, New York, New York, USA.

Rutigliano, F.A., R. D'Ascoli, A. De Marco, and A.V. De Santo. 2002. Soil microbial community as influenced by experimental fires of different intensities. pp. 137-149. *In* L. Trabaud and R. Prodon [eds.]. *Fire and Biological Processes.* Backhuys Publishers, Leiden, The Netherlands.

Saá, A., M.C. Trasar-Cepeda, F. Gil-Sotres, and T. Carballas. 1993. Changes in soil phosphorus and acid phosphatase activity immediately following forest fires. *Soil Biology & Biochemistry*, 25: 1223-1230.

Saá, A., M.C. Trasar-Cepeda, and T. Carballas. 1998. Soil P status and phosphomonoesterase activity of recently burnt and unburnt soil following laboratory incubation. *Soil Biology & Biochemistry*, 30: 419-428.

Sakamoto, K. and Y. Oba. 1994. Effect of fungal to bacterial ratio on the relationship between CO_2 evolution and total soil microbial biomass. *Biology and Fertility of Soils*, 17: 39-44.

Serra, A., E. Mateos, X. Parra, and V. Sarlé. 1992. Estudio de los efectos de un incendio forestal sobre poblaciones de artrópodos edáficos. *Historia Animalium*, 1: 41-62.

Serrasolsas, I. and P.K. Khanna. 1995a. Changes in heated and autoclaved forest soils of S.E. Australia. I. Carbon and nitrogen. *Biogeochemistry*, 29: 3-24.

Serrasolsas, I. and P.K. Khanna. 1995b. Changes in heated and autoclaved forest soils of S.E. Australia. II. Phosphorus and phosphatase activity. *Biogeochemistry*, 29: 25-41.

Sgardelis, S.P. and N.S. Magaris. 1993. Effects of fire on soil microarthropods of a phryganic ecosystem. *Pedobiologia*, 37: 83-94.

Shakesby, R.A. and S.H. Doerr. 2006. Wildfire as a hydrological and geomorphological agent. *Earth-Science Reviews*, 74: 269-307.

Siemann, E., J. Haarstad, and D. Tilman. 1997. Short-term and log-term effects of burning on oak savanna arthropods. *American Midland Naturalist*, 137: 349-361.

Skujinš, J.J. 1967. Enzymes in soil. pp. 371-414. *In* A.D. McLaren and G.H. Peterson [eds.]. *Soil Biochemistry Vol. 1.* Marcel Dekker, New York, New York, USA.

Smith, J.E., D. McKay, G. Brenner, J. McIvers, and J.W. Spatafora. 2005. Early impacts of forest restoration treatments on the ectomycorrhizal fungal community and fine root biomass in a mixed conifer forest. *Journal of Applied Ecology*, 42: 526-535.

Springett, J.A. 1976. The effects of prescribed burning on the soil fauna and on litter decomposition in Western Australian forest. *Australian Journal of Ecology*, 1: 77-99.

Staddon, W.J., L.C. Duchesne, and J.T. Trevors. 1998a. Impact of clear-cutting and prescribed burning on microbial diversity and community structure in a Jack pine *Pinus banksiana* Lamb. clear-cut using Biolog Gram-negative microplates. *World Journal of Microbiology and Biotechnology,* 14: 119-123.

Staddon, W.J., L.C. Duchesne, and J.T. Trevors. 1998b. Acid phosphatase, alkaline phosphatase and arylsulfatase activities in soils from a jack pine (*Pinus banksiana* Lamb.) ecosystem after clear-cutting, prescribed burning, and scarification. *Biology and Fertility of Soils,* 27: 1-4.

Stendell, E.R., T.R. Horton, and T.D. Bruns. 1999. Early effects of prescribed fire on the structure of the ectomycorrhizal fungus community in a Sierra Nevada ponderosa pine forest. *Mycological Research,* 103: 1837-1359.

Swengel, A.B. 2001. A literature review of insect responses to fire, compared to other conservation managements of open habitat. *Biodiversity Conservation,* 10: 1141-1169.

Tabatabai, M. and Bremner, J.M. 1970. Arylsulfatase activity of soils. *Soil Science Society of America Proceedings,* 34: 225-229.

Tateishi, T. and T. Horikoshi. 1995. Microbial biomass in the soils of burned and unburned Japanese red pine forest in the Setouchi district, western Japan. *Bulletin of Japanese Society of Microbial Ecology,* 10: 9-20.

Tateishi, T., T. Horikoshi, H. Tsubota, and F. Takahashi. 1989a. Application of the chloroform fumigation-incubation method to the estimation of soil microbial biomass in burned and unburned Japanese red pine forest. *FEMS Microbiology Ecology,* 62: 163-172.

Theodorou, C. and G.D. Bowen. 1982. Effects of a bushfire on the microbiology of a South Australian low open dry sclerophyll forest soil. *Australian Forest Research,* 12: 317-327.

Tiwari, V.K. and B. Rai. 1977. Effect of soil burning on microfungi. *Plant and Soil,* 47: 693-697.

Torres, P. and M. Honrubia. 1997. Changes and effects of a natural fire on ectomycorrhizal inoculum potential of soil in a *Pinus halepensis* forest. *Forest Ecology and Management,* 96: 189-196.

Úbeda, X., M. Lorca, and S. Bernia. 2002. Changes in some soil parameters in the topsoil after a prescribed burning. *In* D.X. Viegas [ed.]. *Forest Fire Research & Wildland Fire Safety.* Millpress, Rotterdam, The Netherlands.

Ulery, A.L., R.C. Graham, and C. Amrhein. 1993. Wood-ash composition and soil pH following intense burning. *Soil Science,* 156: 358-364.

Valette, J.C., V. Gomendy, J. Maréchal, C. Houssard, and D. Gillon. 1994. Heat transfer in the soil during very low-intensity experimental fires: the role of duff and soil moisture content. *International Journal of Wildland Fire,* 4: 225-237.

Vázquez, F.J., M.C. Acea, and T. Carballas. 1993. Soil microbial populations after wildfire. *FEMS Microbiology Ecology,* 13: 93-104.

Vilariño, A. and J. Arines. 1991. Numbers and viability of vesicular-arbuscular fungal propagules in field soil samples after wildfire. *Soil Biology & Biochemistry,* 23: 1083-1087.

Visser, S. 1995. Ectomycorrhizal succession in jack pine stands following wildfire. *New Phytologist,* 129: 389-401.

Wallwork, J.A. 1970. *Ecology of Soil Animals.* McGraw-Hill, Maidenhead, Berkshire, UK.

Warcup, J.H. and K.F. Baker. 1963. Occurrence of dormant ascospores in soil. *Nature,* 197: 1317-1318.

Wardle, D.A. and A. Ghani. 1995. A critique of the microbial metabolic quotient (qCO_2) as a bioindicator of disturbance and ecosystem development. *Soil Biology & Biochemistry,* 27: 1601-1610.

Wanner, M. and W.E.R. Xylander. 2003. Transient fires useful for habitat-management do not affect to soil microfauna (testate amoebae) – a study on an active military training area in eastern Germany. *Ecological Engineering,* 20: 113-119.

Wellington, E.M.H. and I.K. Toth. 1994. Actinomycetes. pp. 269-290. *In* R.W. Waver, S. Angle, P. Bottomley, D. Bezdicek, S. Smith, A. Tabatabai, and A. Wollum [eds.]. *Methods of Soil Analysis, Part 2. Microbiological and Biochemical Properties.* Soil Science Society of America, Madison, Wisconsin, USA.

Wells, C.G., R.E. Campbell, L.F. DeBano, C.E. Lewis, R.L. Fredicksen, E.C. Franklin, R.C. Froelich, and P.H. Dunn. 1979. Effects of fire on soil: A state-of-knowledge review. General Technical Report WO-7, U.S. Department of Agriculture, Forest Service, Washington D.C., USA.

White, C.S. 1986. Effects of prescribed fire on rates of decomposition and nitrogen mineralization in a ponderosa pine ecosystem. *Biology and Fertility of Soils,* 2: 87-95.

Widden, P. and D. Parkinson. 1975. The effects of a forest fire on soil microfungi. *Soil Biology & Biochemistry,* 7: 125-138.

Wikars, L.O. and J. Schimmel. 2001. Immediate effects of fire severity on soils invertebrates in cut and uncut pine forests. *Forest Ecology and Management,* 141: 189-200.

Wolf, D.C. and H.D. Skipper. 1994. Soil sterilization. pp. 41-50. *In* R.W. Weaver, S. Angle, P. Bottomley, D. Bezdicek, S. Smith, A. Tabatabai, and A. Wollum [eds.]. *Methods of Soil Analysis, Part 2.* Soil Science Society of America, Madison, Wisconsin, USA.

Wright, E. and W.B. Bollen. 1961. Microflora of Douglas-fir soil. *Ecology,* 42: 825-828.

Wüthrich, C.D. Schaub, M. Weber, P. Marxer, and M. Conedera. 2002. Soil respiration and soil microbial biomass after fire in a sweet chestnut forest in southern Switzerland. *Catena,* 48: 201-215.

Yeager, C.M., D.E. Northup, C.C. Grow, S.M. Barns, and C.R. Kuske. 2005. Changes in nitrogen-fixing and ammonia-oxidizing bacterial communities in soil of conifer after wildfire. *Applied and Environmental Microbiology,* 71: 2713-2722.

Yong-Mei, Z.W. Ning, Z. Guo-Yi, and B. Wei-Kai. 2005. Changes in enzyme activities of spruce (*Picea balfouriana*) forest soil as related to burning in the eastern Qinghai-Tibetan Plateau. *Applied Soil Ecology,* 30: 215-225.

Zornoza R. 2006. Evaluación de la calidad ambiental en suelos de la provincia de Alicante: desarrollo y aplicación de diferentes técnicas. Ph.D. thesis, Universidad Miguel Hernández, Elche, Spain.

6

Soil Erosion after Forest Fire

David F. Scott[1]*, *Michael P. Curran*[2], *Peter R. Robichaud*[3] and *Joseph W. Wagenbrenner*[3]

Abstract

Soil erosion is a three-part process consisting of detachment, transport, and deposition of soil particles. All of these processes are controlled by erosive energy and soil properties, which can be either directly or indirectly affected by wildfire. These post-fire changes often lead to erosion rates that are several orders of magnitude greater than in forested environments which have adequate vegetation and forest floor material protecting the mineral soil. Concentrated overland flow or rill flow is one of the main hillslope erosion processes after wildfires. Since the forest floor is often consumed, overland flow velocities increase due to the reduced surface roughness, allowing for greater sediment transport capacity. The result is both increased overland flow and erosion. Erosion of hillslopes, the formation of rills, and the incision of gullies and stream channels produces high concentrations of sediment that can result in sediment-rich flood events such as debris floods and debris flows. However, the erosional response to wildfires is highly variable and reflects the specific site characteristics, such as watershed area, topography, and rainfall characteristics (e.g., intensity and duration) in combination with the severity of the fire and its effects on the forest floor and soil properties. Therefore, the change in the amount of erosion resulting from a wildfire will depend on the soil burn severity including: the proportion of exposed mineral soil and the quality of any remaining forest floor; the degree of soil heating and presence of soil water repellency that may increase runoff; the physiographic factors such as drainage and slope characteristics that may influence erosion energy and concentration times of surface water runoff; and the inherent erosivity of the site, in particular the probability of high

[1] *Watershed Management, University of British Columbia Okanagan, Kelowna, British Columbia, Canada.*

[2] *British Columbia Ministry of Forests and Range, Kootenay Lake Forestry Centre, Nelson, British Columbia, Canada.*

[3] *U.S. Department of Agriculture, Forest Service, Rocky Mountain Research Station, Moscow, Idaho, USA.*

* *Corresponding author: David F Scott, Watershed Management, The University of British Columbia Okanagan Kelowna, British Columbia V1V 1V7 Canada, Tel. 250 807 8755, Fax 250 807 8004, e-mail: david.scott@ubc.ca*

intensity rainstorms in the first years after fire. Some forms of gravity-driven erosion, namely dry ravel and mass failures, are also important in some areas. Wind erosion is less common, though it may be significant in drier environments with erodible soils.

INTRODUCTION

For many years fires have been known to cause increases in flooding and erosion. Early work in this area arose from concern over the destructive effects of flooding and sediment after some chaparral and forest fires in California (Hoyt and Troxell 1934, Colman 1951, Rowe et al. 1951). Extreme flooding and erosion have been recorded in forested landscapes, especially where fire frequency is low (e.g., Brown 1972, Leitch et al. 1983). In other instances, little post-fire increase in erosion was measured (e.g., Colman 1951, Ronan 1986, Prosser and Williams 1998). It is clear from a review of the literature that the degree of erosion following fire in forests varies considerably from place to place and from fire to fire; the environmental setting and fire characteristics make each event unique. In this chapter we explore the factors and circumstances that lead to changes in erosion processes and the general observation of large increases in soil erosion after fires. This provides an explanation of the variability in the post-fire erosion processes and why some fires produce devastating flooding and erosion while others do not. This information should also shed light on the level of erosion risk that a site may face after a fire and provide some insight into effective mitigation practices (covered in detail in Section II–Rehabilitation and Restoration Strategies).

Soil erosion is a three-part process consisting of detachment, transport, and deposition of soil particles. Working from this simple starting point, we first need to consider those factors that determine the soil's vulnerability to detachment and movement, or its erodibility. The erodibility of the site often increases as a result of wildfire. Second, but not of lesser importance, are those factors that determine the amount of energy that is available to detach and move soil particles, the erosive forces or erosivity of the site. The erosive force is provided by the environmental setting of the hillslope – either by rainfall, runoff, wind, or gravity – and these forces are not changed by fire. However, the net effect of some of these forces on the soil can be changed as a result of the consumption of organic matter by fire and soil heating.

FACTORS THAT AFFECT EROSION PROCESSES

Soil Erodibility

Soil erosion is often described as the movement of soil particles via the energy provided by water (rain splash, overland flow), gravity, or wind. All of these processes can be affected by wildfire, and the effects of fire and relative magnitudes of post-fire changes in erosion rates can be considerable. In general terms, all of these types of erosion are affected by post-fire changes to the ground cover that protects the soil and to changes in the soil itself.

Erodibility is the term that describes the inherent vulnerability of a soil to erosion due to soil cohesion and particle mass. Soil cohesion is the ease with which particles become detached from their neighbors, and the lack of cohesion is exemplified by sandy or gravely soils. Erodibility also incorporates the inertia of particles or the ease with which particles are moved and this is largely a function of their mass. Hence, larger particles are less vulnerable to erosion as more energy would be needed to transport these particles. To some extent these two factors work to counter each other because fine-textured soils, having the smallest and lightest particles, are also those that have the highest cohesion. More specifically, the clay content of any soil is crucial in stabilizing a soil as it is likely to increase cohesion within the soil body, and thereby reduce the risk of particles becoming detached from each other.

Organic Matter

Soil aggregation results from the cohesion between soil particles, often by organic material, such as humus, clay and humus complexes, or organic structures such as fungal hyphae and fine roots. Soils frequently are coarser near the surface, with less clay content than deeper in the soil profile, as a result of weathering of mineral particles and leaching of the fine fraction during soil formation, or due to the deposition of coarser materials such as volcanic ash (tephra) or loess on the surface soil. The role of organic matter in aiding aggregation of soils is thus important in reducing the erodibility of surface soils. Another expression for this resistance to erosion is aggregate stability, a measure of a soil's inherent resistance to the breakdown of aggregation.

Geological characteristics may also play a part in soil erodibility through their contribution to soil cohesiveness. Coarse-textured and gravelly soils, such as those typically derived from coarse-grained granitic rocks, are often poorly aggregated and vulnerable to erosion by flowing water. Meanwhile, well-sorted sandy or silty soils are vulnerable to erosion because of their lower cohesion, particularly when exposed to flowing water.

Forest Floor

In addition to aiding aggregation, organic matter can provide protection to the soil surface. Litter and duff protect the soil surface from direct impact by rain drops, thus reducing the amount of energy applied by rainfall (or erosivity), and reducing the amount of rain splash erosion. Also, the organic layers on top of the mineral soil increase the roughness of the surface, thus impeding overland flow. This results in lower transport capacities. Rock fragments also will provide soil protection, reducing the amount of energy applied by rainfall and increasing the roughness for overland flow. An analogous situation occurs in wind erosion, where litter and duff interrupt the boundary layer, add roughness, and reduce wind velocities and carrying capacities. The root systems of living vegetation also add stability to the soil profile, further increasing the resistance to detachment by water, wind, or gravity, and help to hold the soil mantle in place in situations where mass failure might otherwise occur.

Post-fire Changes to Soil Protective Layers

In a natural forest environment, vegetation and forest floor (litter and duff) protect the surface soil, maintaining surface soil erosion at negligible levels. If the vegetation or forest floor is removed and/or the soil is disturbed, the risk of soil erosion is increased. Once the soil is exposed there is a greater risk of erosion from all three energy sources (water, gravity, and wind), and elimination of the protective surface cover can lead to orders-of-magnitude increases in erosion in the first years following fire. When the soil burn severity is either moderate or low, the forest floor is not completely consumed so some soil protection is maintained and erosion levels will not change as much as when the forest floor is completely removed.

It has been found that high soil burn severity fires lead to greater erosion, and that the amount of bare ground is positively related to erosion rates (Johansen et al. 2001, De Luis et al. 2003). A study of erosion after wildfire in the Front Range of Colorado showed that 81 percent of the observed variability in sediment yields was explained by the amount of ground cover (Benavides-Solorio and MacDonald 2001). At the other end of the mineral soil exposure spectrum, an assessment of long-term erosion in boreal forests of northeast Canada concluded that there was no association with fire because the substantial forest floor in these cooler, northern forests typically is not fully consumed in fires (Carcaillet et al. 2006).

Post-fire Physical and Chemical Changes to the Soil

The passage of a wildfire is clearly not going to change the particle size distribution of a soil. However, the aggregation of soils that is produced by organic matter may be affected if soil temperatures are high enough to combust or transform the organic matter (DeBano 1981, Andreu et al. 2001, Kunze and Stednick 2006). The result is that these soils lose their aggregation and are consequently more erodible (McNabb and Swanson 1990). Erodibility has seldom been measured directly but it has been deduced from the measured loss of soils off hillslope plots (e.g., DeBano and Rice 1973).

Hosking (1938) showed that substantial losses of organic matter are possible at temperatures below 200°C while most organics are consumed before temperatures reach 300°C. Moody et al. (2005) measured critical shear stress in relation to temperatures and found that soils exposed to temperatures above 275°C were more easily detached. Such temperatures may be produced in soils during wildfires once the insulating layer of litter and duff is consumed and soil water has been driven off. In other words, where the forest floor is combusted and the soils are dry, the organic matter that aids soil aggregation is likely to be consumed during the fire (Ryan 2002), leaving the soil less aggregated and more erodible. But this only occurs over the depth to which soil is heated sufficiently for combustion of the organics to occur. The fire-induced temperatures in a soil profile decrease quickly with depth because of the low thermal conductivity of dry soil (Van Wijk and De Vries, 1963). Thus the loss of aggregation is typically confined to the surface soils, usually less than the top 50 mm, though this depth is dependent on the degree and duration of soil heating (DeBano et al. 1977, Certini 2005, Shakesby and Doerr 2006).

In the field, evidence of such heating is provided by the ashed or blackened soil where organics have been oxidized or charred (see Chapter 4 for a complete discussion). The severely burned soils themselves often have a powdery feel, with ash and mineral soil being mixed and lacking aggregation and in severe cases developing a reddish or orange color due to the oxidation of iron minerals. Beneath these charred soils, the temperatures during the fire were lower, and the soils have a normal color and degree of aggregation. In these soils, however, fire may cause changes in the water repellency (Doerr et al. 2006), which can adversely affect erosion by water.

Water repellency is often found in unburned forest soils as a result of leaching of organic compounds from the vegetative litter into the mineral soil (DeBano 2000). In many environments fire will change the water repellent layer of the soil. It has been shown that when litter is consumed at a sufficiently high temperature, the water repellent layer will be affected, usually moving slightly deeper (20 to 50 mm) into the soil profile, and often becoming more severe so that infiltration capacity decreases and overland flow increases (DeBano 1981; see Chapters 3 and 7 for further discussion).

POST-FIRE EROSION PROCESSES

Water-driven (Fluvial) Erosion

Rainfall and overland flow are the most commonly occurring agents of erosion following wildfires in forests, and therefore this form of erosion has received the most attention in reported studies. Rainfall erosivity expresses the energy available to dislodge and move sediments (Renard et al. 1997). This value is highly variable in both time (from storm to storm and year to year) and space. Although the erosivity of a given rainstorm will not be affected by wildfire, where a wildfire has consumed the forest floor the soil surface will be unprotected from the erosive energy of raindrops, and more energy will be transmitted to the soil surface. Also, increased soil erodibility following wildfires makes soil particles more susceptible to detachment by rain splash (Wells et al. 1979, Terry and Shakesby 1993). This results in the redistribution of mineral soil and ash on the surface. The soil surface is likely to be re-arranged downslope at a micro-scale, and pedestals will typically result if gravels are present (Fig. 1). Numerous authors have shown that water-repellent soils can reduce infiltration capacities and increase the risk of overland flow on burned soils (see reviews by DeBano (1981) and Shakesby and Doerr (2006)). In addition, detached soil particles and ash can block pores and thereby seal the soil surface, further impeding infiltration and increasing the likelihood of overland flow (Andreu et al. 2001). Resultant increases in overland flow greatly increase the erosive power of water. Inter-rill (sheetwash) erosion occurs when water flows downslope without concentrating. Convex slopes that become steeper down slope will increase the energy of flowing water and thereby increase erosion, and when the hillslope has a concave shape, flow concentrates and rill erosion usually occurs more quickly. Water's erosive power increases as the depth increases, therefore rill erosion rates generally are much greater than sheetwash or rainsplash erosion rates. In recent studies, rill and channel erosion has accounted for up to 80 percent of measured post-fire sediment yields (Moody and Martin 2001a). Similarly, given sufficient overland flow volume, gullies will form and often quickly incise (Fig. 2).

Fig. 1 Pedestals formed in coarse surface soils by rain splash erosion on frequently burned sites in southern Sudan where high intensity rain storms are common.

Fig. 2 A square gully created by post-fire surface flows in an area that normally does not experience surface runoff, Kelowna, British Columbia, Canada. Note how the plant roots have been exposed by the erosion. Photo: Don Dobson, Dobson Engineering, Kelowna, British Columbia, Canada.

In an unburned forest, the organic forest floor has a very high porosity and water-holding capacity and, if not saturated, can absorb sudden water inputs and then release the water slowly, allowing steady infiltration into the mineral soil. This infiltration occurs at much lower energy levels than a typical rain storm, since water either maintains contact with the litter until reaching the soil or, if drops form they fall less than a few centimeters from the litter layer to the soil surface. In addition, soils that develop under thick forest floor layers tend to have a porous and friable surface structure with a high infiltration capacity. As a result, overland flow is relatively infrequent in most forested environments. Also, the litter on the forest floor provides surface roughness which reduces the energy of overland flow and thus limits the ability of water to detach or transport soil particles. The litter also creates opportunities for trapping and infiltration of runoff and thereby deposition of eroded material.

When wildfire consumes the overlying vegetation and forest floor layers, there is less interception and more precipitation reaches the (altered) soil surface that may also have reduced infiltration capacity from the physical and chemical changes that fire can cause. Also, since rain can fall directly on the soil surface the net erosive force (erosivity) on the burned soil is greater. Finally, removal of the litter reduces the surface roughness of the overland flow path, which allows the runoff to maintain its full energy and a greater sediment transport capacity than if the litter layer were at least partially intact. The result is increases in both overland flow and erosion. There will be a greater relative increase in post-fire runoff and erosion in soils that had a thick overlying forest floor before burning than in soils where the overlying litter layer was thin or absent (Curran et al. 2006).

The eroded materials may be re-deposited where flows are slowed because of reductions in slope or where obstacles to flow are encountered. Such obstacles may be surface litter, stone, surface roughness or cavities in the soil, such as holes where roots have been burned. However, re-entrainment of eroded material commonly occurs if the amount of transported sediment is greater than available storage.

Several studies from around the world have provided quantitative values of post-fire erosion (Table 1). The relative increases in sediment production vary by climate, topography, soil type, burn severity and method and scale of measurement. In these studies maximum measured erosion rates generally ranged from 2 to 20 ton ha^{-1} (Table 1). The larger values were typically measured in mountainous areas with erodible soils and high intensity convective rainfall. In one study of wildfire related erosion in Portugal, Terry and Shakesby (1993) found that, on recently burned sites, rain-splash detachment rates were an order of magnitude and soil losses two orders of magnitude higher than for 'old' or previously burned sites, and both were two orders of magnitude higher than for unburned sites. These erosion effects were associated with water repellent soils.

After the 1998 fires around Yellowstone National Park (Wyoming, USA), suspended sediment yields increased up to 473 percent (Ewing 1996). In the same area, Troendle and Bevenger (1996) measured four times the average annual sediment yield in a burned watershed as compared to an unburned watershed.

The results from hillslope- and watershed-scale studies in the Colorado Front Range suggest that convective precipitation is one of the primary controls on post-fire runoff and erosion rates. Maximum intensity for a 30 minute interval (I_{30}) has been used to explain the peak discharge response of burned watersheds in the Front Range (Moody and Martin 2001b). Similarly, I_{30} explained better than 70 percent of the observed storm runoff and 80 percent of sediment yield variability after a wildfire in Colorado Front Range conifer forest (Kunze and Stednick 2006).

At the watershed scale, fire will likely result in greatly increased sediment yields relative to the pre-fire condition, but in per-unit-area terms sediment yields typically will be less than those measured on the hillslope. The degree to which this occurs will depend on the sediment delivery potential (or more specifically, the delivery ratio) which is the proportion of eroded material that ends up in the channel. The survival of riparian vegetation or litter (natural sediment traps) at the foot of slopes or along streams may greatly reduce the delivery of sediments to streams. Scott and van Wyk (1990) estimated a

Table 1 Some notable examples of fluvial erosion rates measured after forest fires. Unless otherwise noted, the data represent annual erosion rates in high soil burn severity sites in the first year following wildfire.

Location	Forest type; Fire type	Erosion rate (normalized to $t\ ha^{-1}\ yr^{-1}$)	Comments	Reference
Agueda, Portugal	Eucalyptus; Wildfire	176	Burned area plowed for replanting	Shakesby et al. 2002
Bitterroot Valley, Montana, USA	Ponderosa pine; Wildfire	0.2-63	Hillslope plots (100 m^2)	Spiegel and Robichaud 2007
Several locations, Western USA	Coniferous; Wildfire	0.2-24	Catchments (1-13 ha)	Robichaud et al. 2008
Warburton, Victoria, Australia	Mixed eucalypt; Wildfire	22	Sediment from single storm	Leitch et al. 1983
North central Washington, USA	Subalpine fir; Wildfire	20	Hillslope plots (36 m^2); Steep slopes	Robichaud et al. 2006
Colorado Front Range, USA	Ponderosa pine; Wildfires	13	Simulated rainfall 1 m^2 plots	Benavides-Solorio and MacDonald 2002
Colorado Front Range, USA	Ponderosa pine; Wildfires	2-10	Large hillslope plots (<1 ha)	Benavides-Solorio and MacDonald 2005
Colorado Front Range, USA	Ponderosa pine; Wildfires	9.5	Large hillslope plots (<1 ha)	Wagenbrenner et al. 2006
Narrabeen Lagoon, New South Wales, Australia	Dry sclerophyll forest; Wildfire	2.5-8.2	Small hillslope plots (8 m^2)	Blong et al. 1982
La Concordia, Valencia, Spain	Scrubland; Experimental fires	5.1	Hillslope plots (80 m^2)	Campo et al. 2006
Kiewa River, Southeast Australia	Mixed eucalypt/ alpine ash; Wildfire	up to 3.0	Gauged catchments; Up to 50% of in-stream sediment from one large storm	Lane et al. 2006
Colorado Front Range, USA	Ponderosa and lodgepole pine; Wildfire	1.0	Watershed (4 km^2); Suspended sediment from one large storm	Kunze and Stednick 2006

Contd.

Table 1 *Contd.*

Location	Forest type; Fire type	Erosion rate (normalized to t ha^{-1} yr^{-1})	Comments	Reference
Kakadu NP, Australia	Mixed eucalypt; Experimental fires	0.06	Watersheds (~7 km^2); In-stream maximum suspended solids	Townsend and Douglas 2000
Northern New Mexico, USA	Ponderosa pine; Wildfire	0.08 per mm of rain	Rainfall simulation; 25 times more sediment from burned plots than from unburned plots	Johansen et al. 2001
Mt. Carmel, Israel	Pine; Wildfire	0.9-3.7	Hillslope plots (400-500 m^2); 100-500 times more sediment from burned plots than from unburned plots	Inbar et al. 1997
Coastal New South Wales, Australia	Eucalyptus; Wildfire	not measured	Increased sediment measured from hillslope plots, but very small increases in sediment yield at the catchment (3-5 ha) scale.	Prosser and Williams 1998

delivery ratio of around 50 percent in a severely burned plantation watershed where almost no vegetation remained unburned. Prosser and Williams (1998) found that although erosion on hillslopes was greatly increased following wildfires in a eucalypt forest in the vicinity of Sydney, Australia, there were only small increases in sediment yields at the watershed scale because of numerous trapping and storage opportunities on the less than severely burned forest floor.

With an increased occurrence of overland flow and surface wash, the time of concentration is shortened in burned catchments. As a result, channels, gullies and streams carry more water, and the flowing water will have more energy, be more sediment-rich, and be delivered downstream faster than before the fire (Fig. 2). These effects lead to increased stormflow volumes and higher peak flows in burned watersheds as compared to unburned watersheds of similar size. Scouring of gullies and stream channels may occur where stream gradients are sufficient to maintain the sediment transport capacity, and greater sedimentation rates will occur in lower-slope channels, lakes, or reservoirs. As Kunze and Stednick (2006) point out, after wildfire rain storms of a high recurrence frequency (i.e., normally unremarkable precipitation events) can lead to both localized flooding and erosion, with peak storm discharge being an order of magnitude, or more, higher in the first year or two after fire. Some examples of this are Brown (1972) in a burned eucalypt forest in Australia, Helvey (1980) in a Pacific Northwest conifer forest and Scott and van Wyk (1990) and Scott and Schulze (1993) in burned pine and eucalypt plantations, respectively, in South Africa.

Conventional experience would suggest that large-scale erosion and flooding may be considered more likely during the rainy season. However, it is the intensity of any individual storm in the post-fire period, particularly during the first post-fire year, that will determine the erosion event, and these often occur during convective summer storms (Kunze and Stednick 2006, Robichaud et al. 2008). Understanding the risk of large and intense rainfall events and their spatial extent is important in determining the risk of erosion and flooding. Prediction of these precipitation events, which often control post-wildfire erosion, may be difficult because they can be highly localized and vary widely in severity. The assessment of this risk will depend on the quality of precipitation records, which may be sparse in mountainous or remote areas where large wildfires often occur. Even with good records, the current trend towards more variable weather and climate change suggests that more intense rainfall events may be expected over larger areas in the future (however, the predictability of these events is uncertain; Houghton et al. 2001).

The nature of precipitation events plays a role in the influence of soil wettability on a catchment's hydrologic response after wildfire. Water repellency is weakened as soil water content increases (DeBano 1981) and some authors have found a specific threshold wetness above which soils are effectively wettable (MacDonald and Huffman 2004). If initial rains after a wildfire are particularly gentle and prolonged, or if wetting results from snowmelt, then water repellency is likely to be negated as a factor. Therefore, the greatest risk lies with large, intense rain events, typically convective storms that occur when the soils are dry and water repellency is most pronounced. Environments where such storms are typical during or immediately after the dry season will have the greatest risk of post-fire erosion. Generally, little erosion occurs as a result of low intensity rainfall or snow melt (Benavides-Solorio and MacDonald 2005, Wagenbrenner et al. 2006, Robichaud et al. 2008).

The first storms after a wildfire are typically the most erosive because this is when the sites are most vulnerable (Wagenbrenner et al. 2006, Spigel and Robichaud 2007). Eroded material will typically include a large percentage of ash and organic material, much of it possibly charred. However, numerous studies have shown a marked decrease in erosion rates after the first post-fire year; an order of magnitude reduction in sediment yields each year being a 'rule of thumb' that is frequently used but is sometimes misleading (Robichaud et al. 2008). Factors that contribute to this observed reduction in erosion rates over time include: 1) an increase in ground cover due to litter or revegetation which reduces the amount of erosive energy applied to the soil surface, increases rainfall interception and reduces the energy of overland flow; 2) a decrease in soil water repellency and an increase in vegetation which lead to greater infiltration capacity; 3) an armoring of the soil surface after wash caused by the initial events; 4) a reduced supply or availability of sediment; and 5) a reduced erodibility of sediment because of increases in organic matter and its aggregating effects. These changes are time, climate, and vegetation dependent and so time of recovery will vary by the fire's location.

Gully and Channel Erosion and Mass Failure

In gullied terrain and mountainous areas, very large debris floods or flows can result from increased runoff that did not cause much hillslope erosion (as was the case in Fig. 3). The very large increases in energy that result from such increases in stormflows are what drive accelerated channel and bank erosion. Erosion of channels and transport of sediments eroded from higher slopes can be classified as sediment-rich flood events: debris floods occur where concentrations of sediment and debris are very high but flow is water driven, while debris flows are where the flow is dominated by solids rather than water (Fig. 3). Some authors include debris floods and debris flows, which are channel erosion events, as mass failures (Cannon and Reneau 2000) while others (Sidle et al. 2006) classify mass failures as gravity-driven events. The latter are more commonly termed landslides, and involve the movement of a soil body *en masse*. Cannon et al. (submitted to Geological Society of America Bulletin) have developed an empirically based model to predict post-fire debris flows based on an extensive survey of fire-related debris flows in the western United States.

Mass Failures

While the processes of erosion by flowing water are fairly well understood and have been well studied, the mechanisms of mass erosion after wildfire are more complex and are more difficult to predict (Meyers et al. 2001; see discussion in Chapter 2). However, in terms of amounts of material moved these mass failures are particularly important.

The root systems of trees are often cited as a stabilizing network on steep slopes, providing strength to the slope. As living roots often are not burned during wildfires, this strengthening factor is not affected by fire in the short

Fig. 3 A debris flow fan that was associated with wildfire in the mountains directly above Kuskanook, British Columbia, Canada. Approximately 60,000 m^3 was deposited on the existing alluvial fan. Photo: Mike Curran, British Columbia Ministry of Forests and Range, Nelson, British Columbia, Canada.

term. The root systems must first start to decompose, a process that is likely to take numerous years, before their role in stabilizing slopes is lost (e.g., 4 to 15 years is often cited in the Pacific Northwest of North America, Chatwin et al. 1994). However, the large reduction in canopy and forest floor interception and the loss of transpiration will cause soils to be wetter. These additional inputs of water can be considerable, depending on the pre-fire forest type and the rainfall regime. The higher water content will add to the stress on slopes and will serve to reduce shear strength in the soil body by reducing friction and increasing pore water pressure. After very large rain or rain-on-snow events the slopes may fail as a result of the balance of strength and stresses shifting, and it is through this mechanism that fires may increase the risk of mass failures on steep hillslopes. Meyers et al. (2001) recorded numerous slide failures off slopes that were burned eight years earlier in west-central Idaho; the erosion being triggered by intense rains on melting snow. These slides led to debris flows and sediment-charged flooding further downstream.

Gravity-driven Erosion (Dry Ravel)

Dry ravel is a form of gravity-driven soil creep that sometimes occurs in steep terrain. On slopes exceeding the angle of repose for the soil particles, the particles may move downhill under the pull of gravity. The term was first used in reference to post-wildfire erosion in the early descriptions of wildfire effects in California (Wells et al. 1979) and may be widespread in some areas. In fact, the phenomenon has long been recognized and described in steep terrain where surface soils are coarse or gravelly, such as the scarp slopes of alluvial terraces or active gravel pits. During wildfire, slopes are destabilized by the combustion of wood and roots that have become incorporated into the soil surface and helped maintain soil resistance to the force of gravity. Movement occurs until a new, stable position is reached and this typically occurs on gentler slopes below the angle of repose, or until re-vegetation helps to anchor the soil. Dry ravel can be recognized in the field by bare patches of soil and evenly-sloped, small fans of coarser material forming below steep slope sections. Anderson et al. (1959) recorded dry ravel amounting to 0.2 to 4.3 ton ha^{-1}, while Krammes (1960) found that dry ravel can be increased nine-fold in the first post-fire year. This transported material can supply water-driven erosion in subsequent wet seasons. Some risk assessment guidance documents provide a relative rating of dry ravel potential for angular or rounded surface materials (e.g., Curran et al. 2000). See Chapter 19 for discussion of dry ravel in California, USA.

Wind-driven (Eolian) Erosion

Forests generally occur in humid regions where wind is rarely considered a significant erosive force. In some cases there is a window of opportunity for wind-driven erosion immediately following fire when the cohesiveness of surface soils are altered and the soils are exposed. Also, with consumption of the forest litter and vegetation during wildfire there is less protection of the soil surface and therefore more erosive force can be applied to the soil surface. Finally, wind velocities may be higher because of the removal of the drag effects provided by the forest floor and forest and understory canopies. On all high soil burn severity sites, some degree of wind erosion of ash and mineral soil is likely to occur until the soil surface is dampened. Where conditions are generally drier and there is poor soil cohesion wind erosion may be important, as seen in inter-forest dunes in eastern Canada (Carcaillet et al. 2006). Whicker et al. (2006) measured wind erosion and dust plume transport of sediments after the Cerro Grande wildfire in the Los Alamos National Laboratory, New Mexico, USA. In the first year after burning, the horizontal dust flux in the moderate soil burn severity sites (0.95 g m^{-2} d^{-1} at the 0.25 m elevation) was as much as double to that of the unburned site (0.45 g m^{-2} d^{-1} at the 0.25 m elevation), while the dust flux in the high burn severity sites (5.5 g m^{-2} d^{-1}) was 10 times the unburned. These dust concentrations were elevated for at least three years after the fire, and this was attributed to continuing drought conditions.

SUMMARY

Soil erosion processes – by water, gravity, and wind – are all affected by fire. In each case the loss of organic matter reduces the aggregation of the soil and makes it less cohesive. The consumption of the protective forest floor layer makes the soil more exposed to the erosive energy of water (rain drops, overland flow), gravity, and wind. Soil water repellency and sealing of pores in the soil inhibit infiltration leading to increases in overland flow. Soil erosion by gravity and mass failure is less widespread than erosion by water or wind, but the magnitude of erosion from these processes can be significant. The duration of increased risk is shortest with gravity-driven erosion, on the order of weeks or months after a fire, depending on revegetation rates, while it is greatest for mass failure, where the risk may not increase for several post-fire years. The durations of risk for erosion by both water (and wind in some instances) are on the order of several years, with some reduction in erosion rates likely each post-fire year.

The passage of a wildfire does not change the erosive forces *per se*, but it may make the site much more vulnerable to those same erosive forces by reducing the soil protective layers and altering the soil structure. Current strategies in post-fire rehabilitation aim to increase the soil protective cover, and these strategies are discussed in Section II, Rehabilitation and Restoration Strategies.

In summary, the risk of erosion after wildfire will depend on the combination of a number of risk factors. These are

- the soil burn severity,
- the degree of reduced soil cohesion and increased erodibility,
- the proportion of mineral soil that is newly exposed to erosive forces and how quickly this is recovered with litter or revegetation,
- the degree of water repellency in the burned soils,
- the risk of large and intense rain or wind events, particularly in the first year after the fire and while the soils are dry,
- site factors, including soil erodibilty and surface roughness (e.g., litter and vegetation that provide sediment or storage opportunities at the micro-site scale),
- topographic factors at the hillslope scale, including slope gradient, length and evenness, and the drainage pattern (i.e., the degree to which erosive forces are concentrated or dissipated in the channel system).

References

Anderson, H., G. Coleman, and P. Zinke. 1959. Summer slides and winter scour dry-wet erosion in southern California mountains. Technical Paper 36. U.S. Department of Agriculture Forest Service, Pacific Southwest Forest and Range Experiment Station, Albany, California, USA.

Andreu, V., A.C. Imeson, and J.L. Rubio. 2001. Temporal changes in soil aggregates and water erosion after a wildfire in a Mediterranean pine forest. *Catena,* 44: 69-84.

Benavides-Solorio, J.D. and L.H. MacDonald. 2001. Post-fire runoff and erosion from simulated rainfall on small plots, Colorado Front Range. *Hydrological Processes,* 15: 2931-2952.

Benavides-Solorio, J.D. and L.H. MacDonald. 2002. Erratum to 'Post-fire runoff and erosion from simulated rainfall on small plots, Colorado Front Range. *Hydrological Processes,* 16: 1131-1133.

Benavides-Solorio, J.D. and L.H. MacDonald. 2005. Measurement and prediction of post-fire erosion at the hillslope scale, Colorado Front Range. *International Journal of Wildland Fire,* 14: 457-474.

Blong, R., S. Riley, and P. Crozier. 1982. Sediment yield from runoff plots following bushfire near Narrabeen Lagoon, NSW. *Search,* 13: 36-38.

Brown, J. 1972. Hydrologic effects of a bushfire in a catchment in south-eastern New South Wales. *Journal of Hydrology,* 15: 77-96.

Campo, J., V. Andreu, E. Gimeno-Garcia, O. Gonzalez, and J.L. Rubio. 2006. Occurrence of soil erosion after repeated experimental fires in a Mediterranean environment. *Geomorphology,* 82: 376-387.

Cannon, S.H. and S.L. Reneau. 2000. Conditions for generation of fire-related debris flows, Capulin Canyon, New Mexico. *Earth Surface Processes and Landforms,* 25: 1103-1121.

Cannon, S.H., J.E. Gartner, M.G. Rupert, J.A. Michael, A.H. Rea, and C. Parrett. Submitted. Predicting the probability and volume of post-wildfire debris flows in recently burned basins in the inter-mountain western U.S. *Geological Society of America Bulletin.*

Carcaillet, C., P.J.H. Richard, H. Asnong, L. Capece, and Y. Bergeron. 2006. Fire and soil erosion history in East Canadian boreal and temperate forests. *Quaternary Science Reviews,* 25: 1489-1500.

Certini, G. 2005. Effects of fire on properties of forest soils: a review. *Oecologia,* 143: 1-10.

Chatwin, S.C., D.E. Howes, J.W. Schwab, and D.N. Swanston. 1994. A Guide for Management of Landslide-Prone Terrain in the Pacific Northwest – Second Edition. *Land Management Handbook* No. 18. BC Ministry of Forests, Victoria, British Columbia, Canada.

Colman, E. 1951. Fire and water in southern California's mountains. Miscellaneous, Paper number 3. U.S. Department of Agriculture, Forest Service, California Forest and Range Experiment Station, Albany, California, USA.

Curran, M.P., I. Davis, and B. Mitchell. 2000. Silviculture prescription data collection field handbook: interpretive guide for data collection, site stratification, and sensitivity evaluation for silviculture prescriptions. *Land Management Handbook* No. 47. BC Ministry of Forests, Victoria, British Columbia, Canada.

Curran, M.P., B. Chapman, G.D. Hope, and D. Scott. 2006. Large-scale Erosion and Flooding after Wildfires: Understanding the Soil Conditions. Technical Report 030. BC Ministry of Forests and Lands, Research Branch, Victoria, British Columbia, Canada.

De Luis, M., J.C. Gonzalez-Hidalgo, and J. Raventos. 2003. Effects of fire and torrential rainfall on erosion in a Mediterranean gorse community. *Land Degradation and Development,* 14: 203-213.

DeBano, L. 1981. Water repellent soils: a state-of-the-art. General Technical Report PSW-46. U.S. Department of Agriculture, Forest Service, Pacific Southwest Research Station, Berkeley, California, USA.

DeBano, L. 2000. Water repellency in soils, a historical overview. *Journal of Hydrology,* 231-232: 4-32.

DeBano, L. and R. Rice. 1973. Water repellent soils: Their implications in forestry. *Journal of Forestry,* 71: 220-223.

DeBano, L., P. Dunn, and C. Conrad. 1977. Fire's effect on physical and chemical properties of chaparral soils. pp. 65-74. *In* H.A. Mooney and C.E. Conrad [technical eds.]. Proceedings of the Symposium on the Environmental Consequences of Fire and Fuel Management in Mediterranean Ecosystems. General Technical Report WO-3. U.S. Department of Agriculture, Forest Service, Washington, D.C., USA.

Doerr, S.H., R.A. Shakesby, W.H. Blake, G.S. Humphreys, C. Chafer, and P.J. Wallbrink. 2006. Effects of contrasting wildfire severity on soil wettability in Australian eucalypt catchments. *Journal of Hydrology,* 319: 295-311.

Ewing, R. 1996. Post-fire suspended sediment from Yellowstone National Park, Wyoming. *Water Resources Bulletin,* 32: 605-627.

Helvey, J.D. 1980. Effects of a north central Washington wildfire on runoff and sediment production. *Water Resources Bulletin,* 16: 627-634.

Hosking, J. 1938. The ignition at low temperatures of the organic matter in soils. *Journal of Agriculture Science,* 38: 393-400.

Houghton, J.T., Y. Ding, D.J. Griggs, M. Noguer, P.J. van der Linden, X. Dai, K. Maskell, and C.A. Johnson [eds.]. 2001. *Climate Change 2001: The Scientific Basis.* Edited for the Intergovernmental Panel on Climate Change. Cambridge University Press, Cambridge, UK.

Hoyt, W.G. and H.C. Troxell. 1934. Forests and streamflow. *American Society Civil Engineers Transactions,* 99: 1-30.

Inbar, M., L. Wittenberg, and M. Tamir. 1997. Soil erosion and forestry management after wildfire in a Mediterranean woodland, Mt. Carmel, Israel. *International Journal of Wildland Fire,* 7: 285-294.

Johansen, M.P., T.E. Hakonson, and D.D. Breshears. 2001. Post-fire runoff and erosion from rainfall simulation: contrasting forests with shrublands and grasslands. *Hydrological Processes,* 15: 2953-2965.

Krammes, J.S. 1960. Erosion from mountainside slopes after fire in southern California. Research Note PSW-171. U.S. Department of Agriculture, Forest Service, Pacific Southwest Research Station, Berkeley, California, USA.

Kunze, M.D. and J.D. Stednick. 2006. Streamflow and suspended sediment yield following the 2000 Bobcat fire, Colorado. *Hydrological Processes,* 20: 1661-1681.

Lane, P.N.J., G.J. Sheridan, and P.J. Noske. 2006. Changes in sediment loads and discharge from small mountain-catchments following wild-fire in south eastern Australia. *Journal of Hydrology,* 331: 495-510.

Leitch, C., D. Flinn, and R. Van De Graaff. 1983. Erosion and nutrient loss resulting from Ash Wednesday (February 1983) wildfires: a case study. *Australian Forest Research,* 46: 173-180.

MacDonald, L.H. and E.L. Huffman. 2004. Post-fire soil water repellency: Persistence and soil moisture thresholds. *Soil Science Society of America Journal,* 68: 1729-1734.

McNabb, D.H. and F.J. Swanson. 1990. Effects of fire on soil erosion. pp.159–176. *In* J.D.Walstad, S.R. Radosevich and D.V. Sandberg [eds.]. *Natural and Prescribed Fire in Pacific Northwest Forests*. Oregon State University Press: Corvallis, Oregon, USA.

Meyers, G., J. Pierce, S. Wood, and A. Jull. 2001. Fire, storms, and erosional events in the Idaho batholith. *Hydrological Processes,* 15: 3025-3038.

Moody, J.A. and D.A. Martin. 2001a. Initial hydrologic and geologic response following a wildfire in the Colorado Front Range. *Earth Surface Processes and Landforms,* 26: 1049-1070.

Moody, J.A. and D.A. Martin. 2001b. Post-fire, rainfall intensity–peak discharge relations for three mountainous watersheds in the western USA. *Hydrological Processes,* 15: 2981-2993.

Moody, J.A., J.D. Smith, and B.W. Ragan. 2005. Critical shear stress for erosion of cohesive soils subjected to temperatures typical of wildfires. *Journal of Geophysical Research,* 110: 13.

Prosser, I.P. and L. Williams. 1998. The effect of wildfire on runoff and erosion in native Eucalyptus forest. *Hydrological Processes,* 12: 251-265.

Renard, K.G., G.R. Foster, G.A. Weesies, D.K. McCool, and D.C. Yoder. 1997. Predicting soil erosion by water: a guide to conservation planning with the revised universal soil loss equation (RUSLE). Agricultural Handbook No. 703, U.S. Department of Agriculture, Washington, D.C., USA.

Robichaud, P.R., T.E. Lillybridge, and J.W. Wagenbrenner. 2006. Effects of post-fire seeding and fertilizing on hillslope erosion in north-central Washington, USA. *Catena,* 67: 56-67.

Robichaud, P.R., J.W. Wagenbrenner, R.E. Brown, P.M. Wohlgemuth and J.E. Beyers. 2008. Evaluating the effectiveness of contour-felled log erosion barriers as a post-fire runoff and erosion mitigation treatment in the western United States. *International Journal of Wildland Fire,* 17: 255-273.

Ronan, N. 1986. The hydrological effects of fuel reduction burning and wildfire at Wallaby Creek. Report No. MMBW-W-0015. Melbourne and Metropolitan Board of Works, Melbourne, Australia.

Rowe, P.B., H. Storey, and E.L. Hamilton. 1951. Some results of hydrologic research. pp. 19-29. *In* Some aspects of watershed management in southern California watersheds. Miscellaneous Paper 1. U.S. Department of Agriculture, Forest Service, California Forest and Range Experiment Station, Albany, California, USA.

Ryan, K.C. 2002. Dynamic interactions between forest structure and fire behavior in boreal ecosystems. *Silva Fennica,* 36(1): 13-39.

Scott, D.F. and D.B. van Wyk. 1990. The effects of wildfire on soil wettability and hydrological behaviour of an afforested catchment. *Journal of Hydrology,* 121: 239-256.

Scott, D.F. and R.E. Schulze. 1993. The hydrological effects of a wildfire in a eucalypt afforested catchment. *South African Forestry Journal,* 160: 67-74.

Shakesby, R.A. and S.H. Doerr. 2006. Wildfire as a hydrological and geomorphological agent. *Earth-Science Reviews,* 74: 269-307.

Shakesby, R.A., C.O.A. Coelho, A.J.D. Ferreira, and R.P.D. Walsh. 2002. Ground-level changes after wildfire and ploughing in eucalyptus and pine forests. Portugal: Implications for soil microtopographical development and soil longevity. *Land Degradation and Development,* 13: 111-127.

Sidle, R.C., A.D. Ziegler, J.N. Negishi, A.R. Nik, R. Siew, and F. Turkelboom. 2006. Erosion processes in steep terrain – truths, myths, and uncertainties related to forest management in Southeast Asia. *Forest Ecology and Management,* 224: 199-225.

Spigel, K.M. and P.R. Robichaud. 2007. First-year post-fire erosion rates in Bitterroot National Forest, Montana. *Hydrological Processes,* 21: 998-1005.

Terry, J.P. and R.A. Shakesby. 1993. Soil hydrophobicity effects on rainsplash: simulated rainfall and photographic evidence. *Earth Surface Processes and Landforms,* 18: 519-525.

Townsend, S.A. and M.M. Douglas. 2000. The effect of three fire regimes on stream water quality, water yield and export coefficients in a tropical savanna (northern Australia). *Journal of Hydrology,* 229: 118-137.

Troendle, C.A. and G.S. Bevenger. 1996. Effect of fire on streamflow and sediment transport Shoshone National Forest, Wyoming. pp. 43-52. *In* J. Greenlee [ed.]. Proceedings of the Second Biennial Conference on the Greater Yellowstone Ecosystem. The Ecological Implications of Fire in Greater Yellowstone. International Association of Wildland Fire, Fairfiled, Washington, USA.

Van Wijk, W.R. and D.A. De Vries. 1963. Periodic temperature variations in a homogeneous soil. pp.102-143. *In* W.R. Van Wijk [ed.]. *Physics of the Plant Environment.* North Holland Publishing Company, Amsterdam, The Netherlands.

Wagenbrenner, J.W., L.H. MacDonald, and D. Rough. 2006. Effectiveness of three post-fire rehabilitation treatments in the Colorado Front Range. *Hydrological Processes,* 20: 2989-3006.

Wells, C.G., R.E. Campbell, L.F. DeBano, C.E. Lewis, R.L. Frederiksen, E.C. Franklin, R.C. Forelich, and P.H. Dunn. 1979. Effects of fire on soil: a state of knowledge review. General Technical Report WO-7. U.S. Department of Agriculture, Forest Service, Washington, D.C., USA.

Whicker, J.J., J.E. Pinder, III, and D.D. Breshears. 2006. Increased wind erosion from forest wildfire: implications for contaminant-related risks. *Journal of Environmental Quality,* 35: 468-478.

7

Soil Water Repellency: A Key Factor in Post-fire Erosion

Stefan H. Doerr[1]*, *Richard A. Shakesby*[1] and *Lee H. MacDonald*[2]

Abstract

Soil water repellency (hydrophobicity) prevents water from wetting or infiltrating dry soil. This condition has been documented under a wide range of vegetation types and climates and particularly following forest fires. Water repellency is of considerable interest to land managers, hydrologists and soil scientists because i) it can be induced, enhanced or destroyed during burning and ii) its presence can cause a marked reduction in infiltration rate. This reduction in infiltration is commonly presumed to be the primary cause of the increases in runoff and erosion that are often observed at a range of scales following forest fires. The goal of this chapter is to provide a basic understanding of soil water repellency, its measurement, the effects of burning on soil water repellency, and its relative importance in runoff and erosion processes at different scales.

It is widely accepted that water repellency is caused by the presence of organic compounds with hydrophobic properties on soil particle surfaces. During burning such substances in the litter and topsoil can be volatized and condensed in the soil, inducing or intensifying water repellency. During very hot fires, however, these compounds can be destroyed and the soil surface is rendered wettable. In many cases fire increases repellency, which tends to be confined to the top centimeters of the soil, and often is highly variable spatially, temporally and in its degree. It is typically most pronounced under dry conditions and reduced or absent following prolonged wet conditions. The duration and amount of wetting needed to reduce or eliminate soil water repellency, however, varies with soil type, burn severity, and the persistence of soil water repellency prior to wetting.

[1] *Institute of Environmental Sustainability, School of the Environment and Society, Swansea University, Swansea, UK.*

[2] *Department of Forest, Rangeland, and Watershed Stewardship, Colorado State University, Fort Collins, Colorado 80523-1472, USA.*

* *Corresponding author: Stefan Doerr, Institute of Environmental Sustainability, School of the Environment and Society, Swansea University, Singleton Park, Swansea SA2 8PP, UK. Tel: +44 1792 295147, Fax: +44 1792 295955, e-mail: s.doerr@swan.ac.uk*

Burning also induces a series of other changes to soils and the vegetative cover that may be just as, or possibly even more, important in causing the observed increases in runoff and erosion following fire. These factors make it challenging to assess the role of water repellency in post-fire hydrology and erosion processes. This is particularly so at larger scales due to the high spatial variability of many factors involved and the difficulty in characterizing the counteracting role of wettable soil patches, ash, bioturbation, soil cracks, and burned-out tree roots in reducing the surface runoff engendered by strongly water repellent patches.

Research to date has demonstrated that water repellency can strongly affect post-fire runoff and erosion processes and the factors that can enhance or reduce the relative impact of water repellency in burned landscapes have been reasonably well established. Quantifying and predicting the relative contribution role of water repellency in post-fire erosion processes, however, remains a major challenge particularly at larger scales. Additional manipulative experiments and more detailed monitoring are needed to provide a better knowledge base on the extent of the impact of soil water repellency on runoff and erosion in the field under natural rainfall events at different scales.

INTRODUCTION

Soil water repellency refers to the inability of water to wet or infiltrate soil, and this phenomenon has been documented in a wide range of vegetation types and climates (Doerr et al. 2000, Dekker et al. 2005). Soil water repellency is of considerable interest to soil scientists and land managers because of its implications for increasing runoff and erosion. Much of the research on soil water repellency has focused on the effects of wild and prescribed fires, as numerous studies have suggested that soil water repellency is the primary cause of reduced infiltration rates after burning. This reduction in infiltration is commonly presumed to be the primary cause of the observed increases in post-fire runoff and erosion at the plot, hillslope and watershed scales (e.g., Sartz 1953, DeBano et al. 1970, Swanson 1981, Scott and Van Wyk 1990, Inbar et al. 1998, Robichaud et al. 2000, Shakesby and Doerr 2006).

The goal of this chapter is to provide a basic understanding of soil water repellency, the effects of burning on soil water repellency, and the effects of this soil water repellency on overland flow and erosion rates at different scales. The first part of this chapter provides an overview of the origin, occurrence and measurement of soil water repellency. We then summarize the effects of burning on the strength and persistence of soil water repellency in different vegetation types, and highlight some of the difficulties in accurately characterizing soil water repellency. The next section reviews current knowledge with respect to the role of post-fire soil water repellency in

Fig. 1 Water drops resting on a highly repellent organic-rich soil (photo by Erik van den Elsen).

increasing runoff, increasing erosion, and causing land degradation. The understanding provided in the first part of the chapter is used to help explain some of the apparent disparities in the literature. The reader is referred to a series of recent reviews for more detailed information on most of the topics that are covered in this chapter (Neary et al. 1999, 2005, DeBano 2000, Doerr et al. 2000, Shakesby et al. 2000, Letey 2001, Doerr and Moody 2004, DeBano et al. 2005, Shakesby and Doerr 2006).

ORIGIN, OCCURRENCE AND MEASUREMENT OF SOIL WATER REPELLENCY

Origin of Water Repellency

It is commonly assumed that soils wet readily under rainfall or irrigation, but an increasing body of literature indicates that this is often not the case. Many soils behave in a water-repellent manner at low or moderate moisture contents under both burned and unburned conditions (Fig. 2). Schreiner and Schorey (1910) were amongst the first to document soil water repellency, as they described a soil in California that *"could not be wetted, either* [sic] *by man, by rain, irrigation, or the movement of water from the subsoil"*. Another early study showed that soil water repellency reduced the productivity of citrus orchards in Florida (Jamison 1942). Numerous other examples of soil water repellency in unburned areas can be found in the literature (Doerr et al. 2000), and during

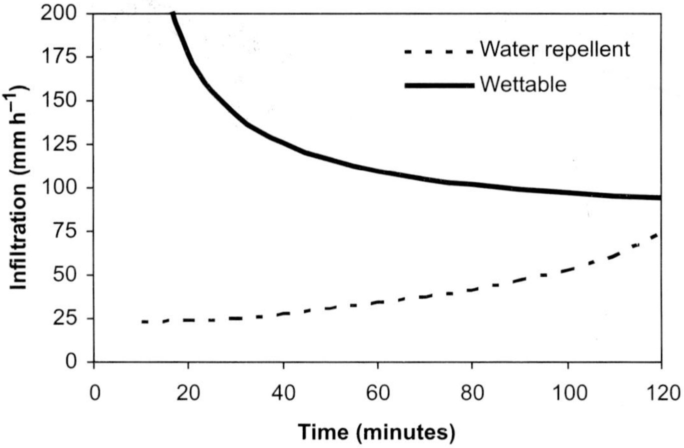

Fig. 2 Infiltration rates into water repellent and wettable soil (modified from Letey et al. 1962).

the 1990s and early 2000s it became evident that soil water repellency is widespread and not restricted to burned areas or a narrow set of other conditions.

Water repellency in unburned soils has been reported from all continents except Antarctica, for climates that range from seasonal tropical to subarctic, for soils that range from coarse- to fine-textured, and for many of land uses, including plowed cropland, grasslands, shrublands, and a wide range of forest types (e.g., Wallis and Horne 1992, Bauters et al. 2000, Doerr et al. 2000, 2006a). In some industries, such as horticulture and turf grass, wetting agents are widely used to increase soil wettability (e.g., Cisar et al. 2000).

The different techniques used to assess soil water repellency measure either the 'strength' or the 'persistence' of soil water repellency (see following section). Both properties can vary from extremely high, such as under eucalyptus plantations in Portugal (Doerr et al. 1998), to only being detectable with a purpose-built micro-infiltrometer as reported for some agricultural soils in Scotland (Hallett and Young 1999). Most scientists working on soil water repellency agree that "*water repellency in soils is the norm rather than the exception, with its degree being highly variable*" (Wallis et al. 1991). In general, soil water repellency is confined to the top centimeters or decimeters of soil where organic molecules with hydrophobic properties are present on the surfaces of soil pores. Water repellency occurs when the hydrophobic 'ends' of these molecules are oriented towards the pore space.

Molecules with hydrophobic properties are ubiquitous in the environment. Plants produce these compounds to protect leaf surfaces from desiccation, and to help repel insects or microbes. These hydrophobic molecules are relatively resistant to physical or chemical degradation, so they are common in vegetated soils and are believed to be the cause of soil water repellency (Doerr et al. 2000). Unfortunately the degree of soil water repellency cannot be predicted reliably from soil organic matter content or the amount of hydrophobic compounds (Doerr et al. 2005a). Hence it is still not completely clear why some soils exhibit soil water repellency and others do not. Specific combinations of organic compounds (Morley et al. 2005) and their inter-molecular arrangement in response to environmental conditions (Roy and McGill 2000) seem to be critical.

A review of the literature indicates that certain soil and vegetation combinations are particularly likely to develop strong soil water repellency. Coarse-textured soils are more susceptible to the development of soil water repellency than finer textured soils, and this is thus generally attributed to the much smaller particle surface area and hence the number of potential adsorption sites for organic molecules. Soils under vegetation types with oil- or wax-rich leaves, such as sclerophyllous shrubs, conifers, and eucalypts, are much more prone to develop strong water repellency than under broad-leaved deciduous forests (Doerr et al. 2000 and references therein). Under unburned conditions this soil water repellency is typically strongest at the surface of the mineral soil and drops off rapidly with depth (e.g., Huffman et al. 2001). Many of the vegetation and soil types that are likely to exhibit strong soil water repellency under unburned conditions are also particularly susceptible to wildfires. As discussed below, burning can greatly affect the magnitude and depth of soil water repellency, and burning is more likely to enhance soil water repellency in vegetation types that are prone to soil water repellency under unburned conditions.

Measurement and Classification of Water Repellency

There are many techniques for measuring and classifying soil water repellency, and these are summarized in Tschapek (1984), Wallis and Horne (1992) and Letey et al. (2000). The two primary techniques for assessing soil water repellency are the 'Water Drop Penetration Time' (WDPT) test (Van't Woudt 1959) and the 'Molarity of an Ethanol Droplet' (MED) test (also known as the 'Percentage Ethanol' or 'Critical Surface Tension' test) (Letey et al. 2000).

In the WDPT test water drops are applied to the soil being tested and the investigator simply notes how long it takes until these water drops are absorbed into the soil. A longer duration indicates stronger water repellency, and the WDPT is best considered as measuring the *persistence* of soil water repellency. In the MED test, drops with an increasing concentration of ethanol are applied to the soil to measure indirectly the apparent soil surface tension. This effectively determines how strongly the water is repelled, and this property is considered the *strength* or *severity* of soil water repellency (Letey et al. 2000). The persistence and strength of soil water repellency are often related, but the relationship is not always clear or consistent (Dekker and Ritsema 1994). Some of the more recent literature relating to these tests includes Dekker et al. (1998), Doerr (1998), Roy and McGill (2002), and Shirtcliffe et al. (2006).

Both the WDPT and the MED tests provide quantitative data, but the subsequent classification or characterization of these data vary with the investigator's objectives and perception of what constitutes low or high soil water repellency (Table 1). WDPT thresholds as short as 1 (Roberts and Carbon 1971) and 5 sec (Bisdom et al. 1993) have been used to distinguish between wettable and water repellent soils. The U.S. Department of Agriculture, Forest Service (1995) uses only 3 categories, as a WDPT of less than 5 sec indicating no water repellency, 5 to 40 sec as moderately water repellent, and a WDPT greater than 40 sec is characterized as strongly water repellent. The different definitions of the various water repellency classes used can hinder the comparison of results among studies. One widely used set of water repellency classes for the WDPT test is presented in Table 1.

The comparability and statistical analyses of WDPT data also are complicated because the maximum time of observation varies among studies, and the maximum time of observation effectively truncates the data. For

Table 1 Water Drop Penetration Time (WDPT) class intervals in seconds (upper limit) and associated repellency persistence rating.

WDPT interval	<=5	10 30 60	180 300 600	900 1800 3600	18000 >18000
Persistence rating [1]	-	slight	strong	severe	extreme

[1] based on Bisdom et al. (1993)

practical reasons most studies do not make observations for more than 600 sec. In many studies WDPTs of less than 600 sec have been reported, however, higher values are also commonly found in burned and also unburned soils (e.g., Dekker et al. 2001, Doerr et al. 2006a, b).

The reported values also can vary with the exact methodology used to collect and analyze the data. Most studies use multiple drops to ascertain the WDPT for a given sample, but the reported value can vary depending on whether one uses the maximum observed value, the mean, or the median. The median is generally considered to provide the most accurate index of soil water repellency because the mean can be greatly affected by one or a few drops with very long penetration times. Furthermore, a few very high values may not be of much practical significance because of the typically high spatial variability in soil water repellency.

The MED test is less commonly used by field practitioners but is preferred by many researchers because the observation times are much shorter and it may exhibit less variability (e.g., Huffman et al. 2001). Table 2 provides the molarity and surface tension values for the most commonly used volumetric concentrations of ethanol, and two different schemes for converting the MED values to a categorical classification of soil water repellency.

The analysis of MED data is complicated because one typically uses a set of predetermined solutions to assess soil water repellency as indicated in

Table 2 Ethanol concentrations, Molarity of an Ethanol Droplet (MED) values, apparent surface tension (γ), and two associated descriptive classifications used in water repellency testing.

% Ethanol (vol.)	Molarity (MED)	γ (mNm^{-1})	Severity rating [1]	Severity rating [2]
0	0	72.1		
1	0.17	66.9	none	
3	0.51	60.9		low
5	0.85	56.6	slight	
8.5	1.45	51.2	moderate	
13	2.22	46.3	strong	moderate
18	3.07	42.3	very strong	severe
24	4.09	38.6		very severe
36	6.14	33.1	extreme	

[1]after Doerr (1998); [2]after King (1981)

Table 2. Since only certain concentrations are used, the MED data are discrete rather than continuous and should be analyzed using nonparametric statistics. It also is important to note that the surface tension of water decreases quite markedly as temperature increases, so the surface tension values in Table 2 are only applicable at 20°C.

Both the WDPT and the MED tests provide useful characterizations of soil water repellency, but it is very difficult to use either of these measurements to predict infiltration rates and the wetting behavior of bulk soil material directly (Dekker et al. 1999, Lewis et al. 2005, Doerr et al. 2006b, Cerdà and Doerr 2007). More studies are needed to link the measured soil water repellency to the infiltration rate as measured at the point scale with infiltrometers, the small plot scale using rainfall simulators, and runoff rates at the small watershed scale.

Changes to Water Repellency during Burning

Burning induces or enhances soil water repellency by volatilizing the hydrophobic organic compounds in the litter and topsoil. The simultaneous development of a pressure gradient in the layer being heated causes some of these compounds to be driven upwards into the atmosphere while some are forced downwards. The decline in soil temperature with depth means that these compounds will condense onto cooler soil particles at or below the soil surface (DeBano et al. 1976). The heat generated by burning, in addition to redistributing and concentrating the naturally-occurring hydrophobic substances in the soil and litter, is also thought to make these compounds more hydrophobic by pyrolysis and conformational changes in their structural arrangement (Doerr et al. 2005b). Burning also is believed to facilitate the bonding of these substances to soil particles (Savage et al. 1972,

Giovannini 1994). Laboratory studies show that soil water repellency is intensified at soil temperatures of 175 to 270°C, but is destroyed at temperatures above 270 to 400°C. The duration of heating can also affect the degree of soil water repellency with longer heating during influencing the temperature at which these changes occur (e.g., DeBano et al. 1976, Doerr et al. 2004). When there is insufficient oxygen, the combustion of the hydrophobic compounds and hence the temperature at which soil water repellency is destroyed it may rise to 500 to 600°C (Bryant et al. 2005).

These principles mean that the effects of burning on soil water repellency can be highly variable, as fires can induce soil water repellency in soils that were largely non-repellent, and either enhance or reduce pre-existing water repellency. The effect of burning depends primarily on the amount and type of organic matter consumed, the duration and amount of soil heating, and the amount of oxygen available during burning (DeBano and Krammes 1966, Robichaud and Hungerford 2000, Doerr et al. 2004, Bryant et al. 2005). Figure 3 illustrates some of the different soil wettability scenarios due to burning, which include:

i) burning can induce a low level of soil water repellency in a formerly non-repellent surface soil (e.g., Reeder and Jurgensen 1979, Cerda and Doerr 2005);
ii) a weakly water repellent soil can develop a stronger water repellent layer at or near the soil surface after burning; this has been observed after moderate and high severity fires in pine forests in the western USA (e.g., Huffmann et al. 2001, Woods et al. 2007; Fig. 3a);
iii) a high surface heating can destroy a strong or weak surface water repellency while creating increased repellency a few centimeters below the surface; this has been reported for hot chaparral fires in the western USA (e.g., DeBano 1971; Fig. 3b); and eucalypt fires in Australia (Doerr et al. 2006b; Fig. 3c);
iv) a soil that is strongly water repellent in both the surface and subsurface layers may have similar or weaker water repellency after burning when the heating was not sufficient to destroy the surface repellency and does not enhance a strong pre-existing subsurface water repellency. This has been observed in relatively wet eucalyptus stands in Portugal (Doerr et al. 1998) and in conifer forests of the northwestern USA (Doerr et al., submitted).

Numerous other scenarios can fall between these, and different scenarios can occur within the same fire depending on the conditions prior to the fire and fire behavior.

The apparent effects of burning on soil water repellency can also vary with the methodology used. For example, the position of the soil surface may be defined as the surface of the residual ash or litter after burning or the top of the mineral soil. In the former case the surface is most likely to be characterized as non-repellent because ash is typically hydrophilic rather

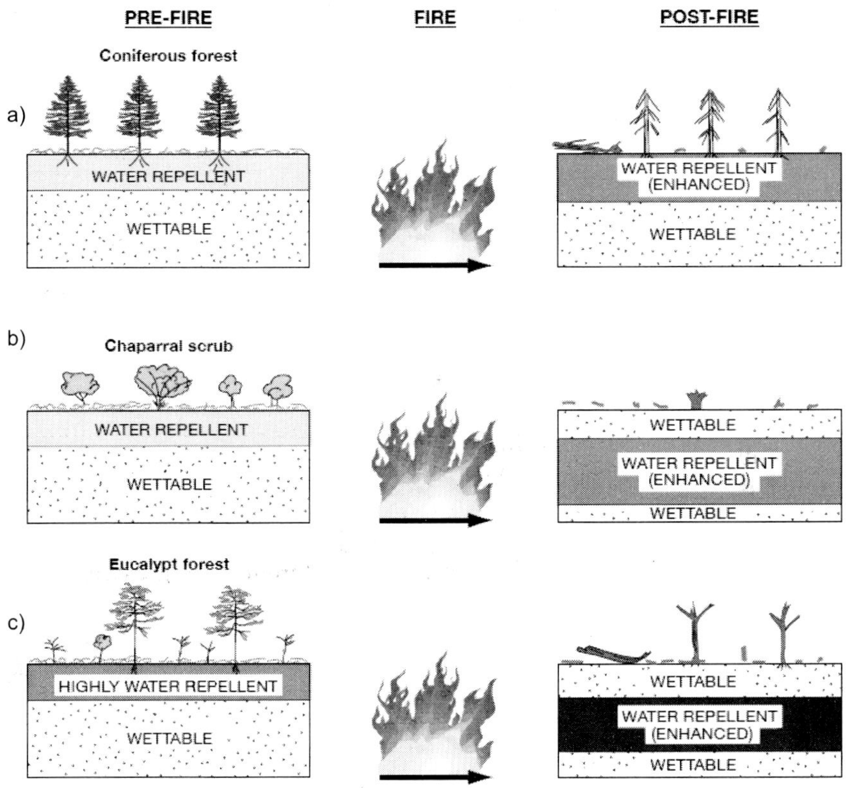

Fig. 3 Soil water repellency changes following fire for moderate or high soil burn severity conditions in: a) coniferous forest in the northwestern USA; b) Californian chaparral; and c) Australian eucalypt forest. Darker shading represents more severe repellency.

than hydrophobic. If the ash and residual litter is first swept away, the surface is much more likely to be characterized as water repellent. The identification of the mineral soil surface can also be problematic when there is a gradual boundary between the organic and mineral layers. The surface repellency can also change quite quickly as the wettable ash and any wettable mineral layers are removed by wind or overland flow. The relationship between burn severity and soil water repellency can vary because of differences in how different investigators characterize burn severity and soil water repellency (see previous section). Finally, short-term changes in soil moisture can greatly affect soil water repellency as discussed below.

Changes to Water Repellency in the Post-fire Period

The effects of burning on soil water repellency follow directly from the combustion of the organic matter and the associated soil heating. However, soil water repellency also can change very rapidly in response to changes in soil moisture, and somewhat slower as the fire-induced changes in soil water repellency decay towards pre-fire conditions or a new status according to the amount and type of post-fire vegetation. Any effort to predict the effects of burning on infiltration and erosion must clearly distinguish between this longer-term recovery and the shorter-term changes due to variations in soil moisture.

Both burned and unburned soils become less repellent or completely lose their water repellency as soil moisture increases. A water repellent soil can resist wetting for periods ranging from a few seconds to days or even months (e.g., King 1981, Dekker and Ritsema 1994, Doerr et al. 2006b), but the strong pressure gradient and the presence of macropores or other preferential flow paths means that water will eventually enter the soil. At a certain soil moisture content (i.e., critical threshold) the soil changes from being water repellent to wettable (Dekker et al. 2001). This soil moisture threshold can be less than 5 percent (per volume) for dune sands with a low organic matter content to more than 30 percent for finer textured soils (Doerr and Thomas 2000, Dekker et al. 2001, Benavides-Solorio and MacDonald 2005). One study has suggested that the soil moisture threshold increases with increasing burn severity, but the mechanism for this relationship is unknown (MacDonald and Huffmann 2004).

Once a water-repellent soil dries out, the soil water repellency is gradually or immediately re-established. Observations by one of the authors in Colorado suggest that in burned areas this repellency may be slightly weaker than prior to wetting, so a series of wetting and drying cycles may eventually eliminate the soil water repellency induced by burning (L.H. MacDonald, unpublished data). The exact processes involved in this are not fully understood, but changes in the spatial configuration of the hydrophobic molecules such as those suggested by Roy and McGill (2000) and Morley et al. (2005) are thought to be important.

The longer-term changes in soil water repellency after burning are due to a variety of physical and chemical processes. The duration of fire-induced increases in soil water repellency is an important concern for resource managers, but relatively little is known about the factors that control the changes in post-fire soil water repellency over time. There are several reasons for this, including: i) the relative paucity of longer-term studies on post-fire soil water repellency; ii) the large variability in results amongst those studies that have been conducted; iii) the short-term changes in soil water repellency due to changes in soil moisture are not always separated from the 'true' recovery to pre-fire conditions; iv) the variation amongst investigators in terms of what constitutes soil water repellency; and v) the difficulty of identifying and characterizing the effect of the different processes on the longevity of post-fire soil water repellency.

Most studies indicate that the increase in soil water repellency due to burning will break down within a few months to a couple of years. The longest documented duration of post-fire soil water repellency was in a severely burned pine forest in Oregon, USA, where the fire-induced soil water repellency apparently persisted for 6 years (Dyrness 1976). In less severely burned sites the post-fire soil water repellency persisted for 3 to 4 years (Dyrness 1976). On the other hand, 65 percent of the soil area that was water repellent after a fire in a mixed species forest in Michigan, USA was wettable a year later (Reeder and Jurgensen 1979). A low-severity fire in mixed chaparral in California almost doubled the frequency of measurements indicating moderate to extreme surface repellency, but the frequency of water repellency returned to pre-fire values in less than two months (Hubbert and Oriol 2005).

The longevity of post-fire soil water repellency can be highly variable, even for a particular vegetation type and geographic area. Huffmann et al. (2001) reported that fire-induced repellency persisted for at least 22 months in pine stands in Colorado, USA, while in a nearby location affected by a high-severity wildfire, the soil water repellency returned to pre-fire conditions within one year (MacDonald and Huffman 2004). In a Sardinian scrubland, the moderate pre-fire surface repellency was reduced after an experimental burn, but recovered to pre-fire levels within three years (Giovannini et al. 1987). Similarly, a severe fire in a *Pinus halepensis* stand in Spain destroyed the pre-fire soil water repellency, but this returned within three years (Cerda and Doerr 2005). A moderate to severe fire in an Australian eucalypt forest resulted in a patchy destruction of, or increase in, surface soil water repellency and increased the already high subsurface water repellency. Subsequent measurements showed no significant decline in the area of wettable surface soil one and two years after the fire, but there was a progressive decline in extreme water repellency in both the surface and subsurface soil (Doerr et al. 2006b).

Relatively little is known about the different processes that control the changes in post-fire soil water repellency over time. Observations after the 2002 Hayman wildfire in Colorado showed that the fire-induced soil water repellency was longer lived at 3 cm depth than at the soil surface (MacDonald and Rough, 2005). The faster breakdown of soil water repellency at the soil surface was attributed to the physical disruption caused by freeze-thaw cycles, but it also could be due to the greater biological activity associated with vegetative regrowth or the armoring of the soil surface with larger particles. One study in a shrubland in Idaho, USA, showed that the differences in solar insolation with slope aspect affects soil moisture, and the resulting differences in soil moisture will affect both the soil water repellency and post-fire vegetative regrowth (Pierson et al. 2002). More detailed studies are needed to determine i) the duration of fire-induced soil water repellency in different vegetation types and ii) the relative roles of physical, chemical, and biological factors in breaking down post-fire soil water repellency.

EFFECTS OF SOIL WATER REPELLENCY ON HYDROLOGY AND EROSION

The primary hydrologic and erosional effects of soil water repellency include: i) lower infiltration rates (Fig. 2) and a corresponding increase in the likelihood and amount of infiltration-excess (Hortonian) overland flow; ii) more spatial variability in infiltration and soil moisture fluxes, causing an uneven distribution of soil moisture; iii) increased surface erosion due to the increase in overland flow; and iv) increased susceptibility to wind erosion due to drier soil conditions and reduced cohesion of soil particles. The reduction in infiltration can also have secondary effects, such as hindering the germination and growth of vegetation, which can prolong fire impacts on runoff and erosion rates (see reviews by DeBano 2000, Doerr et al. 2000, Shakesby et al. 2000, and Shakesby and Doerr 2006).

Soil water repellency is of particular concern in burned areas because it can be much stronger and more persistent (in terms of WDPT) than in unburned areas (Fig. 4). Water repellency is typically implicated as the main cause of the increase in overland flow and erosion after burning (DeBano 2000, Martin and Moody 2001), but the effect of soil water repellency on runoff and erosion is complicated by other fire-induced changes such as the loss of the protective litter layer, the change in soil structure and cohesion due to the loss of soil organic matter, the potential reduction in infiltration due to soil sealing, and the reduction in interception due to the loss of overstory vegetation (e.g., Shakesby et al. 1993). This is illustrated in Figure 5, where all these factors may have contributed to the occurrence of this overland flow event in post-fire terrain. The hydrological and erosional effects of soil water repellency in unburned and burned areas are discussed in more detail in the following sections.

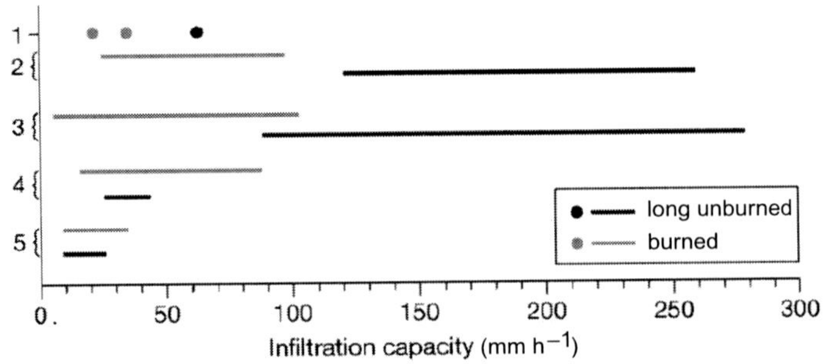

Fig. 4 Infiltration capacities measured after wildfire and on comparable unburned terrain according to various authors. Lines represent ranges of values and points represent individual values. 1) pine, Arizona, USA (Campbell et al. 1977); 2) pine and mixed conifer, Washington, USA (Martin and Moody 2000); 3) pine and eucalypt, Portugal (Shakesby et al. 1993); 4) pine and oak scrub, Israel (Kutiel et al. 1995); and 5) oak scrub, Spain (Imeson et al. 1992) (from Shakesby and Doerr 2006).

Fig. 5 Overland flow transporting burned soil, ash and charred debris during intense rain following wildfire in eucalypt forest in the Victorian Alps, southeast Australia in 2003 (photo by Rob Ferguson).

Hydrological Effects of Soil Water Repellency

One of the most striking consequences of soil water repellency is the reduction in the infiltrability (or infiltration rate); extremely water repellent soils may show very little wetting during rainstorms. For example, 40 to 46 mm h^{-1} of simulated rain caused minimal wetting of some strongly water repellent forest soils in Portugal (Walsh et al. 1998), even though the infiltration capacities of these same soils were around 80 mm h^{-1} in the laboratory when rendered non-repellent (Doerr et al. 2003). In laboratory experiments these soils remained dry despite having a ponded water layer for more than three weeks (Doerr and Thomas 2000). DeBano (1971) found that in the first five minutes the infiltration into a water-repellent soil was only one percent of the value when wettable; the maximum infiltration rate was 25 times less than a similar soil rendered hydrophilic by heating (see Fig. 2). Empirical data indicate a positive and significant relationship between soil water repellency and the amount of runoff from rainfall simulations (Robichaud 2000, Benavides-Solorio and MacDonald 2001, 2002), but few studies have rigorously isolated the effect of soil water repellency on infiltration and runoff. A recent study found that the application of wetting agents caused a 16-fold increase in infiltration from simulated rainfall on bare soils in eucalypt stands with strong soil water repellency (Leighton-Boyce et al. 2007).

A moderate reduction in infiltration rates due to soil water repellency would be expected to have minimal effect if the infiltration rate is still greater than rainfall intensities. In areas with distinct wet and dry seasons, an increase in soil water repellency may only be important for the first few storms, as once the soil wets up the water repellency is eliminated until the soil dries out. In snowmelt-dominated areas, soil water repellency is rarely a problem because snowmelt rates tend to be low relative to rainfall intensities, and the initial snowmelt would typically be expected to wet the soil beyond the critical soil moisture threshold. This explains why there is virtually no surface runoff or erosion during the winter and spring from severely burned areas in the central and northern Rocky Mountains, USA (Benavides-Solorio et al. 2005). Soil water repellency can have a much greater effect on infiltration and runoff under dry conditions when the soil moisture is below the critical threshold (Shahlaee et al. 1991, Walsh et al. 1994, Soto and Díaz-Fierros 1998, Doerr and Thomas 2000, Dekker et al. 2001). In an Australian eucalypt forest, exceptionally dry conditions were reported to enhance repellency and cause an increase in the overland flow coefficient from 5 to 15 percent (Burch et al. 1989).

Comparison of infiltration capacities from burned and unburned sites in Fig. 4 show the possible range in response. In two of the five studies, burning had no apparent effect on infiltration capacities, while burning greatly decreased infiltration rates in the other three studies. Post-fire soil water repellency is of greatest concern when it is sufficient to cause a shift in the dominant runoff process from subsurface stormflow to overland flow. In most unburned forests and shrublands the infiltration capacity is greater than rainfall intensities, and storm runoff is dominated by subsurface stormflow or saturation overland flow (Dunne and Leopold 1978, Hewlett 1982). In certain vegetation types burning can reduce the infiltration rate to the extent that even moderate rainstorms may exceed the infiltration rate. In the lower montane forests in Colorado, for example, summer thunderstorms with intensities of 60 to 65 mm h^{-1} generate no surface runoff or erosion, but in the first two years after burning, rainfall intensities of only 8 to 10 mm h^{-1} generate extensive surface runoff and erosion (Moody and Martin 2001, Benavides-Solorio and MacDonald 2005, Kunze and Stednick 2006). This remarkable difference, however, can not only be attributed to post-fire water repellency, but also to other fire effects such as vegetation removal as outlined in more detail in the next section. Overland flow generated as a result of post-fire soil water repellency is generally characterized as infiltration-excess or Horton overland flow (Horton 1933), but the presence of a wettable surface layer above a strongly repellent layer can induce a shallow layer of saturation overland flow (Walsh et al. 1994, Doerr et al. 2006b).

The reduction in infiltration after burning is exacerbated by the concomitant loss of interception and storage losses in the vegetation and litter, and this also acts to increase the amount of runoff. In high-severity fires, there also is a loss of surface roughness, and this increases the velocity of the overland flow and hence the size of peak flows. The exposure of the mineral soil surface to rainsplash, sheetwash, and rill erosion also may induce soil sealing, which further reduces the infiltration rate (Shakesby and Doerr 2006).

The net effect of these changes is that the size of peak flows can increase by one or more orders of magnitude (Robichaud et al. 2000, Neary et al. 2005). For example, there was a 5- to 15-fold increase in summer peakflows after burning for a chaparral site in Arizona (Robichaud et al. 2000), and a comparable increase was observed after burning for summer convective storms in a ponderosa pine forest in Colorado (Kunze and Stednick, 2006). Scott (1993) also noted an increase of up to two orders of magnitude in peak flows following burning in South Africa. In a burned pine forest in South Africa stormflows were 7.5 percent of precipitation as compared to 2.2 percent in unburned areas, and this difference was attributed to saturation overland flow in the wettable surface layer (Scott and Van Wyk 1990).

Other studies have shown little increase in runoff after burning because the change in infiltration was not sufficient to induce infiltration-excess overland flow (see, for example, Anderson et al. 1976). In oak woodlands in northeast Spain, burning increased the water-holding capacity of moderately water repellent soils by stabilizing water-retaining pores (Imeson et al. 1992). After several years these pores were lost by compaction and the infiltration capacity was reduced, but overall there was little difference in infiltration rates between the burned and unburned soils.

Burning can also affect the amount of runoff when much of the overstorey vegetation is killed, as this will reduce interception and transpiration rates. In areas with at least 450 to 500 mm of annual precipitation, annual water yields will increase. This increase, however, is typically smaller than the potential change in the size of peak flows because annual water yields are driven primarily by changes in evapotranspiration while peak flows are driven by the change in runoff processes on a storm by storm basis (Anderson et al. 1976). Similar to what is observed following forest harvest, the fire-induced increases in annual water yields tend to be greatest in humid ecosystems with denser pre-fire vegetation and high pre-fire evapotranspiration rates (Anderson et al. 1976, Robichaud et al. 2000).

Conclusive investigations into the larger-scale effects of soil water repellency are hindered by the high spatial variability in soil water repellency and infiltration rates. Intensive sampling of soil water repellency along burned transects in Colorado and Montana indicated an autocorrelation length of only about 2 m (Woods et al. 2007), and this high spatial variability is consistent with the results of most other studies (e.g., Huffman et al. 2001, Hubbert et al. 2006). In unburned soils the probability distribution of infiltration rates typically follows a lognormal distribution and also shows a relative low level of spatial correlation (e.g., Loague and Gander 1990). This suggests that the generation of overland flow is spatially heterogeneous, and as spatial scale increases there is a greater likelihood of encountering patches at the extreme end of the probability distribution where there are relatively high infiltration rates. In burned areas the overland flow generated from severely water repellent patches may infiltrate farther downslope in less

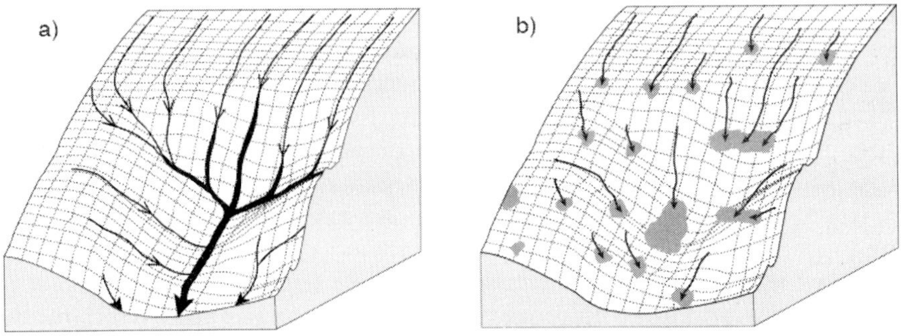

Fig. 6 The possible influence of variation in the spatial contiguity of water repellency on Hortonian overland flow in a catchment with a) a uniformly repellent soil and b) a repellent soil interrupted by 'sink' areas in the form of wettable patches or macropores. Overland flow (arrows) generated on repellent areas is shown intercepted by sinks (shaded areas) (modified from Shakesby et al. 2000).

severely burned patches, through macropores created by burned-out roots, or in areas that were burned at a similar severity but are less water repellent (Shakesby et al. 2000, Woods et al. 2007). Figure 6 compares the runoff pathways on a uniformly water repellent hillslope with a hillslope that has a water repellent soil interrupted by hydrologic 'sinks' in the form of macropores or wettable patches. In the first case, the volume and depth of overland flow increases uniformly in the downslope direction, while in the latter case some of the runoff only flows as far as the nearest sink. The continuity and connectivity of water repellent areas is believed to be an important control on hydrological impacts of soil water repellency in both burned and unburned areas (Doerr and Moody 2004, Hubbert et al. 2006, Woods et al. 2007).

The temporal and spatial variability of soil water repellency is only one reason for the difficulty in determining the effects of soil water repellency on runoff rates at the watershed (catchment) scale as compared to the point and plot scales. In many cases, it is very difficult to accurately measure runoff at the watershed scale after burning because of the very high sediment and debris loads. There can also be problems in characterizing the precipitation and snowmelt inputs at the watershed scale, as well as the spatial variation in other controlling factors, such as soil depth, soil type, and vegetation (Miller et al. 2003). The difficulty of measuring soil water repellency at the watershed scale means that the hydrologic effects of soil water repellency usually have to be inferred by comparing runoff rates under different conditions. A comparison of storm hydrographs from similar rainstorms on an unburned forested watershed in Portugal showed that peak and total runoff did not significantly differ between dry antecedent (i.e., repellent) and moderately moist antecedent conditions (less or non-repellent), but the time to peak was considerably reduced (Doerr et al. 2003). Data from burned ponderosa pine forests in Colorado indicate that surface runoff and erosion rates can be nearly as high one year after burning as immediately after burning (Benavides-Solorio and MacDonald 2005, Kunze and Stednick 2006), even though the soil water repellency was substantially weaker a year after burning. These results suggest that soil water repellency can increase storm runoff at the watershed scale as well as at the point and plot scales, but soil water repellency is only one of the many factors that affect the amount and timing of runoff. The complexity of runoff responses precludes simple generalizations, and better procedures are needed to predict the effects of burning on post-fire soil water repellency and infiltration (Doerr and Moody 2004). A general review of post-fire hydrologic changes at the watershed scale can be found in Shakesby and Doerr (2006).

Effects of Soil Water Repellency on Soil Erodibility and Erosion

In general, the greatest influence of soil water repellency on erosion is its potential for increasing overland flow. As the amount of overland flow increases, so does its depth and velocity, and hence the ability of the water to scour and transport particles by sheetwash (Meeuwig 1970). The concentration of overland flow into small rivulets can initiate rill erosion (e.g., Doehring 1968, Wells et al. 1987, Benavides-Solorio and MacDonald 2005), and the topographic convergence of water at larger scales can result in gully, bank, and channel erosion (e.g., Moody and Martin 2001). Pre- and post-fire data from the Colorado Front Range show that the first storms after a high-severity fire caused an extensive rill network to develop in previously unchanneled swales, and sediment yields to increase from zero to around 10 Mg ha^{-1} yr^{-1} (Libohova 2004).

Soil water repellency can also directly affect erosion rates by altering the erodibility of the soil by either wind or water action. Laboratory tests have shown that raindrops on water repellent soils produce fewer, slower-moving ejection droplets than raindrops on wettable soils, but the droplets from water repellent soils had more sediment (Terry and Shakesby 1993). With successive drops, the surface of the water repellent soil remained dry and non-cohesive, and the soil particles could be displaced by rainsplash despite the overlying film of water. In contrast, the simulated rainfall caused the surface of the wettable soil to become sealed and compacted, and this increased the resistance of the soil to detachment by rainsplash. These results have been replicated by simulated rainfall experiments on long-unburned, water repellent soils in the laboratory (Doerr et al. 2003) and in the field (Leighton-Boyce et al. 2007), though they have not yet been verified for natural rainfall on bare, newly burned surfaces.

In burned areas, it is very difficult to determine the effect of soil water repellency on surface erosion rates because moderate and high severity fires also remove the protective litter layer and expose the mineral soil surface to rainsplash. By definition, high severity fires consume some of the organic matter in the surface layer of the mineral soil, and the resulting disaggregation of the soil particles greatly increases the susceptibility of the surface soil to rainsplash, sheetwash, and rill erosion. The removal of the litter layer also reduces the surface roughness and increases the velocity of overland flow, which will further increase the surface erosion rates. In steeper terrain that burned at high severity, a wettable surface soil layer can become saturated, and the increased pore pressures will decrease the shear strength and lead to the downslope movement of soil by mass failure (DeBano 2000) and miniature debris flows (e.g., Wells 1981, Gabet 2003). Detailed measurements in the Colorado Front Range indicate that about 80 percent of the sediment produced from 0.1 to 0.5 ha plots can be attributed to rill incision (Pietraszek 2006). The concentration of flow necessary to produce rills can also be facilitated by anthropogenic modifications such as roads and paths (e.g., Atkinson 1984, Scott and Van Wyk 1990, Zierholz et al. 1995), skid trails or cable rows resulting from post-fire salvage logging (Chase 2006), or improperly installed contour-felled logs (Wagenbrenner et al. 2006).

Wetting agents are perhaps the best way to isolate the unique role of soil water repellency in hydrology and erosion, and Osborn et al. (1964) found that post-fire rills developed only on plots that were untreated (i.e., water repellent). On the other hand, removing the litter cover on hillslope-scale plots in a ponderosa pine forest in Colorado caused similar amounts of rilling and surface erosion as a high severity fire (L.H. MacDonald, unpublished data). This, together with the high correlation between surface cover and post-fire erosion rates (Pietraszek 2006), suggests that the amount of vegetation cover is a more important control on post-fire erosion rates in the Colorado Front Range than fire-induced soil water repellency.

In some environments and soil types rill networks may not develop, and this can be due to several reasons. First, the shear stress exerted by flowing water is several orders of magnitude lower than the force exerted by rainsplash (Hudson 1995, also see Chapter 16) and may not exceed the critical shear stress for particle entrainment. Once concentrated flow develops, the depth of water may protect the soil surface from rainsplash and erosion rates may be higher in the interrill areas. In other cases rill networks may not develop on water repellent burned soils due to the interception of overland flow by cracks, burned-out root holes, and burrows from insects and animals (Burch et al. 1989, Booker et al. 1993, Shakesby et al. 2003).

At larger hillslope scales the role of soil water repellency in increasing erosion is uncertain. There is no shortage of studies that document large increases in post-fire channel erosion and sediment yields in small to moderate-sized watersheds (see review by Shakesby and Doerr 2006), and these increases are readily attributable to the increase in runoff. However, there is considerable uncertainty with respect to the specific effect of soil water repellency in increasing surface runoff as opposed to other factors, and this

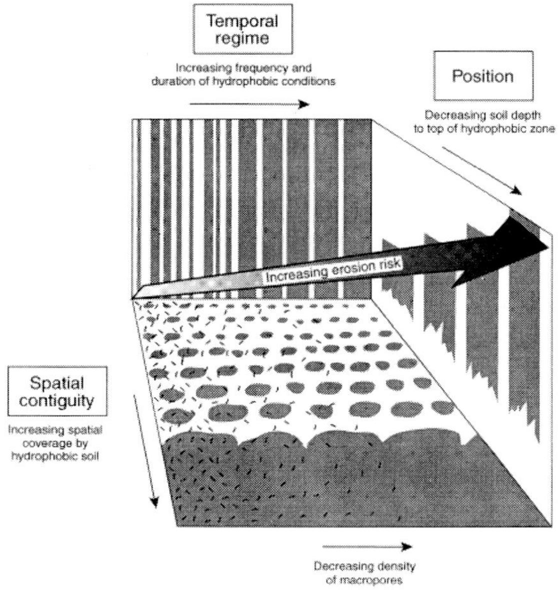

Fig. 7 A conceptual model of the effect of three soil water repellency attributes (temporal variation, spatial contiguity and thickness of wettable soil overlying water repellent soil of undefined depth) on erosion risk. Grey tones represent repellent conditions and unshaded areas represent wettable conditions. Other obvious factors that have a positive relationship with erosion risk, but are not included in this figure to retain its clarity, are the persistence and severity of soil water repellency, the intensity and duration of rainstorms, vegetation cover and slope angle. These are assumed to be unvarying here (modified from Shakesby et al. 2000).

uncertainty also applies to the effect of soil water repellency on larger-scale erosion rates.

Conceptually, in addition to the amount of protective vegetative cover, the effect of soil water repellency on surface erosion rates depends on not only the strength and persistence of soil water repellency, but also its spatial and temporal frequency and its spatial contiguity (Fig. 7). Several of these characteristics depend on other factors, such as soil texture, the seasonal timing of precipitation and its effect on soil moisture, and the dominant cause of runoff (i.e., snowmelt or rainfall).

The final issue is the effect of soil water repellency on wind erosion. As in the case of water-driven surface erosion, burning can greatly increase wind erosion rates, and much of this increase can be attributed to the removal of the protective vegetation and litter cover. Soil water repellency can also play a role in wind erosion in both burned and unburned areas, as it reduces the surface soil moisture, which reduces soil particle cohesion and lowers the threshold wind velocity for particle detachment and entrainment (Whicker et al. 2002).

CONCLUSIONS

A reasonably detailed knowledge base now exists on the origin and characteristics of post-fire water repellency. We know that post-fire soil water repellency is:

i) spatially, temporally and in its degree highly variable;
ii) common amongst certain vegetated soils irrespective of burning;
iii) often enhanced, but in some cases unaffected or eliminated following fire depending on the degree of soil heating;
iv) confined to the top few centimeters or decimeters of soil; and
v) typically most pronounced under dry conditions, but reduced or absent after prolonged rainfall, which in turn varies with soil type, burn severity, and the degree of soil water repellency prior to wetting.

Soil water repellency is a common characteristic of post-fire soils, and in some cases it is much stronger and more persistent than in the same soils prior to burning. The observed decreases in infiltration after burning suggest that post-fire soil water repellency plays a major role in causing the large increases in peak flows and surface erosion that are observed after high severity fires. The problem is that burning induces a series of other changes to the surface soils and vegetative cover that may be just as, or possibly even more, important in causing the observed increases in runoff and erosion. It is also much more difficult to determine the role of soil water repellency at larger scales due to the high spatial variability and the difficulty of characterizing the counteracting role of wettable soil patches, ash, bioturbation, soil cracks, and burned-out tree roots for reducing the surface runoff engendered by strongly water-repellent patches. Additional manipulative experiments and more detailed monitoring are needed to determine the extent to which soil water repellency has a measurable impact on runoff and surface erosion in the field under natural rainfall events at different scales.

Acknowledgements

We wish to acknowledge the grants supporting our research on soil water repellency and wildfire from the European Union (grants EV4V-0106-C TT, EV5V-0041 and FAIR 6CT98-4027), Natural Environment Research Council (Urgency Grant NER/A/S/2002/00143 and Advanced Fellowship NER/J/S/200200662), U.S. Environmental Protection Agency, U.S. Department of Agriculture, Forest Service, and the U.S. Department of Interior and Department of Agriculture, Forest Service, Joint Fire Science Program. We thank Anna Ratcliffe and Nicola Jones for producing the line drawings and Erik van den Elsen and Rob Ferguson for kindly providing images (Figs. 1 and 5). We are also indebted to the many colleagues, including a number of the authors of this book, for their fruitful collaboration and informative discussions over many years.

References

Anderson, H.W., M.D. Hoover, and K.G. Reinhart. 1976. Forests and water: effects of forest management on floods, sedimentation, and water supply. General Technical Report PSW-18, U.S. Department of Agriculture, Forest Service, Pacific Southwest Forest and Range Experimental Station, Berkeley, California, USA.

Atkinson, G. 1984. Erosion damage following bushfires. *Journal of the Soil Conservation Service of NSW*, 40: 4-9.

Bauters, T.W.J., T.S. Steenhuis, D.A. DiCarlo, J.L. Nieber, L.W. Dekker, C.J. Ritsema, J.-Y. Parlange, and R. Haverkamp. 2000. Physics of water repellent soils. *Journal of Hydrology*, 231-232: 233-243.

Benavides-Solorio, J. and L.H. MacDonald. 2001. Post-fire runoff and erosion from simulated rainfall on small plots, Colorado Front Range. *Hydrological Processes*, 15: 2931-2952.

Benavides-Solorio, J. and L.H. MacDonald. 2002. Correction to post-fire runoff and erosion from simulated rainfall on small plots, Colorado Front Range. *Hydrological Processes*, 16: 1131-1133.

Benavides-Solorio, J.D. and L.H. MacDonald, 2005. Measurement and prediction of post-fire erosion at the hillslope scale, Colorado Front Range. *International Journal of Wildland Fire*, 14: 457-474.

Bisdom, E.B.A., L.W. Dekker, and J.F. Schoute. 1993. Water repellency of sieve fractions from sandy soils and relationships with organic material and soil structure. *Geoderma*, 56: 105-118.

Booker, F.A., W.E. Dietrich, and L.M. Collins. 1993. Runoff and erosion after the Oakland firestorm. *California Geology*, 46: 159-173.

Brunsden, D. and J.B. Thornes. 1979. Landscape sensitivity and change. *Transactions of the Institute of British Geographers*, New Series. 4: 463-484.

Bryant, R., S.H. Doerr, and M. Helbig. 2005. The effect of oxygen deprivation on soil hydrophobicity during heating. *International Journal of Wildland Fire*, 14: 449-455.

Burch, G.J., I.D. Moore, and J. Burns. 1989. Soil hydrophobic effects on infiltration and catchment runoff. *Hydrological Processes*, 3: 211-222.

Campbell, R.E., M.B. Baker, P.F. Fiolliott, F.R. Larson, and C.C. Avery. 1977. Wildfire effects on a ponderosa pine ecosystem: an Arizona case study. Research Paper RM-191. U.S. Department of Agriculture, Forest Service, Rocky Mountain Forest and Range Experimental Station, Fort Collins, Colorado, USA.

Cerdà, A. and S.H. Doerr. 2005. Influence of vegetation recovery on soil hydrology and erodibility following fire: an eleven-year investigation. *International Journal of Wildland Fire*, 14: 423-437.

Cerdà, A. and S.H. Doerr. 2007. Soil wettability, runoff and erosion responses for major dry-Mediterranean land use types on calcareous soils. *Hydrological Processes*, 21: 2325-2336.

Chase, E. 2006. Effects of salvage logging in post-fire sediment production rates on the Star Fire in Central Sierra Nevada, CA. M.S. Thesis, Colorado State University, Fort Collins, Colorado, USA.

Cisar, J.L., K.E. Williams, H.E. Vivas, and J.J. Haydu. 2000. The occurrence and alleviation by surfactants of soil-water repellency on sand-based turfgrass systems. *Journal of Hydrology*, 231-232: 352-358.

DeBano, L.F. 1971. The effect of hydrophobic substances on water movement in soil during infiltration. *Soil Science Society of America Proceedings*, 35: 340-343.

DeBano, L.F. 2000. The role of fire and soil heating on water repellency in wildland environments: a review. *Journal of Hydrology*, 231-232: 195-206.

DeBano, L.F. and J.S. Krammes. 1966. Water repellent soils and their relation to wildfire temperatures. *Bulletin of the International Association of Hydrological Sciences*, 2: 14-19.

DeBano, L.F., L.D. Mann, and D.A. Hamilton. 1970. Translocation of hydrophobic substances into soil by burning organic litter. *Soil Science Society of America Proceedings*, 34: 130-133.

DeBano, L.F., S.M. Savage, and D.A. Hamilton. 1976. The transfer of heat and hydrophobic substances during burning. *Soil Science Society of America Proceedings*, 40: 779-782.

DeBano, L.F., D.G. Neary, and P.F. Ffolliott. 2005. Soil physical properties. pp. 29-51. *In* D.G. Neary, K.C. Ryan, and L.F. DeBano [eds.]. Wildland Fire in Ecosystems: Effects of Fire on Soil and Water. General Technical Report RMRS-GTR-42, vol. 4, U.S. Department of Agriculture, Forest Service, Rocky Mountain Research Station, Fort Collins, Colorado, USA.

Dekker, L.W. and C.J. Ritsema. 1994. How water moves in a water-repellent sandy soil. 1. Potential and actual water-repellency. *Water Resources Research*, 30: 2507-2517.

Dekker, L.W., C.J. Ritsema, K. Oostindie, and O.H. Boersma. 1998. Effect of drying temperature on the severity of soil water repellency. *Soil Science*, 163: 780-796.

Dekker, L.W., C.J. Ritsema, O. Wendroth, N. Jarvis, K. Oostindie, W. Pohl, M. Larsson, and J.P. Gaudet. 1999. Moisture distributions and wetting rates of soils at experimental fields in The Netherlands, France, Sweden and Germany. *Journal of Hydrology*, 215: 4-22.

Dekker, L.W., S.H. Doerr, K. Oostindie, A.K. Ziogas, and C.J. Ritsema. 2001. Actual water repellency and critical soil water content in a dune sand. *Soil Science Society of America Journal*, 65: 1667-1675.

Dekker, L.W., K. Oostindie, and C.J. Ritsema. 2005. Exponential increase of publications related to soil water repellency. *Australian Journal of Soil Research*, 43: 403-441.

Doehring, D.O. 1968. The effect of fire on geomorphic processes in the San Gabriel Mountains, California. pp. 43-65. *In* R.B. Parker [ed.]. Contributions to Geology. University of Wyoming, Laramie, Wyoming, USA.

Doerr, S.H. 1998. On standardising the "Water Drop Penetration Time" and the "Molarity of an Ethanol Droplet" techniques to classify soil hydrophobicity: a case study using medium textured soils. *Earth Surface Processes and Landforms*, 23: 663-668.

Doerr, S.H. and A.D. Thomas. 2000. The role of soil moisture in controlling water repellency: new evidence from forest soils in Portugal. *Journal of Hydrology*, 231-232: 134-147.

Doerr, S.H. and J. Moody. 2004. Hydrological impacts of soil water repellency: on spatial and temporal uncertainties. *Hydrological Processes*, 18: 829-832.

Doerr, S.H., R.A. Shakesby, and R.P.D. Walsh. 1998. Spatial variability of soil hydrophobicity in fire-prone eucalyptus and pine forests, Portugal. *Soil Science,* 163: 131-324.

Doerr, S.H., R.A. Shakesby, and R.P.D. Walsh. 2000. Soil water repellency, its characteristics, causes and hydro-geomorphological consequences. *Earth Science Reviews,* 51: 33-65.

Doerr, S.H., A.J.D. Ferreira, R.P.D. Walsh, R.A. Shakesby, G. Leighton-Boyce, and C.O.A. Coelho. 2003. Soil water repellency as a potential parameter in rainfall-runoff modelling: experimental evidence at point to catchment scales from Portugal. *Hydrological Processes,* 17: 363-377.

Doerr, S.H., W.H. Blake, G.S. Humphreys, R.A. Shakesby, F. Stagnitti, S.H. Vuurens, and P. Wallbrink. 2004. Heating effects on water repellency in Australian eucalypt forest soils and their value in estimating wildfire soil temperatures. *International Journal of Wildland Fire,* 13: 157-163.

Doerr, S.H., C.T. Llewellyn, P. Douglas, C.P. Morley, K.A. Mainwaring, C. Haskins, L. Johnsey, C.J. Ritsema, F. Stagnitti, G. Allinson, A.J.D. Ferreira, J.J. Keizer, A.K. Ziogas, and J. Diamantis. 2005a. Extraction of compounds associated with water repellency in sandy soils of different origin. *Australian Journal of Soil Research,* 43: 225-237.

Doerr, S.H., P. Douglas, R. Evans, C.P. Morley, N. Mullinger, R. Bryant, and R.A. Shakesby. 2005b. Effects of heating and post-heating equilibration times on soil water repellency. *Australian Journal of Soil Research,* 43: 225-237.

Doerr, S.H., R.A. Shakesby, L.W. Dekker, and C.J. Ritsema. 2006a. Occurrence, prediction and hydrological effects of water repellency amongst major soil and land use types in a humid temperate climate. *European Journal of Soil Science,* 57: 741–754.

Doerr, S.H., R.A. Shakesby, W.H. Blake, G.S. Humphreys, C. Chafer, and P.J. Wallbrink. 2006b. Effects of contrasting wildfire severity on soil wettability in Australian eucalypt catchments. *Journal of Hydrology,* 319: 295-311.

Doerr, S.H., S.W. Woods, D.A. Martin, and M. Casimiro (submitted). "Natural" soil water repellency in conifer forests of western USA: its prediction and relationship to wildfire occurence. *Journal of Hydrology.*

Dunne, T. and L.B. Leopold. 1978. *Water in Environmental Planning.* W.H. Freeman and Company, New York, New York, USA.

Dyrness, C.T. 1976. Effect of wildfire on soil wettability in the High Cascade of Oregon. pp. 444-447. *In* Research Paper PNW-202, U.S. Department of Agriculture, Forest Service, Pacific Northwest Forest and Range Experimental Station, Portland, Oregon, USA.

Gabet, E.J. 2003. Post-fire thin debris flows: sediment transport and numerical modelling. *Earth Surface Processes and Landforms,* 28: 1341-1348.

Giovannini, G. 1994. The effect of fire on soil quality. pp. 15-27. *In* M. Sala and J.L. Rubio [eds.]. *Soil Erosion as a Consequence of Forest Fires.* Geoforma Ediciones, Logroño, Spain.

Giovannini, G., S. Lucchesi, and M. Giachetti. 1987. The natural evolution of a burnt soil: a 3-year investigation. *Soil Science,* 143: 220-226.

Hallet, P.D. and Young, I.M. 1999. Changes to water repellency of soil aggregates caused by substrate induced microbial activity. *European Journal of Soil Science,* 50: 35-40.

Hewlett, J.D. 1982. *Principles of Forest Hydrology*. University of Georgia Press, Athens, Georgia, USA.

Horton, R.E. 1933. The role of infiltration in the hydrologic cycle. *Transactions of the American Geophysical Union*, 14: 446-460.

Hubbert, K.R. and V. Oriol. 2005. Temporal fluctuations in soil water repellency following wildfire in chaparral steeplands, southern California. *International Journal of Wildland Fire*, 14: 439-447.

Hubbert, K.R., H.K. Preisler, P.M. Wohlgemuth, R.C. Graham, and M.G. Narog. 2006. Prescribed burning effects on soil physical properties and soil water repellency in a steep chaparral watershed, southern California, USA. *Geoderma*, 130: 284-298.

Hudson, N.W. 1995. *Soil Conservation. 3rd edition*, Batsford Ltd., London, UK.

Huffmann, E.L., L.H. MacDonald, and J.D. Stednick. 2001. Strength and persistence of fire-induced hydrophobicity under ponderosa and lodgepole pine, Colorado Front Range. *Hydrological Processes*, 15: 2877-2892.

Imeson, A.C., J.M. Verstraten, E.J. van Mulligen, and J. Sevink. 1992. The effects of fire and water repellency on infiltration and runoff under Mediterranean type forest. *Catena*, 19: 345-361.

Inbar, M., M. Tamir, and L. Wittenberg. 1998. Runoff and erosion processes after a forest fire in Mount Carmel, a Mediterranean area. *Geomorphology*, 24: 17-33.

Jamison, V.C. 1942. The slow reversible drying of surface sandy soils beneath citrus trees in central Florida. *Soil Science Society of America Proceedings*, 10: 25-29.

King, P.M. 1981. Comparison of methods for measuring severity of water repellence of sandy soils and assessment of some factors that affect its measurement. *Australian Journal of Soil Research*, 19: 275-285.

Kunze, M.D. and J.D. Stednick. 2006. Streamflow and suspended sediment yield following the 2000 Bobcat fire, Colorado. *Hydrological Processes*, 20: 1661-1681.

Kutiel, P., H. Lavee, M. Segev, and Y. Benyamini. 1995. The effect of fire-induced surface heterogeneity on rainfall-runoff-erosion relationships in an eastern Mediterranean ecosystem, Israel. *Catena*, 25: 77-87.

Leighton-Boyce, G., S.H. Doerr, R.A. Shakesby, and R.D.P. Walsh. 2007. Quantifying the impact of soil water repellency on overland flow generation and erosion: a new approach using rainfall simulation and wetting agents on in situ soils. *Hydrological Processes*, 21: 2337-2345.

Letey, J. 2001. Causes and consequences of fire-induced soil water repellency. *Hydrological Processes*, 15: 2867-2875.

Letey, J., J. Osborn, and R.E. Pelishek. 1962. The influence of the water-solid contact angle on water movement in soil. *Bulletin of the International Association of Hydrological Sciences*, 3: 75-81.

Letey, J., M.L.K. Carrillo, and X.P. Pang. 2000. Approaches to characterize the degree of water repellency. *Journal of Hydrology*, 231-232: 61-65.

Lewis, S.A., J.Q. Wu, and P.R. Robichaud. 2005. Assessing burn severity and comparing soil water repellency, Hayman Fire, Colorado. *Hydrological Processes*, 20: 1-16.

Libohova, Z. 2004. Effects of thinning and a wildfire on sediment production rates, channel morphology, and water quality in the Upper South Platte River watershed. M.S. Thesis. Colorado State University, Fort Collins, Colorado, USA.

Loague, K.M. and G.A. Gander. 1990. R-5 revisited: 1. Spatial variability of infiltration on a small rangeland catchment. *Water Resources Research*, 26: 957-971.

MacDonald, L.H. and E.L Huffman. 2004. Post-fire soil water repellency: persistence and soil moisture thresholds. *Soil Science Society of America Journal*, 68: 1729-1734.
MacDonald, L., D. Rough, and Z. Libohova. 2005. Effects of forest fires on the strength and persistence of soil water repellency in the Colorado Front Range. *Geophysical Research Abstracts*, Vol. 7, 08613. SRef-ID: 1607-7962/gra/EGU05-A-08613.
Martin, D.A. and J.A. Moody. 2001. Comparison of soil infiltration rates in burned and unburned mountainous watersheds. *Hydrological Processes*, 15: 2893-2903.
Meeuwig, R.O. 1970. Sheet erosion on intermountain summer ranges. Research Paper INT-85, U.S. Department of Agriculture, Forest Service, Intermountain Forest Range Experimental Station, Ogden, Utah, USA.
Miller, J.D., J.W. Nyhan, and S.R. Yool. 2003. Modeling potential erosion due to the Cerro Grande Fire with a GIS-based implementation of the Revised Universal Soil Loss Equation. *International Journal of Wildland Fire*, 12: 85-100.
Morley, C.P., K.A. Mainwaring, S.H. Doerr, P. Douglas, C.T. Llewellyn, and L.W. Dekker. 2005. Identification of organic compounds at different depths in a water repellent soil. *Australian Journal of Soil Research*, 43: 239-249.
Neary, D.G., C.C. Klopatek, L.F. DeBano, and P.F. Ffolliott. 1999. Fire effects on belowground sustainability: a review and synthesis. *Forest Ecology and Management*, 122: 51-71.
Neary, D.G., P.F. Ffolliott, and J.D. Landsberg. 2005. Fire and streamflow regimes. pp. 107-118. *In* D.G.Neary, K.C. Ryan, and L.F. DeBano [eds.]. Wildland Fire in Ecosystems: Effects of Fire on Soil and Water. General Technical Report RMRS-GTR-42, vol. 4, U.S. Department of Agriculture, Forest Service, Rocky Mountain Research Station, Fort Collins, Colorado, USA.
Osborn, J.R., R.E. Pelishek, J.S., Krammes, and J. Letey. 1964. Soil wettability as a factor in erodibility. *Soil Science Society of America Proceedings*, 28: 294-295.
Pierson, F.B., D.H. Carlson, and K.E. Spaethe. 2002. Impacts of wildfire on soil hydrological properties of steep sagebrush-steppe rangeland. *International Journal of Wildland Fire*, 11: 145-151.
Pietraszek, J.H. 2006. Controls on post-fire erosion at the hillslope scale, Colorado Front Range. M.S. Thesis. Colorado State University, Fort Collins, Colorado, USA.
Reeder, C.J. and M.F. Jurgensen. 1979. Fire-induced water repellency in forest soils of upper Michigan. *Canadian Journal of Forest Research*, 9: 369-373.
Roberts, F.J. and B.A. Carbon. 1971. Water repellency in sandy soil of southwestern Australia: I. Some studies related to field occurrence. Australian Commonwealth Scientific and Research Organization (CSIRO), Division of Plant Industry, Field Station Record 10: 13-20.
Robichaud, P.R. 2000. Fire effects on infiltration rates after prescribed fire in Northern Rocky Mountain Forests, USA. *Journal of Hydrology*, 231-232: 220-229.
Robichaud, P.R., J.L. Beyers, and D.G. Neary. 2000. Evaluating the effectiveness of postfire rehabilitation treatments. General Technical Report RMRS-GTR-63, U.S. Department of Agriculture, Forest Service, Rocky Mountain Research Station, Fort Collins, Colorado, USA.
Roy, J.L. and W.B. McGill. 2000. Flexible conformation in organic matter coatings: An hypothesis about soil water repellency. *Canadian Journal of Soil Science*, 80: 143-152.

Roy, J.L. and W.B. McGill. 2002. Assessing soil water repellency using the molarity of ethanol droplet (MED) test. *Soil Science*, 79: 83-97.

Sartz, R.S. 1953. Soil erosion on a fire-denuded forest area in the Douglas-fire region. *Journal of Soil and Water Conservation*, 8: 279-281.

Savage, S.M., J. Osborn, J. Letey, and C. Heton. 1972. Substances contributing to fire induced water repellency in soils. *Soil Science Society of America Proceedings*, 36: 674-678.

Schreiner, O. and E.C. Shoery. 1910. Chemical nature of soil organic matter. *U.S. Department of Agriculture Soils Bulletin*, 74: 1-48.

Scott, D.F. 1993. The hydrological effects of fire in South African mountain catchments. *Journal of Hydrology*, 150: 409-432.

Scott, D.F. and D.B. Van Wyk. 1990. The effects of wildfire on soil wettability and hydrological behaviour of an afforested catchment. *Journal of Hydrology*, 121: 239-256.

Shahlaee, A.K., W.L. Nutter, E.R. Burroughs Jr., and L.A. Morris. 1991. Runoff and sediment production from burned forest sites in the Georgia Piedmont. *Water Resources Bulletin*, 27: 485-493.

Shakesby, R.A. and S.H. Doerr. 2006. Wildfire as a hydrological and geomorphological agent. *Earth Science Reviews*, 74: 269-307.

Shakesby, R.A., C. de O.A. Coelho, A.D. Ferreira, J.P. Terry, and R.P.D. Walsh. 1993. Wildfire impacts on soil erosion and hydrology in wet Mediterranean forest, Portugal. *International Journal of Wildland Fire*, 3: 95-110.

Shakesby, R.A., S.H. Doerr, and R.P.D. Walsh. 2000. The erosional impact of soil hydrophobicity: current problems and future research directions. *Journal of Hydrology*, 231-232: 178-191.

Shakesby, R.A., C.J. Chafer, S.H. Doerr, W.H. Blake, G.S. Humphreys, P. Wallbrink, and B.H. Harrington. 2003. Fire severity, water repellency characteristics and hydrogeomorphological changes following the Christmas 2001 forest fires. *Australian Geographer*, 34: 147-175.

Shakesby, R.A., P.J. Wallbrink, S.H. Doerr, P.M. English, C. Chafer, G.S. Humphreys, W.H. Blake, and K.M. Tomkins. 2007. Distinctiveness of wildfire effects on soil erosion in south-east Australian eucalypt forests assessed in a global context. *Forest Ecology and Management*, 238: 347-364.

Shirtcliffe, N.J., G. McHale, F.B. Pyatt, M.I. Newton, and S.H. Doerr. 2006. Critical conditions for the wetting of soils. *Applied Physics Letters*, 89: 094101.

Soto, B. and F. Díaz-Fierros. 1998. Runoff and soil erosion from areas of burnt scrub: a comparison of experimental results with those predicted by the WEPP model. *Catena*, 31: 257-270.

Swanson, F.J. 1981. Fire and geomorphic processes. pp. 401-421. *In* H.A. Mooney, T.M. Bonnicksen, N.L. Christiansen, J.E. Lotan, and W.A. Reiners [eds.]. Fire Regime and Ecosystem Properties. General Technical Report WO-26. U.S. Department of Agriculture, Forest Service, Washington D.C., USA.

Terry, J.P. and R.A. Shakesby. 1993. Soil water repellency effects on rainsplash: simulated rainfall and photographic evidence. *Earth Surface Processes and Landforms*, 18: 519-525.

Tschapek, M. 1984. Criteria for determining the hydrophilicity-hydrophobicity of soils. *Zeitchrif fur Planzenernahrung and Bodenkunde*, 147: 137-149.

U.S. Department of Agriculture, Forest Service. 1995. Water-Repellent Soils – Article 23.31. *In* Burned-Area Emergency Rehabilitation Handbook, Section FSH 2509.130 of the Forest Service Handbook.

Van't Woudt, B.D. 1959. Particle coatings affecting the wettability of soils. *Journal of Geophysical Research*, 64: 263-267.

Wagenbrenner, J.W., L.H. MacDonald, and D. Rough. 2006. Effectiveness of three post-fire rehabilitation treatments in the Colorado Front Range. *Hydrological Processes*, 20: 2989-3006.

Wallis, M.G. and D.J. Horne. 1992. Soil water repellency, Volume 20, pp. 91-146. *In* B.A. Stewart [ed.]. *Advances in Soil Science*. Springer, New York, USA, 20: 91-146.

Wallis, M.G., D.R. Scotter, and D.J. Horne. 1991. An evaluation of the intrinsic sorptivity water repellency index on a range of New Zealand soils. *Australian Journal of Soil Research*, 29: 353-362.

Walsh, R.P.D., D.J. Boakes, C.O.A. Coelho, A.J.B. Gonçalves, R.A. Shakesby, and A.D. Thomas. 1994. Impact of fire-induced water repellency and post-fire forest litter on overland flow in northern and central Portugal. Vol. II, pp. 1149-1159. *In* Proceedings of the Second International Conference on Forest Fire Research, University of Coimbra, Coimbra, Portugal.

Walsh, R.P.D., C. de O.A. Coelho, A. Elmes, A.J.D. Ferreira, A.J.B. Gonçalves, R.A. Shakesby, J.L. Ternan, and A.G. Williams. 1998. Rainfall simulation plot experiments as a tool in overland flow and soil erosion assessment, north-central Portugal. *Geookodynamik*, 19: 139-152.

Wells, W.G. 1981. Some effects of brushfires on erosion processes in coastal southern California, pp. 305-342. *In* Proceedings of Symposium on Erosion and Sediment Transport in Pacific Rim Steeplands. Christchurch, New Zealand, Publication 132, International Association of Hydrological Sciences, Oxford, UK.

Wells, W.G. 1987. The effect of fire on the generation of debris flows. *In* J.E. Costa. and G.F. Wieczorek [eds.]. Debris flows/avalanches. Geological Survey of America, *Reviews in Engineering Geology*, 7: 105-114.

Whicker, J.J., D.D. Breshears, P.T. Wasiolek, T.B. Kirchner, R.A. Tavani, D.A. Schoep and J. C. Rodgers. 2002. Temporal and spatial variation of episodic wind erosion in unburned and burned semiarid shrubland. *Journal of Environmental Quality*, 31: 599-612.

Woods, S.W., A. Birkas, and R. Ahl. 2007. Spatial variability of soil hydrophobicity after wildfires in Montana and Colorado. *Geomorphology*, 86: 465-479.

Zierholz, C., P. Hairsine, and F. Booker. 1995. Runoff and soil erosion in bushlands following the Sydney bushfires. *Australian Journal of Soil and Water Conservation*, 8: 28-37.

8

Effects of Fire on Forest Nutrient Cycles

R. John Raison[1*], *Partap K. Khanna*[1], *Kris L.S. Jacobsen*[1], *Joan Romanya*[2] and *Isabel Serrasolses*[3]

Abstract

Forest fires can have a major impact on nutrient cycling, with significant consequences for many other important ecosystem processes. The critical cycling fraction of nutrients is affected by fire in a number of ways: 1) by loss of nutrients during and after fires through direct (gaseous and particulate during combustion) and indirect (leaching or erosion of ash and soil material) means; 2) by transforming some nutrients from organic to inorganic forms and soil heating, which initiates chemical and biotic processes affecting decomposer activity and nutrient uptake by vegetation; 3) by affecting vegetation composition, structure, and growth rate, and thus the uptake and turnover of nutrients; 4) by changing the chemical environment of plant roots (pH, exchangeable cations, soil solution composition, solubility and composition of minerals); 5) by affecting other soil processes via changes in hydrology and temperature regimes; and 6) by triggering vegetation succession (e.g., stimulation of N-fixing understory vegetation).

These effects of fire on nutrient cycling are caused by heat, ash additions, altered microclimate, and changed vegetation dynamics and may be of short- or long-term duration. Changed rates of nutrient cycling have important implications for forest management, including the application of prescribed burning regimes and actions taken to assist rehabilitation of ecosystems processes after wildfire. Care needs to be taken in making broad assumptions and in generalizing about the effects of fire on nutrient cycling in particular forest systems. While there are some well established principles that are discussed in this chapter, local studies are required to refine understanding of the relationship between fire and nutrient cycling and to guide fire management.

[1] *CSIRO Sustainable Ecosystems, Kingston, Australia Capital Territory, Australia.*
[2] *Department of Natural Products, Plant Biology and Soil Science, Universitat de Barcelona, Barcelona, Spain.*
[3] *Unitat d'Ecologia, Facultad de Biociencies, Universitat Autònoma de Barcelona, Barcelona, Spain.*
* *Corresponding author: R. John Raison, CSIRO Sustainable Ecosystems, P.O. Box E 4008, Kingston, Australia Capital Territory, ACT 2604, Australia. Tel: 61 2 6281 8280, Fax: 61 2 6281 8312, e-mail: John.Raison@csiro.au*

INTRODUCTION

Fire impacts forest ecosystems by changing many important soil and plant processes. Among these, the effect of fire on nutrient budgets and short- and longer-term nutrient cycling processes are important (Raison et al. 1993a). Fires result in direct losses of nutrients during combustion and create the potential for accelerated losses after a fire. Soil nutrients are mobilized by heating and ash deposition in the shorter-term and N fixation is often stimulated after fire. Decomposition processes in the forest floor and soils can be changed for many years. These processes alter rates of nutrient cycling and can affect forest productivity which in turn controls most other properties and processes in the forest (e.g., carbon and water cycles, biodiversity values). Fire can also lead to significant emissions of greenhouse gases, especially CO_2, methane and nitrous oxide. The processes and magnitude of these are not discussed here, but are described in IPCC (2006).

Fires in forests can be described in terms of three key variables: severity, frequency, and season of burn. Various combinations of these variables define the fire regime to which the forest is exposed. The fire regime, in combination with the characteristics of the site and species making up the forest, will determine the effects of fire on nutrient cycles and other properties and processes of the forest ecosystem. The following are broad fire categories and typical forest responses:

- Low-severity burns that consume only ground litter and understory vegetation. The dynamics and species composition and structure of the understory can be affected. The overstory is not defoliated, and any effects on the dynamics of the tree layer only occur in the longer-term. Direct impacts of a single fire on the soil are small. Cumulative effects of fire on the nutrient cycle are likely to be more significant than those of a single fire.
- Moderate- to high-severity fires that defoliate but do not kill the overstory trees. Trees recover either by sprouting from buds on branches or the trunk. The understory dynamics are markedly changed. The soil may be bared and subjected to significant heating and input of ash, and have changed microclimate for 1 to 2 yr after the fire while the understory vegetation and tree canopy recovers.
- High-severity fires that kill most of the vegetation which initiates a myriad of changes that leads to stand replacement. N fixing understory vegetation may be markedly stimulated as part of the successional process. The soil environment is significantly changed in the short-term and over many years as the forest recovers.
- High-severity fire in forest residues (slash) produced during timber harvesting or clearing prior to conversion of forest to agriculture. The soil will be subjected to considerable heating and ash input, especially under

piles of woody slash (leading to the creation of 'ashbeds'). Fire impacts on the soil will be highly spatially variable because of variations in heating and ash input. Vegetation dynamics and soil microclimate are changed over the time scale of years to decades.

The effect of fire on forest nutrient cycles has been the subject of many earlier reviews (Raison 1979, Walker et al. 1986, Raison et al. 1993a, Fisher and Binkley 2000, Wan et al. 2001, Certini 2005, Smithwick et al. 2005, and others). Available information is best examined within a framework that defines the key variables that can affect change. The framework considers the direct (short-term) and indirect (longer-term) effects of fire (Table 1), the interacting effects of heat, ash, and changed microclimate, and the importance of the fire regime which determines the spatial impact of fire in a forested landscape. All these factors are important and they interact. For example in the short-term, fire affects the state and rate of recovery of vegetation and thus becomes a critical factor of longer-term impacts because recovery of vegetation helps to stabilize soil processes and to re-establish nutrient cycling.

Table 1 Important direct and indirect effects of burn severity on forest nutrient cycles. Note that the fire regime (combination of burn severity, frequency, and season of burn) is critical, and that the cumulative effects of repeated fires need to be considered.

Direct effects	Indirect effects
Loss of nutrients to the atmosphere during combustion	Loss of fire-mobilized nutrients in ash/soil by erosion (wind/water) or leaching
Heating of soil	Changes in micro-climate in soil and forest floor
Deposition of ash	Altered soil mineralization rates
Removal or killing of vegetation	Changed N fixation rates Changed nutrient uptake and storage by vegetation, and modified litter input and decomposition

We use the framework from Table 1 to provide a synthesis of the current understanding of the effects of fire on forest nutrient cycles by building on the considerable body of work that has been conducted during the last half century. Emphasis is placed on identifying emerging principles and on defining key gaps in understanding, especially in relation to the longer-term effects of fire, which are much less well understood (Raison et al. 1993a, Wan et al. 2001), but which are critically important to the functioning of forest ecosystems. Two case studies describing the effects of contrasting fire types in Australian eucalypt forests are used to illustrate some of the key principles.

METHODOLOGY FOR ASSESSMENT OF NUTRIENT LOSSES AND CHANGES IN NUTRIENT CYCLING

The measurement and assessment of changes in nutrient cycling processes due to forest fire present several methodological problems. These are primarily related to the highly spatially variable distribution of fuel available for burning, its nutrient composition and varying fire severity. This causes high spatial variability in soil heating and ash production which creates sampling and assessment problems. A retrospective approach is often adopted in field studies where samples are collected at different periods after fire and any changes are assigned to fire effects. Such studies assume the uniformity of sites prior to fire and similarity in the burning conditions. In most studies the validity of these assumptions is not assessed. Results from such studies should be interpreted with caution.

In other field studies a different method is used where the nutrients in the fuel and surface soil are measured (pre-fire conditions) and compared with the nutrients remaining in the post-fire residues and surface soil. From the differences in these measurements, it is then possible to study by differences the amount of nutrients lost during fire and changes in soil nutrients can be determined. Results from such studies can provide reliable information, as long as the collection of fire residues is quantitative and the depth of soil sample taken represents the fire-affected zone (1 to 5 cm depth). Inclusion of unaffected soil (e.g., 5 to 20 cm depth) will increase the possibility of incorrectly finding 'no-effect'.

Selection of appropriate soil parameters is another pre-requisite for a successful study. Often the selected soil parameters are either insensitive to small changes which may be caused by fire, or studies are not of sufficient duration. In order to distinguish between the effect of fire and the other factors affecting temporal variation, it is necessary to carry out studies over several climatic events and preferably over several years. The objective in most studies should be to ensure that the active biological sites, labile nutrient fractions and soil physical components of the surface soils are adequately sampled. It is important to realize that any soil treatment prior to laboratory analysis may mask the effects of fire that are occurring in the field. For example, if disturbed, mixed, or dried soil samples are moistened to a fixed level and incubated to follow a fixed temperature regime to study the effects of fire on N mineralization, the results may not resemble those obtained *in situ*.

A number of dynamic processes, such as the accelerated death and decay of roots, may complicate interpretation of soil C and nutrient changes after fire. This point is illustrated as part of Case Study 2 described below.

Further, the effects of single fires should be carefully distinguished from the cumulative effects of previous fires. Fire history must be considered when interpreting the results of an experimental fire treatment.

NUTRIENT LOSS MECHANISMS

Losses of elements during forest fires are expected to have major effects on nutrient cycling processes because for many elements the losses are a significant fraction of the pool of actively cycling nutrients. There are direct losses during the burning of forest fuels due to gaseous and particulate transfers. Indirect losses may occur following the fire when ash is removed by the action of wind and water, sometimes together with surface soil due to lack of protection by vegetation. Fire-mobilized nutrients can also be leached, especially from coarse-textured soils. The amount of nutrients lost during a fire will vary markedly for the range of fire categories (Table 2). In high severity fires, nutrient loss from atmospheric transfers is high, large amounts of ash are deposited, and the risks from erosion and leaching are greater. There are no simple direct ways of measuring site nutrient losses due to fire, but it is possible to apply some general principles to make reasonable estimates of losses and these are discussed below.

Atmospheric Transfers

Fisher and Binkley (2000) identified three processes that contribute to atmospheric transfer of nutrients in fire:

- Oxidation of carbon (C), nitrogen (N) and sulfur (S) leading to gaseous loss.
- Volatilization of phosphorus (P) and potassium (K) leading to gaseous loss.
- Convection of ash which is important for loss of calcium (Ca), P and K in particular.

The gaseous transfers are considered to be losses from the site, whereas some of the particulate transfers may be re-deposited as ash fall-out. Raison et al. (1985a) proposed the following approaches to examine the mechanisms and magnitude of atmospheric transfer of elements:

- Loss of C and N occurs mostly in gaseous form and is directly related to fuel consumption. If the fraction of fuel consumed during the fire and the N contents of the fuel are known, atmospheric losses of N can be assessed with sufficient accuracy.
- Particulate losses of elements through convection can be assessed by considering the Ca balance between the fuel and the fire residues. Loss of Ca occurs only by particulate transfer.
- Non-particulate losses of P, K, and other elements can be calculated from the changes in the ratio of the given element to Ca (Ca is not lost in non-particulate form).

The concepts advanced by Raison et al. (1985a) were confirmed in later studies by Gillion and Rapp (1989) in a Mediterranean pine forest, and by

Table 2 Characteristics of forest fires relevant to potential direct losses of nutrients.

	Land clearing, slash and burn	Forest regeneration	Wildfire	Prescribed-low severity
Fuel components	Mostly trunks, crown and understory slash	Tree crowns, woody residues, shrubs and litter	Tree foliage, shrubs, woody and fine litter	Shrubs and fine litter
Fuel consumption (t ha^{-1})	100–400+	50–300+	20–60	8–15
Fire intensity* (kW m^{-1})	70000–100000+ (extreme)	7001–70000 (very high)	501–7000 (moderate to high)	<500 (low)
Maximum temperature (°C) at 2 cm depth	500	200	100	<60
Time (hr) for which soil is heated above 80% of the maximum temp.	>24	0.5–2	0.1–1	0.1–0.3
Fire frequency (yr)	30–100+	15–30+	20–100+	5–15
Ash residues produced	Very high on part of the area	Medium to high on most of the area	Low to medium on most of the area	Low on patches of the area
N losses (kg ha^{-1})	500–2000+	350–800	>200	50–150
Vegetation effects	Most vegetation removed, recovery depends on management goals	Much vegetation removed, recovery on decadal timescale	Defoliation and possible mortality of understory and trees	Part removal of understory, trees little affected

*Fire intensity is described by an index as defined by Bryam (1959)

Gillion et al. (1995) in controlled laboratory combustion studies. The reduction in the quantity of N in fuels during fire is linearly correlated with the fuel consumption percentage (Raison et al. 1985a, Gillion and Rapp 1989, Wan et al. 2001). About 5 to 10 kg of N are lost for each ton of fuel consumed (Fisher and Binkley 2000, Serrasolses and Vallejo 1999).

A number of studies have shown that K and P are lost by both volatilization and particulate mechanisms. Volatile losses of P are of special significance with losses of 27 to 50 percent of that in burned fuel (Raison et al. 1985b, DeBano and Klopatek 1988, Mackenson et al. 1996), depending upon the temperature and the type of fuel involved. P is only very slowly replenished by atmospheric inputs and soil weathering in most natural forest ecosystems.

Season of burn may be an important factor determining the amount of nutrients lost. Boring et al. (2004) reported that more nitrogen was lost from growing season burning than during the dormant period burning in a longleaf pine ecosystem. This was related to high N levels in the foliage burned during the growing season.

• Losses of nutrients during combustion are low in savanna woodlands where the major fuels are grass and small amounts of woody litter (Cook 1994), but can be very large in intense forest fires and land conversion fires, especially for N which is readily oxidized (Table 2). For example, Kauffman et al. (1995) estimated losses of up to 1600 kg N ha^{-1} and 20 kg P ha^{-1} during burning of slashed primary tropical moist forest.

Erosion

Erosion of ash and surface soil can have profoundly negative effects on soil nutrient pools and on soil fertility (Wright and Bailey 1982, Raison et al. 1993a, Gimeno-García et al. 2000, De Luís et al. 2003; Chapter 10). Wind can rapidly remove a large fraction of nutrients contained in ash (Ewel et al. 1981, Kauffman et al. 1993, Giardina et al. 2000). Erosional losses can dominate the effects of fire on site nutrient budgets, especially for elements that are less subject to losses by oxidation or volatilization. Repeated fire is likely to lead to progressive incremental degradation of soil fertility if significant erosion occurs. Clearly, strong efforts need to be made to minimize erosion rates after wildfires or managed burns, and this is the subject of several other chapters in this volume.

Leaching

Leaching losses of N are small compared to gaseous losses of N in fire (Belillas and Feller 1998, Johnson et al. 2004). Murphy et al. (2006) observed an increase in ammonium (NH_4) and nitrate (NO_3) (and also of P and S) concentrations in soil solutions during the first winter after a wildfire, but the gaseous losses from the forest floor were much higher than the leaching losses. Exceptions may occur after an intense fire on very coarse-textured soils (Fisher and Binkley 2000). However, leaching can lead to significant loss of cations after an intense fire, with the accompanying anions often being sulfate and bicarbonate (see discussion in Khanna and Raison 1986). If erosion is increased after fire, it will be a much more significant nutrient loss mechanism than leaching.

CHANGED SOIL NUTRIENT AVAILABILITY

There are four main sets of processes affecting nutrient cycling in forest ecosystems:

a) Nutrient inputs, which include atmospheric inputs of a number of elements, fixation of atmospheric N, and release of many elements by mineral weathering.
b) Nutrient outputs by leaching, erosion, atmospheric transport, and removal of biomass.
c) Nutrient uptake and retention by the overstory and understory vegetation and also by microbial and faunal populations.
d) Nutrient turnover through above-ground and below-ground litter.

Depending upon the severity of fire, all of these processes undergo changes for a varying duration. While some of these changes are easy to measure and have been extensively studied, some changes can be more subtle but are of long-term significance. For example, recent studies by Mabuhay et al. (2004) of burned forest 2 mon and 3, 6, 9, and 25 yr after fire, showed that forest wildfires had a long-term effects on microbial biomass, abundance, and diversity in the soil.

Nutrient Changes due to Soil Heating, Ash Additions, and Heat-Ash Interactions

Deposition of ash adds variable quantities of nutrients to the soil, and much of this is in plant-available forms. Ash addition can make some elements available to vegetation at faster rates than would occur from biological decomposition. Ash usually has a low nitrogen (N), sulfur (S) and boron (B) content, but is rich in many other nutrients. In comparison with unburned eucalypt fuels, concentrations of calcium (Ca), magnesium (Mg), and phosphorus (P) in ash increased by 10 to 50-fold, 10 to 35-fold and 10-fold respectively (Raison et al. 1985a), with very high values in grey ash.

Not all of the nutrients present in the ash are easily available for transport in soils and uptake by plants. Using dissolution experiments on ash derived from forest fuels, Khanna et al. (1994) grouped the elements into three categories:

a) Most of the potassium (K), sulfur (S), and boron (B) occurred in water soluble form but about 30 percent of the amount remained insoluble, probably occluded in glass-like substances.
b) Much of the calcium (Ca), magnesium (Mg), silica (Si), and iron (Fe) were in sparingly soluble forms and dissolved progressively as more water was added to the ash.
c) Phosphorus (P) was mostly insoluble in water and required protons for dissolution.

These results describe most of the patterns observed in many studies after fire for both soil solutions and deeply leached waters. It also indicates why P in ash becomes available to plants only when the ash is mixed with acid soils. When ash remains on the soil surface it can take a long time to dissolve, depending upon the proton inputs to the soil surface. In their long-term (3 yr) study of soil solutions after fire, Ludwig et al. (1998) reported that magnesium calcite, which is commonly formed during fires, may be changed into insoluble forms as a result of changes in the chemistry of surface layers of the ash. This leads to its long-term retention on the soil surface.

Ash is highly alkaline and will increase the pH of the soil and its capacity to buffer protons. An increase in pH may also increase the cation exchange capacity of soils and help retain some of the cations released from the ash (Humphreys and Lambert 1965, Khanna et al. 1986). Changes in cation composition of soil and soil pH after ash addition are controlled by the soil buffering ranges described in detail by Walker et al. (1986). The salt content of ash is high and may initially inhibit seedling growth, but plants around an ashbed benefit from the nutrients made available in the short-and long term, causing the so-called 'ashbed effect'.

With respect to heating effects, the peak temperature during a fire is more important than the amount of heat to which soil and plant components are subjected. Gimeno-Garcia et al. (2004) reported that the peak temperatures (higher than 600°C) measured on the soil surface were independent of the amount of fuel burned, however the mean values differed depending upon the amount of fuel consumed (439° and 232°C lasting for 36 min and 17 min for heavy and light fuel loads, respectively). Heating causes many changes in soil properties, especially the organic and biological components and these are discussed in Chapters 4 and 5 of this volume. The effects of changes in organic matter and other soil properties on long-term nutrient availability remain poorly understood.

Heat and ash interactions affect many biological soil processes which affect cycling of nutrients via changes in mineralization of organic matter and nutrient uptake. Fire often enhances the subsequent decomposition of organic matter and the mineralization of organically bound elements. This may result from changed microbial populations, altered substrate quality (additional amount of available carbon (C) for respiration), and more favorable environmental factors (temperature, moisture, and pH; Walker et al. 1986). For example, in laboratory experiments where ash was added to different soils, enhanced respiration rates, especially in soils with high organic matter content, were observed during 6 wk of incubation (Khanna et al. 1994) until the easily mineralizable soil C was respired. Nitrogen mineralization rates and the amount of nitrate produced usually increase following a fire and in some cases may even lead to denitrification losses (Khanna et al. 1994). This may be related to changes in microbial populations. Bauhus et al. (1993) proposed that autotrophic nitrifiers might replace heterotrophic nitrifiers after fire due to changes in soil pH associated with the addition of ash. Similarly,

Hart et al. (2005b) noted that the relative abundance of bacteria to fungi increased in soils following fire. Very little is known of denitrification rates after fire, and this requires additional research effort.

There is sufficient evidence in the literature to show that soil microbial biomass decreases after fire (Bauhus et al. 1993) and it can take many years to achieve the pre-fire levels again. In addition to changes in microbial biomass and the microbial populations, enzyme activities vary significantly after fire (Serrasolses and Khanna 1995).

Changes in Soil Carbon Dynamics

Soil carbon (C) stocks and turnover affect nutrient cycles and many other soil properties and processes that control soil fertility and forest functioning (Raison and Rab 2001). The effects of fire on soil C dynamics and on C stocks are complex and poorly researched and understood. Fire can significantly affect inputs of C to the soil from above-ground litter and from root systems (live and fire-killed roots), as well as rates of loss of soil C directly from soil heating or from stimulated soil respiration after fire. It is the magnitude of change to C input and output processes that determines overall soil C dynamics. These changes are likely to be ecosystem-specific, so that generic responses are unlikely.

Reviewers of the literature often claim either little effect of fire on soil C, or sometimes an increase in soil C (e.g., Johnson and Curtis 2001, Gonzalez et al. 2004). However, many studies included in the above reviews are flawed because of the following methodological problems:

- Failure to account for both soil and root C when comparing 'soil' C stocks before and after fire. In soils sampled before fire, roots are mostly removed during sieving and thus root C is not measured as part of the soil C pool. Where fire kills vegetation, some dead and decaying roots can pass through the sieve, and are thus measured as part of the soil C pool. This renders pre- and post-fire comparisons of total soil C invalid because different C pools are being measured and compared before and at different times after fire. The input of C from dead roots can be high in forests that have high root biomass and where fire kills much of the vegetation. Soil C stock changes with time after fire of contrasting severity in the coastal eucalypt forests are described in Case Study 2. The input of C from killed roots on the strongly heated ashbeds during the initial years after forest harvesting can be clearly seen (Fig. 1).
- Failure to recognize that fires produce significant quantities of inert charcoal (char, black carbon), and that this progressively accumulates in soils, sometimes contributing 30 to 50 percent of soil C (Ponomarenko and Anderson 2001, Forbes et al. 2006). Thus, while total soil C may not change much, the biologically active pools may be more sensitive to fire. This is shown for soil N reserves and N mineralization rates in the sub-alpine eucalypt forests described in Case Study 1 (Fig. 1).

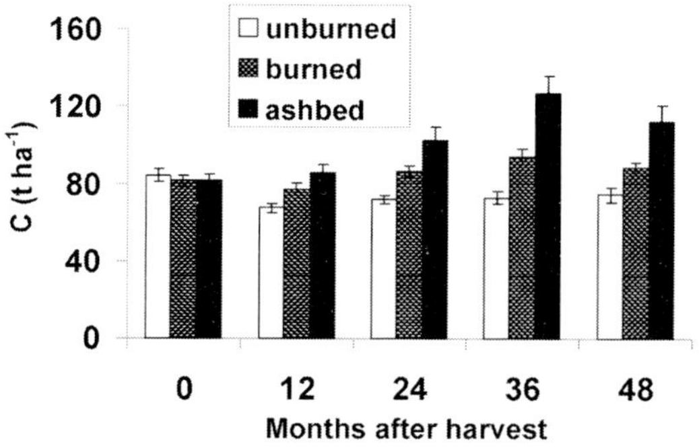

Fig. 1 Change in soil C (t ha^{-1}, 0-40 cm depth) during the first four yr after clearfell harvesting of coastal eucalypt forest followed by fire of differing severity (see Case Study 2 for details).

Future research must consider the above issues, and focus on the effects of fire on the key processes affecting soil C dynamics, and the consequences of these for nutrient cycling and soil fertility.

Nitrogen

The pools of ammonium nitrogen (NH_4-N) after fire in surface soil increase with the severity of fire, and this is mainly an effect of the denaturation of proteins by soil heating (Walker et al. 1986). Ash deposition can increase soil mineral nitrogen pools mainly as NH_4-N. Pools of nitrate nitrogen (NO_3-N) in ash are low and may greatly vary depending upon the type of the burned vegetation (Romanyà et al. 2001). Ash adds only small quantities of N to the soil, but ash can stimulate the mineralization of soil organic N (Khanna et al. 1994, Grogan et al. 2000), especially in soils that have been heated (Walker et al. 1986, Klopatek et al. 1990).

Many of the above changes are of short-term duration after fire. However, Grady and Hart (2006) observed a long-term decrease in N mineralization rates measured in the field 6 to 15 yr after a prescribed fire. This decrease was associated to low aboveground C inputs. However, in stands after high-severity wildfires, an increase in N availability was observed which was related to changes in soil microclimatic conditions. Changes in the understory components after fire may change the quantity and especially the quality of litter inputs to the forest floor and thus rates of nutrient turnover.

Redistribution of ash by wind or water after fire may increase spatial heterogeneity of soil N availability and increase patchiness of vegetation in fire-prone ecosystems (Grogan et al. 2000). Similarly, erosion and deposition of nutrient-rich surface soil after fire can create strong gradients in soil fertility.

Fire, by enhancing short-term rates of N cycling, has the potential to increase gaseous losses of N resulting from nitrification and denitrification. In particular, wetting of burned surfaces can enhance such losses in savannahs (Levine et al. 1996, Swap et al. 1996).

Wan et al. (2001) conducted a meta-analysis to examine the relationships between fire and N pools and dynamics across terrestrial ecosystems. They found that fire increased soil ammonium and nitrate pools. Soil ammonium pools declined exponentially with time after fire, and on average returned to pre-burn levels after 3 yr. Soil nitrate pools on average peaked about 1 yr after fire, and returned to pre-fire levels after about 2 yr. It is important to recognize however that mineral-N pools merely represent the residual between N mineralization and plant uptake, and are not a good indicator of soil N mineralization rates and thus the availability of N to vegetation.

Phosphorus

Many of the world's forests that are subject to recurrent fire grow on phosphorus-deficient soils. Despite the fact that fire can result in significant losses of P in smoke, or by erosion of ash or surface soil, and that fire can mobilize P in the soil, there have been relatively few studies of the effects of fire on P cycling in forests. Following slash burning on a harvested eucalypt forest site, Romanyà et al. (1994) estimated that about 8 kg P ha^{-1} was deposited as ash. The ashbeds (18 percent of the area) showed an increase in P sorption capacity in the 0 to 5 cm soil layer, but the sorbed P was not tightly bound to the solid phase. Solubility of P deposited in ash is related to the availability of protons as discussed above.

Fire, by changing soil P availability, may indirectly change the rates of N fixation by native legume systems which are typically constrained by low P availability. Addition of P to a range of Australian eucalypt forests, for example, increased rates of N fixation by native legumes several-fold (Raison et al. 1993a). Addition of P can dramatically increase legume growth, and after several years this is reflected in enhanced rates of soil N mineralization (Table 6). Likewise, N fixation rates are typically increased for several years after fire disturbance, and this is likely to be partly due to the short-term enhancement of P availability after fire, as well as to the regeneration (heat often stimulates the germination of 'hard' seeds) of dense actively growing stands of understory legumes. We can speculate that frequent fire might lead to P depletion and a reduction in long-term P availability, with consequent negative effects for N fixation rates. This coupled with the cumulative combustion losses of N that can lead to reduction in rates of litter and soil N mineralization (Table 3) and may create significant N limitations in re-currently burned forests.

CHANGES IN VEGETATION DYNAMICS AFFECTING NITROGEN (N) FIXATION, NUTRIENT UPTAKE AND NUTRIENT TURNOVER

Vegetation Dynamics

Fire can exert major effects on ecosystem nutrient cycling and vegetation structure and composition over the short and long-term. Fire has an important

Table 3 Percentage reduction in indices of N cycling caused by five repeated low-severity fires applied at three year intervals, compared with the absence of fire for 14 years (after Raison et al. 1993a).

N cycling indicator	Reduction (%)
N mineralization – soil	50(a)
– litter	57
N concentration in expanded leaves	13
N stock in foliage	17
N in annual leaf fall	17

(a) 35% reduction after two fires, each 7 yr apart.

role in the maintenance of landscape diversity and often restricts the dominance of climax ecosystems (Rovira et al. 2004). For instance, in the Mediterranean landscapes, wildfires at 25 to 50 yr intervals are thought to be necessary to maintain shrubland communities intermingled with *maquias* or forested areas (Trabaud 1991, Specht 1981). However, the long-term impacts of fire regimes on ecosystems are complex and not well understood.

Post-fire plant regeneration induces important changes in the soil system and become the drivers of the transformations occurring in the medium and long-term after fire. With plant growth, soil becomes less vulnerable to erosion. According to Thornes (1990) soil protection against erosion is ensured when at least 30 percent plant cover is achieved. With plant recovery, nutrient availability is controlled again by micro-organisms (organic matter mineralization, immobilization), plants (nutrient uptake, litter fall, organic matter quality) and plant-microbial mutualisms (symbiotic N-fixing plants, mycorrhizae; Hart et al. 2005b).

In fire-prone ecosystems often there is a quick recovery of plant cover after fire. The main reproductive strategies for regeneration in response to fire are resprouting from fire-resistant structures and germination of fire-protected seeds. In the Mediterranean ecosystems most of the post-fire recovery is carried out through auto-succession or direct regeneration resulting from a rapid recovery of the pre-fire plant species. This is the case of the Californian chaparral, the *Quercus coccifera garrigues*, and other *Quercus* spp. forests, or the Aleppo pine (*Pinus halepensis)* forests. However other plant formations such as some pine forests do not follow direct regeneration after fire; for example, *P. nigra* (Trabaud and Campant 1991), *P. sylvestris* and *P. pinea* forests (Retana et al. 2002, Rodrigo et al. 2004) because of a lack of serotinous cones or failure in germination, or high mortality of seedlings. Even Aleppo pine and *P. pinaster*, which usually recover well after fire, may disappear if a fire recurs in less than 10 to 15 yr, due to the absence of viable seeds. These pine formations change to other plant communities such as shrublands or *Quercus* spp. forests growing from individuals already present in the pre-fire understory.

Thus, in some forests, fire can lead to a dramatic loss of species, or change species composition and landscape structure, which in turn may change the litter nutrient composition and turnover rates. In some cases, poor plant regeneration and low plant cover may lead to soil degradation and erosion. Short-term plant regeneration strategies are important for re-establishing plant cover and, thus, for reducing the potential for erosion and leaching of nutrients after fire.

Studies of the long-term effects of fire on nutrient cycling are scarce and often the effects of the fire regime, land use history and forest aging are confounded. In the California chaparral, a decrease in net vegetation productivity and nutrient assimilation has been observed after one or two decades, as nutrients become immobilized in vegetation, litter, and soil organic matter (Schlesinger and Gill 1980). In South African fynbos, the period of fire recurrence was found to be shorter than the time for litter accumulation to reach the steady-state (Mitchell et al. 1986). A high fire frequency can also deplete the productivity of the Mediterranean *Quercus coccifera* garrigue in southern areas while in northern areas, with more rain, this ecosystem has shown higher resilience (Delitti et al. 2005). Belowground nutrient reserves are responsible for the quick recovery of aboveground N and P stocks in the garrigue (Ferran et al. 2005), but after repeated fires, nutrient losses may deplete these reserves and produce a loss of biomass and productivity. Cobo and Carreira (2003) suggest that when fire frequency is abnormally high, the ecosystem nutrient capital does not fully recover, and the long-term balance between fire-related nutrients outputs and succession-related nutrients inputs is upset.

By separating the effects of previous land cultivation from fire, Duguy et al. (2007) established that both fire and long-term cultivation decreased soil organic matter and C/N ratio, and increased P availability.

In forest soils, soil C increases with the age of the forests while nutrient availability often decreases. In mature forests there is also a higher presence of allelochemicals such as tannins, phenols and terpenoids that among other effects may inhibit N mineralization (White 1991). Polglase et al. (1992a, b) found a large decrease in N mineralization and labile P in *Eucalyptus regnans* forests 250 yr after fire as compared to forests of age 80 yr.

Fire Effects on N Fixing Systems

There are a range of mechanisms (symbiotic and non-symbiotic N fixation, rainfall accession) that can replenish the losses of N in fires.The activity of N-fixing systems can be of major importance after fire, in particular N fixation by understory plants is significant. Rates of N fixation are affected by canopy dynamics and the availability of soil resources such as water and phosphorus, and these are affected by patterns of vegetation development during post-fire plant succession.

N fixation by woody understory species is a common mechanism in forests and woodlands, and this is often stimulated by fire as a consequence of the germination of 'hard' seeds after soil heating, rapid resprouting from rootstocks, or improved availability of water and soil P, which limits N fixation rates in many forest environments.

Nitrogen-fixing systems can range from symbiosis or associations with some vascular plant species, mainly legumes, to free living organisms such as cyanobacteria, or heterotrophic N-fixing free living bacteria (see Marschner 1995). N-fixing systems depend on soils resources, except for the autotrophic cyanobacteria that are constrained by the availability of labile organic energy sources.

Fisher and Binkley (2000) concluded that only changes to symbiotic N-fixation rates are likely to be of ecological significance because non-symbiotic rates are so low as to generally only be important over time frames of decades to centuries. However, in the N-limited savannah of southern Africa, Aranibar et al. (2003) concluded that atmospheric N depositions, and N-fixation by forbs and soil cyanobacteria, also appear to be important. Cook (1994) concluded that in the savannah of northern Australia, N inputs from known mechanisms were inadequate to replace the loss of N resulting from the usual annual fires that occur naturally in that part of the world.

N-fixation rates of free living heterotrophic bacteria are generally low and highly dependent on the inputs of fresh organic matter. The amount of N fixed by heterotrophic bacteria living in the forest floor vary greatly (from 0.18 to 4 kg N ha^{-1} yr^{-1}) (Vitousek and Hobbie 2000). No study on the effects of fire on N-fixation by these bacteria have been found, but it can be assumed that their fixing capacity may be lowest after fire, when the availability of fresh organic matter will be low.

Autotrophic N-fixation normally occurs in microbiotic crusts on the soil surface in which cyanobacteria associate with mosses, liverworts, algae, lichens, fungi or other bacteria. While these crusts have been found to be highly vulnerable to fire (Evans and Johansen 1999), some authors have found, by incubating soils in the laboratory, that cyanobacteria can exhibit rapid growth in fire-disturbed soils (Vazquez et al. 1993) and enhance nutrient cycling (Acea et al. 2003). In a boreal forest, a study of a chronosequence of burned sites ranging in age from 35 to 355 yr after fire, showed that autotrophic N fixation, in this case associated with mosses, increased with the age of the forest (Zackrisson et al. 2004). Most studies suggest N-fixation rates for such systems to be much less than 18 kg N ha^{-1} yr^{-1} (Evans and Johansen 1999). Furthermore, ammonia (NH_3) volatilization and denitrification have also been found to increase in these microbiotic crusts suggesting lower net N inputs.

The woody legumes are one of the most successful plant groups in tropical and sub-tropical areas (Crews 1999), and unlike in the temperate and cold regions are present in many mature plant communities. In temperate, boreal and tundra ecosystems, herbaceous legumes are abundant in the early successional floras of most communities including Mediterranean ecosystems where legumes have been identified as an important component of post-fire successional communities (Crews 1999, Naveh 1967, Bell and Koch 1980). In non-tropical areas there are some studies showing that legumes are reduced in number and plant cover as succession progresses (Arianoutsou and Thanos 1996, Newland and DeLuca 2000). In spite of the high abundance of woody legumes in many mature tropical and subtropical ecosystems, rates of N-fixation are often low. The rate of N-fixation by understory acacias can be increased after fire (Guinto et al. 2000), but these authors found slightly reduced fixation rates on dry sites with frequent burning.

It has been suggested that the pulse of growth and N-fixation following fire can be attributed to the removal of shading and to the enhancement of P availability (Vitousek and Field 1999). Rastetter et al. (2001) used modeling to conclude that legume N-fixation can be more economical than soil N uptake when the canopy is open, the soil is well exploited by roots as occurs in resprouting communities, soil inorganic N concentration is low and other soil resources such as P and water are readily available. Some studies show enhanced N-fixation after fire when soil N availability is low (Newland and DeLuca 2000). However, in Mediterranean grasslands and shrublands significant N-fixation by legumes occurs soon after the fire in spite of increased post-fire soil mineral-N (Casals et al. 2005). In that study, N-fixation 2 to 4 mo after the fire was high for both seedlings and resprouters, suggesting that N-fixing systems were soon operative after fire and were independent of the age of the rooting systems. Post-fire N fixation seems to depend mainly on soil conditions rather than on the legume regeneration strategies. Studies in frequently burned longleaf pine savannahs found that the timing of fire did not have much impact on N fixation rates in spite of the changes that occurred in plant growth patterns (Hiers et al. 2003).

CHANGES IN PATTERNS OF LITTER INPUT, QUALITY, DECOMPOSITION AND NUTRIENT MINERALIZATION

Fire can disrupt the cycling of nutrients through litter that are critical for the on-going nutrition of forests. Fire can lower the pools of nutrients in the forest floor, change the rates of litter input by affecting vegetation dynamics and modify rates of litter decomposition.

Fire can markedly change the micro-climate of the forest floor and soil, especially higher severity fires that defoliate the tree canopy or result in stand death and replacement. Fire can result in elevation of soil temperatures, and sometimes soil moisture because of reduced water use by vegetation, with effects persisting for months to years depending on the rate of recovery of the vegetation. These changes will favor increased soil mineralization rates. However, the forest floor can become drier in warm climates, and this can inhibit litter decay rates after fire (e.g., Raison et al. 1986).

The pattern of accumulation of fuels, and the nutrients they contain, after a fire event is a crucial variable affecting the impacts of fire on nutrient cycles.

The rate of build-up of fuel determines the development of fire risk and the likely frequency and severity of fire, as well as the impacts of fire on nutrient cycles as a consequence of nutrient transfers to the atmosphere, the nature of ash inputs, and impacts on forest successional processes including the abundance and activity of N-fixing species. Forest systems in which fuels (and fuel nutrient content) re-accumulate quickly after a previous fire are generally more susceptible to fire-induced depletion of nutrients (Raison et al. 1993a).

In the Mediterranean area, forest fires usually burn L and F layers of the forest floor but the H horizon burns only slightly, or does not burn, and it is enriched by the ashes and charred remains of the biomass and the overlaying layers (Ferran et al. 1992a, b, Serrasolses and Vallejo 1999). The climatic conditions just after the fire in relation to the degree of revegetation and soil erodibility will determine the intensity of the erosive processes. In areas with high stoniness in the soil surface, the H layer remains protected from fire and from erosion, as this is the case of garrigues growing in soils developed over limestone (Ferran 1996), and the remaining H layer decomposes slowly over some years. Field experiments have shown that soils with high stone content show higher moisture content and biological activity in the H layer than soils without stones (Casals et al. 2000). When erosive processes are high, the H layer remaining after fire moves down slope and it is quickly mineralized or mixed with the mineral soil (Ferran 1996, Serrasolses and Vallejo 1999).

The replacement of the organic matter lost during fire or erosion events occurs via long-term plant production and litter inputs. The forest floor is reconstructed along with the re-development of vegetation, related mostly to increasing plant cover. Litterfall from resprouting *Quercus coccifera* garrigue was highest 3 yr after fire and then decreased slightly to values of the mature vegetation (Ferran 1996). The *Q. coccifera* litter falling during the initial year after fire had a high nutrient concentration, related to the high nutrient availability, and consequently litter decomposition was rapid in 1-yr-old burned stands and decreased during the succession. Forest floor reconstruction of regenerating garrigue occurred gradually and stabilized around 6 to 10 yr after fire. In a resprouting coppice of holm oak (*Quercus ilex*), litterfall recovers quickly and the litter layer (L horizon) recovers in about 2 yr. A longer period was required for recovery of the F layer (18 to 23 yr), and it was estimated that the H layer may take several decades to be completely reconstructed (Ferran and Vallejo 1992b). During the first stages of succession, plant cover was dominated by different species of seeder shrubs with abundance of N-fixing species that produced litter with a low C/N ratio.

In some areas, fire recurrence does not permit the development of the whole forest floor profile. In an Aleppo pine forest in the Mediterranean, Eugenio et al. (2006) found that fire recurrence decreased the thickness and dry mass of the forest floor by comparing once and twice-burned forests, 9 yr after the last fire. In *Quercus coccifera* garrigues growing in dry Mediterranean areas Ferran et al. (2005) found that the forest floor contains a high proportion of the total soil nutrient pools, especially N. Thus combustion of aboveground biomass and forest floor may produce significant losses of N and P. However, significant belowground nutrient reserves may account for the quick recovery of *Quercus coccifera* aboveground biomass and forest floor L horizon. Successive fires often do not permit the development of the whole forest floor present in unburned garrigues. Under these conditions, recurrent fires can deplete soil nutrient reserves and produce a loss of biomass and productivity of this species (Ferran et al. 2005). Forest floor accumulation in Mediterranean garrigues often increases with mean annual rainfall. This may result from both low plant productivity and/or higher fire frequency in drier areas. Higher nutrient accumulation that occurs in garrigues in areas with more rain suggest that the single fire impact may be larger in these areas although their resilience may be also higher due to high nutrient reserves in the mineral soil and root systems.

CONTRASTING EFFECTS OF FIRE ON NUTRIENT CYCLING PROCESSES

Case Study 1. Effects of Repeated Low-severity Prescribed Burning on N Cycling in Sub-alpine Eucalyptus Forest

Australian sub-alpine eucalypt forests are subjected to low-severity prescribed fires that aim to reduce fuel loads and the risk of subsequent wildfires (Raison et al. 1986). These fires consume part of the understory vegetation and litter layer, but flame heights are usually <1 m so that the overstory is largely unaffected. Figure 2 shows a *Eucalyptus pauciflora* (snowgum) forest about 10 yr since a previous low severity fire that has a well developed understory of the woody shrub *Davesia mimosoides* and a litter layer of mass about 15 t ha^{-1}. Figure 3 shows a low severity prescribed burn and Figure 4 shows the recently burned forest. Only very small (<10 cm dbhob) trees that have very thin bark are killed or damaged by these low severity fires. The effects of such fires, applied at a range of frequencies, on N budgets and N cycling has been studied over several decades in a *Eucalyptus pauciflora* forest near Canberra, Australia (Raison et al. 1985a, b, Raison et al. 1993a).

The forest is characterized by a rapid re-accumulation of fuel (and fuel N content) after fire. N stocks in the leguminous understory plus ground litter reach about 120 and 180 kg ha^{-1}, respectively, 3 and 10 yr after a previous low-intensity fire (Raison et al. 1993a). Low severity fire oxidizes 50 to 75 percent of the N in the fuel (Raison et al. 1985b). The rate of N input to these forests, mostly derived from symbiotic N fixation, is ~10 kg $ha^{-1}yr^{-1}$. Figure 5(a) summarizes N losses by oxidation in a fire in fuels of increasing age (and fuel N content), and for situations where either 50 or 75 percent of the fuel N is lost. Figure 5(b) shows the effect of differing fire frequency and percentage loss of N from fuels on rates of N accretion or depletion in the forest.

Fig. 2 Snowgum forest unburned for ten years.

Fig. 3 Low severity prescribed burn in snowgum forest.

Fig. 4 State of understory and forest floor soon after a low severity prescribed burn in snowgum forest.

Very frequent (three yearly) fires lead to N depletion rates of 8 to 17 kg $ha^{-1}yr^{-1}$. If fire frequency or spatial coverage results in <50 percent of fuel N loss, site N balance will be maintained where fires occur about every 9 yr. If 75 percent of fuel N is lost, inter-fire periods of about 15 yr are required to maintain N balance.

Changes in the N budget are reflected in rates of N cycling, because fire has the most effect on the more biologically labile N components (those contained in litter, understory, and surface soil organic matter) in the forest ecosystem. Table 3 compares the effects of repeated burning, with the absence of fire on several indicators of N cycling in the snowgum forest.

The snowgum forest contains large amounts of N in soil organic matter, and despite the fact that the five fires would have removed <5 percent of the soil N capital, there was a 50 percent decline in soil N mineralization rate (Table 3). This study demonstrates that the labile pools of soil and litter N can be rapidly depleted by fire. Similar findings have been reported for a range of forest types around the world (Raison et al. 1993a), demonstrating that a minimum inter-fire period is required to maintain N cycling. The appropriate fire regime will vary, and needs to be established by regional research studies to provide guidance for sustainable fire management practices.

Case Study 2. Effects of Slash Burning on P and N Cycling After Harvest of Coastal Mixed Species *Eucalyptus* Forest

This case study summarizes the effects of clearfell harvesting followed by slash fire of differing intensity on nutrient dynamics and forest growth in

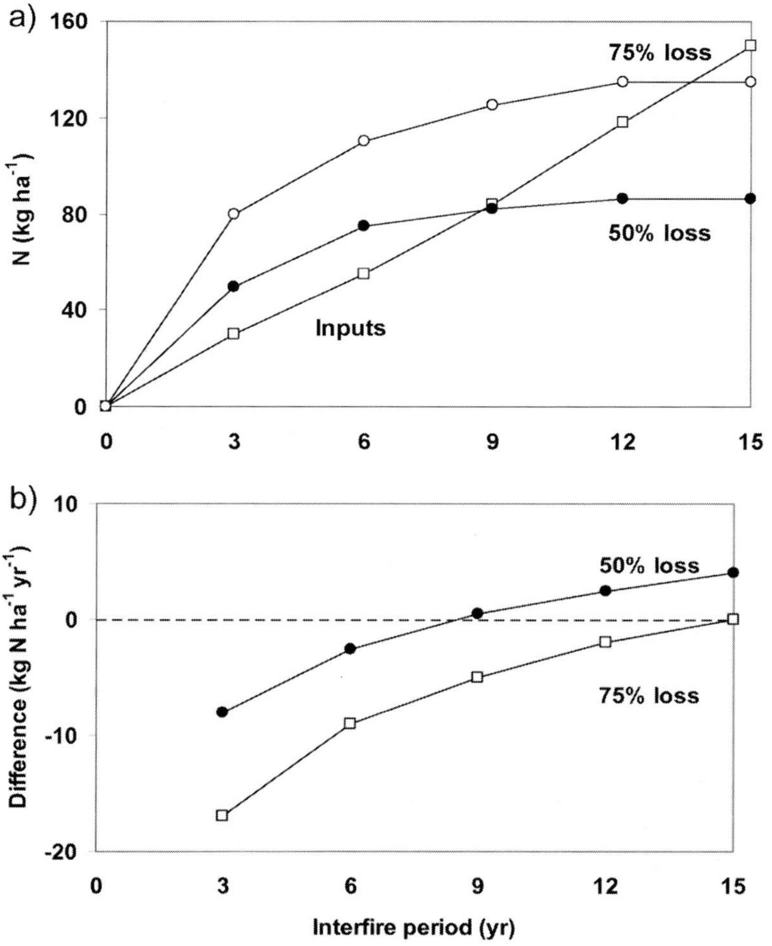

Fig. 5 Predicted losses of N (kg ha^{-1}) in low severity fires in snowgum forest assuming either 50 or 75 percent loss of N from fuel components, for periods up to 15 yr after the previous low severity fire a), and the effects of the length of the interfire period (fire frequency) on rates of depletion or accretion of N b). Negative numbers indicate net depletion of N.

coastal mixed-species eucalypt (see Raison et al. 1993b and Romanyà et al. 1994 for further details) during the initial 8 yr of forest regeneration near Orbost, in East Gippsland, Australia. Figure 6 shows the slash distribution after harvesting, and Figure 7 the nature of the slash burn. Figure 8 shows the spatial variation in fuel consumption and fire severity that has led to formation of a range of microsites (unburned areas, moderately burned areas, and 'ashbeds' where accumulations of woody slash have burned resulting in intense heating of the soil and large inputs of ash). Supplementary work

Fig. 6 Recently harvested coastal Eucalyptus forest.

Fig. 7 Slash burn.

describing the response of the forest to small inputs of P fertilizer (50 kg P ha^{-1} applied as superphosphate to lightly burned areas 6 mo after the slash burn) that are useful in interpreting the effects of fire on nutrient dynamics is also discussed.

The sandy duplex soil supporting the forest is low in total P (<50 mg kg^{-1} in the surface 0 to 10 cm layer). Forest growth is markedly stimulated by the addition of small amounts of P fertilizer (Raison et al. 1993b; Table 6). Fire

Fig. 8 Immediately after slash fire showing a range of microsites experiencing contrasting burn severities and modification of soil properties.

also stimulates soil P availability and the uptake of P and early growth rate of vegetation, especially on ashbeds (Romanyà et al. 1994; Fig. 9).

Fire severity varies greatly across the burned forest coupe (Fig. 8), in the study reported below, 35 percent of the area remained unburned, 41 percent was burned by fire of moderate severity, and 24 percent was 'ashbed'. Removal of vegetation leads to soils becoming wetter and warmer for several years after clearfelling and burning, with changes diminishing over time as the vegetation re-develops.

Clearfelling and slash burning greatly accelerates soil N mineralization rates during the initial 3 yr of the forest rotation (Table 4). By the third year soil N mineralization rates are declining, at the same time as N demand by the regenerating forest is increasing. During the initial 2 yr after burning, N mineralization rates were lowest on the 'ashbeds' (Table 4) as a consequence of direct losses of soil N by heating, and a period of net immobilization caused by high C input from fire-killed roots to the soil (Raison et al. 1993b). In the third year, mineralization rates had recovered on the ashbeds, probably due

Table 4 Changes in *in situ* N mineralization rates (kg $ha^{-1}yr^{-1}$) during the first three years after clearfelling and burning mixed-species Eucalyptus forest

Site	Mean of years 1 and 2	Year 3
Uncut forest	30	15
Cut – unburned	60	26
– burned	55	36
– ashbed	40	37

Fig. 9 Rapid regeneration of vegetation, especially acacias near an ashbed one year after the slash burn.

to enhanced inputs of N from acacias that respond rapidly to increased P availability (Table 6) and which grew very rapidly on the ashbeds (Fig. 9).

The addition of a small amount of P stimulated plant growth to a level comparable to that on ashbeds, and greatly increased the concentrations of P in the biomass components of acacias and eucalypts (Table 5). P addition significantly increased the concentration of N in acacias despite a 10-fold increase in acacia biomass (Table 6) and suggests increased rates of N fixation by acacia. Two years after P addition, the concentrations of N in leaves of the eucalypts had declined because of dilution as a consequence of rapid growth response to added P. At this stage the transfer of N fixed by acacias was probably still small, but soil N mineralization rates were high (~60 kg N $ha^{-1}yr^{-1}$; Table 4) and this was sufficient to enable a significant eucalypt growth response to added P. Between years four and five, N mineralization rates were 40 percent higher on the +P plots (Table 6) presumably because of the turnover of N previously fixed by acacias.

Table 5 Concentrations of N (%) and P (mg kg^{-1}) in foliage of 2.5-yr-old regenerating eucalypts and acacias in unfertilized plots and in areas fertilized 2 yr previously with 50 kg P ha^{-1}. The experimental area had only been lightly burned by the regeneration burn.

	P Acacia	P Eucalypt	N Acacia	N Eucalypt
–P	175	300	1.1	1.0
+P	310	430	1.4	0.7
+P/–P	1.8	1.4	1.3	0.7

Table 6 Effect of fertilization with P, on the growth of naturally regenerated acacias and eucalypts, and on rates of *in situ* soil N mineralization.

	Age	–P	+P	+P/–P
Acacia biomass (t ha^{-1})	2	3.8	37.4	9.8
Eucalypt volume (cm^3 $tree^{-1}$)	7	2210	7200	3.3
Soil N mineralization (kg $ha^{-1}yr^{-1}$)	4–6	74	102	1.4

Growth of both acacias and eucalypts was increased by P addition (Table 6). Growth rates of eucalypts on P fertilized plots were similar to those on ashbeds, and acacias grew vigorously on both areas. By about age 5 yr, eucalypts began to overtop the dense biomass of acacias on both areas leading to suppression of their growth and inducing some mortality and greater input of N-rich litter. Soil N mineralization rates were increased by about 40 percent in the +P plots during years four to six, and this coupled with good availability of P (Table 5) led to a three-fold increase in volume growth by eucalypts (Table 6).

This study illustrates some important principles:

- The effects of fire on the nutrient cycle and on plant growth are highly spatially variable; for example, by age 8.5 yr, 60 percent of the volume production by eucalypts on the harvested and burned area was on the 'ashbeds' which occupied only 24 percent of the area.
- Fire severity greatly affects nutrient losses, soil heating and ash inputs. In P-limited forests, fire-mobilized P can greatly stimulate symbiotic N fixation and help re-build N capital and maintain N cycles.
- P limitation can be important in mature forests, and further loss of P in high severity fires is significant. Replacement of P losses by natural processes will be very slow. Fire may result in longer-term decline in P cycling (and as a consequence, of N cycling) especially on P-poor sites. The longer-term effects of fire regimes on P and N cycles needs to be better understood because the availability of these elements to vegetation is a primary determinant of ecosystem productivity which affects many other forest values.

CONCLUSIONS

A cautious approach needs to be taken in generalizing the effects of fire on nutrient budgets in particular forest systems so that broad assumptions are not incorrectly made. This is demonstrated by the work of Johnson et al. (2005, 2007) who compared measured losses of C and nutrients after a wildfire in a mixed conifer forest, with losses estimated using the following commonly adopted assumptions:

- All of the C and N contained in the foliage and forest floor is combusted.
- All of the P, K, Ca, and Mg is converted to ash and retained on site.
- No soil C or N is lost by direct oxidation.
- Fire accelerated leaching is minimal in comparison to other fluxes.

They found that losses based on these assumptions underestimated actual losses of C and N by about 50 percent, and that for other elements the results were variable.

Fire can have significant impacts on P and N cycles in forests. Long-term depletion of stocks or rates of cycling of these elements is clearly undesirable, so management regimes need to be tailored to avoid this. Appropriate fire regimes will vary depending on the capacity of ecosystems to buffer and replenish nutrient losses. Characteristics of more vulnerable ecosystems are described in Table 7, and also discussed in Case Study 1 describing the effects of repeated fire on N cycling in a sub-alpine eucalypt forest. Where fire is used as a management tool, it should not be applied in situations where the risks of accelerated soil erosion are high. Applying prescribed fire when the surface soil and lower litter layers are moist is desirable because atmospheric transfer of nutrients from fuels are reduced, direct loss of organic matter and N from surface soils is prevented, and a layer of partly combusted forest floor remains that can help reduce erosion (Raison et al. 1986).

There are many studies worldwide demonstrating that frequent fires can have deleterious effects on forest nutrition and ecosystem function (Raison et al. 1993a, Fisher and Binkley 2000, Wan et al. 2001, Smithwick et al. 2005). A recent example is the detailed studies by Hart et al. (2005a) who showed

Table 7 Situations where fire may pose a significant risk to the maintenance of forest nutrient cycles.

- Where forests are already nutrient-limited and where natural inputs of nutrients, especially of N and P, are small.
- Where a high proportion of site nutrient pools are held in combustible vegetation or dead organic matter.
- Where fuel mass (and fuel nutrient content) re-accumulate rapidly after burning, and there is frequent fire.
- Where fire intensity and the mass of fuel combusted are high.
- Where fire effects induce increased erosion of ash and/or surface soil (typically occurs where there are erodible soils, steep slopes, and high intensity rain events).
- Where fire reduces rates of N fixation (e.g., by changing the composition of understory vegetation).
- Where the frequency of fire is increased in comparison to historical rates, due to climate change, population pressure, and/or to changed fire management policies.

that 2 yr interval burning in Ponderosa pine forests negatively impacted on fine root and mycorrhizal dynamics and nutrient cycling processes.

Tree growth rates are sometimes changed after fire (often increased in the short-term) and can be the result of many factors (changed nutrient and water availability, altered tree density, effects on competing vegetation). Very few studies have tried to separate out the contribution of these factors (Fisher and Binkley 2000). Care should be taken not to assume that short-term increase in growth after fire will be maintained in the long-term, and that the effects of fire on nutrient cycling are always positive. In order for long-term forest productivity to be maintained, the long-term nutrient cycling must also be maintained. This requires the application of appropriate fire regimes for particular forest types and site conditions.

References

Acea, M.J., A. Prieto-Fernández, and N. Diz-Cid. 2003. Cyanobacterial inoculation of heated soils: effect on microorganisms of C and N cycles and on chemical composition in soil surface. *Soil Biology and Biochemistry,* 35: 513-524.

Aranibar, J.N., S.A. Macko, A.C. Anderson, A.L.F. Potgeiter, R. Sowry, and H.H. Shugart. 2003. Nutrient cycling responses to fire frequency in Kruger National Park (South Africa) as indicated by stable isotope analysis. *Isotopes Environmental Health Studies,* 39: 141-158.

Arianoutsou, M. and C.A. Thanos. 1996. Legumes in the fire-prone Mediterranean regions: An example from Greece. *International Journal of Wildland Fire,* 280: 119-125.

Bauhus, J., P.K. Khanna, and R.J. Raison. 1993. The effect of fire on carbon and nitrogen mineralization and nitrification in an Australian forest soil. *Australian Journal of Soil Research,* 31: 621-639.

Belillas, C.M. and M.C. Feller. 1998. Relationship between fire severity and atmospheric and leaching nutrient losses in British Columbia's coastal western hemlock zone forests. *International Journal of Wildland Fire,* 8: 87-101.

Bell, D.T. and J.M. Koch. 1980. Post-fire succession in the northern jarrah forest of western Australia. *Journal of Ecology,* 5: 9-14.

Boring, L.R., J.J. Hendrick, C.A. Wilson, and R.J. Mitchell. 2004. Season of burn and nutrient losses in a longlead pine ecosystem. *International Journal of Wildland Fire,* 13: 443-453.

Byram, G.M. 1959. Combustion of forest fuels. pp. 61-89. *In* K.P. Davis [ed.]. *Forest Fire: Control and Use.* McGraw-Hill, New York, USA.

Casals, P., J. Romanya, and V. Vallejo. 2005. Short-term nitrogen fixation by legume seedlings and resprouts after fire in Mediterranean old fields. *Biogeochemistry,* 76: 477-501.

Casals, P., J. Romanyà, J. Cortina, P. Bottner, M.M. Coûteaux, and V.R. Vallejo. 2000. CO_2 efflux from a Mediterranean semi-arid forest soil. I. Seasonality and effects of stoniness. *Biogeochemistry,* 48: 261-281.

Certini, G. 2005. Effects of fire on properties of forest soils: A review. *Oecologia,* 143: 1-10.

Choromanska, U. and T.H. DeLuca. 2002. Microbial activity and nitrogen mineralization in forest mineral soils following heating: evaluation of post-fire effects. *Soil Biology Biochemistry,* 34: 263-271.

Cobo, M.C. and J.A. Carreira. 2003. Fire recurrence thresholds and desertification in Mediterranean semiarid shrublands: the role of soil nutrient capital depletion. *Arid Lands Newsletter*, 54.

Cook, G.D. 1994. The fate of nutrients during fires in a tropical savannah. *Austral Ecology*, 19: 359-365.

Crews, T.E. 1999. The presence of N fixing legumes in terrestrial communities: Evolutionary versus ecological considerations. *Biogeochemistry*, 46: 233–246.

DeBano, L. and J. Klopatek. 1988. Phosphorus dynamics of pinyon-juniper soils following simulated burning. *Soil Science Society of America Journal*, 52: 271-277.

Delitti, W., A. Ferran, L. Trabaud, and V.R. Vallejo. 2005. Effects of fire recurrence in *Quercus coccifera* L. shrubland of the Valencia Region (Spain) I. Plant composition and productivity. *Plant Ecology*, 177: 57-70.

De Luis, M., J.C. González-Hidalgo, and J. Raventós. 2003. Effects of fire and torrential rainfall on erosion in a Mediterranean gorse community. *Land Degradation and Development*, 14: 203-213.

Duguy, B., P. Rovira, and V.R. Vallejo. 2007. Land-use history and fire effects on soil fertility in eastern Spain. *European Journal of Soil Science*, 58(1): 83-91.

Eugenio, M., F. Lloret, and J.M. Alcañiz. 2006. Regional patterns of fire recurrence effects on calcareous soils of Mediterranean *Pinus halepensis* communities. *Forest Ecology and Management*, 221: 313-318.

Evans, R.D. and J.R. Johansen. 1999. Microbiotic crusts and ecosystem processes. *Critical Reviews in Plant Sciences*, 18(2): 183-255.

Ewel, J., B. Berish, N. Brown, N. Price, and J. Raich. 1981. Slash and burn impact on a Costa Rican wet forest site. *Ecology*, 62: 816-829.

Ferran, A., I. Serrasolses, and R. Vallejo. 1992a. Soil evolution after fire in *Quercus ilex* and *Pinus halepensis* forests. pp. 397-404. *In* A. Teller, P. Mathy, and N.R.S. Jeffers [eds.]. *Responses of Forest Ecosystems to Environmental Changes*. Elsevier Applied Science, London, UK.

Ferran, A. and V.R. Vallejo. 1992b. Litter dynamics in post-fire successional forests of *Quercus ilex*. *Vegetation*, 99/100: 239-246.

Ferran, A. 1996. La fertilitat de sols forestalls en la regeneració després del foc de diferents ecosistemes mediterranis. Ph.D. thesis, Universitat Barcelona, Barcelona, Spain.

Ferran, A., W. Delitti, and V.R. Vallejo. 2005. Effects of fire recurrence in *Quercus coccifera* L. shrubland of the Valencia Region (Spain) I. Plant and soil nutrients. *Plant Ecology*, 177: 71-83.

Fisher, R.F. and D. Binkley. 2000. Fire effects. pp. 241-262. *In* R.F. Fisher, D. Binkley, and W.L. Pritchett [eds.]. *Ecology and Management of Forest Soils*. John Wiley and Sons, Inc., New York, USA.

Forbes, M.S., R.J. Raison, and J.O. Skjemstad. 2006. Formation, transformation and transport of Black Carbon (charcoal) in terrestrial and aquatic ecosystems. *Science of the Total Environment*, 370: 190-206.

Giardina, C.P., R.L. Sanford Jr., and I.C. Dockersmith. 2000. Changes in soil phosphorus and nitrogen during slash-and-burn clearing of a dry tropical forest. *Soil Science Society of America Journal*, 64: 399-405.

Gillon, D. and M. Rapp. 1989. Nutrient losses during a winter low-intensity prescribed burn in a Mediterranean forest. *Plant and Soil*, 120: 69-77.

Gillon, D., V. Gomendy, C. Houssard, J. Marechal, and J.C. Valette. 1995. Combustion and nutrient losses during laboratory burns. *International Journal of Wildland Fire,* 5: 1-12.

Gimeno-García, E., V. Andreu, and J.L. Rubio. 2000. Changes in organic matter, nitrogen, phosphorus and cations in soil as a result of fire and water erosion in a Mediterranean landscape. *European Journal of Soil Science,* 51: 201-210.

Gimeno-García, E., V. Andreu, and J.L. Rubio. 2004. Spatial patterns of soil temperature during experimental fires. *Geoderma,* 118: 17-38.

González-Perez, J.A., F.J. González-Vila, G. Almendros, and H. Knicker. 2004. The effect of fire on soil organic matter – A review. *Environment International,* 30: 855-870.

Grady, K.C. and S.C. Hart. 2006. Influences of thinning, prescribed burning, and wildfire on soil processes and properties in southwestern ponderosa pine forests: A retrospective study. *Forest Ecology and Management,* 234: 123-135.

Grogan, P., T.D. Bruns, and F.S. Chapin III. 2000. Fire effects on ecosystem nitrogen cycling in a California bishop pine forest. *Oecologia,* 122: 537-544.

Guinto, D. F., Z. Xu, A.P.N. House, and P.G. Saffigna. 2000. Assessment of N_2 fixation by understorey acacias in recurrently burnt eucalypt forests of subtropical Australia using 15N isotope dilution techniques. *Canadian Journal of Forest Research,* 30(1): 112-121.

Hart, S.C., A.T. Classen, and R.J. Wright. 2005a. Long-term interval burning alters fine root and mycorrizal dynamics in a ponderosa pine forest. *Journal of Applied Ecology,* 42: 752-761.

Hart, S.C., T.H. DeLuca, G.S. Newman, M.D. MacKenzie, and S.I. Boyle. 2005b. Post-fire vegetative dynamics as drivers of microbial community structure and function in forest soils. *Forest Ecology and Management,* 220: 166-184.

Hiers, J.K., R.J. Mitchell, L.R. Boring, J.J. Hendricks, and R. Wyatt. 2003. Legumes native to longleaf pine savannas exhibit capacity for high N_2-fixation rates and negligible impacts due to timing of fire. *New Phytologist,* 157: 327-338.

Humphreys, F.R. and M.J. Lambert. 1965. An examination of a forest site which has exhibited the ashbed effect. *Australian Journal of Soil Research,* 3: 81-94.

Intergovernmental Panel on Climate Change. 2006. Generic methodologies applicable to multiple land-use categories. pp. 2.1-2.59. *In* 2006 IPCC Guidelines for National Greenhouse Gas Inventories. IPCC Secretariat, Geneva, Switzerland.

Johnson, D.W. and P.S. Curtis. 2001. Effects of forest management on soil C and N storage: Meta analysis. *Forest Ecology and Management,* 140: 227-238.

Johnson, D.W., R.B. Susfalk, T.G. Caldwell, J.D. Murphy, W.W. Miller, and R.F. Walker. 2004. Fire effects on carbon and nitrogen budgets in forests. *Water, Air, and Soil Pollution Focus,* 4: 263-275.

Johnson, D.W., J.D. Murphy, R.B. Susfalk, T.G. Caldwell, W.W. Miller, R.F. Walker, and R.F. Powers. 2005. The effects of wildfire, salvage logging, and post-fire N fixation on the nutrient budgets of a Sierran Forest. *Forest Ecology and Management,* 220: 155-165.

Johnson, D.W., J.D. Murphy, R.F. Walker, D.W. Glass, and W.W. Miller. 2007. Wildfire effects on forest carbon and nutrient budgets. *Ecological Engineering,* 31: 183-192.

Kauffman, J.B., R.L. Sanford Jr., D.L. Cummings, I.H. Salcedo, and E.V.S. Sampio. 1993. Biomass and nutrient dynamics associated with slash fires in neotropical dry forest. *Ecology,* 74: 140-151.

Kauffman, J.B., D.L. Cummings, D.E. Ward, and R. Babbitt. 1995. Fire in the Brazilian Amazon: 1. Biomass, nutrient pools, and losses in slashed primary forests. *Oecologia,* 104: 397-408.

Khanna, P.K. and R.J. Raison. 1986. Effect of fire intensity on solution chemistry of surface soil under a *Eucalyptus pauciflora* forest. *Australian Journal of Soil Research,* 24: 423-434.

Khanna, P.K., R.J. Raison, and R.A. Falkiner. 1986. Exchange characteristics of some acid organic-rich forest soils. *Australian Journal of Soil Research,* 24: 67-80.

Khanna, P.K., R.J. Raison, and R.A. Falkiner. 1994. Chemical properties of ash derived from *Eucalyptus* litter and its effects on forest soils. *Forest Ecology and Management,* 66: 107-125.

Klopatek, J., C.C. Klopatek, and L.F. Debano. 1990. Potential variation of nitrogen transformations in pinyon-juniper ecosystems resulting from burning. *Biology and Fertility of Soils,* 10: 35-44.

Levine, J.S., E.L. Winstead, D.A.B. Parsons, M.C. Scholes, R.J. Scholes, W.R. Cofer, D.R. Cahoon, and D.I. Sebacher. 1996. Biogenic soil emissions of nitric oxide (NO) and nitrous oxide (N_2O) from savannahs in South Africa: The impact of wetting and burning. *Journal of Geophysical Research,* 101: 115-122.

Ludwig, B., P.K. Khanna, R.J. Raison, and K.L. Jacobsen. 1998. Modelling cation composition of soil extracts under ashbeds following an intense slashfire in a eucalypt forest. *Forest Ecology and Management,* 103: 9-20.

Mackensen, J., D. Holscher, D. Klinge, and H. Folster. 1996. Nutrient transfer to the atmosphere by burning debris in eastern Amazonia. *Forest Ecology and Management,* 86: 121-128.

Marschner, H. 1995. *Mineral Nutrition of Higher Plants.* Academic Press, New York, New York, USA.

Mitchell, D.T., P.G.F. Coley, S. Webb, and N. Allsop. 1986. Litterfall and decomposition processes in the coastal fynbos vegetation, south-western Cape, South Africa. *Journal of Ecology,* 74: 977-993.

Murphy, J.D., D.W. Johnson, W.W. Miller, R.F. Walker, E.F. Carroll, and R.R. Blank. 2006. Wildfire effects on soil nutrients and leaching in a Tahoe Basin watershed. *Journal of Environmental Quality,* 35: 479-489.

Naveh, Z. 1967. Mediterranean ecosystems and vegetation types in California and Israel. *Ecology,* 48: 445-459.

Newland, J.A. and T.H. DeLuca. 2000. Influence of fire on native nitrogen-fixing plants, and soil nitrogen status in Ponderosa pine/Douglas-fir forests in western Montana. *Canadian Journal of Forest Research,* 30: 274-282.

Polglase, P.J., P.M. Attiwill, and M.A. Adams. 1992a. Nitrogen and phosphorus cycling in relation to stand age of *Eucalyptus regnans* F. Muell. II. N mineralization and nitrification. *Plant and Soil,* 142: 167-176.

Polglase, P.J., P.M. Attiwill, and M.A. Adams. 1992b. Nitrogen and phosphorus cycling in relation to stand age of *Eucalyptus regnans* F. Muell. III. Labile inorganic and organic P, phosphatase activity and P availability. *Plant and Soil,* 142: 177-185.

Ponomarenko, E.V. and D.W. Anderson. 2001. Importance of charred organic matter in Black Chernozem soils of Saskatchewan. *Canadian Journal of Soil Science,* 81: 285-297.

Raison, R.J. 1979. Modification of the soil environment by vegetation fires, with particular reference to nitrogen transformations: A review. *Plant and Soil,* 51: 73-108.

Raison, R.J., P.K. Khanna, and P.V. Woods. 1985a. Mechanisms of element transfer to the atmosphere during vegetation fires. *Canadian Journal of Forestry Research,* 15: 132-140.

Raison, R.J., P.K. Khanna, and P.V. Woods. 1985b. Transfer of elements to the atmosphere during low-intensity prescribed fires in three Australian sub-alpine eucalypt forests. *Canadian Journal of Forestry Research,* 15: 657-664.

Raison, R.J., P.V. Woods, and P.K. Khanna. 1986. Decomposition and accumulation of litter after fire in sub-alpine eucalypt forests. *Australian Journal of Ecology,* 11: 9-19.

Raison, R.J., A.M. O'Connell, P.K. Khanna, and H. Keith. 1993a. Effects of repeated fires on nitrogen and phosphorus budgets and cycling processes in forest ecosystems. pp. 347-363. *In* L. Trabaud and R. Prodon [eds.]. Fire in Mediterranean Ecosystems. Report No. 5 Ecosystem Research Report Series, Commission of the European Communities, Brussels, Belgium.

Raison, R.J., K.L. Jacobsen, M.J. Connell, P.K. Khanna, H. Keith, S.J. Smith, and P. Piotrowski. 1993b. Nutrient cycling and tree nutrition. pp. 8-89. *In* Collaborative research in regrowth forests of East Gippsland between CSIRO and the Victorian department of Conservation and Natural Resources. CSIRO Division of Forestry, Canberra, Australia.

Raison, R.J. and M.A. Rab. 2001. Guiding concepts for the application of indicators to interpret change in soil properties and processes in forests. pp. 231-258. *In* R.J. Raison, A.G. Brown, and D.W. Flinn [eds.]. Criteria and Indicators for Sustainable Forest Management. International Union of Forest Research Organizations (IUFRO) Research Series, No. 7. CABI, Wallingford, UK.

Rastetter, E.B., P.M. Vitousek, C. Field, G.R. Shaver, D. Herbert, and G.I. Agren. 2001. Resource optimization and symbiotic nitrogen fixation. *Ecosystems,* 4(4): 369-388.

Retana, J., J.M. Espelta, A. Habrouk, J.L. Ordoñez, and F. Solà-Morales. 2002. Regeneration patterns of three Mediterranean pines and forests changes after a large wildfire in northeastern Spain. *Ecoscience,* 9: 89-97.

Rodrigo, A., J. Retana, and F.X. Pico. 2004. Direct regeneration is not the only response of Mediterranean forests to large fires. *Ecology,* 85: 716-729.

Romanyà, J., P.K. Khanna, and R.J. Raison. 1994. Effects of slashburning on soil phosphorus fractions and sorption and desorption of phosphorus. *Forest Ecology and Management,* 65: 89-103.

Romanyà, J., P. Casals, and V.R. Vallejo. 2001. Short term-effects of fire on soil nitrogen availability in Mediterranean grasslands and shrublands growing in old fields. *Forest Ecology and Management,* 147: 39-53.

Rovira, P., J. Azcón-Bieto, I. Fleck, and V.R. Vallejo. 2004. Carbon sequestration in Mediterranean ecosystems: critical aspects related to plant respiration, wildfires and nitrogen budget. *Contributions to Science,* 2(4): 467-482.

Serrasolses, I. and P.K. Khanna. 1995. Changes in heated and autoclaved forest soils of SE Australia: II. Phosphorus and phosphatase activity. *Biogeochemistry,* 29: 25-41.

Serrasolses, I. and V.R. Vallejo. 1999. Soil fertility after fire and clearcutting. *In* F. Rodà, C.A. Gracia, J. Retana, and J. Bellot [eds.]. Ecology of Mediterranean Evergreen Oak Forests. *Ecological Studies,* 137: 316-328.

Schlesinger, W.H. and D.S. Gill. 1980. Biomass, production, and changes in the availability of light, water, and nutrients during the development of pure stands of chaparral shrubs, *Ceanothus megacarpus*, after fire. *Ecology*, 61: 781-789.

Smithwick, E.A.H., M.G. Turner, M.C. Mack, and F.S. Chapin III. 2005. Postfire soil N cycling in northern conifer forests affected by severe stand-replacing wildfires. *Ecosystems*, 8: 163-181.

Specht, R.L. 1981. Response of fires in heathlands and related shrublands. pp. 395-345. *In* A.M. Gill, R.H. Groves, and I.R. Noble [eds.]. *Fire and the Australian Biota.* Surrey Beatty and Sons, Chipping Norton, Sydney, Australia.

Swap, R.J., M. Garstang, S.A. Macko, P.D. Tyson, and P. Kallberg. 1996. Comparison of biomass burning emissions and biogenic soil emissions to the tropical South Atlantic. pp. 396-402. *In* J.S. Levine [ed.]. *Biomass Burning and Global Change.* MIT Press, Cambridge, Massachusettes, USA.

Thornes, J.B. 1990. The interaction of erosional and vegetation dynamics in land degradation: spatial outcomes pp. 41-53. *In* J.B. Thornes [ed.]. *Vegetation and Erosion.* John Wiley and Sons, Chichester, UK.

Trabaud, L. 1991. Le feu est-il un facteur de changement pour les sistèmes écologiques du bȧsin méditérraneen. *Sécherese*, 2: 163-174.

Trabaud, L. and C. Campant. 1991. Problem of naturally colonizing the salzmann pine, *Pinus nigra arn ssp salzmannii (dunal) franco* after a fire. *Biological Conservation*, 58: 329-343.

Vázquez, F.J., M.J. Acea, and T. Carballas. 1993. Soil microbial populations after wildfire. *FEMS Microbiology Letters*, 13: 93-104.

Vitousek, P.M. and C.B. Field. 1999. Ecosystem constraints to symbiotic nitrogen fixers: a simple model and its implications. *Biogeochemistry*, 46 (1-3): 179-202.

Walker, J., R.J. Raison, and P.K. Khanna. 1986. Fire. pp. 185-216. *In* J.S. Russell and R.F. Isbell [eds.]. *Australian Soils: The Human Impact.* University of Queensland Press, Queensland, Australia.

Wan, S., D. Hui, and Y. Luo. 2001. Fire effects on nitrogen pools and dynamics in terrestrial ecosystems: A meta analysis. *Ecological Applications*, 11: 1349-1365.

White, C.S. 1991. The role of monoterpenes in soil nitrogen cycling processes in Ponderosa pine. *Biogeochemistry*, 12: 43-68.

Wright, H.A. and A.W. Bailey, 1982. *Fire Ecology: United States and Southern Canada.* John Wiley and Sons, New York, New York, USA.

Zackrisson, O., T.H. DeLuca, M.C. Nilsson, A. Sellstedt, and L.M. Berglund. 2004. Nitrogen fixation increases with successional age in boreal forests. *Ecology*, 85: 3327-3334.

9

The Effect of Forest Fire on Vegetation

Francisco Lloret[1*] and *Paul H. Zedler*[2]

Abstract

Fire events are common disturbances in many ecosystems. The role of fire in ecosystems is encompassed in the concept of 'fire regime', which includes both temporal and spatial aspects as well as the physical characteristics of the fire. Vegetation recovery mechanisms determine ecosystem resilience, or the capacity to return to pre-fire conditions. In general, ecosystems submitted to frequent fires tend to be more resilient. The new environment produced by fire often determines successional trajectories. Thus, species that were not dominant in the pre-fire conditions will eventually be replaced by shade-tolerant species. However, this successional scheme is not universal. In some forests, such as fire-prone Mediterranean forests, direct regeneration occurs resulting in post-fire communities that are similar to pre-fire communities. This case supports the fire regime-resilience coupled model. However, under certain conditions, this coupling is disrupted and results in decreased resilience to fire. For example, new climatic conditions, change in the species pool (planted species or invasive species moving into an area), or policies modifying fire regimes, may result in vegetation dynamics that move toward new communities.

Forest resilience may be changed by management strategies designed to shape the fire regime and modify fuel amounts and types. Fire suppression strategies may lead to the dominance of late successional low resilient species at the expense of more resilient species, while frequent fires may precipitate a shift from forest to more open communities. Enhancing more resilient species will meliorate resilience of the whole community.

General patterns of post-fire vegetation dynamics are reviewed for humid tropical, dry deciduous tropical, Mediterranean, temperate deciduous, and boreal forest ecosystems. The focus is on fire regime characteristics and

[1] *CREAF (Center for Ecological Research and Forestry Applications), Departament de Biologia Animal, Biologia Vegetal i Ecologia, Universitat Autònoma de Barcelona, Barcelona, Spain.*

[2] *Nelson Institute for Environmental Studies and the Arboretum, University of Wisconsin–Madison, Madison, Wisconsin, USA.*

* *Corresponding author: Francisco Lloret, CREAF (Center for Ecological Research and Forestry Applications), Department de Biologia Animal, Biologia Vegetal i Ecologia, Universitat Autònoma de Barcelona, 08193 Bellaterra, Barcelona, Spain. Tel: + 49 35 812700, e-mail: Francisco.Lloret@uab.cat*

ecological processes that determine the forest resilience. The potential causes of reduction or loss of resilience are presented in relation to the fire regime, as the interaction of these two factors can lead to abrupt shifts in vegetation dynamics.

INTRODUCTION

Fire Regime in Ecosystems

It is now universally agreed that fire is a natural component of ecosystem dynamics. This ubiquity is easily explained. In an oxygen-rich atmosphere accumulated organic biomass always has the potential to become fuel for a fire. It is only necessary that the biomass be dry and dense enough for the fire to propagate when ignited. Lightning is the most important non-human ignition source, but volcanism, falling rocks, and even spontaneous combustion also can provide the necessary spark. These conditions are most reliably present when periods favorable for plant growth are followed by extended periods where a large fraction of this vegetation either dies or is stressed with reduced moisture levels. This sequence occurs in many regions of the world. In temperate zones, the onset of winter can cause the death of plants and plant parts and lead to an accumulation of burnable biomass. In arid and semi-arid regions, drought has an even more pronounced effect. Therefore, wildfires are a main driver of vegetation and landscape dynamics in a large number of ecosystems across the world: coniferous forests in boreal and temperate regions, deciduous temperate forests, Mediterranean evergreen forests and shrublands, and tropical savannahs and deciduous forests. On the other hand, fire is rarely present in other systems, such as rainforest tropical forests or deserts with low fuel continuity.

The role of fire in an ecosystem is encompassed in the concept of 'fire regime' – the totality of fire behavior for a given region or ecosystem (Agee 1993). It is usually taken to include both temporal and spatial aspects as well as the actual physical characteristics of the fire (e.g., intensity, duration, maximum temperature). Knowing the historical fire regime of an ecosystem is the key to understanding its resilience (its natural capacity to recover) from fire. Data on waiting times to recurrence tell us how often the species in the ecosystem are subject to direct fire effects (such as heat caused mortality) and how long the system has to recover, which, for example, determines the opportunities of fire-sensitive species to reinvade a burned site. Unfortunately, accurate characterization of a fire regime requires long periods of observation over large areas. Detailed paleoecological studies (charcoal records, for example Tinner et al. 1999, Asselin and Payette 2005; see Chapter 1) are valuable, but costly and often difficult to translate directly into a fire regime. As a result, our ability to make definitive statements about the connection between ecosystem resilience and fire regime is limited.

Fire intensity, which is critically important to plant response, is difficult to assess. Fire researchers define intensity as the heat release per unit time per unit length of the fire front (Whelan 1995). A less direct way to assess fire intensity is to characterize fires by the degree to which they consume or kill the biomass present, as this will greatly determine post-fire vegetation response. Some fires burn through the entire depth of the vegetation and are referred to as 'ground-to-crown fires'. In them, nearly everything above the ground is either consumed by the fire or killed by the heat of the fire. Ground-to-crown fires may or may not have a large effect on the species composition and in many systems, subject to these fires the same individuals will recover and reassume dominance. In contrast 'understory fires' or 'ground fires' consume only the surface fuels, such as plant litter and standing dead and live herbaceous and low shrubby plants, and leave the overstory plants (almost exclusively woody) unscathed or only slightly damaged. Over longer periods of time, frequent understory fires may also significantly affect the dynamics of the community by modifying the patterns of recruitment of overstory species. The distinction between these two fire types is well known to fire scientists and practitioners (e.g., Biswell 1989); however, the failure to apply this well-known distinction to decision-making has caused confusion as to the natural role of fire and appropriate fire management prescriptions.

Whether a given fire will be a ground-to-crown fire or an understory fire is strongly dependent on the structure of the vegetation, the moisture content of the plant materials at the time of the fire, and the weather. In certain ecosystems, most notably grasslands and dense shrublands of arid and semi-arid regions, 'understory fires' are rare or absent. In these ecosystems, if there is enough heat energy to allow a fire to propagate, there will usually be enough energy to burn through the entire depth of the vegetation. In other ecosystem types, the lower layers of vegetation usually cannot generate enough heat fast enough to ignite the layers above them, and fires typically remain ground fires. But, under extreme conditions, intensity can be great enough to carry the fire into the canopies converting an understory fire into a ground-to-crown fire.

Fire regimes are dependent on the spatial pattern, or completeness, of the burn, which is determined by the proportion of area within the burn perimeter where combustion occurred. Completeness of burn within a perimeter depends on a number of factors (topography, wind and weather), but the most important factor is fuel continuity. Fires in non-desertic grasslands typically have nearly complete burning within the fire boundary because the fuel typically is continuous and uniformly combustible. Unburned patches can occur where fuel continuity is interrupted as where fossorial rodents have exposed the bare ground or around rocky outcrops. In contrast, within the boundaries of a single forest fire, there may be ground-to-crown fire in some patches, but only understory fires in others, or more rarely, crown fires without ground fires (e.g., Turner and Romme 1994). Spatial extent and pattern of fires also interact with fire frequency because recurrent fires can reduce fuel connectivity, making fires smaller, or patchier, or both (Swetnam 1993).

The size of a fire is another spatial aspect of the fire regime. In all fire regimes, small fires are more abundant than large fires (e.g., Schoenberg et al. 2002). This pattern is the result of time since the last fire affecting fuel connectivity and the seasonal cycle. After a fire, it generally takes more than a single year (except in many grasslands) for fuel connectivity to reach a level that will permit a large fire. Over the course of the year, small patches may be able to burn (say an equator-facing slope) when another nearby (a pole-facing slope) may not. Thus, conditions that allow a high-energy ignition event (lightning) to start a fire and burn a small area are much more common than conditions that allow a fire to spread across all landscape elements. It is important to stress, however, that greater number of small fires does not mean that most of the area is burned in such fires. On the contrary, the larger area of big fires make it much more likely that any particular point on the landscape will be burned in a large fire than in a median-sized fire, and being burned in a medium-sized fire is more likely than in a small fire.

The time between fires is the feature of a fire regime that most directly affects plant demography, especially in ground-to-crown fire systems. This is because times of recurrence determine how frequently the mature plants are at risk of fire mortality, the number of times recruitment from seed is possible, and other related critical demographic transitions. The empirical frequency distribution of historical times between fires is related to the synthetic concept of the fire interval probability density distribution, which in turn relates to the general concept of 'fire frequency' as the "probability of an element burning per unit time" (Johnson and Gutsell 1994). A great deal of attention has been directed to the relation between fire frequency and fire type. Low frequency, and therefore long periods between fires, can increase fuel contiguity in both the horizontal and vertical dimensions favoring larger fires and ground-to-crown fires in circumstances where understory fires would be common if intervals between fires were shorter.

Resilience to Fire

Fire produces important physical-chemical and biological changes which include the loss of aerial biomass and transformation of the litter and the upper few centimeters of soil. However, shortly after fire, vegetation regrowth and the ecosystem functioning are potentially restored. Throughout this chapter, questions related to post-fire vegetation recovery, and particularly the circumstances in which fire can impair the natural capacity to recover (i.e., resilience), are discussed. How long is required for vegetation to reach pre-fire conditions? Are new communities replacing the pre-fire communities? Can fires cause irreversible changes in the vegetation and ecosystem features in relation to the pre-fire situation? To what extent are human actions affecting forest resilience? How do human activities alter characteristics of fire occurrence, which, in turn, modifies the distribution, structure, and composition of plant communities before and after fires?

The ability of vegetation to recover after fire is determined by the impact of the disturbance on plant population and by the ability of the populations to reestablish in the new post-fire conditions. In disturbance ecological terminology, *resistance* describes the ability to endure disturbance, while the capacity to restore previous conditions (often estimated as a function of the time needed to achieve pre-disturbance features) is called *resilience*. A useful approach to determine the vegetation response to fire is to consider the performance at the species level. In other words, those communities with a larger number of resilient species will recover faster than communities dominated by low resilient species. Consequently, it is useful to explore the mechanisms that allow species to survive fires and to reestablish their populations, as well as the factors limiting these processes.

THE BIOLOGICAL BASIS OF REGENERATION

After fire, regeneration of the vegetation is based on the individual behavior of the populations of the different species. Some low intensity fires do not have enough heat to kill many individual plants of some poplulations. This occurs in understory fires where flames and high temperatures do not affect the tree canopy, or when there is limited continuity of the canopies and only some of the trees experience partial charring. In many species, the bark provides thermal protection and improves the ability to survive low intensity surface fire. Therefore, canopy recovery from understory fires is mostly dependent on axillary or epicormic regrowth, just as it is after other disturbances like insect infestation or drought breakdowns (Bond and Midgley 2001).

When the standing individuals are mostly charred or destroyed (high intensity fire), there are two main mechanisms by which populations may establish in a burned site – vegetative regrowth of pre-existing individuals or establishment of new plants from seeds. In some species, meristems are able to survive and reactivate after burning producing new shoots and leaves. Since below ground organs experience much lower temperatures than aerial parts, resprouting is, in most cases, from stumps, rhizomes, bulbs, root crowns, roots or lignotubers (Fig. 1). Lignotubers, swollen structures located between stems and roots, act as resource storage sites and as bud banks (James 1984, Canadell and Zedler 1995). In some species, the thermal isolation of stems allows existing buds within the branches to survive resulting in a quick recovery of the canopy. This thermal isolation may be provided by the bark, as in the Mediterranean cork-oak (*Quercus suber*), the California live oak (*Q. agrifolia*), the Canarian pine (*Pinus canariensis*), the Douglas-fir (*Pseudotsuga menziesii, P. macrocarpa*), and South American species of *Araucaria* (Climent et al. 2004, Pausas 1997, Gonzalez et al. 2005). Thermal isolation may also be

Fig. 1 a) Serotinous fruits in South African Proteaceae; b) seedling post-fire germination of Mediterranean Cistaceae; c) canopy resprouting of Australian *Eucalyptus;* and d) resprouting from below-ground tissues in Mediterranean *Quercus*.

provided by the leaves as in palms and other monocots (Reinhart and Menges 2004). Resprouting allows species to persist through fires and thereby confers high resilience to fire.

Resprouting is a response to any disturbance that removes aboveground parts, and it is widely distributed, both taxonomically and geographically (Bond and Midgley 2001, Vesk and Westoby 2004). Therefore, resprouting can hardly be considered as an adaptation uniquely associated with frequent fires. But the capacity to resprout after fire may be enhanced by structures or features that may well have been fire selected, such as the protective cork or deep bark that is present in many species of frequently burned vegetation (Climent et al. 2004). Although the ultimate explanation of the capacity to resprout may be elusive, it certainly is part of the reason certain species achieve dominance in fire-prone communities. In a subtropical shrubland of Central Mexico, where fires were not historically present, Lloret et al. (1999) found that resprouting ability was present in taxa sharing their biogeographical origin with taxa from the Californian chaparral (where fire is a natural disturbance) and in neotropical taxa without historical contact with fire. However, the group related to the Californian species exhibited better resprouting performance than the neotropical group.

The degree of resprouting not only varies between species, but also is highly variable within species (Bond and Midgley 2001). There are three major factors contributing to the resprouting performance – individual characteristics (age and size), fire regime and frequency, and post-fire conditions. Although bigger plants are likely to produce more and larger resprouts, this relationship is not universal and depends on the ontogenic stage, the plant growth form, and the disturbance type (Vesk 2006). The number of buds from arising new shoots may be positively related to the extent of the underground organs, for example the lignotuber surface, but this number may diminish in old plants due to the depletion of the bud bank (Zammit 1988). Plants of larger pre-fire size are expected to have a more extensive root system and potentially have greater resource uptake ability from the soil and/or from stored stocks. However, at sites where larger plants provide greater fuel loads, fire intensity may be higher and cause more damage. In addition, the physiological and resource storage state of the individuals are influenced by the competitive interactions existing before fire. Thus, saplings growing under a close canopy and subjected to intense light competition have been observed to have worse regeneration than those growing in more open areas (Marod et al. 2004, Dokrar et al. 2004).

Higher fire intensity, determined by fire temperature and duration, diminish resprouting success (Lloret and López-Soria 1993). The season of the fire will impact resprouting because of the seasonal effect on fire intensity. In addition, fires that occur at times when the stored resources are low are potentially more severe than those occurring after the growing season (Konstantinidis et al. 2006). Fire frequency may also negatively affect resprouting response by depleting the bud bank or preventing replenishment of resources stocks (Zammit 1988, Canadell and López-Soria 1998). Also, each fire causes some degree of mortality within populations that accumulate with successive fires. Therefore, population density will gradually decline unless the intervals between fires are long enough and favorable enough for the establishment of new individuals. Post-fire conditions including temperature stress, water, and nutrient availability, and biotic interactions, such as herbivory, are the final determinants of vegetative regrowth of pre-fire vegetation (López-Soria and Castell 1992, Bailey and Whitham 2002).

Many species find that the burned site is advantageous for the establishment of new populations. The lack of competition from previously established plants and the increase of available resources (light from the canopy opening and the post-fire flush of available nutrients) provide an advantage. However, seeds must be available by dispersal from unburned populations or from a seed bank that has been able to survive the high temperatures of fire. Therefore, colonist species with long distance dispersal mechanisms and the propensity to establish in open habitats are typically good candidates to initiate the regeneration process. Fire extent and severity are thus two major limitations for the achievement of this post-fire colonization.

Some species produce seed banks that are fire resistant. In many cases seeds survive in the soil, because the soil affords protection from the heat, with only the upper few centimeters experiencing temperatures in the lethal range (Bradstock et al. 1992). In forests, however, accumulations of wood, as in artificially produced slash piles, can result in hot fires with long residence time that can virtually sterilize the soil (Korb et al. 2004). High temperatures stimulate seed germination of some fire resistant species by breaking seed coats, and in some species smoke and other chemical cues produced by the fire may induce germination (Keeley and Fotheringham 2000, Dixon et al. 1995; Fig. 1). The evolutionary significance of fire on the acquisition of these traits is not clear as fire-related germination cues seem narrowly coupled to selective forces associated with fires. Seed coats may, however, be broken by a variety of agents associated with disturbance situations other than fire such as soil gaps formed by fossorial rodents (e.g., DeSimone and Zedler 1999).

In some species, seedling establishment is fire-dependent (Zammit and Zedler 1994, Greene and Johnson 1999). This strategy insures that seedling establishment occurs under the most advantageous conditions of resource availability and reduced competition. It requires, however, investment in structures that provide long-term protection for seeds from predators and decay. Obviously this strategy requires a positive synchronization between fire regime and life cycle—fire intervals that are too short may prevent an adequate seed bank from accumulating, while fire intervals that are too long may result in the death of the established seed-producing individuals and the gradual loss of seed bank viability. The extreme case of this strategy is represented for species with serotinous fruits that delay the seed release until fire occurs (Lamont et al 1991; Fig. 1). In fact, the level of serotiny is variable within and between *Pinus* species (Keeley and Zedler 1998) and will have a large influence on the timing and the amount of seedling recruitment. Thus, trees with high levels of serotiny will retain high numbers of seeds and will be capable of establishing high numbers of seedlings after fire.

As in resprouting, the characteristics of the fire regime and the post-fire conditions will determine the successful establishment of seedlings for both fire-dependent and non-fire dependent plants. High burn severity often results in loss of seed viability even in those species with fire-stimulated germination (Habrouk et al. 1999). Fire interval is a vital factor in restoring seed banks and maintaining seed sources (Eugenio et al. 2006b, Quintana-Ascencio et al. 2003). Fire season may also influence post-fire seedling establishment as the number of available diaspore in the seed bank fluctuates in ecosystems with seasonal patterns of seed production and release (Domínguez et al. 2002, Kerns et al. 2006). The size of the area burned is a factor in seedling establishment of those species that depend on dispersal from unburned seed sources (Greene and Johnson 2000, Elmqvist et al. 2002). Although several mechanisms exist for post-fire vegetation establishment, the fire regime characteristics and the post-fire environmental conditions ultimately determine its success.

FIRE AND VEGETATION DYNAMICS

The Successional Canon

Succession after fire is a classic example of ecological succession – the predictable replacement over time of 'early successional' species by other species better able to establish and grow in shadier, more competitive environments, which leads to a quasi-equilibrium condition with dynamics regulated by small-scale processes (e.g., 'gap phase' replacement). This general model applies to forest systems that experience fire at long intervals relative to the life span of the species (Fig. 2). The new post-fire environment (with fertilized soils, higher radiation, and reduced competition) favors species that were not dominant or perhaps not even present in the pre-fire environment. For example, in the Great Lakes region of North America, there are extensive areas of conifer and broadleaf forest mixtures. If burned at an advanced stage of succession with an intense crown fire, the dominant species present at the time of the fire are greatly reduced in abundance and rapidly growing species of high light requirement (e.g., *Populus* spp.) replace them. These, in turn, are replaced by more shade-tolerant species (e.g., *Pinus strobus*). This intermediate stage can persist for long periods if the dominants are capable of reaching a large size and have great longevity (as is the case for *P. strobus*). But the absence of fire over long intervals will favor the increase of very shade-tolerant species capable of surviving in a shaded understory (e.g., *Acer saccharum, Tsuga canadensis;* Stearns and Likens 2002). In such a system the resilience is dependent, in part, on larger-scale phenomena that allow all of the species that participate in the sequence to persist in the region at the same time. Thus, these species can recover by resprouting on the site or they can invade from adjacent areas at a different state of succession or from small-unburned patches.

This general pattern is also found in tropical, boreal, and temperate forests. Tropical forests are dominated by fire-sensitive species that in many

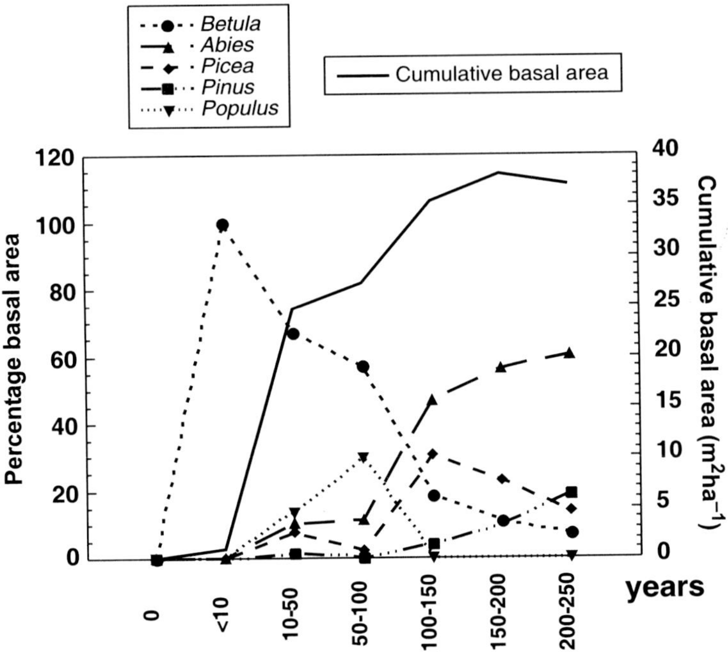

Fig. 2 Basal area of tree species as related to successional age after fire in the dark taiga of Central Siberia. *Betula: B. pendula, Abies: A. sibirica, Picea: P. obovata, Pinus: P. sibirica, Populus: P. tremula.* The case illustrates successional species replacements after fire. Data from Schulze et al. (2005).

cases disappear after fire. Fire promotes the establishment of opportunistic species that eventually can be replaced by shade-tolerant more competitive trees (Goldammer and Seibert 1990, Holdsworth and Uhl 1997). In the absence of fire, transitions from grasslands and woodlands to forests have also been documented (King et al. 1997, McCoy et al. 1999, Russell-Smith et al. 2004). In boreal forests, the colonizing stages of normal succession are represented by pulses of aspen recruitment, which is mostly vegetative after a disturbance (Greene and Johnson 1999). These species are eventually replaced by long-lived shade tolerant tree species (Whelan 1995, Archambault et al. 1997, De Grandpre et al. 2000, Clark et al. 2003, Schulze et al. 2005). The competitive replacement of fire-resilient species by fire-sensitive ones occurs in long periods without fire. This has been observed in Tasmania, where the mesophytic rainforest invaded the eucalyptus forest, and in southeastern North America, where hardwood trees supplant pine forests and scrublands (Whelan 1995).

Fires interact with regional climate so that the successional trajectories will vary depending on where they occur along regional gradients. In mesic sites of the boreal mixed wood forest region of Quebec, fire eliminates shade-intolerant species. Consequently, the vegetation composition should evolve toward potential climax vegetation based on topology (elevation): the sugar maple-yellow birch type on upper slopes; the balsam fir-yellow birch type on midslopes; and the balsam fir-yellow birch-cedar type on lower slopes (Archambault et al. 1997).

Fire may also promote the coexistence of species, some of which would be competitively excluded if succession proceeded without disturbances. Lodgepole pines and spruce species occur together in mature stands in boreal North American forests where the current fire cycle (ca. 80 years) is not long enough to permit the more shade-tolerant spruce to exclude pine competitively (Johnstone and Chapin 2003). In austral forests of South America, fire allows the co-existence between the vigorously sprouting, shade-intolerant *Nothofagus antarctica* and *Araucaria araucana*, that partly survives fire and is more shade-tolerant (Burns 1993). In the first years after a fire, root resprouts or seedlings of *A. araucana* establish beneath the resprouted *N. antarctica* canopy. *Araucaria araucana* trees will eventually overtop *N. antarctica* if no disturbance occurs for more than 150 years. However, the suppression of *N. antarctica* may be interrupted by fires which are common in this region. The result is the existence of stands with clustered trees of *A. araucana* over a shorter *N. antarctica* sub-canopy. In the Siberian taiga (Uemura et al. 1990): *Betula* resprouts after fire, creating conditions that favor shade-tolerant, later successional species of *Larix*. *Picea koraiensis* is less tolerant of fire because it has a thinner bark and a tendency to burn as a crown fire. However, understory species of *Picea* forests are more fire resistant because the wetter conditions of these forests. Thus, the position of the site along a moisture gradient will determine which successional path a site will follow, and which stage is likely to be most abundant over time.

Alternatives to the Classical Successional Model

Vegetation changes after wildfires do not always follow an equilibrium model in which successional series start from short-lived well dispersed species and lead to a pre-determined stable state dominated by more competitive, low growing, shade-tolerant species. For example, the vegetation may directly regenerate (i.e., be self-replaced) after fire. This means that the early stages, dominated by shade intolerant species, are short-lived or that herbaceous dominance may be transient and give way to shrub lands or woodlands (Calvo et al. 2002). This woody vegetation becomes dominant a few decades after the disturbance and is essentially the same as existed before the fire. Such truncated successional sequences are common in arid regions where the capture of space is more about obtaining access to reliable soil moisture than gaining access to light. In such circumstances the first generation of woody plants creates understory conditions that are unfavorable for the establishment of species that are both shade tolerant and capable of growing up through the canopies of the established individuals. The fact that few species can be both shade and drought tolerant severely limits successional possibilities, and the succession 'stalls' at the first generation. Though some woody species do establish seedlings to form a 'seedling bank' the rate of replacement is slow (relative to the probable time of fire recurrence) permitting the light-demanding dominants to reassert their control of a site. This means that shade tolerant species are able to regenerate after fire, or alternatively, that the frequency or intensity of disturbances do not allow a permanent fire-intolerant community to establish. The combination of different types of disturbances may contribute to the latter situation, as is observed in the Mediterranean Basin where there is a long history of burning and land transformation. There is little post-fire regeneration of plantation species and native populations are too depleted to initiate new succession. Rodrigo et al. (2004) reported important shifts after a single fire burned extended areas dominated by planted *Pinus nigra* populations in northeastern Spain.

Post-fire self-replacement is typical in evergreen Mediterranean forests, such as those dominated by *Quercus ilex* and *Q. suber* in the Mediterranean Basin. These communities regenerate well after fire and maintain their composition due to the resprouting ability of many species or the existence of fire-resistant seed banks (Ojeda et al. 1996, Rodrigo et al. 2004). A similar behavior is found in many woodlands and shrublands (e.g., Mediterranean garriga, maquis and frygana, Californian chaparral, South African fynboss, Australian mallee and kwongan, Chilean matorral), which are considered climax communities in Mediterranean type-ecosystems across the world.

Fires may also promote the coexistence of different permanent communities at the landscape level within the same climatic region by transforming the environmental conditions. This may also happen when burning modifies the conditions for regeneration (e.g., consumption of soil duff in boreal systems). In conifer forests from Alaska, Johnstone and Kasischke (2005) reported the effects of burn severity (estimated by the amount of organic layer consumed) on post-fire regeneration. Areas that had burned at high severity favored aspen (*Populus tremuloides*) recruitment, whereas, black spruce (*Picea mariana*) establishment was negatively associated to burn severity. Therefore, increased burn severity is likely to shift successional trajectories away from simple conifer self-replacement towards a trajectory of mixed conifer and deciduous dominance.

Finally, fires may promote important vegetation shifts or even lead to new stable states that can last for many years in the absence of fire. A particular vegetation type may be able to persist despite climatic change, but be susceptible to sudden change when burned. After a fire a different successional trajectory may be set in motion leading towards a new climax community. Johnstone and Chapin (2003) reported that in Canadian boreal forests fire favored the expansion of *Pinus contorta* var. *latifolia* in the northern edge of its distribution range at the expense of *Picea mariana* and *P. glauca*, but this replacement was not observed in the interior portions of the range. The authors suggest that these changes are a step in the northwards expansion of the distribution area of lodgepole pine since the last glacial period. The mechanism of these transitions may be related to the provision of microhabitat that allows establishment and growth of shade tolerant species. The forest is able to remain even under a suboptimal climate until fires destroy these microhabitats. This is the explanation given for the transformation of boreal forests that abruptly shifted to open krummholz after a fire in 1568 AD, when colder climate inhibited post-fire regeneration of the forest (Arseneault and Payette 1997). In a situation of non-equilibrium between climate and vegetation, wildfires can often trigger the impending change.

FIRE REGIME AND RESILIENCE

The Fire Regime-Resilience Coupled Model

The time period in which systems are susceptible to ignition and fire propagation vary from very large (semi-arid grasslands of the North American prairie, South American pampas, African savannah, and Eurasian steppes) to almost zero (tropical rain forest). Systems more prone to fire are expected to contain more resilient species as a result of species sorting during the process of community assembly along the successive fire cycles. As explained above, the mechanisms allowing high levels of resilience may be the result of pre-adaptations to disturbances in general or by selective pressures exercised by fire along evolutionary time-scales. Additionally, positive feedbacks between vegetation providing fuel load and fire may appear. Often highly resilient species, such as grasses, are also highly flammable and promote the existence of frequent fires. It has been suggested that the existence of selective forces promoting acquisition of flammable traits, such as high levels of standing dead branches and bark, would benefit high fire-resilient species and eliminate competitor neighbors that are less resilient (Bond and Midgley 1995). The result would be the occurrence of 'steady-states' where fire regime and vegetation are highly interdependent, such as those found in grass driven-systems or in some Mediterranean communities.

Thus, if a fire regime remains in place for a long period, some kind of relationship is expected to link species resilience and fire regime. At a biogeographical scale, the concept is generally observed within the direct post-fire regeneration patterns in Mediterranean and grass-dominated systems and with successional trajectories observed in boreal, temperate, and tropical rainforests. In these successional dynamics, longer times are needed for forests with more fire-sensitive species, such as tropical rainforests, to recover the pre-fire communities.

At a more local scale evidence of a coupling between fire regime and resilience is also found. In a study of the transition between dry Afromontane forests and *Acacia* woodlands in southern Ethiopia, Eriksson et al. (2003) reported that fire propagates readily in the *Acacia* woodland as a result of large, continuous fuel loading; however, in the Afromontane forests, sustained combustion is unlikely to occur due to the poor aeration of the densely packed leaf litter and its direct contact with the forest floor. Fire survival of trees in these communities is tightly linked to bark thickness, which tends to be smaller in Afromontane forest species than in *Acacia* woodland species. Also, fire breaks seed dormancy in *Acacia* species, favoring its establishment after fire. This scenario of coupling between fire likelihood and fire resistance and resilience may be modified by cutting and browsing in the Afromontane forests, which results in a drier microclimate and greater abundance of grasses and herbs. This eventually would increase fire spread potential and the shift to a more fire-tolerant woodland (Fig. 3).

In areas where 'zones of tension' between more and less fire resilient vegetation exist, the landscapes may show considerable overall resilience (capacity to prevent soil loss and sustain annual primary productivity) because of the intermingling of more and less fire resilient species. Human influence, however, can alter disturbance regimes over large areas, thereby exceeding the capacity of landscapes to recover and increasing the need for

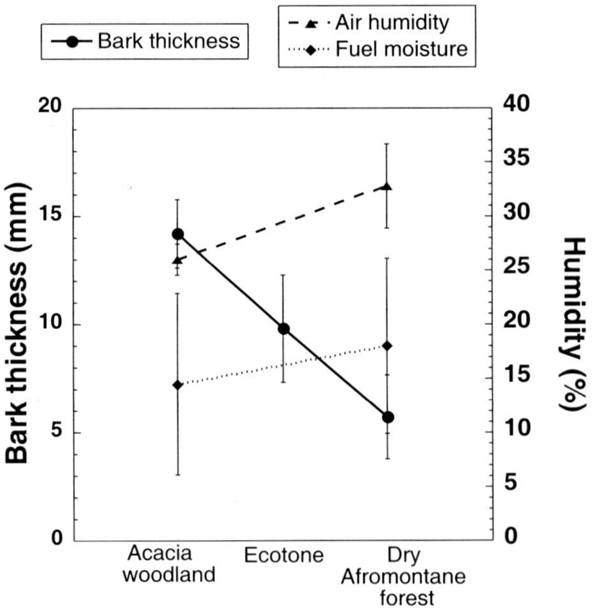

Fig. 3 Mean bark thickness of species growing in Acacia woodland (n = 4), ecotone (n = 9) and dry Afromontane forest (n = 13) in southern Ethiopia, and air relative humidity (%) and fuel moisture (%) measured during experimental fires in the Acacia woodland and the dry Afromontane forest. Mean bark thickness is correlated to species resilience to fire, while air and fuel moisture are inversely correlated to fire likelihood. The case illustrates a coupling between fire regime and resilience after fire. Data from Eriksson et al. (2003).

restoration. There are a variety of factors that make the fire regime-resilience coupled model highly sensitive to changes.

Failure of Resilience

Climate, ignitions, and fuel basically control wildfires. Humans modify the natural fire regime by increasing fire ignitions, by suppressing fires, or by changing the amount and distribution of fuel. At the global scale, climate change driven by human activity is also increasing climatic fire risk in some regions (Piñol et al. 1998, Westerling et al. 2006). Significant changes in forest plant communities are introduced when deliberate planting of preferred but less resilient species occurs, or accidental introduction and the spread of exotic species. As a result, plant communities that were coupled to a given fire regime experience a new situation with potential loss of resilience (Fig. 4).

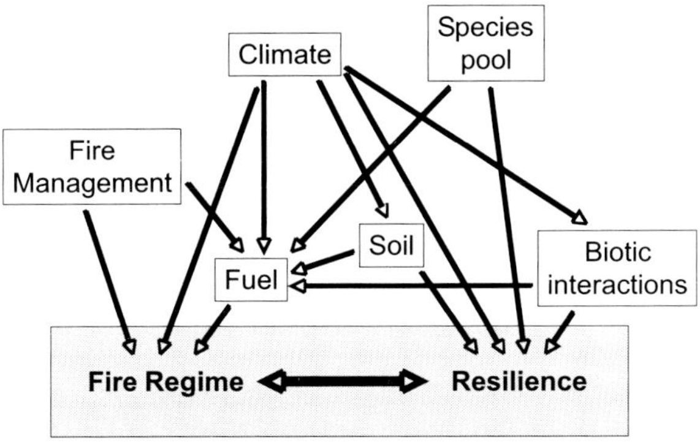

Fig. 4 Fire regime and resilience are often positively correlated, but this relationship may be altered by a set of interacting factors that often are determined by humans at different levels (from global to local).

The increase of fire ignitions is a recognized cause of the recent transformation of low resilient tropical forests to savannah-type formations (Swaine 1992, Fensham et al. 2003, Vieira and Scariot 2006). In more resilient ecosystems, such as in the Mediterranean region, there has been an increase in the number and extent of fires since the last decades of the past century. Some studies show that high fire recurrences may result in a loss of resilience in terms of the time needed to restore plant cover (Díaz-Delgado et al. 2002). Such loss of resilience would be the result of changes in species composition or lower release of nutrients by fire. In both cases, shorter intervals between fires may preclude the establishment of fully reproductive populations (Eugenio et al. 2006b) or reduce the nutrient capital of a site (i.e., loss of fire fertilization effect; Eugenio et al. 2006a; Table 1).

In ecosystems where fires naturally occur at some frequency, it has been argued that fire suppression policies have produced a homogeneous landscape dominated by an accumulation of continuous fuel that is conducive

Table 1 Effect of fire recurrence in Mediterranean *Pinus halepensis* communities from Catalonia (Spain). Paired stands experienced one and two fires in the last two decades were sampled about 10 years after fire. The case illustrates loss of resilience as a consequence of high frequency of fires in Mediterranean communities where direct regeneration is the rule. Number of stands and statistical significance are indicated in n and p columns. Standard errors are given between brackets. Data from Eugenio and Lloret (2004) and Eugenio et al. (2006a, b).

	Once burned	Twice burned	n	p
Overall canopy				
Cover (%)	87.9 (2.2)	77.9 (2.7)	15	0.002
Height (cm)	32.9 (4.2)	13.4 (3.0)	28	<0.0001
Tree layer				
Cover (%)	32.9 (4.2)	13.4 (3.0)	28	<0.0001
Height (cm)	149.6 (7.3)	92.0 (17.4)	28	<0.0001
Shrub layer				
Cover (%)	53.2 (5.0)	60.9 (4.7)	28	<0.0001
Height (cm)	107.9 (7.2)	70.0 (4.5)	28	<0.0001
Herb layer				
Cover (%)	43.8 (4.6)	33.2 (4.7)	28	0.025
Height (cm)	35.5 (3.7)	25.4 (2.3)	28	0.012
Pinus halepensis				
Density (10^3 ha^{-1})	25.4 (12.7)	14.1 (10.1)	14	0.001
Height (cm)	168.8 (10.9)	142.0 (13.0)	14	0.014
Diameter (mm)	25.9 (1.5)	23.3 (2.0)	14	0.043
Soil organic horizons				
LF dry mass (Mg ha^{-1})	16.9 (2.0)	10.8 (1.8)	15	0.006
H dry mass (Mg ha^{-1})	1.0 (0.6)	0.2 (0.1)	15	0.015

to catastrophic, high intensity crown fires (Minnich 2001). Then resilience after fire may be diminished by the severity of fires and by the absence in the landscape of species able to restart a new succession. In fact, contemporary evidence suggests that where fire becomes rare, less fire resilient species expand at the expense of more fire resilient species (Abrams and Downs 1990, Brose et al. 2001, Ruffner and Groninger 2006). However, the analysis of fire records does not provide conclusive evidences of a relationship between fire suppression practices and frequency and extent of fires. Instead land use changes and extreme climate events provide more satisfactory explanations of changes in fire regime (Johnson et al. 1998, Keeley et al. 1999).

Therefore, in some sparsely populated regions, limited fire suppression may be an alternative to avoid catastrophic fires. However, this is much less advisable in densely populated areas where temperate and Mediterranean forests grow. In these areas, prescribed burning has been introduced as a landscape management practice. In many countries of the Mediterranean Basin, prescribed burning is not socially accepted and fuel reduction by clipping is preferred. Vegetation is mainly cleared in strategic sites, like roadsides, that could act as barriers to fire spread and facilitate the action of fire fighters. In deciduous and coniferous forests, prescribed fires are generally low intensity surface fires. They eliminate litter and favor the establishment of seedlings (Brose and Van Lear 1998). In Mediterranean shrublands of California and Australia, prescribed fires often are moderate or high intensity crown fires. To minimize risks, prescribed burns are not ignited in the driest season when most natural fires occur. Consequently, prescribed fires used for fuel reduction do not mimic natural fire regimes. For example, the prescribed burns are done in a different season than most natural fires and fires may become very frequent at some sites. Kerns et al. (2006) reported that prescribed fires occurring in the 'off ' season (in the wetter spring season rather than in the dryer fall season), may increase the occurrence of exotic species in *Pinus ponderosa* forests, and the community resulting from these prescribed fires may be quite different from the natural ones. Following the fire regime-resilience coupled model, these communities (often dominated by grasses after successive burnings) are more prone to burn again (although with low intensity). A critical point of prescribed burning is that it must be applied at relatively short intervals. Although some studies do not detect important changes in soil properties (Murphy et al 2006), long-term application of burning may negatively affect fine roots and mycorrhizal symbionts with resulting changes in nutrient cycling (Hart et al. 2005).

As post-fire burn severity increases, resilience tends to decrease. Since pre-fire populations are more deeply affected, there are more opportunities for the establishment of new species (Fig. 5). Broncano et al. (2005) found that the transition from a mixed Mediterranean forest of *Pinus halepensis* and *Quercus ilex* to monospecific forests was higher in more high burn severity stands. Thus, the intrinsic variability of fire behavior may result in a forced modeling of the landscape into a mosaic of forest types according to the differential sensitivity of species to fire, as described for austral forests of *Araucaria* and *Nothofagus* from South America (González et al. 2005). Fires may also increase erosion risks by destroying the protection provided by vegetation, increasing runoff, and destroying soil aggregates. These effects increase with burn severity (Giovannini and Lucchesi 1997) and cause dramatic consequences when combined with heavy rain before the post-fire vegetation is able to provide cover (Conedera et al. 2003).

A trend toward drier climatic conditions in some areas has increased the likelihood of fires in regions where fires were, until recently, a relatively rare disturbance. This is the case of increasing fire climatic risk in deciduous temperate forests of central Europe where an increase of drier periods have been observed in recent decades (Reinhard et al. 2005). These areas have had several crown fires with dramatic consequences due to the inability of the dominant species to regenerate after fires (Delarze et al. 1992). Paleohistorical records have shown that such fire-driven vegetation shifts have occurred in the past (Tinner et al. 1999; Fig. 6). Climate also influences patterns of

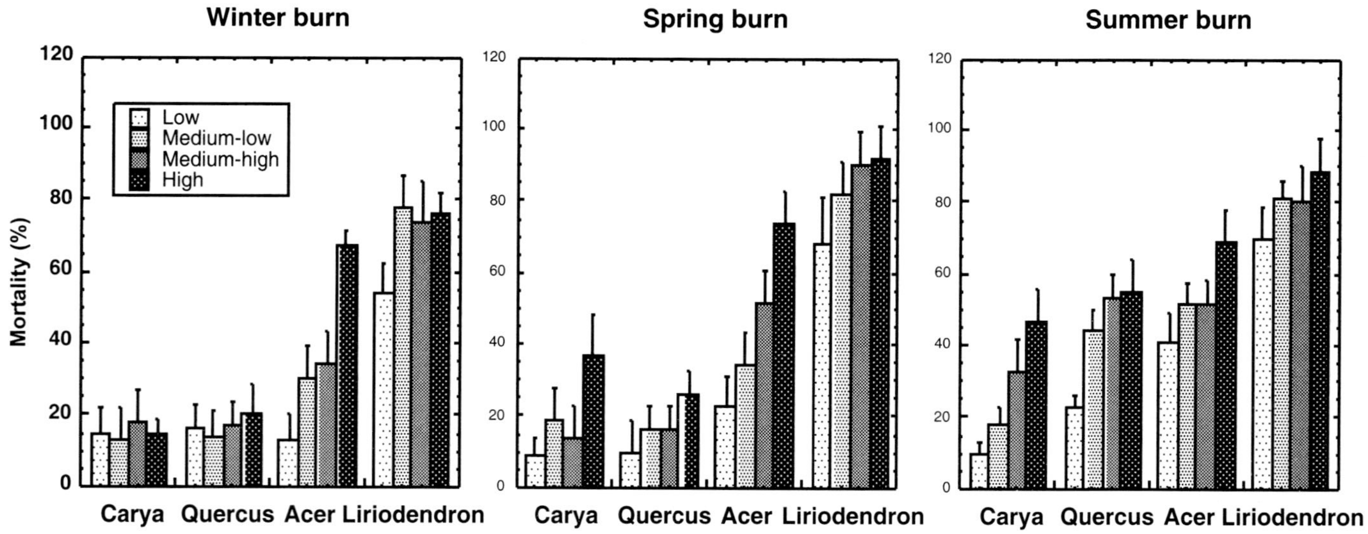

Fig. 5 Percent mortality (mean percentage of stems failing to sprout by the end of the first post-burn season) after prescribed burnings of different intensity carried out in winter, spring, and summer in eastern hardwood forests of North America. Bars indicate standard error. The case illustrates prescribed fires improving oak regeneration respective to other hardwood species and effects of intensity and seasonality on post-fire resilience. Data from Brose and Van Lear (1998).

establishment and growth after fire (Hogg and Wein 2005). Therefore, under new climate scenarios, wildfires may be the catalyst for changes in the vegetation-climate relationship. From paleohistorical records, vegetation shifts that differ from successional trajectories have been observed after fires in boreal regions, where a single fire in a region without a history of fires resulted in a shift from lichen-spruce krummholz to lichen tundra (Arseneault and Payette 1992).

Human-induced changes in species composition (e.g., by plantations) may influence fuel characteristics and resilience properties at the community level. Resilience properties of planted species may differ from the native vegetation, altering the historical fire regime-resilience coupling (Rodrigo et al. 2004). Vegetation shifts induced by fires may also occur through the expansion of exotic species (Keeley 2006). These shifts may involve changes in species composition, structure, and function as has been seen in the invasion of European herbs into frequently burned shrublands in California (Keeley et al. 2005) and in the pine populations invading South African and New Zealand shrublands (Versfeld and Van Wilgen 1986, Richardson and Higgins 1998). Williams and Wardle (2005) have reported that low frequency fires exacerbate the invasion of the exotic serotinous *Pinus radiata* from plantations into *Eucalyptus* woodlands of Australia. Similarly, Simberloff et al. (2003) pointed out that prevention of catastrophic fires is the best way to maintain native coniferous forests (South America) from invasions of exotic

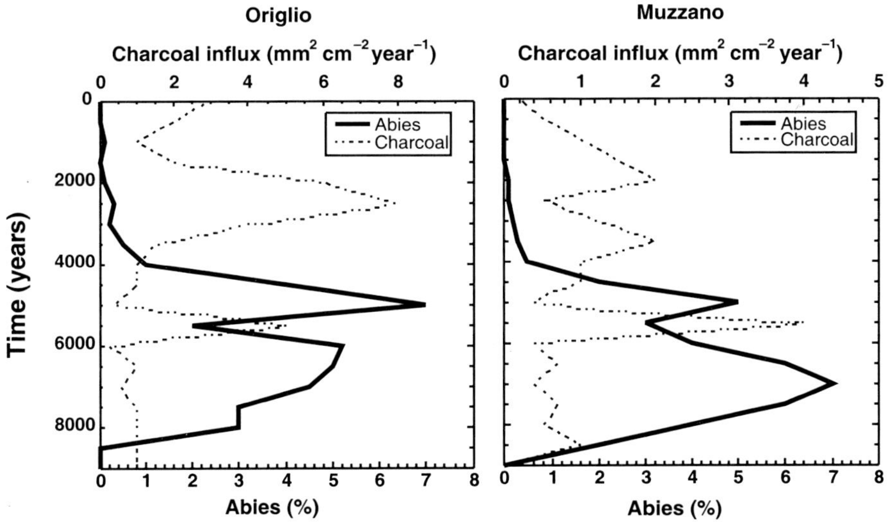

Fig. 6 Diagrams of *Abies alba* pollen percentage and charcoal influx (mm^2 cm^{-2} × yr^{-1}) from sediments of two lakes in southern Switzerland (Origlio and Muzzano). Note the negative correlation of charcoal and *A. alba* occurrence, which is currently absent in the area. The case illustrates vegetation shifts related to the occurrence of fire in the paleohistorical record. Data from Tinner et al. (1999).

trees. Interestingly, positive feedbacks between exotics species and fire have been reported when the exotic species promote fire occurrence and then are able to establish successfully after the fire (Litton and Santelices 2002, Hoffmann et al. 2004). The result may be a completely new system controlled by the intruder.

Biotic interactions may also alter resilience patterns. Granivory may substantially decrease post-fire seedling establishment, contributing to the variability of regeneration at the landscape scale (Ordoñez and Retana 2004). O'Dowd and Gill (1984) argued that the post-fire mass release of seeds by *Eucalyptus* species might be critical for seedling establishment because only a massive seed input will allow some to escape ant predation. It follows that any fire regime that failed to produce these strong pulses might jeopardize the capacity of a population to sustain itself. An instance of the interaction of fire and granivores has been reported by Greene et al. (2004) in the boreal forest region where lightly burned sites also generally had higher predatory pressure. In several Mediterranean countries, grazing by sheep and goats is forbidden in the first few years after fire due to the impact on young resprouts and seedlings. Also, the combined effect of fire and an insect pest has been reported to induce a shift from forest to lichen woodland in boreal regions of North America (Payette et al. 2000).

Management Strategies to Improve Forest Resilience

Improvement of forest resilience is the general goal of most fire-focused management strategies. These actions operate directly, like fire suppression (putting fires out), or indirectly, like prescribed burning (starting fires), to alter the fire regime. Fuel management strategies also impact the fire regime by decreasing or enhancing the probability of fire occurrence in its potential intensity. As discussed above, reduced fire frequencies (consequences of fire suppression policies) often lead to dominance by late successional, low resilient species (Fig. 5) and frequent fires (from prescribed burning) can cause the loss of forests and replacement by high resilient communities, such as grasslands. Management strategies attempt to find the middle ground between the expansion of resilient non-forest communities and old forests dominated by species with a depleted ability to regenerate after fire.

In theory, the fire resilience of almost any system could be enhanced simply by the frequent application of fire, simulating the historical process by which fire resilient systems have developed. There are many reasons why this simplistic prescription is not widely adopted, despite the high probability that it in the long run it would be successful at least to some degree. The main problem, even in regions with long histories of fire, is that an unacceptably unstable intermediate situation can occur before the system settles into a new strongly fire-resilient configuration. Another concern is the possibility of accelerated erosion, as the gaps left by the death of less resilient forms are not filled or are replaced by species with a lower erosion reduction capacity. The fire may also open the system to aggressive invaders that can be difficult or impossible to remove from the system, effectively shifting it into a new 'basin of attraction.' To avoid these and other problems, prescribed fire should be judiciously applied. Historical fire patterns, even where they are known, may not be a very good guide for the management regime that is required in the initial stages of shifting the vegetation towards greater resilience. These patterns may not apply once the system stabilizes because of changes in fuel conditions, such as the introduction of exotic species, or to changes in soils, such as erosional losses. There may not be a clear natural fire regime to use as a reference. In many ecosystems humans have modified the fire regime many centuries ago and any attempt to restore a natural fire regime may introduce elements that lead to unpredictable new states (Keeley 2006). In some cases, managers may want to maintain the low resilient systems, such as a tropical rain forests or late successional patches of forest, and will include fire suppression as a management strategy. Finally, fire suppression will always be the management goal when human life is at risk and fire suppression usually supercedes other goals when properties infrastructure are at risk for damage from fire.

Species-focused strategies are aimed at managing species composition and community structure. For example, the goal of managers may be to establish a forest of resilient species or to favor the establishment of less resilient species after disturbance. Actions require knowledge of individual species resilience and potential interactions of the factors involved (Fig. 4). Climate is not amenable to management, but actions on soil properties, the species pool (including plantations and control of non-natives), and biotic interactions (such as herbivory, granivory, and pests) may minimize resilience failure at the species level. Species-focused management to improve the fire resilience of forests is difficult because species composition and vegetation structure are closely related, but to a degree, independent. It is possible to have the needed resilient species, yet have a vegetation community structure that can lead to fires that cause catastrophic change. It is also possible to alter the structure of one part of the system to the desired state, but still have non-resilient system because another less readily manipulated part of the system is not appropriate. For example, in pine forests of the southwestern U.S., thinning to reduce tree density does little to improve fire resilience if the understory is not made up of fire-resilient grasses and shrubs to carry the frequent fires that are needed to sustain the modified structure (Friederici 2003).

Management goals should be based on the resilience characteristics of the system and the implementation objectives will impact vegetation composition, structure, and function of the ecosystem. Due to the many uncertainties involved in manipulation of fuel and species composition (cutting, thinning, planting of desirable species, chemical control of undesirable exotic species) and in prescribed burning, management strategies to improve fire resilience of forests requires an experimental adaptive approach.

FIRE EFFECTS ON VEGETATION IN REGIONAL FOREST TYPES

Tropical Rainforests

In tropical forests, there exists a continuum of fire frequency decreasing from evergreen tropical forests to savannahs, with seasonal deciduous forests occupying an intermediate position. Human activity modifies this gradient by clearing, slashing, and burning forests. Therefore, increasing fire frequency has favored the expansion of savannahs and open forests with grasses and other clonal plants in areas where less fire-prone communities would exist in the absence of humans (Swaine 1992, Fensham et al. 2003, Vieira and Scariot 2006). This dynamic is particularly intense along the forest-savannah ecotone and limited in areas located far from grass-dominated communities (Saha and Howe 2003), which supports the hypothesis that this substitution is also limited by propagule availability. At the opposite extreme, fire suppression may enhance the expansion of evergreen forests in areas where historically fire facilitated the expansion of dry forests, such as protected areas of southeastern Asia (Stott et al. 1990).

Rainforests were considered to have been historically safe from wildfires, but more recent information has shown that burning has been present (albeit with low frequency) since the end of the Pleistocene (Goldammer and Seibert 1990). Fast decomposition avoids the accumulation of a continuous horizontal fuel layer with large amounts of standing dead fuel; however, widely scattered patches of flammable plant materials may be present. Usually, fires within the rainforest are low intensity ground fires mostly consuming dry leaf litter. Most trees are, however, relatively thin barked and fire-sensitive, so that tree survival is strongly dependent on stem size with larger trees having thicker barks and greater fire resistance (Cochrane et al. 1999, Pinard et al. 1999, Eriksson et al. 2003). The mortality produced by these fires may result in an accumulation of woody debris favoring later fires of higher intensity. Also, the canopy opening favors the proliferation of herbaceous vines and lianas that impede tree regeneration and supply additional fine fuels. In contrast to trees, vines are prone to resprout (Pinard et al. 1999). Therefore, previously burned forests have been reported to be much more likely to burn with high intensity than unburned ones (Cochrane et al. 1999), which generates positive feedback and accelerates the transformation to an unforested landscape.

Although some rainforest species are able to resprout vigorously and may achieve dominance in burned sites (Goldammer and Seibert 1990), generally mortality is high (Holdsworth and Uhl 1997). Overall it has been reported that fire reduces the number of trees and tree species per unit area, especially the dominant tree species (Slik et al. 2002, Cleary and Priadjati 2005). Although tree density tends to recover, the number of species declines and a shift in species composition appears. Therefore, there is no trend for recovery to pre-disturbance conditions, and the burned forest may remain severely degraded for a long period of time (Cleary and Priadjati 2005).

Regeneration after burning is supported by gap-opportunistic pioneer species that can establish abundant seedlings from the seed bank (Goldammer and Seibert 1990, Holdsworth and Uhl 1997). When fires are small due to patchy fuel distribution, the spatial distribution of disturbed patches is similar to that produced by tree death, and the gap-replacement dynamics are similar. Late-successional species may also establish in the small burned patches. However, not all fires are small and patchy. Drought periods and fuel accumulation due to large-scale disturbances, such as hurricanes, may result in fires to a large extent (Goldammer and Seibert 1990, Nepstad et al. 1999). Simulataneous die-off of bamboos in some forest systems will result in large fires (Keeley and Bond 1999). Analogously, logging can create large gaps that increase fire vulnerability, which can be reduced by the use of low-impact logging techniques (Holdsworth and Uhl 1997). Since most tropical rainforest species do not form a seed bank, establishment in large burned areas is highly dependent on long-distance dispersal mechanisms and the existence of remnant trees that can provide a favorable microhabitat for seedlings of late-successional species (Elmqvist et al. 2002). Therefore, both fire intensity and fire size, and the completeness of the burn within the fire perimeter play important roles in vegetation community recovery.

If fire is excluded from unforested areas and appropriate growing conditions persist, a new succession towards greater tree dominance may begin if seeds of suitable species disperse to the site. This conversion to rainforest has been documented in the forest-savannah ecotone of central Africa (King et al. 1997), the eucalypt-dominated woodland savannah of northeastern Australia (Russell-Smith et al. 2004), and in shrublands ('maquis') of New Caledonia that were originated and maintained by fire (McCoy et al. 1999).

Dry-Deciduous Tropical Forests

Although not as frequent as in savannahs, natural occurrence of fires in tropical deciduous seasonal dry forests is not uncommon in the dry season. A number of causes contribute to natural ignitions, but humans have a long history of increasing fire frequency (Stott et al. 1990). Most fires are surface, low intensity fires, although ground-to-crown high intensity fires may occur. Fires that burn only in the crowns are more unusual because of the low continuity of the tree canopy.

Although post-fire regeneration is provided by both seedlings and resprouts in deciduous tropical forests (Miller and Kauffman 1998), several studies have shown seeds to be fire sensitive with low post-fire germination rates (Miller 1999, Kennard et al. 2002, Saha and Howe 2003). Resprouting, however, is a widespread strategy, including species producing suckers from root buds, root-crowns, or branches from meristems protected from high heat by structures in the aerial parts. In addition, the resources stored in below ground organs may provide more reserves than provided by seeds, which in this type of forest are generally small, and wind-dispersed. These general resprouting and seeding patterns can be modified by characteristics of the fire regime and pre-fire forest structure.

Dry forests have fire resistant species and species that benefit from fire; however, frequent fires generally simplify community species composition. In central India, Saha and Howe (2003) documented that a century of annual understory burning in an otherwise fire-free, deciduous tropical forest has favored tree species that produce suckers from root buds over species that produce sprouts basally from root crowns. Since root-sprouting offers a means of occupying new ground with clonal ramets away from the original parental base, over time, forests may become dominated by clonal root-sprouters, in contrast to historical records of forest dominated by root-crown resprouters.

Burn severity may influence the differential success of seedlings and resprouts. Kennard et al. (2002) reported that in Bolivian dry forests, there existed a greater dominance of seedlings in high burn severity areas. This is due to the sensitivity of sprouting tissues and underground organs to fires with high soil heating over long durations. At the same time, resprouting mechanisms are differentially affected by burn severity – stem resprouting was common after low severity fires while species success after high severity fires favored resprouting from belowground organs. Pre-fire distribution of gaps within the forest may also influence the regeneration patterns. Seedlings growing in gaps are more likely to survive surface fires by resprouting because they were able to accumulate more resources in the belowground organs than seedlings growing below the closed canopy (Marod et al. 2004, Dokrar et al. 2004). In general, after high severity fires seedlings of shade-intolerant species are plentiful. These results suggest that low to moderate severity fires (most common in these forests) increases resilience and favors self-replacing dynamics, whereas high severity fires induce a different dynamic, in which shade-intolerant, long-distance dispersed colonizers contribute to the initiation of a new succession.

Mediterranean Forests

Mediterranean ecosystems are considered to be fire prone due to the dry, hot season which characterizes the climate. In addition, dead branches and leaves from vegetation and standing shoots accumulate on the forest floor because decomposition rates are low. This material has a high surface area to volume ratio and living leaves are rich in volatile, highly flammable compounds both contribute to fire spread. However, the historical incidence of fires has not been identical in the different Mediterranean regions of the world. Most fires are naturally ignited by lightning, but human activities have modified the natural fire regime. In many regions, the current increase in large, high intensity fires constitutes a matter of great concern (Lavorel et al. 1998, Keeley et al. 1999).

Fires in Mediterranean ecosystems are characterized by high intensity burning of most standing biomass, with understory fires being rare. However, the ability of Mediterranean communities to self-replace after fire has been well documented (Whelan 1995, Bond and van Wilgen 1996). Although a wave of ephemeral, transient herbs commonly appear immediately after fire (Calvo et al. 2002), most species are fire resilient. Some of them resprout from aerial or underground structures (lignotuber, root crown, and rhizomes) and the establishment of their new cohorts often occurs gradually between fires (Keeley 1992, Lloret and Zedler 1992). Some resprouting species, such as the Californian *Adenostoma fasciculatum*, germinate abundantly after fire and achieve dominance. Others develop seed banks that can resist fire in the soil or in standing fruits, and germination may be stimulated after the break of the seed coat or by smoke cues (Dixon et al. 1995, Keeley and Fotheringham 2000). Some of these seeder species, called obligate seeders, apparently need fire to complete their life cycle since little or no recruitment is observed in the absence of this disturbance. Particularly, in some regions, such as South Africa, Australia, and California, serotinous species occur within a few genera (Le Maitre and Midgley 1992). This character is considered to be an adaptation to frequent fires by concentrating all the progeny of a population at the time when conditions for germination and establishment are optimal. Thus, the seed predation and light competition that occur between fires are avoided. In other cases fire-dependent recruitment appears in species accumulating a long-term soil seed bank (Zammit and Zedler 1994). Finally, some seeder species, like short shrubs from the Californian and the Mediterranean basin coastal scrublands, may take advantage of gap opening or other disturbances even in the absence of fire (DeSimone and Zedler 1999).

Overall, in Mediterranean ecosystems there is a strong relationship between community resilience and the probability of fires. In fact, fire-sensitive species are rare in these communities (Lloret et al. 2005). However, the ability of these communities to regenerate after fire is not unlimited, particularly when changes in the fire regime occur as a consequence of human activity. Fire regime modifications have been shown to effect Mediterranean forests by modifying the resilience properties of the system (Zedler et al. 1983).

Increasing burn severity may result in a decrease of resprouting ability by depletion of the bud bank (Zammit 1988) or the resources stored in belowground organs (Canadell and López-Soria 1998). In species depending on seed germination, short fire intervals may prohibit replenishment of the seed bank as seen in *Pinus halepensis*, an active post-fire colonist in the western Mediterranean Basin, which declines if the fire interval becomes short enough to prohibit populations from reaching a reproductive age (Eugenio et al. 2006a, b; Table 1). As in tropical ecosystems, high frequency fire favors grasses and short-living species, which can cause a transition to short shrublands, grasslands, or savannah-like community structures (Grove and Rackham 2001, Keeley 2002). In turn these communities are more prone to burn and positive feed-backs between vegetation and fire may appear (Grigulis et al. 2005). Also, non-native herbaceous species may invade and become dominant in frequently burned shrublands (Keeley et al. 2005). Díaz-Delgado et al. (2002) used remote sensing techniques to show that vegetation recovered more slowly after recurrent fires. On the other hand, decreasing fire frequency, as a result of fire suppression policies, may risk renewal of fire-dependent species, such as the serotinous *Pinus torreyana* of southern California (McMaster and Zedler 1981). Although seeder and resprout species often coexist in Mediterranean communities, the relative abundance of each group is sensitive to change in the fire regime, and particularly to fire frequency. In northeast Spain, Lloret et al. (2005) observed an increase of seeder species in burned forests, and these species are generally more abundant in Mediterranean conditions. In the Californian chaparral, Franklin et al. (2004) found that resprouter species tended to increase in unburned stands, while obligate seeders decreased; the opposite pattern was observed in frequently burned sites.

Fire severity also determines resilience of Mediterrean plant communities. It has been shown that the resprouting ability decreases with burn severity (Lloret and López-Soria 1993). Although high soil temperature stimulates germination of some species, seed viability declines when temperature or heat duration reach specific thresholds (Habrouk et al. 1999). As a result, community composition and structure resilience may be negatively affected by severity (Broncano et al. 2005, Díaz-Delgado et al. 2003).

The season of fire occurrence also determines the impact on the vegetation. If fire occurs before the replenishment of the seed soil bank, which for many species occurs in late spring or summer, there is a decline in germination the following autumn (Domínguez et al. 2002). Resprouting is also expected to be more effective if fire occurs when the level of stored resources in underground organs is high (Konstantinidis et al. 2006); however, direct links between resources stored in below ground organs and resprouting vigor have not been clearly established (Cruz et al. 2003).

Temperate Deciduous Forests

In temperate deciduous forests, fires are usually understory fires. In spite of their low intensity they may play an essential role in the dynamics of these forests by determining the regeneration of the populations. Two main mechanisms would be involved in this control. First, the transformation of the habitat by fires, which includes: a) the opening of gaps by reducing the density of understory and overstory tree density and allowing the establishment of shade-intolerant species (Signell et al. 2005), and b) the generation of a mineral soil seedbed that promotes germination. The second controlling mechanism is the ability of different species to resprout or to germinate after fire (Brose and Van Lear 1998; Fig. 5). Therefore, in temperate deciduous forests, changes in the fire frequency may induce important changes in species dominance.

During the last centuries, humans in eastern North America have modified the fire regime. American Indians performed frequent low-intensity surface fires which allowed oaks to be the dominant species. Since the 19th century, intense exploitation and stand-replacing fires were introduced. After the 1950s, fire suppression policies were implemented. The result has been a recent recovery of forests but a decrease of oak regeneration and a loss of species that are replaced by mesophytic species across many landscapes (Abrams and Downs 1990, Brose et al. 2001, Ruffner and Groninger 2006). For example, Brose and Van Lear (1998) studied regeneration of hardwoods in Virginia where disturbances often stimulate regeneration of fast-growing intolerant species (*Betula lenta, Liquidambar styraciflua, Liriodendron tulipifera*) that would be successionally replaced by shade-tolerant species (i.e., *Fagus grandiflora, Acer rubrum, A. saccharum*). In this sequence, surface fires maintain oak (*Quercus alba, Q. rubra, Q. velutina, Q. coccinea*) stands on productive sites (Fig. 5). Fire would also reduce competition from the recruits of fast-growing species that are more sensitive to surface fires than oak recruits. Currently, prescribed burning is being examined as a method to improve oak regeneration (Brose and Van Lear 1998); however, the results may depend on the ability of other understory species to resprout and close the gaps opened by fires (Chiang et al. 2005). In northern Europe, natural fires allowed deciduous stands to exist within the matrix of coniferous forests; but, after 1900, fire suppression and the direct elimination of hardwood species in favor of coniferous species became widespread. Recently prescribed burning after clear-cutting is being used to favor the restoration of deciduous stands *in Pinus sylvestris*-dominated areas (Axelson et al. 2002).

As a result of increasing climatic hazard, temperate deciduous forests fires may shift to ground-to-crown fires that could rapidly change species composition due to the poor regeneration capacity of the dominant species after complete burning. Pollen records indicate that this transition occurred in northern and central Europe during Holocene (Tinner et al. 1999), and prompts this concern given the current increase in wildfires associated with a drier climate (Delarze et al. 1992). In some cases, crown fires trigger changes due to interactions with exotic species. Although the deciduous *Nothofagus glauca* of south-central Chile are able to resprout after fire, fires have been absent in this region until very recently. Crown fires in stands of *Nothofagus glauca* open the door to the establishment of exotic invasive species, including trees such as *Pinus radiata*, whose plantations are often the origin of the fires (Litton and Santelices 2002). This fire-mediated interaction of coexisting species with different post-fire regeneration capabilities also occurs in the North America southeastern coast. There, the native longleaf pine (*Pinus palustris*), promotes intense crown fires that cause mortality of the smaller resprouter turkey oak (*Quercus laevis*; Rebertus et al. 1989). The establishment of pine seedlings is favored by the reduction of competition from resprouting oaks.

Boreal Forests

Fires are a natural component of the dynamics of boreal ecosystems, resulting in landscapes shaped by a mosaic of burned areas at different stages of regeneration. Crown fires ignited by lightning strikes are common in these forests (Gromstev 2002). The intensity of the crown fire determines individual survival and seed availability. Ground level fire intensity is related to reduction of the inhibiting effect of deep organic layers on seedling establishment (Schimmel and Granström 1996, Greene et al. 2004). Since the distribution of canopy and ground fuels may exhibit different spatial patterns, the burn severity in the canopy and on the ground may not be coupled.

In boreal and subarctic forests, most understory vascular plants regenerate by sprouting from buried vegetative parts (Sirois 1995, Schimmel and Granström 1996, Kemball et al. 2005). The regeneration of tree canopy generally occurs shortly after fire (Greene et al. 2004). Some species produce serotinous (*Pinus banksiana*) or semi-serotinuous (*Picea mariana*) fruits and their populations readily recover after fire; conversely, in the absence of fires, these species do not persist (Johnson 1992, Sirois 1995, Greene and Johnson 1999). Other trees (white spruce *Picea glauca*, balsam fir *Abies balsamea*, paper birch *Betula papyrifera*, and aspen *Populus tremuloides*) must disperse seeds into a burned area from surviving individuals or from adjacent unburned stands. Therefore, post-fire regeneration may be enhanced by synchronization with masting years, as reported for *Picea glauca* (Peters et al. 2005). Paper birch and aspen are particularly good colonists since their copious vegetative reproduction by stem resprouts or root suckers, respectively allows them to quickly assume dominance of an area opened by fire (Greene and Johnson 1999). However, white spruce and balsam fir are fire-sensitive and may become locally extinct in the absence of efficient dispersal (Johnson 1992). These different responses to fire, together with longevity, shade tolerance, and niche regeneration may determine successional trajectories of replacing species (Gromstev 2002) or vegetation shifts. The result is a landscape mosaic reflecting fire, climate, soil type, and moisture.

In sub-boreal forests of North America, most recruitment occurs a few years after the fire, followed by decades of low or no recruitment (Johnstone et al. 2004, Greene et al. 2004). Thus, variations in forest composition are largely determined by factors affecting recruitment immediately after fire (Sirois and Payette 1989, Purdy et al. 2002). The colonist populations often are self-thinning (Johnstone et al. 2004), which can be particularly effective in species with a high-density of recruits due to vegetative resprouting (e.g., aspen)

(Greene and Johnson 1999). In the absence of shade-intolerant competitors, gap dynamics may maintain near-pure aspen stands for a long time (Cumming et al. 2000). Late successional species will eventually competitively replace the species that dominated burned stands in the first years after fire (Archambault et al. 1997). Often, late successional species will establish quickly after fire and sustain their populations for long periods in the absence of further disturbance (Clark et al. 2003, Schulze et al. 2005). In British Columbia, the early colonizers are aspen (*Populus tremuloides*) and the conifer *Pinus contorta*, which is replaced by *Picea glauca* and *Picea mariana* (Johnstone et al. 2004), and eventually dominated by the long-lived *Abies lasiocarpa* (De Grandpre et al. 2000, Clark et al. 2003). In the dark taiga of Siberia, the succession after stand replacing (crown) fires is characterized by an early dominance of *Betula pendula* and *Populus tremula*, which will be replaced by long-lived, taller *Abies sibirica* (able to resprout), *Picea sibirica* (bird-dispersed) and *Pinus sibirica* (Schulze et al. 2005). These conifers achieve successive dominance according to their longevity: first *Abies*, with life spans of 70 to 150 years, and later *Picea*, which lives for 150 to 200 years (Fig. 2). Although late successional species are often able to establish shortly after fire, in some cases these successional replacements may be due to facilitation processes, as in the Siberian taiga where *Betula* resprouts after fire, and then favors the establishment of shade-tolerant, later successional species of *Larix* (Uemura et al. 1990). In contrast, in North American forests dense stands of aspen (*Populus tremuloides*), which are favored by severe fires, may inhibit the establishment of previously dominant species, such as black spruce (*Picea mariana*; Johnstone and Kasischke 2005). The distribution of stands dominated by these species across the landscape is not only controlled by fire history, but also by factors such as soil type, moisture, and topographic position (Larsen 1997, Archambault et al. 1997, De Grandpre et al. 2000).

Burn severity is important in determining tree survival. It is particularly important for fire-sensitive species that must regenerate from seed dispersed from unburned trees (Greene and Johnson 2000, Asselin et al. 2001). However, burn severity also effects serotinuous species, which disperse seeds at intermediate levels of flame duration (de Groot et al. 2005). Initial differences in regeneration patterns due to variation of crown burn severity will be reflected in long term differences on forest structure and composition (Arseneault 2001).

Burn severity at the ground level determines the survival of underground organs and soil seed banks (Schimmel and Granström 1996), as well as the availability of microsites were seedlings can establish (Johnstone and Kasischke 2005). In the unburned ground of boreal forests, the deep thick layer of forest floor organic matter (duff) inhibits seedling establishment, and duff consumption by fire reduces this effect (de Groot et al. 2004, Johnstone and Kasischke 2005, Johnstone and Chapin 2006). Thus seedling establishment is stimulated by fires (Kuuluvainen and Rouvinen 2000), and indeed increasing ground level burn severity increases seedling establishment. However, the complete destruction of the duff layer would also destroy most of the seed bank. Therefore, species that regenerate from soil seed banks have greater regeneration in sites with moderate duff burn depth, while species regenerating from post-fire seed dispersal have greater regeneration with complete consumption of the duff layer (Schimmel and Granström 1996). Thus, the spatial heterogeneity of ground fire burn severity has a strong influence on the density and composition of the forest stands (Johnstone and Chapin 2006).

Fire frequency may also influence post-fire regeneration patterns by promoting coexistence of forested and unforested patches across the landscape (Asselin et al. 2006). Model simulations have shown that the serotinous jack pine (*Pinus banksiana)* may expand its populations after short fire intervals (50 years), but will become locally extinct after long fire intervals (220 years). The opposite trend is expected for the semi-serotinous late successional black spruce (*Picea mariana*; Le Goff and Sirois 2004). The relationship between fire frequency and dominant species may also reflect the indirect effects of other factors. Larsen (1997) found that fire frequency was inversely correlated to site soil moisture, estimated as the distance to water-related firebreaks (lakes, rivers or streams), and jack pine and aspen forests (associated with high fire frequency) tended to be found in drier sites.

Different components of fire regime, as fire frequency and intensity may interact with unfavorable climate during and after fires to produce changes in the dominant species (Lavoie and Sirois 1998). Under the extreme conditions, fire may interact with boreal forest climate variability to produce important shifts after a single fire or delay the expected vegetation-climate equilibrium by favoring some species. In northwestern Quebec, the use of small low intensity fires in areas with many lakes has been proposed to maintain mixed wood forests over the coniferous forests (Bergeron et al. 2004). Control of vegetation type by fire is particularly prone to occur near the boreal forest ecotones with tundra or mixed wood forests. Historical data shows that fires reduced black spruce (*Picea mariana*) populations of forest-tundra uplands of Quebec in the late Holocene (Asselin and Payette 2005). More recently, extensive fires in the 1950s resulted in expansion of the forest tundra, with a characteristically patchy distribution of forests stands and scattered trees, into the upper boreal forest (Sirois and Payette 1991).

CONCLUSION

Vegetation recovery after fire depends on the ability of plant populations to endure fire and to regrow from surviving tissues or establish from viable seeds that remained in the soil or canopy seed banks or dispersed from unaffected populations. This recovery is determined to a certain extent by the fire regime, both temporal and spatial aspects, and the physical characteristics of the fires.

In general, ecosystems that sustain frequent fires tend to be more resilient. Both the selective pressures of fire acting on species evolution and fire events sorting species composition in communities contribute to this coupling between the fire regime and resilience.

The new environment produced by fire with fertilized soils, greater solar radiation, and reduced competition, as well as the differential ability to establish after fire often determines successional trajectories. Thus, species that were not dominant in the pre-fire conditions will eventually be replaced by shade-tolerant species. This successional pattern, often found in tropical, temperate, and boreal forests, results in a mosaic of forests stands whose structure and species composition reflect the fire history at the landscape level. However, this successional scheme is not universal, particularly when fire does not occur at long intervals relative to the life span of the species.

Several factors, that are becoming more common, may disrupt this coupling and decrease fire resilience. New climatic conditions, changes in species pool (such as those promoted by humans by exploitation, plantation or introduction of non-natives species), and/or policies modifying fire regimes and fuel accumulation may result in vegetation dynamics conducting to new communities.

Management may improve forest resilience by shaping the fire regime, modifying fuel characteristics, or favoring more resilient species. However, in most situations, it is difficult to determine what the 'natural' fire regime may have been, making it problematic to re-establish alleged natural regimes. Current experiences indicate that fire suppression strategies may lead to the loss of early successional resilient species, while too frequent fires may result in 'type conversion' to more open communities (i.e., savannah and grassland). Therefore, management decisions should consider the limits of resilience of individual species and the characteristics of fire regime-resilience coupling.

Acknowledgments

This work has partly been financed by the Spanish government SINREG project (REN2003-07198-C02-01).

References

Abrams, M.D. and J.A. Downs. 1990. Successional replacement of old-growth white oak by mixed mesophytic hardwoods in Southwestern Pennsylvania. *Canadian Journal Forest Research,* 20: 1864-1870.

Agee, J.K. 1993. *Fire Ecology of Pacific Northwest Forests.* Island Press, Washington, D.C., USA.

Archambault, L., J. Morissette, and M. Bernier-Cardou. 1997. Forest succession following a fire in a balsam fir yellow birch stand in the Bas-Saint-Laurent region of Quebec. *Forestry Chronicle,* 73: 702-710.

Arseneault, D. 2001. Impact of fire behavior on postfire forest development in a homogeneous boreal landscape. *Canadian Journal of Forest Research,* 31: 1367-1374.
Arseneault, D. and S. Payette. 1992. A postfire shift from lichen-spruce to lichen-tundra vegetation at tree line. *Ecology,* 73: 1067-1081.
Arseneault, D. and S. Payette. 1997. Reconstruction of millenial forest dynamics from tree remains in a subarctic tree line peatland. *Ecology,* 78: 1873-1883.
Asselin, H. and S. Payette. 2005. Late Holocene deforestation of a tree line site: estimation of pre-fire vegetation composition and black spruce cover using soil charcoal. *Ecography,* 28: 801-805.
Asselin, H., M.J. Fortin, and Y. Bergeron. 2001. Spatial distribution of late-successional coniferous species regeneration following disturbance in southwestern Quebec boreal forest. *Forest Ecology and Management,* 140: 29-37.
Asselin, H., A. Belleau, and Y. Bergeron. 2006. Factors responsible for the co-occurrence of forested and unforested rock outcrops in the boreal forest. *Landscape Ecology,* 21: 271-280.
Axelson, A.L., L. Östlund, and E. Hellberg. 2002. Changes in mixed deciduous forests of boreal Sweden 1866-1999 based on interpretation of historical records. *Landscape Ecology,* 17: 403-418.
Bailey, J.K. and T.G. Whitham. 2002. Interactions among fire, aspen, and elk affect insect diversity: Reversal of a community response. *Ecology,* 83: 1701-1712.
Bergeron, Y., S. Gauthier, M. Flannigan, and V. Kafka. 2004. Fire regimes at the transition between mixedwood and coniferous boreal forest in Northwestern Quebec. *Ecology,* 85(7): 1916-1932.
Biswell, H.H. 1989. *Prescribed Burning in California's Wildlands Vegetation Management.* University of California Press, Berkeley, California, USA.
Bond, W.J. and J.J. Midgley. 1995. Kill the neighbour – an individualistic argument for the evolution of flammability. *Oikos,* 73: 79-85.
Bond, W.J. and B.W. van Wilgen. 1996. *Fire and Plants. Population and Community.* Chapman & Hall, London, UK.
Bond, W.J. and J.J. Midgley. 2001. Ecology of resprouting in woody plants: the persistence niche. *Trends in Ecology and Evolution,* 16: 45-51.
Bradstock, R.A., T.D. Auld, M.E. Ellis, and J.S. Cohn. 1992. Soil temperatures during bushfires in semiarid, Mallee shrublands. *Australian Journal of Ecology,* 17: 433-440.
Broncano, M.J., J. Retana, and A. Rodrigo. 2005. Predicting the recovery of *Pinus halepensis* and *Quercus ilex* forests after a large wildfire in northeastern Spain. *Plant Ecology,* 180: 47-56.
Brose, P.H. and D.H. Van Lear. 1998. Responses of hardwood advance regeneration to seasonal prescribed fires in oak-dominated shelterwoods stands. *Canadian Journal Forests Research,* 28: 331-339.
Brose, P.H., T. Schuler, D. Van Lear, and J. Berst. 2001. Bringing fire back – The changing regimes of the Appalachian mixed-oak forests. *Journal of Forestry,* 99: 30-35.
Burns, B.R. 1993. Fire-induced dynamics of *Araucaria araucana–Nothofagus antarctica* forests in Southern Andes. *Journal of Biogeography,* 20: 669-685.
Calvo, L., R. Tarrega, and E. de Luis. 2002. Secondary succession after perturbations in a shrubland community. *Acta Oecologica,* 23: 393-404.
Canadell, J. and P.H. Zedler. 1995. Underground structures of woody plants in

Mediterranean ecosystems of Australia, California, and Chile. pp. 177-210. *In* M. T. Kalin Arroyo, P.H. Zedler, and M.D. Fox. [eds.]. *Ecology and Biogeography of Mediterranean Ecosystems in Chile, California, and Australia*. Springer Verlag. New York, New York, USA.

Canadell, J. and L. López-Soria. 1998. Lignotuber reserves support regrowth following clipping of two Mediterranean shrubs. *Functional Ecology,* 12: 31-38.

Chiang, J.M., M.A. Arthur, and B.A. Blankenship. 2005. The effect of prescribed fire in an oak forest understory on the Cumberland Plateau. *Journal Torrey Botanical Society,* 132: 432-441.

Clark, D.F., J.A. Antos, and G.E. Bradfield. 2003. Succession in sub-boreal forests of west-central British Columbia. *Journal of Vegetation Science,* 14: 721-732.

Cleary, D.F.R. and A. Priadjati. 2005. Vegetation responses to burning in a rain forest in Borneo. *Plant Ecology,* 177: 145-163.

Climent, J., R. Tapias, J.A. Pardos, and L. Gil. 2004. Fire adaptations in the Canary Islands pine (*Pinus canariensis*). *Plant Ecology,* 171: 185-196.

Cochrane, M.A., A. Alencar, M.D. Schulze, C.M. Souza, D.C. Nepstad, P. Lefebvre, and E.A. Davidson. 1999. Positive feedbacks in the fire dynamic of closed canopy tropical forests. *Science,* 284: 1832-1835.

Conedera, M., L. Peter, P. Marxer, F. Forster, D. Rickenmann, and L. Re. 2003. Consequences of forest fires on the hydrogeological response of mountain catchments: a case study of the riale Buffaga, Ticino, Switzerland. *Earth Surface Processes and Landforms,* 28: 117-129.

Cruz, A., B. Perez, and J.M. Moreno. 2003. Plant stored reserves do not drive resprouting of the lignotuberous shrub *Erica australis*. *New Phytologist,* 157: 252-261.

Cumming, S.G., F.K.A. Schmiegelow, and P.J. Burton. 2000. Gap dynamics in boreal aspen stands: Is the forest older than we think? *Ecological Applications,* 10: 744-759.

de Groot, W.J., P.M. Bothwell, S.W. Taylor, B.M. Wotton, B.J. Stocks, and M.E. Alexander. 2004. Jack pine regeneration and crown fires. *Canadian Journal of Forest Research,* 34: 1634-1641.

DeSimone, S.A. and P.H. Zedler. 1999. Shrub seedling recruitment in unburned Californian coastal sage scrub and adjacent grassland. *Ecology,* 80: 2018-2032.

Giovannini, G. and S. Lucchesi. 1997. Modifications induced in soil physico-chemical parameters by experimental fires at different intensities. *Soil Science,* 162: 479-486.

Gromtsev, A. 2002. Natural disturbance dynamics in the boreal forests of European Russia: A review. *Silva Fennica,* 36: 41-55.

De Grandpre, L., J. Morissette, and S. Gauthier. 2000. Long-term post-fire changes in the northeastern boreal forest of Quebec. *Journal of Vegetation Science,* 11: 791-800.

Delarze, R., D. Caldelari, and P. Hainard. 1992. Effects of fire on forest dynamics in southern Switzerland. *Journal of Vegetation Science,* 3: 55-60.

Díaz-Delgado, R., F. Lloret, X. Pons, and J. Terradas. 2002. Satellite evidence of decreasing resilience in Mediterranean plant communities after recurrent wildfires. *Ecology,* 83: 2293-2303.

Díaz-Delgado, R., F. Lloret, and X. Pons. 2003. Influence of fire severity on plant regeneration by means of remote sensing imagery. *International Journal Remote Sensing,* 24: 1751-1763.

Dixon, K.W., S. Roche, and J.S. Pate. 1995. The promotive effect of smoke derived from burnt vegetation on seed germination of Western Australian plants. *Oecologia,* 101: 185-192.

Dokrar, M., K. Utis, H. Tanaka, and T. Nakashizuka. 2004. Effects of drought and fire on seedling survival and growth under contrasting light conditions in a seasonal tropical forest. *Journal of Vegetation Science,* 15: 691-700.

Dominguez, L., L. Calvo, and E. Luis. 2002. The impact of wildfire season on regeneration of *Quercus pyrenaica* forest and *Pinus* sp. stands. *Journal of Mediterranean Ecology,* 3: 47-54.

Elmqvist, T., M. Wall, A.L. Berggren, L. Blix, A. Fritioff, and U. Rinman. 2002. Tropical forest reorganization after cyclone and fire disturbance in Samoa: Remnant trees as biological legacies. *Conservation Ecology,* 5: Art. No. 10.

Eriksson, I., D. Teketay, and A. Granström. 2003. Response of plant communities to fire in an *Acacia* woodland and a dry Afromontane forest, southern Ethiopia. *Forest Ecology and Management,* 177: 39-50.

Eugenio, M. and F. Lloret. 2004. Fire recurrence effects on the structure and composition of Mediterranean *Pinus halepensis* communities in Catalonia (northeast Iberian Peninsula). *Ecoscience,* 11: 446-454.

Eugenio, M., F. Lloret, and J.M. Alcañiz. 2006a. Fire recurrence effects on the soil organic horizons of Mediterranean *Pinus halepensis* communities. *Forest Ecology Management,* 221: 313-318.

Eugenio, M., I. Verkaia, F. Lloret, and J.M. Espelta. 2006b. Density and growth decline in *Pinus halepensis* populations after recurrent wildfires in Catalonia (NE Iberian Peninsula). *Forest Ecology and Management,* 231: 47-54.

Fensham, R.J., R.J. Fairfax, D.W. Butler, and D.M.J.S. Bowman. 2003. Effects of fire and drought in a tropical eucalypt savanna colonized by rain forest. *Journal of Biogeography, 30*: 1405-1414.

Franklin, J., C.L. Coulter, and S.J. Rey. 2004. Change over 70 years in a southern California chaparral community related to fire history. *Journal of Vegetation Science,* 15: 701-710.

Friederici, P. 2003. Introduction. pp. xv-xxii. *In* P. Friederici [ed.]. *Ecological Restoration of Southwestern Ponderosa Pine Forests*. Island Press, Washington, D.C., USA.

Giovannini G. and S. Lucchesi. 1997. Modifications induced in soil physico-chemical parameters by experimental fires at different intensities. *Soil Science,* 162: 479-486.

Goldammer, J.G. and B. Seibert. 1990. The impact of droughts and forest fires on tropical lowland rain forest of East Kalimantan. pp. 19-31. *In* J.G. Goldammer [ed.]. *Fire in the Tropical Biota: Ecosystems Processes and Global Changes*. Springer-Verlag, Berlin, Germany.

Gonzalez, M.E., T.T. Veblen, and J.S. Sibold. 2005. Fire history of *Araucaria-Nothofagus* forests in Villarrica National Park, Chile. *Journal of Biogeography,* 32: 1187-1202.

Greene, D.F. and E.A. Johnson. 1999. Modeling recruitment of *Populus tremuloides, Pinus banksiana,* and *Picea mariana* following fire in the mixedwood boreal forest. *Canadian Journal of Forest Research,* 29: 462-473.

Greene, D.F. and E.A. Johnson. 2000. Tree recruitment from burn edges. *Canadian Journal of Forest Research,* 30: 1264-1274.

Greene, D.F., J. Noel, Y. Bergeron, M. Rousseau, and S. Gauthier. 2004. Recruitment of *Picea mariana, Pinus banksiana,* and *Populus tremuloides* across a burn severity gradient following wildfire in the southern boreal forest of Quebec. *Canadian Journal of Forest Research,* 34: 1845-1857.

Grove, A.T. and O. Rackham. 2001. *The Nature of Mediterranean Europe: and Ecological History.* Yale University Press, New Haven, Connecticut, USA.

Grigulis, K., S. Lavorel, I.D. Davies, A. Dossantos, F. Lloret, and M. Vilà. 2005. Landscape-scale positive feedbacks between fire and expansion of the large tussock grass, *Ampelodesmos mauritanica* in Catalonian shrublands. *Global Change Biology,* 11: 1042-1053.

Habrouk, A., J. Retana, and J.M. Espelta. 1999. Role of heat tolerance and cone protection of seeds in the response of three pine species to wildfires. *Plant Ecology,* 145: 91-99.

Hart, S.C., A.T. Classen, and R.J. Wright. 2005. Long-term interval burning alters fine root and mycorrhizal dynamics in a ponderosa pine forest. *Journal of Applied Ecology,* 42: 752-761.

Hoffmann, W.A., V.M.P.C. Lucatelli, F.J. Silva, I.N.C. Azeuedo, M.D. Marinho, A.M.S. Albuquerque, A.D. Lopes, and S.P. Moreira. 2004. Impact of the invasive alien grass *Melinis minutiflora* at the savanna-forest ecotone in the Brazilian Cerrado. *Diversity and Distribution,* 10: 99-103.

Hogg, E.H. and R.W. Wein. 2005. Impacts of drought on forest growth and regeneration following fire in southwestern Yukon, Canada. *Canadian Journal of Forest Research,* 35: 2141-2150.

Holdsworth, A.R. and C. Uhl. 1997. Fire in Amazonian selectively logged rain forest and the potential for fire reduction. *Ecological Applications,* 7: 713-725.

James, S. 1984. Lignotubers and burls – their structure, function and ecological significance in Mediterranean ecosystems. *Botanical Review,* 50: 225-266.

Johnson, E.A. 1992. *Fire and Vegetation Dynamics. Studies from the North American Boreal Forest.* Cambridge University Press, Cambridge, UK.

Johnson, E.A. and S.L. Gutsell. 1994. Fire frequency models, methods, and interpretations. *Advances in Ecological Research,* 25: 239-287.

Johnson, E.A., K. Miyanishi, and J.M.H. Weir. 1998. Wildfires in the western Canadian boreal forest: Landscape patterns and ecosystem management. *Journal of Vegetation Science,* 9: 603-610.

Johnstone, J.F. and F.S. Chapin. 2003. Non-equilibrium succession dynamics indicate continued northern migration of lodgepole pine. *Global Change Biology,* 9: 1401-1409.

Johnstone, J.F., F.S. Chapin, J. Foote, S. Kemmett, K. Price, and L. Viereck. 2004. Decadal observations of tree regeneration following fire in boreal forests. *Canadian Journal of Forest Research,* 34: 267-273.

Johnstone, J.F. and F.S. Chapin. 2006. Effects of soil burn severity on post-fire tree recruitment in boreal forest. *Ecosystems,* 9: 14-31.

Johnstone, J.F. and E.S. Kasischke. 2005. Stand-level effects of soil burn severity on postfire regeneration in a recently burned black spruce forest. *Canadian Journal of Forest Research,* 35: 2151-2163.

Keeley, J.E. 1992. Recruitment of seedlings and vegetative sprouts in unburned chaparral. *Ecology,* 73: 1194-1208.

Keeley, J.E. 2002. Native American impacts on fire regimes of the California coastal ranges. *Journal of Biogeography,* 29: 303-320.

Keeley, J.E. 2006. Fire management impacts on invasive plants in the western United States. *Conservation Biology,* 20: 375-384.

Keeley, J.E. and P.H. Zedler. 1998. Evolution of life histories in *Pinus*. pp. 219-250. *In* D.M. Richardson [ed.]. *Ecology and Biogeography of Pinus*. Cambridge University Press, Cambridge, UK.

Keeley, J.E. and W.J. Bond. 1999. Mast flowering and semelparity in bamboos: The bamboo fire cycle hypothesis. *American Naturalist*, 154: 383-391.

Keeley, J.E. and C.J. Fotheringham. 2000. Smoke-induced seed germination in Californian chaparral. *Ecology*, 79: 2320-2336.

Keeley, J.E., C.J. Fotheringham, and M. Morais. 1999. Reexamining fire suppression impacts on brushland fire regimes. *Science*, 284: 1829-1832.

Keeley, J.E., M. Baer-Keeley, and C.J. Fotheringham. 2005. Alien plant dynamics following fire in Mediterranean-climate California shrublands. *Ecological Applications*, 15: 2109-2125.

Kemball, K.J., G.G. Wang, and Q.L. Dang. 2005. Response of understory plant community of boreal mixed wood stands to fire, logging, and spruce budworm outbreak. *Canadian Journal of Botany*, 83: 1550-1560.

Kennard, D.K., K. Gould, F.E. Putz, T.S. Fredericksen, and F. Morales. 2002. Effect of disturbance intensity on regeneration mechanisms in a tropical dry forest. *Forest Ecology and Management*, 162: 197-208.

Kerns B.K., W.G. Thies, and C.G. Niwa. 2006. Season and severity of prescribed burn in ponderosa pine forests: Implications for understory native and exotic plants. *Ecoscience*, 13: 44-55.

King, J., J.B. Moutsinga, and G. Doufoulon. 1997. Conversion of anthropogenic savanna to production forest through fire-protection of the forest-savanna edge in Gabon, Central Africa. *Forest Ecology and Management*, 94: 233-247.

Konstantinidis P., G. Tsiourlis, and P. Xofis. 2006. Effect of fire season, aspect and pre-fire plant size on the growth of *Arbutus unedo* L. (strawberry tree) resprouts. *Forest Ecology and Management*, 225: 359-367.

Korb, J.E., N.C. Johnson, and W.W. Covington. 2004. Slash pile burning effects on soil biotic and chemical properties and plant establishment: recommendations for amelioration. *Restoration Ecology*, 12: 52-62.

Kuuluvainen, T. and S. Rouvinen. 2000. Post-fire understory regeneration in boreal *Pinus sylvestris* forest sites with different fire histories. *Journal of Vegetation Science*, 11: 801-812.

Lamont, B.B., D.C. LeMaitre, R.M. Cowling, and N.J. Enright. 1991. Canopy seed storage in woody plants. *Botanical Review*, 57: 277-317.

Larsen, C.P.S. 1997. Spatial and temporal variations in boreal forest fire frequency in northern Alberta. *Journal of Biogeography*, 24: 663-673.

Lavoie, L. and L. Sirois. 1998. Vegetation changes caused by recent fires in the northern boreal forest of eastern Canada. *Journal of Vegetation Science*, 9: 483-492.

Lavorel, S., J. Canadell, S. Rambal, and J. Terradas. 1998. Mediterranean terrestrial ecosystems: research priorities on global change effects. *Global Ecology Biogeography Letters*, 7: 157-166.

Le Goff, H. and L. Sirois. 2004. Black spruce and jack pine dynamics simulated under varying fire cycles in the northern boreal forest of Quebec, Canada. *Canadian Journal of Forest Research*, 34: 2399-2409.

Le Maitre, D.C. and J.J. Midgley. 1992. Plant reproductive ecology. pp. 135-174. *In* R.M. Cowling [ed.]. *The Ecology of Fynboss: Nutrients, Fire and Diversity*. Oxford University Press, Oxford, UK.

Litton, C.M. and R. Santelices. 2002. Early post-fire succession in a *Nothofagus glauca* forest in the Coastal Cordillera of south-central Chile. *International Journal of Wildland Fire,* 11: 115-125.

Lloret, F. and P.H. Zedler. 1991. Recruitment pattern of *Rhus integrifolia* populations in periods between fire in chaparral. *Journal of Vegetation Science,* 2: 217-230.

Lloret, F. and L. López-Soria. 1993. Resprouting response of *Erica multiflora* after experimental fire treatments. *Journal of Vegetation Science,* 4: 367-374.

Lloret, F., M. Verdú, N. Flores-Hernández, and A. Valiente-Banuet. 1999. Fire and resprouting in Mediterranean ecosystems: insights from an external biogeographical region, the mexical shrubland. *American Journal of Botany,* 86: 1655-1661.

Lloret, F., H. Estevan, J. Vayreda, and J. Terradas. 2005. Fire regenerative syndromes of forest woody species across fire and climatic gradients. *Oecologia,* 146: 461-468.

López-Soria, L. and C. Castell. 1992. Comparative genet survival after fire in woody Mediterranean species. *Oecologia,* 91: 493-499.

McMaster, G.S. and P.H. Zedler. 1981. Delayed seed dispersal in *Pinus torreyana* (Torrey pine). *Oecologia,* 51: 62-66.

McCoy, S., T. Jaffre, F. Rigault, and J.E. Ash. 1999. Fire and succession in the ultramafic maquis of New Caledonia. *Journal of Biogeography,* 26: 579-594.

Marod, D., U. Kutintara, H. Tanaka, and T. Nakashizuka. 2004. Effects of drought and fire on seedling. *Journal of Vegetation Science,* 15: 691-700.

Miller, P.M. 1999. Effects of deforestation on seed banks in a tropical deciduous forest of western Mexico. *Journal of Tropical Ecology,* 15: 179-188.

Miller P.M. and J.B. Kauffman. 1998. Seedling and sprout response to slash-and-burn agriculture in a tropical deciduous forest. *Biotropica,* 30: 538-546.

Minnich, R.A. 2001. An integrated model of two fire regimes. *Conservation Biology,* 15: 1549-1553.

Murphy, J.D., D.W. Johnson, W.W. Miller, R.F. Walker, and R.R. Blank. 2006. Prescribed fire effects on forest floor and soil nutrients in a Sierra Nevada forest. *Soil Science,* 171: 181-199.

Nepstad, D.C., A. Veríssimo, A. Alencar, C. Nobre, E. Lima, P. Lefebvre, P. Sclesinger, C. Potter, P. Moutinho, E. Mendoza, M. Cochrane, and V. Brooks. 1999. Large-scale impoverishment of Amazonian forests by logging and fire. *Nature,* 398: 505-508.

O'Dowd, D.J. and A.M. Gill. 1984. Predator satiation and site alteration following fire: mass reproduction of alpine ash (*Eucalyptus delegatensis*) in southeastern Australia. *Ecology,* 65: 1052-1066.

Ojeda, F., T. Marañon, and J. Arroyo. 1996. Postfire regeneration of a Mediterranean heathland in southern Spain. *International Journal of Wildland Fire,* 6: 191-198.

Ordoñez, J.L. and J. Retana. 2004. Early reduction of post-fire recruitment of *Pinus nigra* by post-dispersal seed predation in different time-since-fire habitats. *Ecography,* 27: 449-458.

Payette, S., N. Bhiry, A. Delwaide, and M. Simard. 2000. Origin of the lichen woodland at its southern range limit in eastern Canada: the catastrophic impact of insect defoliators and fire on the spruce-moss forest. *Canadian Journal of Forest Research,* 30: 288-305.

Pausas, J.G. 1997. Resprouting of *Quercus suber* in NE Spain after fire. *Journal of Vegetation Science,* 8: 703-706.

Peters, V.S., S.E. MacDonald, and M.R.T. Dale. 2005. The interaction between masting and fire is key to white spruce regeneration. *Ecology,* 86: 1744-1750.

Pinard, M.A., F.E. Lutz, and J.C. Licona. 1999. Tree mortality and vine proliferation following a wildfire in a subhumid tropical forest in eastern Bolivia. *Forest Ecology and Management,* 116: 247-252.

Piñol, J., J. Terradas, and F. Lloret. 1998. Climate warming, wildfire hazard, and wildfire occurrence in coastal eastern Spain. *Climatic Change,* 38: 345-357.

Purdy, B.G., S.E. Macdonald, and M.R.T. Dale. 2002. The regeneration niche of white spruce following fire in the mixed wood boreal forest. *Silva Fennica,* 36: 289-306.

Quintana-Ascencio, P.F., E.S. Menges, and C.W. Weekley. 2003. A fire-explicit population viability analysis of *Hypericum cumulicola* in Florida rosemary scrub. *Conservation Biology,* 17: 433-449.

Rebertus, A.J., G.B. Williamson, and E.B. Moser. 1989. Longleaf pine pyrogenicity and Turkey oak mortality in Florida xeric sandhills. *Ecology,* 70: 60-70.

Reinhard, M., M. Rebetez, and R. Schlaepfer. 2005. Recent climate change: Rethinking drought in the context of forest fire research in Ticino, south of Switzerland. *Theoretical and Applied Climatology,* 82: 17-25.

Reinhart, K.O. and E.S. Menges. 2004. Effects of re-introducing fire to a central Florida sandhill community. *Applied Vegetation Science,* 7: 141-150.

Richardson, D.M. and S.I. Higgins. 1998. Pines as invaders in the southern hemisphere. pp. 450-473. *In* D.M. Richardson [ed.]. *Ecology and Biogeography of Pinus.* Cambridge University Press, Cambridge, UK.

Rodrigo, A., J. Retana, and F.X. Picó. 2004. Direct regeneration is not the only response of Mediterranean forests to large fires. *Ecology,* 85: 716-729.

Ruffner, C.M. and J.W. Groninger. 2006. Making the case for fire in southern Illinois forests. *Journal of Forestry,* 104: 78-83.

Russell-Smith, J., P.J. Stanton, P.J. Whitehead, and A. Edwards. 2004. Rain forest invasion of eucalypt-dominated woodland savanna, iron range, north-eastern Australia: I. Successional processes. *Journal of Biogeography,* 31: 1293-1303.

Saha, S. and H.F. Howe. 2003. Species composition and fire in a dry deciduous forest. *Ecology,* 84: 3118-3123.

Schimmel, J. and A. Granström. 1996. Fire severity and vegetation response in the boreal Swedish forest. *Ecology,* 77: 1436-1450.

Schoenberg, F., P.R. Peng, and J. Wood. 2002. On the distribution of wildfire sizes. *Environmetrics,* 14: 583-592.

Schulze, E.D., C. Wirt, D. Mollicone, and W. Ziegler. 2005. Succession after stand replacing disturbances by fire, wind throw, and insects in the dark Taiga of Central Siberia. *Oecologia,* 146: 77-88.

Signell, S.A., M.D. Abrams, J.C. Hovis, and S.W. Henry. 2005. Impact of multiple fires on stand structure and tree regeneration in central Appalachian oak forests. *Forest Ecology and Management,* 218: 146-158.

Simberloff, D., M.A. Relva, and M.A. Nuñez. 2003. Introduced species and management of Nothofagus/Austrocedrus forests. *Environmental Management,* 31(2): 263-275.

Sirois, L. 1995. Initial phase of post-fire forest regeneration in two lichen woodlands of northern Québec. *Ecoscience,* 2: 177-183.

Sirois, L. and S. Payette. 1989. Postfire black spruce establishment in subarctic and boreal Quebec. *Canadian Journal of Forest Research,* 19: 1571-1580.

Sirois, L. and S. Payette. 1991. Reduced postfire tree regeneration along a boreal forest forest-tundra transect in northern Quebec. *Ecology,* 72: 619-627.

Slik, J.W.F., R.W. Verburg, and P.J.A. Kessler. 2002. Effects of fire and selective logging on the tree species composition of lowland dipterocarp forest in East Kalimantan, Indonesia. *Biodiversity and Conservation,* 11: 85-98.

Stearns F. and Likens, G.E. 2002. One hundred years of recovery of a pine forest in northern Wisconsin. *American Midland Naturalist,* 148: 2-19.

Stott, P.J., J.G. Goldamme, and W.L. Werner. 1990. The role of fire in the tropical deciduous forests of Asia. pp. 32-43. *In* J.G. Goldammer [ed.]. *Fire in the Tropical Biota: Ecosystems Processes and Global Changes.* Springer-Verlag, Berlin, Germany.

Swaine, M.D. 1992. Characteristics of dry forest in West Africa and the influence of fire. *Journal of Vegetation Science,* 3: 365-374.

Swetnam, T.W. 1993. Fire history and climate change in giant sequoia groves. *Science,* 262: 885-889.

Tinner, E., P. Hubschmid, M. Ehrli, B. Ammann, and M. Conedera. 1999. Long-term forest fire ecology and dynamics in southern Switzerland. *Journal of Ecology,* 87: 273-289.

Turner, M.G. and W.H. Romme. 1994. Landscape dynamics in crown fire ecosystems. *Landscape Ecology,* 9: 59-77.

Uemura, S., S. Tsuda, and S. Hasegawa. 1990. Effects of fire on the vegetation of Siberian taiga predominated by *Larix dahurica. Canadian Journal of Forest Research,* 20: 547-553.

Versfeld, D.B. and B.W. Van Wilgen. 1986. Impact of woody aliens on ecosystem properties. pp. 239-246. *In* I.A.W. Macdonald, F.J. Kruger, and A.A. Ferrar [eds.]. *Ecology and Management of Biological Invasions in Southern Africa.* Oxford University Press, Cape Town, South Africa.

Vesk, P.A. 2006. Plant size and resprouting ability: trading tolerance and avoidance of damage? *Journal of Ecology,* 94: 1027-1034.

Vesk, P.A. and M. Westoby. 2004. Sprouting ability across diverse disturbances and vegetation types worldwide. *Journal of Ecology,* 92: 310-320.

Vieira, D.L.M. and A. Scariot. 2006. Principles of natural regeneration of tropical dry forests for restoration. *Restoration Ecology,* 14: 11-20.

Whelan R.J. 1995. *The Ecology of Fire. Cambridge Studies in Ecology.* Cambridge University Press, Cambridge, UK.

Westerling, A.L., H.G. Hidalgo, D.R. Cayan, and T.W. Swetnam. 2006. Warming and earlier spring increase western US forest wildfire activity. *Science,* 313: 940-943.

Williams, M.C. and G.M. Wardle. 2005. The invasion of two native Eucalypt forests by *Pinus radiata* in the Blue Mountains, New South Wales, Australia. *Biological Conservation,* 125: 55-64.

Zammit, C. 1988. Dynamics of resprouting in the lignotuberous shrub *Banksia oblongifolia. Australian Journal of Ecology,* 13: 311-320.

Zammit, C. and P.H. Zedler. 1994. Organization of the soil seed bank in mixed chaparral. *Vegetatio,* 111: 1-16.

Zedler, P.H., C.R. Gautier, and G.S. McMaster. 1983. Vegetation change in response to extreme events – the effect of a short interval between fires in California chaparral and coastal scrub. *Ecology,* 64: 809-818.

Section II

Rehabilitation and Restoration Strategies

10

Post-fire Stabilization and Rehabilitation

Peter R. Robichaud

Abstract

After a wildfire, potential increases in runoff, flooding, erosion, and sedimentation are a threat to valued resources both within and downstream of the burned area. Post-fire stabilization treatments, applied to hillslopes, roads, and channels, can reduce flooding and erosion for some, but not all, rain events. The decision of where and when to use post-fire stabilization treatments requires an evaluation of soil burn severity, climate, soils, topography, watershed hydrology, and the resources at risk for damage. Post-fire assessment, including the use of predictive modeling tools, is used to evaluate the need for erosion mitigation. In many cases, it is justifiable and cost effective to choose the 'no treatment option'. Natural mulch (conifer needle cast), which provides immediate ground cover, or natural recovery rates (erosion generally decreases by up to an order of magnitude annually as the site recovers) may be sufficient to avoid long-term damage to natural and man-made resources at risk. Since post-fire stabilization treatments are expensive, they should only be applied if unacceptable levels of flooding, erosion, and/or sedimentation are expected within and downstream of the burned area. Monitoring the effectiveness of post-fire rehabilitation treatments is essential to determine if treatments are functioning as desired, compare treatment effectiveness, identify conditions that enhance or limit treatment effectiveness, and evaluate new treatments as they are implemented.

INTRODUCTION

Wildfires not only consume the vegetation in their path, but also affect soil properties, watershed response, and downstream sedimentation. Within a watershed, post-fire hydrologic and sediment responses are often a function of burn severity and the occurrence of rain events. For a wide range of burn

Peter R. Robichaud, U.S. Department of Agriculture, Forest Service, Rocky Mountain Research Station, 1221 S. Main St., Moscow, Idaho 83843, USA. Tel: 1 208 883 2349, Fax: 1 208 883 2318, e-mail: probichaud@fs.fed.us

severities, the impacts on hydrology and soil loss can be minimal in the absence of intense precipitation. However, when a large precipitation event follows a moderate- to high-burn severity fire, impacts can include increased runoff and peakflows, flooding, erosion, and sediment delivery to streams (Anderson et al. 1976, DeBano et al. 1998, Rinne 1996). After the burning of forest vegetation, increased erosion and flooding are the most visible and dramatic impacts of a wildfire (Robichaud et al. 2000). This chapter examines post-fire stabilization and rehabilitation techniques for forest, shrubland, and grassland ecosystems of western North America. However, much of the content is applicable to any area where the effects of wildfires may require mitigation to protect life, property, or habitat.

The protection of life and property is a significant challenge, not only for wildland fire suppression efforts, but also for the mitigation of increased erosion, downstream sedimentation, flooding, and debris flows that often occur after wildfires (DeBano et al. 1998). In the late 1980s, the U.S. Department of Agriculture, Forest Service (USDA Forest Service) enacted policies that formalized the assessment of site conditions following wildfires and, where necessary, implementation of emergency stabilization measures that reduce the threat to life and property, loss of soil and onsite productivity, loss of control of water, and deterioration of water quality (USDA Forest Service Manual 2523, USDA Forest Service Handbook 2509.13). These policies became the basis of an inter-agency Burned Area Emergency Response (BAER) policy that also included monitoring treatment effectiveness and rehabilitation activities such as repairing facilities for safety reasons, stabilizing biotic communities, and preventing unacceptable degradation of known cultural sites and sensitive natural resources. In the past decade, public land management agencies in the United States have spent hundreds of millions of dollars each year on post-fire watershed stabilization measures intended to minimize flood runoff, peakflows, onsite erosion, offsite sedimentation, mud and debris flows, and other damage to natural habitats as well as roads, bridges, reservoirs, and irrigation systems (US General Accounting Office 2003).

Post-fire activities are divided into three categories, emergency stabilization, rehabilitation, and restoration, which are differentiated not only by the type of activities, but also by the timing of these activities (US General Accounting Office 2006). Emergency stabilization treatments (e.g., mulching to prevent soil erosion, installation of water bars to facilitate water passage over roads) are conducted within one year of a fire to stabilize the burned area, protect public health and safety, and reduce the risk of additional damage to valued resources, such as water supply systems, aquatic habitat, and roads. The burned area assessment and emergency stabilization plans are implemented as soon as possible, often before the wildfire is fully controlled, in an effort to have treatments in place before the first rain event (Robichaud et al. 2000). Rehabilitation activities, conducted within three years of a fire, include repairing facilities (e.g., roads, bridges, picnic areas, fences) and mitigating damage to lands unlikely to recover to a desired condition on their own (e.g., tree or grass planting, noxious weed control, and fuel reduction). Restoration activities are a continuation of rehabilitation activities beyond the initial three years, and are differentiated from emergency stabilization and rehabilitation because the funding and oversight of these long-term activities is separate and decentralized within the various land management agencies. Restoration of habitat quality, resilience, and productivity and the repair of facilities for access and recreation can extend for several years as funding permits (US General Accounting Office 2006). As these terms have evolved over the past decade, much of the literature, especially from the USDA Forest Service and post-fire assessment teams, refers to the immediate post-fire stabilization treatments as 'emergency rehabilitation' or 'BAER treatments.'

During the past 15 years, as wildfire size and severity have dramatically increased in the western United States, the public demand and costs for post-fire emergency stabilization and rehabilitation have also increased. The number of people living in the *wildland-urban interface* (forested lands surrounding urban areas) continues to grow, which increases the risk to human life and property in relation to fire suppression and post-fire stabilization and rehabilitation activities (Stewart et al. 2003). Land managers are being compelled to determine the most cost-effective approaches to mitigate the effects of fire on human lives and property, water supplies, water quality, soil productivity, and protected species and habitat.

Although the United States has the most comprehensive program for post-wildfire assessment and emergency stabilization, other countries also have post-fire stabilization and rehabilitation programs. In British Columbia, Canada, recent efforts have resulted in a post-wildfire rehabilitation pre-planning guide that emphasizes early acquisition of information that is essential for assessment of a burned area (Pike and Ussery 2006). The Australian government has a general Emergency Management Plan that emphasizes the use of risk management techniques for all emergency responses, including wildfire, and some individual state plans include wildfire suppression and rehabilitation planning guidelines (e.g., State of Victoria 2006). Raftoyannis and Spanos (2005) reported that the Greek Forest Service and other land management authorities spend millions of Euros on post-fire rehabilitation measures.

Watershed Responses that Impact Post-fire Treatment Choices and Effectiveness

Watersheds with good hydrologic conditions and adequate rainfall sustain stream baseflow conditions for much or all of the year, produce rainfall-to-runoff ratios of two percent or less, and have minimal erosion (Bailey and Copeland 1961). Fire can destroy accumulated forest floor material and vegetation, altering infiltration by exposing soils to raindrop impact or creating water repellent soil conditions (DeBano et al. 1998). When severe fire produces hydrologic conditions that are poor, surface runoff can increase 70 to over 1000 percent (Neary et al. 2005) and erosion can increase by three orders of magnitude (DeBano et al. 1998, Robichaud et al. 2000; Fig. 1).

Within a watershed, sediment and runoff responses to wildfire are a function of burn severity, topography, soil characteristics, vegetative recovery, and the occurrence of hydrologic events. When a major rainfall event follows a large, high burn severity fire, significant hydrological and erosional responses are likely. High intensity rainfall events tend to exceed the average infiltration rates of many soils such that streamflow is dominated by overland flow (Moody and Martin 2001). After fires, high-intensity rainfall has been associated with high stream peakflows and significant erosion events (DeBano et al. 1998, Neary et al. 1999, Moody and Martin 2001).

Fig. 1 Runoff and erosion from a burned hillslope after the 2006 Shaketable Fire in Oregon, USA.

In the first year after a fire, sediment yields of 0.01 to over 110 Mg $ha^{-1} \times yr^{-1}$ have been reported (Robichaud et al. 2000). In studies where hillslope erosion has been directly measured rather than estimated, a range of sediment yields with a maximum value of 63 Mg ha^{-1} yr^{-1} in the first year after a fire have been reported (Spigel and Robichaud 2007). Barring high intensity rain events, these measured sediment yields usually decrease by an order of magnitude with each successive year (Robichaud and Brown 2000, Robichaud et al. 2008). Consequently, if erosion mitigation is required, treatments need to be applied immediately after fire suppression to provide needed protection in the critical first post-fire year.

Recovery rates vary by climate and geographic area as well as size and severity of the burn. DeBano et al. (1996) found that following a southwestern United States wildfire, sediment yields from a low severity fire recovered to normal levels after 3 years, but moderate and high severity burned watersheds required 7 and 14 years, respectively. In contrast, measured erosion rates recovered to no measurable erosion by the fourth post-fire year following high burn severity wildfires in eastern Oregon and western Montana (Robichaud and Brown 2000, Robichaud et al. 2008). In on-going studies, there have been significant sediment yields following high intensity rain events despite several years of recovery. Post-fire recovery to pre-fire hydrologic conditions may require more time than the regrowth of vegetative cover (Robichaud 2005).

DECIDING WHEN AND WHERE TO APPLY STABILIZATION TREATMENTS

Since 1987, post-fire assessments and treatment recommendations have been delegated to temporary teams of specialists. Most post-fire assessment teams include a soil scientist, hydrologist, engineer, and ecologist with additional members chosen as needed for the area (e.g., aquatic ecologist, fisheries biologist, archeologist, forester). These teams use a risk analysis approach to evaluate post-fire conditions and determine if and where emergency stabilization and rehabilitation are needed. The risk of increased runoff and erosion following a fire is dependent on the climate, soil, topography, size and severity of the fire, and the biotic communities within the burned area. This erosion risk is evaluated in relation to 1) the threat to human life and safety; 2) the potential damage that may occur to valued resources such as water quality, structures, roads, and cultural resources; 3) treatment costs; 4) availability of treatment materials; 5) short and long term effects of treatment applications; and 6) the amount of potential runoff and erosion mitigation a specific treatment may provide. The choice to rely on natural recovery processes and not implement any stabilization or rehabilitation treatments is often the preferable alternative.

Unfortunately, quantified information has not been readily available for many of the factors involved in post-fire assessment decisions, and post-fire assessment teams often have had to rely on their collective experience and perceptions to evaluate and compare erosion potential, recovery rates, and treatment choices (Robichaud et al. 2000). In recent years, research has focused on the development of post-disturbance erosion risk prediction tools (Robichaud et al. 2006a; Robichaud et al. 2007a) and scientific evaluation of treatment effectiveness (Bautista et al. 1996, Beyers 2004, Robichaud 2005, Wagenbrenner et al. 2006, Robichaud et al. 2006b, Robichaud et al. 2008). These new tools and data will allow post-fire assessment teams to be less subjective when determining the likelihood of post-fire erosion occurring and the probabilities of treatment success.

Burn Severity Map

Burn severity is a qualitative indicator of the effects of fire on an ecosystem, and reflects the fuel and soil conditions before a fire, energy released during and after combustion, and effects of fire on site resources (Ryan and Noste 1985, Hartford and Frandsen 1992). Several classification systems have been used to assess burn severity based on observable post-fire conditions, such as amount of canopy remaining, degree of char on trees, and amount of bare soil (Jain 2004). Although most systems classify the burn severity of an area as low, moderate, or high, individual classification schemes may be resource-specific and focus on the post-fire characteristics of a specific component. For example, Hungerford (1996) provided a burn severity classification system based on the effects of fire on the soil resource; the observer determines the degree of soil burn severity by the appearance of the litter and soil.

Since areas of high and moderate burn severity are at higher risk for increased runoff and erosion than areas of low severity, one of the first tasks of any post-fire assessment team is to obtain a burn severity map of the burned area. Burned areas are classified by determining the percentage of the total area within the fire perimeter that is unburned and burned at low, moderate, and high burn severities. Generally there is a mosaic of unburned areas interspersed with areas of variable burn severity. The USDA Forest Service Remote Sensing Applications Center (RSAC) and the U.S. Geological Survey (USGS) Earth Resources Observation Systems (EROS) provide Burned Area Reflectance Classification (BARC) products derived from available satellite and airborne imagery, which are used to generate burn severity maps. Landsat imagery is the default choice for burned area mapping, but Satellite Pour l'Observation de la Terre (SPOT), Advanced Spaceborne Thermal Emission and Reflection Radiometer (ASTER), Moderate Resolution Imaging Spectroradiometer (MODIS) and other imagery are important supplements, especially when clouds or smoke obscure satellite images (Lentile et al. 2006). Most satellite sensors 'read' the conditions of the remaining canopy and differentiate green vegetation from char, rock, and brown and black canopy. Since post-fire assessment teams are determining the need for emergency stabilization, a burn severity map which characterizes burn severity in terms of soil would be more relevant than the canopy-related products listed above (Lewis et al. 2006); however, remotely-sensed, post-fire soil data acquisition and analysis are still being tested (Robichaud et al. 2007b). The initial BARC map is corrected with aerial and ground inspection data, although this correction process is often limited by time and available resources. The corrected burn severity map is then used to help identify the areas at greatest risk for increases in runoff, flooding, erosion, sedimentation, and debris flows within the fire perimeter as well as downstream from the burned area.

Predicting Post-fire Runoff and Erosion Risks

Using the burn severity map to identify the most vulnerable areas, the post-fire assessment team applies climate, past hydrologic response, and runoff and erosion prediction models to predict the responses of the burned watershed to future rain events. These models generally use climate, soil, vegetation, topographic, and burn severity inputs to predict potential rainfall excess as well as overland and channel flow routing. The predicted hydrological responses may be used as inputs in separate erosion prediction models; however, many erosion prediction models include algorithms for climate and hydrological processes, which can provide predictions for both the hydrological and erosion responses. By running the models multiple times, the post-fire assessment team predicts potential runoff, peakflows, streamflows, and erosion for representative areas of concern within the burned area. The model predictions, tempered by the teams' collective experience and judgment, are then used to determine areas that may benefit from post-fire stabilization treatment.

Evaluating the potential effects of wildfire on hydrologic responses is an important step in the assessment process. Several models have been used by post-fire assessment teams to predict the potential runoff, peakflows, erosion, and sediment yields within the burned area and the potentially impacted areas outside of the burned area. Hydrologic responses to a selected design storm are usually modeled using estimated reductions in infiltration rates due to loss of protective organic layers and fire-induced soil water repellency (Robichaud et al. 2000). The U.S. Department of Agriculture, National Resource Conservation Service (NRCS) curve number model is often adapted to determine runoff depths from the selected design storm, which are converted to runoff using the triangular unit hydrograph model on each watershed. This approach does not involve any channel routing (Hawkins and Greenberg 1990). Some hydrologists have used the Hydrologic Engineering Center-Hydrologic Modeling System (HEC-HMS), to simulate hypothetical storm hydrographs and compute flood flows through downstream sub-basins (USACE 1982, 1985).

Some post-fire predictions for erosion and sediment delivery rates are based on a measured post-fire erosion rate from a nearby burned area. Other predictions are made using models. For nearly five decades, erosion predictions have been made using the empirically-based Universal Soil Loss Equation (USLE) (Wischmeier and Smith 1978), which was designed as a management tool for agricultural lands. By incorporating new research findings and improved computer capabilities, the formula has been consistently improved. The current versions, which include the Revised USLE (RUSLE) and the Modified USLE (MUSLE), are incorporated into a wide range of models (Renard et al. 1991, Renard et al. 1994). Limitations of USLE models led to the development of physically-based erosion models such as the Water Erosion Prediction Project (WEPP; Nearing et al. 1989) and the EUROpean Soil Erosion Model (EUROSEM; Morgan et al. 1998). Aksoy and Kavvas (2005) reviewed these and several other hillslope and watershed scale erosion and sediment transport models that are currently in use throughout the world.

In the United States, the WEPP technology has been adapted for use in forest and rangelands (http://forest.moscowfsl.wsu.edu/fswepp/). A suite of web-based interfaces (FS-WEPP) have been developed that allow users to access the basic WEPP model and input customized climate, topography, and soil data, and to quickly evaluate erosion and sediment delivery potential from forest roads and hillslopes after disturbances, including prescribed burns and wildland fire (Elliot 2004). Recently added to the FS-WEPP suite, the Erosion Risk Management Tool (ERMiT) provides the probability of a given rain event sediment yield occurring from a burned forest, range, or chaparral hillslope in each of the five years following a wildfire. In addition, ERMiT predicts the potential reduction in sediment delivery as a result of straw mulch application, erosion barrier (contour-felled logs or straw wattles) installation, or seeding (Robichaud et al. 2006a, Robichaud et al. 2007a). The probabilistic ERMiT output is intentionally designed to facilitate the risk analysis approach used by post-fire assessment teams.

Reducing the Risk of Damage through Post-fire Treatments

Post-fire assessment teams use the runoff and erosion predictions to determine if life, property, cultural features, water quality, sensitive species, or soil productivity within or downstream of the burned area are at risk. If so, they identify the available erosion mitigation treatments, and estimate the probability that application of selected treatments will reduce in the risk of damaging erosion occurring. On United States public lands, the post-fire assessment process requires the *no action* (natural, unaided recovery) alternative to be compared to all other treatment plans under consideration.

POST-FIRE TREATMENT OPTIONS

Early post-fire treatments were aimed at stabilizing the burned area by using various existing management techniques on forest, range land, and grass land watersheds to control both storm runoff and erosion. Many of these techniques have been refined, improved, and augmented from other disciplines, such as agriculture and construction, to form a dynamic set of post-fire treatments in use today (Robichaud et al. 2000). Stabilization and rehabilitation treatments are generally classified by where they are applied—on hillslopes, in channels, and on roads and trails within burned areas. Long-term restoration activities are usually identified by the program areas which fund them, such as habitat restoration, forest or grazing management, recreation, road maintenance, and building construction and maintenance.

Hillslope Treatments

Post-wildfire hillslope treatments are intended to reduce surface runoff and keep post-wildfire soil in place and thereby prevent sediment deposition in unwanted areas. These treatments are regarded as a first line of defense against post-fire erosion and sediment movement (Robichaud et al. 2005). Given that hillslopes are the primary source of post-fire sediments and the cost for treating large hillslope areas is high, there has been more recent research on the development and evaluation of hillslope treatments than on channel or road treatments. Nearly all hillslope treatments use one or a combination of three general techniques—broadcast seeding, mulching, and erosion barriers. Although the three types of hillslope treatments are described briefly below, detailed discussions are found in Chapter 11 (*Non-native and Native Seeding*), Chapter 12 (*Using Erosion Barriers for Post-fire Stabilization*), and Chapter 13 (*Post-fire Mulching*).

Seeding

Broadcast seeding of grasses, usually from an aircraft, is the oldest and most common post-fire treatment. It is relatively easy to apply and rapid establishment of vegetation has been regarded as the most cost-effective method to improve water infiltration and hold soil on burned hillslopes. Non-native annual or perennial grasses typically have been used to provide temporary ground cover until native plants are reestablished; however, recent research has shown that seeded grasses compete with native vegetation and often do not effectively reduce erosion (Robichaud et al. 2000, Beyers 2004, Robichaud et al. 2006b). In recent years, native species and sterile cereal grains have increasingly been used for post-fire seeding (Beyers 2004).

Seeding success is highly dependent on rainfall intensity, amounts, and timing, and as a result, seeding does not assure higher vegetative cover during the critical first year after burning (Robichaud et al. 2006b). In nine seeding studies where quantitative ground cover data were provided, the 60 to 70 percent ground cover needed for erosion reduction, was attained in less than a fourth of the treated areas during the first growing season (Robichaud et al. 2000). Beyers (2004), in a recent review of post-fire seeding effectiveness, reported that when post-fire seed growth provides enough cover to substantially reduce erosion, it generally suppresses revegetation by naturally occurring species. Such findings have encouraged managers to attempt to use seeded grasses to thwart the post-fire influx of noxious and invasive weeds. Research to determine the effectiveness of seeding to achieve this goal is in progress.

Mulching

Mulch is any organic material spread over the soil surface that increases the ground cover and reduces raindrop impact and overland flow. Both wet mulch (hydromulch) and dry mulch (wheat straw, rice straw, wood strands, wood fiber, etc.) can be applied from the air or from the ground; however, mulches have only recently been used as a post-fire rehabilitation treatment. In the past, seed germination from grain or straw mulch was regarded as a bonus as this increased the cover on a site; however, the introduction of noxious weeds and other non-native plants is now considered a drawback to the use of straw mulch and post-fire rehabilitation projects usually require certified 'weed-free' straw for post-fire rehabilitation efforts (Beyers 2004). Straw mulch has been shown to reduce erosion rates after wildfires by 50 to 94 percent (Miles et al. 1989, Bautista et al. 1996, Faust 1998, Wagenbrenner et al. 2006). In some burned areas, natural mulch may provide adequate ground cover making the *no treatment* option a practical choice for those areas. In conifer forests, low and moderate severity burned sites often have trees that are lightly charred and only partially consumed by fire, leaving dead needles in the canopy. These needles fall to the ground and provide enough natural mulch ground cover to reduce erosion (Pannkuk and Robichaud 2003).

Erosion barriers

Straw wattles, contour-felled log erosion barriers, and other natural and engineered structures have been used to provide mechanical barriers to overland flow, promote infiltration, and trap sediment on burned hillsides. Erosion barriers can provide benefits immediately after installation; however, the installation process also can disturb and loosen soil making it easier to erode. Contour-felled log erosion barriers can be effective for low to moderate intensity rainfall events; however, during high-intensity rainfall events their effectiveness is greatly reduced. The effectiveness of contour-felled log erosion barriers is highly dependent on the quality of the installation and also decreases over time as the sediment storage areas above the logs become filled and the barrier can no longer trap mobilized sediment (Wagenbrenner et al. 2006, Robichaud et al. 2008).

Road and Trail Treatments

Post-fire road treatments consist of a variety of practices aimed at increasing the water processing capabilities of roads and road structures (e.g., culverts, low water crossings, and bridges) to prevent large failures that would damage the road and add to downstream sedimentation (Robichaud et al. 2000, Napper 2006). The functionality of the road drainage system is generally not affected by fire, but runoff from burned watersheds can overwhelm the system. Road stabilization treatments include:

- *Armoring*—Road armoring usually involves application of rock and/or gravel to the running surface, drainage ditches, cut and fill slopes (Fig. 2) that are likely to have water flowing over them, and/or entrances and exits of culverts and other stream crossings.
- *Flow directors*—Roads may be graveled and graded to improve the insloping or outsloping of the road surface to avoid water flowing down the road, abrading the surface, and creating ruts and gullies. Water bars and rolling dips are often constructed in the roadway to direct flowing water off the road. Concrete structures, such as jersey barriers, can be used to direct water away from a weak slope.
- *Water passage structures*—Construction of elaborate water passage structures are generally not undertaken during post-fire stabilization. However, culverts may be removed or their size increased and combinations of flow directors and armoring may be installed in critical water passage areas (Fig. 3). A trash rack (a structure built across a

Fig. 2 Armoring of a fill slope on a forest road installed after the 2003 Piru Fire in southern California, USA.

Fig. 3 A new, larger culvert and inlet armoring installed after the 2003 Piru Fire in southern California, USA.

channel to catch debris while allowing water to pass through) can be installed upstream of a culvert to prevent blockages. 'Storm patrols' often travel roads to inspect and clear blocked culverts during and immediately after rain events.

Most post-fire road stabilization treatments are standard road building practices for handling water flow and drainage; however, the overall capacity is increased to handle the potentially higher flows that are common after wildfires. Temporary stabilization structures may later be replaced with permanent road improvements during the post-fire rehabilitation and restoration phases.

Trails are occasionally included in post-fire stabilization plans and are often part of rehabilitation and restoration activities. Trail treatments generally mimic road treatments but on a smaller scale. Trail maintenance is labor intensive as the work often must be done by hand with materials that can be carried or brought in on all-terrain vehicles (ATVs).

Channel Treatments

Channel treatments modify sediment and water movement in ephemeral or low-order channels to reduce sediment inputs into perennial streams and to prevent flooding and debris torrents that may affect downstream resources

(Robichaud et al. 2000). Most channel treatments involve some mechanism to slow water flow and thereby reduce down-cutting and allow sediment to settle. Stream bank armoring (installing rip rap, coir cloth, or engineered materials along the cut banks) and channel clearing (removal of large objects that could become mobilized in a flood) also stabilize channels and downstream waterways (Robichaud et al. 2000).

Check dams and channel grade stabilizing structures are constructed of different materials (straw bales, logs, and rocks) and anchored in channels (Fig. 4). Straw bale check dams are inexpensive, easy to install, and effective at trapping small amounts of sediment, but deteriorate rapidly due to climatic conditions, streamflows, or cattle and wildlife disturbance. Collins and Johnston (1995) evaluated the effectiveness of straw bales on sediment retention after the Oakland Hills fire. About 5000 bales were used to construct 440 straw bale check dams and 100 hillslope barriers. Three months after installation, only 45 percent of the check dams remained functional. Log check dams are similar in function to straw bale check dams, but constructed of more durable material, usually nearby small diameter fire-killed trees. Log check dams require more effort and skill to install, but generally last longer. Properly designed and installed rock check dams (also called rock cage dams or gabions) are capable of halting gully development and reducing sediment yields; however, rock cage dams must be properly located, keyed into the stable part of the stream bank, and anchored to stay in place during large runoff events. Check dams, especially straw bale check dams, tend to fail in large storms, adding to the debris being carried down the channel. The effectiveness of any check dam is dependent on maintenance of the structure

Fig. 4 Straw bale check dams were installed in a small swale below a steep slope after the 2002 Hayman Fire in Colorado, USA.

and periodic sediment removal. The sediment removed from the stream by the check dams can increase downstream scour (Chiun-Ming 1985).

Long-term Rehabilitation and Restoration Treatments

Most of the post-fire emergency stabilization and short-term rehabilitation objectives are concerned with physical ecosystem components – soil, water, and post-disturbance hydrologic processes. In contrast, the objectives of long-term rehabilitation and restoration activities, such as tree planting and habitat improvement, are more often focused on biotic components of the ecosystem – natural recovery of native communities and habitat, maintenance of biodiversity, and the critical need for disturbance habitats for some biota (Beschta et al. 2004, Noss et al. 2006). As emergency stabilization and short-term rehabilitation activities can impact future restoration efforts, post-wildfire management plans should include the types of restoration that likely will be used (based on local environment and management goals) so that emergency stabilization activities will possibly enhance, or at least not impede, those potential long-term efforts (Franklin and Agee 2003).

The question of what are appropriate restoration treatments is a contentious topic. Post-fire logging and tree planting are two of the more controversial practices. Franklin and Agee (2003), like many researchers, recommend that the decision to salvage dead or damaged trees from burned areas take into account biological legacies (surviving biological elements passed from the pre-disturbance ecosystem to the regenerating ecosystem) and the ecology of place, as well as economics and fuel management. In addition, the effects of post-fire timber harvest on water quality, soil erosion, sedimentation, nutrient cycling, cavity-tree formation, invasive species, and quality of habitat must also be taken into account (McIver and Starr 2000, 2001, Beschta et al. 2004, Karr et al. 2004, Reeves et al. 2006, Lindenmayer and Noss 2006). In areas where timber production is a primary management objective, tree planting is often part of the post-wildfire restoration plan (Sessions et al. 2004). However, large disturbed areas that are allowed to recover naturally can be important areas for regional biodiversity. Tree planting should not create new problems or perpetuate old ones by establishing dense plantations on burned sites where timber production is only one of many considerations within the management plan (Franklin and Agee 2003).

MONITORING TREATMENT EFFECTIVENESS

Monitoring the effectiveness of post-fire rehabilitation treatments is essential to determine if the treatments are functioning as desired, compare various treatments, and to determine the conditions under which different treatments are effective and establish the limitations of each treatment. Although direct measurement of hillslope runoff and/or erosion can be expensive, complex, and labor-intensive, quantitative data from post-fire treatment monitoring efforts is needed not only to guide future responses to post-fire stabilization and rehabilitation, but also to develop and refine predictive models. Recent scientific efforts have focused on developing and implementing methods that assess the effectiveness and the limitations of specific post-fire rehabilitation treatments through direct measurement of watershed processes, such as runoff, peakflows, erosion, etc. (Robichaud 2005, Robichaud et al. 2008).

Research and monitoring of post-fire rehabilitation treatment effectiveness demands: 1) a quick response after the fire to measure the potentially largest erosion events that occur in the first post-fire year; and 2) a long-term commitment, up to several years, to evaluate treatment effects through the initial recovery period. To evaluate the effectiveness of post-fire treatments, burned but untreated control areas must be available for comparison. These control areas can be used to assess both short- and long-term treatment effectiveness as well as fire effects and recovery for the ecosystem. A small number of untreated areas can serve as the controls for a larger number of treated areas, as long as the controls are similar to the various treated areas. In addition, successful monitoring projects need timely data collection, analyses, and reporting (MacDonald 1994).

Rapid response approaches have been developed and implemented to compare treatment effectiveness by monitoring runoff response and sediment yield from treated and untreated areas at several scales – hillslope plots (20 to 30 m^2; Fig. 5), hillslope swales (0.1 to 0.5 ha; Fig. 6), and small catchments (1 to 10 ha; Fig. 7). Erosion rates are generally measured in mass of eroded sediment per unit area (1 hectare) per unit time (1 year) or per rain event or rain amount (mm). Occasionally the sediment flux rate, the mass of sediment transported across a unit hillslope contour (1 meter) per unit time (1 day) or per rain event, is used to quantify the erosion rate. Comparing erosion rates between studies is confounded by the variability in rainfall, soil type, topography, and other site differences. In addition, the

Fig. 5 A hillslope plot research site with a geotextile sediment fence (Robichaud and Brown 2002) one year after installation. This plot is one of 24 plots established after the 2000 Valley Complex Fires in Montana, USA to measure contour-felled log erosion barrier treatment effectiveness.

Fig. 6 A hillslope swale research site with a double geotextile fence one year after installation. This is one of 24 swales research sites established after the 2000 Bobcat Fire in Colorado, USA to measure contour-felled log erosion barrier, straw mulch, and seeding treatment effectiveness.

Fig. 7 The sheet metal headwall forms a sediment basin at the outlet of a small catchment (Robichaud and Brown 2003), part of a paired watershed study on treatment effectiveness initiated immediately after the 2002 Hayman Fire in Colorado, USA. The photo was taken in 2003 after the removal of sediment trapped during a rain event.

various controlling hydrological processes at different spatial scales greatly complicate measurement and comparison of erosion rates. Erosion measured at the plot scale usually has interrill and rill components, while erosion measured at the catchment scale will include those two processes as well as some channelized flow. As a result, the erosion rates measured at catchment-scale may be greater or less than the erosion rate measured at the plot-scale as a result of incision and/or deposition in the channel.

Monitoring and research results from the past five to six years suggest that several factors influence the effectiveness of any given treatment. These include: 1) burn severity of treated areas; 2) rainfall event characteristics—intensity, duration, and amount; 3) natural recovery rate; 4) amount of ground cover; and 5) topography, climate, and soils. Specifically, monitoring efforts indicate that:

- Although increased erosion is likely on steep slopes with moderate and high burn severity, the greatest erosion occurs in areas of solid (not patchy) high burn severity where a fire-induced water repellent soil layer exists between 1 and 3 cm below the surface (Robichaud 2005, Robichaud et al. 2006a).
- Post-fire treatments may reduce erosion for some rainfall events. However, in many monitoring/research sites, short duration, high intensity rain events caused large sediment yields in both the treated and control sites regardless of the treatment used (Wagenbrenner et al. 2006, Robichaud et al. 2008).
- Natural recovery of native vegetation reduces erosion over time. The greatest erosion usually is measured during the first post-fire year, and the second post-fire year and subsequent years can be an order of magnitude lower (Robichaud and Brown 2000, Pierson et al. 2001, Robichaud et al. 2008). Recovery rates vary by climate and vegetation type.
- The most effective hillslope treatments provide immediate ground cover to reduce raindrop impact on bare soil and shorten overland flow paths (Wagenbrenner et al. 2006). Naturally occurring mulches, such as conifer needle cast, can provide protective post-fire ground cover (Pannkuk and Robichaud 2003).

Current treatment monitoring/research efforts are providing effectiveness and recovery data for an expanding range of treatments and environments. Other efforts are aimed at evaluating the efficacy of new treatment products and methods. These efforts should guide future treatment development and selection by providing insight on the applicability of specific treatments within various environments as well as the limitations of effectiveness.

CONCLUSION

After a wildfire, it is likely that runoff, erosion, and downstream sedimentation will occur at higher rates than before the fire. Currently available post-fire hillslope treatments can reduce runoff and erosion for many, but not all, rain events, and they do not eliminate these effects. It is important to critically evaluate the need for post-fire stabilization treatments, as in many cases, it is cost effective and more ecologically sound to choose the *no treatment* option. Treatments should only be applied if unacceptable levels of flooding, erosion, and sedimentation are expected, and land managers are compelled to reduce that risk to protect valued resources within and downstream of the burned area. Monitoring the effectiveness of post-fire rehabilitation treatments is necessary to compare treatments and establish the limitations of each treatment. Data from treatment monitoring efforts can guide future post-fire stabilization and rehabilitation efforts and be used to develop and refine treatments as well as predictive models.

Acknowledgments

I gratefully acknowledge the U.S. Department of Interior and U.S. Department of Agriculture Forest Service Joint Fire Science Program (JFSP) for their generous support fire-related studies in the United States. Much of the work cited here and throughout this text are the results of JFSP supported projects.

References

Aksoy, H. and M.L. Kavvas. 2005. A review of hillslope and watershed scale erosion and sediment transport models. *Catena,* 64: 247-271.

Anderson, H.W., M.D. Hoover, and K.G. Reinhart. 1976. Forest and water: effects of forest management on floods, sedimentation, and water supply. General Technical Report, PSW-18. U.S. Department of Agriculture, Forest Service, Pacific Southwest Forest and Range Experiment Station, Berkeley, California, USA.

Bailey, R.W. and O.L. Copeland. 1961. Vegetation and engineering structures in flood and erosion control. *In* Proceedings, 13th Congress of International Union of Forest Research Organization, Paper, 11-1, International Union of Forest Research Organization, Vienna, Austria.

Bautista, S., J. Bellot, and V.R. Vallejo. 1996. Mulching treatment for post-fire soil conservation in a semiarid ecosystem. *Arid Soil Research and Rehabilitation,* 10: 235-242.

Beschta, R.L., J.J. Rhodes, J.B. Kauffman, R.E.Gresswell, G.W. Minshall, J.R. Karr, D.A. Perry, F.R. Hauer, and C.A. Frissell. 2004. Post-fire management on forested public lands of the western United States. *Conservation Biology,* 18(4): 957-967.

Beyers, J.L. 2004. Post-fire seeding for erosion control: effectiveness and impacts on native plant communities. *Conservation Biology,* 18(4): 947-956.

Chiun-Ming, L. 1985. Impact of check dams on steep mountain channels in northeastern Taiwan. pp. 540-548. *In* S.A. El-Swaify, W.C. Moldenhauer and A. Lo [eds.]. *Soil Erosion and Conservation.* Soil Conservation Society of America, Ames, Iowa, USA.

Collins, L.M. and C.E. Johnston. 1995. Effectiveness of straw bale dams for erosion control in the Oakland Hills following the fire of 1991. pp. 171-183. *In* J.E Keeley and T. Scott [eds.]. *Brushfires in California Wildlands: Ecology and Resource Management*. International Association of Wildland Fire, Fairfield, Washington, USA.

DeBano, L.F., P.F. Ffolliott, and M.B. Baker Jr. 1996. Fire severity effects on water resources. pp. 77-84. *In* P.F. Ffolliott, L.F. DeBano, M.B. Baker Jr., G.J. Gottfried, G. Solis-Garza, C.B. Edminster, D.G. Neary, L.S. Allen, and R.H. Hamre [tech. coords.]. Effects of Fire on Madrean Province Ecosystems – A symposium proceedings. General Technical Report RM-289, U.S. Department of Agriculture, Forest Service, Rocky Mountain Forest and Range Experiment Station, Fort Collins, Colorado, USA.

DeBano, L.F., D. Neary, and P.F. Ffolliott. 1998. *Fires Effects on Ecosystems*. John Wiley & Sons, New York, New York, USA.

Elliot, W.J. 2004. WEPP internet interfaces for forest erosion prediction. *Journal of the American Water Resources Association*, 40(2): 299-309.

Faust, R. 1998. Lesson plan: Fork Fire soil loss validation monitoring. Unit VIII, Long term recovery and monitoring. *In* Burned Area Emergency Rehabilitation (BAER) techniques, 1998. Course Notebook, U.S. Department of Agriculture, Forest Service, Pacific Southwest Region, San Francisco, California, USA.

Franklin, J.F. and J.K. Agee. 2003. Forging a science-based national forest fire policy. *In* Issues in Science and Technology, Fall 2003, University of Texas at Dallas, Dallas, USA.

Hartford, R.A. and W.H. Frandsen. 1992. When it's hot, it's hot—or maybe it's not (surface flaming may not portend extensive soil heating). *International Journal of Wildland Fire*, 2: 130-144.

Hawkins, R.H. and R.J. Greenberg. 1990. Wildcat4 Flow Model. School of Renewable Natural Resources, University of Arizona, Tucson, Arizona, USA.

Hungerford, R.D. 1996. Soils: Fire in Ecosystem Management Notes: Unit II-I. U.S. Department of Agriculture, Forest Service, National Advanced Resource Technology Center, Marana, Arizona, USA.

Jain, T.B. 2004. Tongue-tied. *Wildfire*, July-August 2004: 22-26.

Karr, J.R., J.J. Rhodes, G.W. Minshall, F.R. Hauer, R.I. Beschta, C.A. Frissell, and D.A. Perry. 2004. The effects of post-fire salvage logging on aquatic ecosystems in the American west. *BioScience*, 54: 1029-1033.

Lentile, L.B., Z.A. Holden, A.M.S. Smith, M.J. Falkowski, A.T. Hudak, P. Morgan, S.A. Lewis, P.E. Gessler, and N.C. Benson. 2006. Remote sensing techniques to assess active fire characteristics and post-fire effects. *International Journal of Wildland Fire*, 15: 319-345.

Lewis, S.A., J.Q. Wu, and P.R. Robichaud. 2006. Assessing burn severity and comparing soil water repellency, Hayman Fire, Colorado. *Hydrological Processes*, 20: 1-16.

Lindemayer, D.B. and R.F. Noss. 2006. Salvage logging, ecosystem processes, and biodiversity conservation. *Conservation Biology*, 20(4): 949-958.

MacDonald, L.H. 1994. Developing a monitoring project. *Journal of Soil and Water Conservation*, 4a(3): 221-227.

McIver, J.D. and L. Starr. [tech. eds.]. 2000. Environmental effects of post-fire logging: literature review and annotated bibliography. PNW-GTR-486. U.S. Department of Agriculture, Forest Service, Pacific Northwest Research Station, Portland, Oregon, USA.

McIver, J.D. and L. Starr. 2001. A literature review of on the environmental effects of post-fire logging. *Western Journal of Applied Forestry,* 16: 159-168.

Miles, S.R., D.M. Haskins, and D.W. Ranken. 1989. Emergency burn rehabilitation: cost, risk, and effectiveness. pp. 97-102. *In* N.H. Berg [tech. coord.]. Proceedings of the symposium on Fire and Watershed Management. General Technical Report PSW-109. U.S. Department of Agriculture, Forest Service, Pacific Southwest Forest and Range Experiment Station, Berkeley, California, USA.

Moody, J.A. and D.A. Martin. 2001. Initial hydrologic and geomorphic response following a wildfire in the Colorado Front Range. *Earth Surface Processes and Landforms,* 26: 1049-1070.

Morgan, R.P.C., J.N. Quinton, R.E. Smith, G. Govers, J.W.A. Poesen, K. Auerswald, G. Chisci, D. Torri, and M.E. Stycezen. 1998. The European soil erosion model (EUROSEM): a dynamic approach for predicting sediment transport from fields and small catchments. *Earth Surface Processes and Landforms,* 23: 527-544.

Napper, C. 2006. Burned Area Emergency Response Treatments Catalog (BAERCAT). 0625 1801P. U.S. Department of Agriculture, Forest Service, San Dimas Technology and Development Center, San Dimas, California, USA.

Nearing, M.A., G.R. Foster, L.J. Lane, and S.C. Finkner. 1989. A process-based soil erosion model for USDA-water erosion prediction project technology. *Transactions of the ASAE,* 32(5): 1587-1593.

Neary, D.G., C.C. Klopatek, L.F. DeBano, and P.F. Ffolliott 1999. Fire effects on belowground sustainability: a review and synthesis. *Forest Ecology and Management,* 122: 51-71.

Neary, D.G., P.F. Ffolliott, and J.D. Landsberg. 2005. Chapter 5: Fire and streamflow regimes pp. 107-118. *In* D.G. Neary, K.C. Ryan, and L.F. DeBano [eds.]. Wildland Fire in Ecosystems: Effects of Fire on Soils and Water. General Technical Report RMRS-GTR-42-vol.4. U.S. Department of Agriculture, Forest Service, Rocky Mountain Research Station, Ogden, Utah, USA.

Noss, R.F., J.F. Franklin, W.I. Baker, T. Schoennagel, and P.B. Moyle. 2006. Ecology and management of fire-prone forests of the western United States. Society for Conservation Biology Scientific Panel on Fire in Western U.S. Forests. Society for Conservation Biology, North American section, Arlington, Virginia, USA. *Online at http://www.conbio.org/sections/namerica/napolicy.cfm* [accessed 9 Sep 2008]

Pannkuk, C.D. and P.R. Robichaud. 2003. Effectiveness of needle cast at reducing erosion after forest fires. *Water Resources Research,* 39(12): 1333-1344.

Pierson, F.B., P.R. Robichaud, and K.E. Spaeth. 2001. Spatial and temporal effects of wildfire on the hydrology of a steep rangeland watershed. *Hydrological Processes,* 15: 2905-2916.

Pike, R.G. and J.G. Ussery. 2006. Key points to consider when pre-planning for post-wildfire rehabilitation. FORREX Forest Research Extension Partnership, FORREX Series 19, Kamloops, Canada. *Online at* http://www.forres.org/publications/FORREXSeries/FS19.pdf [accessed on 9 Sep 2008]

Raftonyannis, Y. and I. Spanos. 2005. Evaluation of log and branch barriers as post-fire rehabilitation treatments in a Mediterranean pine forest in Greece. *International Journal of Wildland Fire,* 14: 183-188.

Reeves, G.H., P.A. Bisson, B.E. Rieman, and L.E. Benda. 2006. Post-fire logging in riparian areas. *Conservation Biology,* 20(4): 994-1004.

Renard, K.G., G.R. Foster, G.A. Weesies, and J.P. Porter. 1991. RUSLE: revised universal soil loss equation. *Journal of Soil and Water Conservation*, January-February: 30-33.

Renard, K.G., G.R. Foster, D.C. Yoder, and D.K. McCool. 1994. RUSLE revisited: status, questions, answers, and the future. *Journal of Soil and Water Conservation*, May-June: 213-220.

Rinne, J.N. 1996. Short-term effects of wildfire on fishes and aquatic macroinvertebrates in the southwestern United States. *North American Journal of Fisheries Management*, 16: 653-658.

Robichaud, P.R. 2005. Measurement of post-fire hillslope erosion to evaluate and model rehabilitation treatment effectiveness and recovery. *International Journal of Wildland Fire*, 14: 475-485.

Robichaud, P.R. and R.E. Brown. 1999, revised 2000. What happened after the smoke cleared: onsite erosion rates after a wildfire in eastern Oregon. pp. 419-426. *In* D.S. Olsen and J.P. Potyondy [eds.]. Proceedings of the Wildland Hydrology Conference, American Water Resource Association, Hernon, Virginia, USA.

Robichaud, P.R. and R.E. Brown. 2002. Silt fences: an economical technique for measuring hillslope soil erosion. General Technical Report RMRS-GTR-94. U.S. Department of Agriculture, Forest Service, Rocky Mountain Research Station, Fort Collins, Colorado, USA.

Robichaud, P.R. and R.E. Brown. 2003. Quick response small catchment monitoring techniques for comparing post-fire rehabilitation treatment effectiveness. pp. 663-667. *In* Proceedings of the First Interagency Conference on Research in the Watersheds, U.S. Department of Agriculture, Agricultural Research Service: Tucson, Arizona, USA.

Robichaud, P.R., J.L. Beyers, and D.G. Neary. 2000. Evaluating the effectiveness of postfire rehabilitation treatments. General Technical Report RMRS-GTR-63. U.S. Department of Agriculture, Forest Service, Rocky Mountain Research Station, Fort Collins, Colorado, USA.

Robichaud, P.R., J.L. Beyers, and D.G. Neary. 2005. Chapter 10: Watershed rehabilitation. pp. 179-197. *In* D.G. Neary, K.C. Ryan, and L.F. DeBano [eds.]. Wildland Fire in Ecosystems: Effects of Fire on Soils and Water. General Technical Report RMRS-GTR-42-vol. 4. U.S. Department of Agriculture, Forest Service, Rocky Mountain Research Station, Ogden, Utah, USA.

Robichaud, P.R., W.J. Elliot, F.B. Pierson, D.E. Hall, and C.A. Moffet. 2006a. Erosion Risk Management Tool (ERMiT) Ver. 2006.01.18. U.S. Department of Agriculture, Forest Service, Rocky Mountain Research Station, Moscow, Idaho, USA. *Online at* http://forest.moscowfsl.wsu.edu/fswepp/ [accessed 9 Sep 2008]

Robichaud, P.R., T.R. Lillybridge, and J.W. Wagenbrenner. 2006b. Effects of post-fire seeding and fertilizing on hillslope erosion in north-central Washington, USA. *Catena*, 67: 56-67.

Robichaud, P.R., W.J. Elliot, F.B. Pierson, D.E. Hall, and C.A. Moffet. 2007a. Predicting post-fire erosion and mitigation effectiveness with a web-based probabilistic erosion model. *Catena*, 71: 229-241.

Robichaud, P.R., S.A. Lewis, D.Y.M. Laes, A.T. Hudak, R.F. Kokaly, and J.A. Zamudio. 2007b. Post-fire soil burn severity mapping with hyperspectral image unmixing. *Remote Sensing of Environment*, 108: 467-480.

Robichaud, P.R., F.B. Pierson, R.E. Brown, and J.W. Wagenbrenner. 2008. Measuring effectiveness of three post-fire hillslope erosion barrier treatments, western Montana, USA. *Hydrological Processes*, 22: 159-170.

Ryan, K.C. and N.V. Noste. 1985. Evaluating prescribed fires. pp. 230-238. *In* J.E. Lotan, B.M. Kilgore, W.C. Fischer, and R.W. Mutch [tech. coords.]. Proceedings, symposium and workshop on Wilderness Fire. General Technical Report INT-182. U.S. Department of Agriculture, Forest Service, Intermountain Research Station, Ogden, Utah, USA.

Sessions, J., P. Bettinger, R. Buckman, M. Newton, and J. Hamann. 2004. Hastening the return of complex forests following fire: the consequences of delay. *Journal of Forestry*, 102: 38-45.

Spigel, K.M. and P.R. Robichaud. 2007. First year post-fire erosion rates in Bitterroot National Forest, Montana. *Hydrological Processes*, 21: 998-1005.

State of Victoria. 2006. Code of Practice for Fire Management on Public Lands. Victorian Government, Department of Sustainability and Environment, Melbourne, Australia. *Online at* http://www.dse.vic.gov.au/fires [accessed 9 Sep 2008]

Stewart, S.I., V.C. Radeloff, and R.B. Hammer. 2003. Characteristics and location of the wildland-urban interface in the United States. CD-Rom track 4.A1. *In* Proceedings of the Second International Wildland Fire Ecology and Fire Management Workshop. American Meteorological Society, Boston, Massachusetts, USA.

[USACE] US Army Corps of Engineers. 1982. HEC-2 Water Surface Profiles User's Manual. U.S. Army Corps of Engineers, Hydrologic Engineering Center, Davis, California, USA.

[USACE] US Army Corps of Engineers. 1985. HEC-1 Flood Hydrograph Package User's Manual. U.S. Army Corps of Engineers, Hydrologic Engineering Center, Davis, California, USA.

USDA Forest Service. 1995. Burned-area emergency rehabilitation handbook. Forest Service Handbook 2509.13-95-6. U.S. Department of Agriculture, Forest Service, Washington, D.C., USA.

USDA Forest Service. 2004. Emergency stabilization – burned-area emergency response. Section 2523 of Chapter 2520, Watershed protection and management. *Forest Service Manual FSM 2500*, Watershed and Air Management. U.S. Department of Agriculture, Forest Service, Washington, D.C., USA.

US General Accounting Office. 2003. Wildland fires: better information needed on effectiveness of emergency stabilization and rehabilitation treatments. GAO-03-430. U.S. General Accounting Office, Washington, D.C., USA.

US General Accounting Office. 2006. Wildland fire rehabilitation and restoration: Forest Service and BLM could benefit from improved information on status of needed work. GAO-06-670. U.S. General Accounting Office, Washington, D.C., USA.

Wagenbrenner, J.W., L.H. MacDonald, and D. Rough. 2006. Effectiveness of three post-fire rehabilitation treatments in the Colorado Front Range. *Hydrological Processes*, 20: 2989-3006.

Wischmeier, H. and D.D. Smith. 1978. Predicting rainfall erosion losses. Agriculture Handbook no 537. U.S. Department of Agriculture, Science and Education Administration, Washington, D.C., USA.

11

Non-native and Native Seeding

Jan L. Beyers

Abstract

Post-fire seeding is used to stabilize burned slopes by increasing plant cover, prevent invasion of burned areas by noxious weeds, replace weedy annual grasses on burned rangelands, and reestablish desirable vegetation including tree species. Fast-growing pasture grasses and forbs have been most widely applied for post-fire stabilization and rehabilitation, but use of native species is increasing. Successful ground cover enhancement with seeded species depends on protection of the seed from predators and desiccation, the amount and timing of growing-season rainfall, and resting the area from grazing until seeded species and natural regeneration are well-established. As these conditions are often not met, grass seeding has a mixed record of success for erosion reduction and rangeland rehabilitation. In cases where post-fire erosion would threaten life or property, a more effective hillslope stabilization method should probably be used. Seeded grasses can displace native herbaceous plants in post-fire succession, and high grass cover can limit recruitment of tree and shrub seedlings on burned sites. Land managers must weigh the potential erosion control and noxious weed reduction benefits against the economic cost of seeding, the ecological cost of native plant suppression, and the possible economic cost to replant timber species.

INTRODUCTION

Seeding after fire is conducted for a variety of reasons. As noted in the Chapter 10, annual grasses are frequently seeded for erosion control as part of emergency stabilization treatments. In addition, perennial grasses and forbs are often used to rehabilitate burned rangelands, either because overgrazing had caused a shift to shrub dominance or because undesirable non-native annual grasses otherwise would predominate after fire. Tree seed spread for reforestation purposes constitutes post-fire restoration. This chapter describes the use of post-fire seeding for these purposes and reviews some of the

U.S. Department of Agriculture, Forest Service, Pacific Southwest Research Station, 4955 Canyon Crest Dr., Riverside, California 92507 USA. Tel: 1 951 680 1527, Fax: 1 951 680 1501, e-mail: jbeyers@fs.fed.us

literature on treatment effectiveness, with a focus on seeding used for slope stabilization. Changing trends in post-fire seeding, primarily in the western U.S.A., are discussed along with suggestions for needed research.

RATIONALE FOR POST-FIRE SEEDING

Sediment production from burned slopes and disturbed sites correlates inversely with vegetative cover (Noble 1965, Orr 1970), making vegetation enhancement a logical post-fire rehabilitation practice. Quick plant establishment is regarded as the most cost-effective method to promote rapid infiltration of water and to keep soil on hillsides (Rice et al. 1965, Miles et al. 1989). In the western USA, grass seeding is the most widely used post-fire hillslope stabilization treatment, typically involving seeds of annual and/or perennial species applied from an aircraft (Robichaud et al. 2000).

Land management agencies in the USA (USDA Forest Service 2004) and elsewhere (e.g., Australia [State of Victoria 2006], Canada [Pike and Ussery 2006], Greece [Raftoyannis and Spanos 2005]) are required to assess burned landscapes and determine whether post-fire erosion and runoff will threaten human life and property, soil productivity, natural and cultural resource values, or the quantity and quality of water produced from public lands. Post-fire assessments usually focus treatment on severely-burned areas, where plants, litter and duff have been removed by the fire. Seeding has long been prescribed as a hillslope stabilization treatment, with the following primary objectives:

- Reduce erosion by increasing ground cover (post-fire assessments often assume that much of the natural soil seed bank was killed by the fire in addition to standing vegetation)
- Promote water infiltration and reduce runoff
- Maintain site productivity by capturing nutrients that would otherwise be lost in runoff and erosion

Seeding with herbaceous species can provide additional benefits in post-fire rehabilitation:

- Prevent invasion by undesirable plants
- Provide forage for wildlife and/or livestock

Seeding with tree species is used to jump-start forest succession where fires were particularly intense or were outside the natural range of variation for the desired plant community, so that natural regeneration would be delayed (see Chapter 14).

TYPES OF SPECIES SEEDED

Many public lands in the USA were severely overgrazed by the late 1890s, leading to disastrous erosion and a need to identify plant species useful for rangeland rehabilitation (Monsen and McArthur 1995). At the same time, lands laid bare by forest fires had poor capacity for herbaceous regeneration. Considerable research was conducted in the 1930s, '40s, and '50s to find grasses and forbs that would grow quickly after fires, hold soil in place, and withstand grazing by wildlife and livestock (e.g., Forsling 1931, Christ 1934, Friedrich 1947a, b, Evanko 1955, Hull and Johnson 1955, McClure 1956). Besides reducing erosion, post-fire plant establishment was regarded as a way to get useful production – beef cattle and/or sheep that could graze on the herbaceous plants – from lands that would not produce timber revenue for decades. Seed mixes were refined for particular areas as germination and establishment success were evaluated. Most contained annual grasses to provide quick cover, perennial grasses to establish long-term protection, and often legumes to add nitrogen to the soil (Klock et al. 1975, Ratliff and McDonald 1987). Seeded grasses and forbs in the USA were typically pasture species from Europe and Asia, although some selections from native species have been developed as well (Monsen et al. 2004).

Seed of forest trees may be broadcast after fire, particularly where high intensity fires burned through young forest stands just regenerating after timber harvest or a previous fire and natural seed sources would be scarce. However, planting seedlings has largely replaced broadcast seeding in most areas of the USA and Canada (Densmore et al. 1999).

Shrublands, such as the California chaparral or Spanish matorral, are often seeded after fires with annual grasses that will hold soil in place for a few years, until shrub cover is re-established, and then die out. Annual ryegrass (*Lolium multiflorum*) has been the species of choice in the USA because it is inexpensive, readily available in large quantities, and can be easily applied from the air (Fig. 1; Barro and Conard 1987). Cereal grains such as barley (*Hordeum vulgare*) or oats (*Avena sativa*) may be used instead of ryegrass. Post-fire seeding with annual ryegrass or other species and subsequent prescribed fire has also been used to reduce shrub density for range improvement and increased water yield (Schultz et al. 1955, Hibbert et al. 1974).

Most post-fire seeding projects in shrub-steppe or other range types have dual objectives of soil stabilization and rangeland rehabilitation (Friedrich 1947b, Pellant 1990). In the intermountain west of the USA, establishment of non-native annual cheatgrass (*Bromus tectorum*) in sagebrush-steppe vegetation has created continuous fuel beds that promote fire spread, leading to shrub death and increased dominance by cheatgrass (Whisenant 1990). Post-fire rehabilitation has generally used aggressive non-native perennial grass species, including crested wheatgrass (*Agropyron cristatum*), intermediate wheatgrass (*Agropyron intermedium* [*Elymus hispidus*]), smooth brome (*Bromus inermis*), orchard grass (*Dactylis glomerata*), hard fescue (*Festuca ovina*), and meadow foxtail (*Alopecuris pratensis*) (Monsen and McArthur 1995). Interest in restoring rather than just rehabilitating rangelands using native species has increased in recent years (Allen 1995).

Fig. 1 Loading annual ryegrass for post-fire seeding in California, USA (Photo credit: USDA Forest Service).

NATIVE VERSUS NON-NATIVE SPECIES

As described above, post-fire seeding has historically been done with commercially available non-native species selected for vigorous germination, rapid establishment, high production, and grazing tolerance. Public interest in ecological restoration has spurred greater use of native plants for stabilization and rehabilitation. In the USA, federal land management agencies are now directed to use native species, when practicable, for revegetation projects, including post-fire rehabilitation (Richards et al. 1998). However, the difficulty of acquiring enough native seed to cover burned areas after large fires, combined with concern about the ecological impacts of using non-local genotypes of native species, has limited the amount of post-fire seeding actually done with native plants (Richards et al. 1998, Goodrich and Rooks 1999, Robichaud et al. 2000). In some cases the 'native' species used for post-fire seeding were not found at the burn site before the fire (Hunter and Omi 2006). Goodrich and Rooks (1999) point to the need to develop native plant materials that are more competitive with cheatgrass for seeding burned, degraded rangelands; they argue that letting concerns about genetic pollution prevent use of non-local native species may be unrealistic in the face of the greater problems caused by continued cheatgrass dominance of vast rangeland areas (see also Jones 2003). Post-fire rehabilitation in the western USA is now likely to be done with a mix of cereal grains and some native species. In Spain, Pinaya et al. (2000) found that seeded native species produced cover faster and survived a drought period better than non-native annual ryegrass, suggesting that the native species are a better choice there for post-fire rehabilitation . More research is needed on the use of native species for post-fire seeding.

HOW WELL DOES SEEDING WORK?

Seeding after wildfire for erosion control has a mixed record of success (see review in Beyers 2004). Robichaud et al. (2000) examined published literature and numerous USDA Forest Service monitoring reports to assess the effectiveness of a wide range of post-fire rehabilitation practices, including seeding. Relatively few studies reported erosion measurements, so most of the assessment was based on recorded vegetation cover. Cover percentage greater than 30 was regarded as partially effective for erosion control, while 60 percent cover was considered effective, based on findings of Noble (1965) and Orr (1970). Robichaud et al. (2000) found that seeding did increase the likelihood of achieving partially or completely effective ground cover (Table 1), but the proportion of sites with at least 60 percent cover in the first year after seeding was still much less than half. A greater proportion of both seeded and unseeded sites had partially effective cover, more so in seeded sites. Cover was more likely to reach 60 percent the second year after a fire (Table 1).

Table 1: Percentage of study sites in publications and monitoring reports reviewed by Robichaud et al. (2000) that had at least 30 percent and at least 60 percent cover by the end of the first and second growing seasons after a fire. All published studies contained data from both seeded and unseeded plots. Monitoring reports did not always contain both treatments. Multiple study sites within one publication or report were counted separately (Modified from Robichaud et al. 2000).

Study sites	Percentage of sites with >30% cover		Percentage of sites with >60% cover	
	Seeded	Unseeded	Seeded	Unseeded
Number	Percentage			
	First Post-fire Year			
Publications				
19	42	26	26	10.5
Reports				
21	74	38	35	8
	Second Post-fire Year			
Publications				
18	78	67	56	17
Reports				
4	75	75	25	50

A flurry of recent research has included erosion measurements, with continued mixed results on seeding effectiveness. Pinaya et al. (2000) found that seeding with a commercial mix including annual ryegrass or with a mixture of native species was equally effective in reducing erosion on granitic soil in Spain, but seeding had no impact on post-fire runoff. Total soil loss from seeded plots was only about 10 percent of that from controls. Seeding with a mixture of native and naturalized species, with or without straw mulch, reduced erosion on calcareous and gypsiferous soils as well (Badía and Martí 2000). In contrast, seeding with white winter wheat (*Triticum aestivum*) produced relatively little cover and had no impact on erosion in central Washington USA (Robichaud et al. 2006). Wagenbrenner et al. (2006) also reported no reduction in erosion due to seeding at a site in Colorado, USA. Precipitation, especially during the summer, was below average in the latter two studies, affecting seeded grass establishment. Robichaud et al. (2006) also looked at the effectiveness of post-fire fertilization, a common practice in the Pacific Northwest of the USA, for increasing growth of seeded grasses. They found that fertilizers had no impact on seeded grass cover or erosion during the four years of the study.

The efficacy of grass seeding for reducing populations of undesirable species has a similar mixed record of success. Evans and Young (1978) found that seeding must be done immediately after fire to effectively reduce abundance of cheatgrass, and the seeded perennials must be protected from grazing to establish successfully. Goodrich and Rooks (1999) recorded less cheatgrass, yellow salsify (*Tragopogon dubius*), and musk thistle (*Carduus nutans*), a noxious weed, growing six seasons after fire on a Utah, USA site seeded with aggressive non-native perennial grasses (crested wheatgrass, intermediate wheatgrass, orchardgrass, and smooth brome). Squirreltail (*Elymus elymoides*), a native perennial grass, was more abundant on the seeded plots. On the other hand, Ratzlaff and Anderson (1995) found no difference in the amount of cheatgrass on seeded compared to unseeded plots in an Idaho, USA sagebrush rangeland; however, cheatgrass had not been particularly abundant before the fire, and establishment of seeded species was poor because of low precipitation the first year after fire. Floyd et al. (2006) found lower densities of several non-native invasive species on plots in areas seeded after fire with various mixtures of native perennial grasses at Mesa Verde National Park (Colorado, USA), but cheatgrass abundance was not affected by the treatments. In forested areas, researchers have found lower abundance of non-native species in seeded areas compared to non-seeded plots (Schoennagel and Waller 1999, Barclay et al. 2004, Keeley 2004).

Broadcast seeding of tree species can be done successfully (Densmore et al. 1999, Zagas et al. 2004), though planted container-stock seedlings have better survival and grow faster than those germinating from seed. Some type of site preparation enhances the germination rate and establishment of trees (see below).

Factors Affecting Seeding Success and Effectiveness

The quantity of seed applied can obviously affect the likelihood of successful establishment. Target seeding rates of 400 to 600 seeds per m^2 are commonly used, taking into account percent viability of the seed stock. Very high cover levels have been observed where large amounts of seed were applied (Keeley 2004). More importantly, grass establishment and persistence are affected by the amount, timing, and intensity of rainfall (Ratzlaff and Anderson 1995, Robichaud et al. 2000, 2006, Barclay et al. 2004), as is establishment of trees from seeding (Zagas et al. 2004). If the first rains that fall after seeding are not gentle and regular, little seeded grass will be established. Heavy rains wash seed off of steep hillslopes and into channels (Wagenbrenner et al. 2006), and strong wind events before rains fall may blow seed away, necessitating reseeding (Keeley et al. 1995). Second year cover, especially for annual species, depends on favorable conditions for growth and seed-set during the first growing season.

Seed predation by birds and rodents can be significant, especially for large-seeded species like cereal grains (personal observation). Seeding into ash or just before snowfall help minimize seed loss to animals. Covering broadcast seed with mulch can greatly improves plant establishment and the success of the seeding treatment (Dean 2001). Mulch both protects seed from predators and helps retain soil moisture. While Badía and Martí (2000) did not observe greater seeded species cover with mulch added, plant weight was significantly enhanced on mulched plots. Seeds lying on the soil surface may germinate, but their roots can have a hard time penetrating the ground (Fig. 2). Most species used for rangeland rehabilitation do best if planted with a rangeland drill or otherwise covered with a small amount of soil (Monsen et al. 2004). Soil scarification before broadcast seeding may improve tree seeding success by providing favorable microsites for seedling establishment (Densmore et al. 1999).

Evans and Young (1978) found that attempted rangeland rehabilitation without protection from grazing was unsuccessful. Similarly, rehabilitation team members interviewed by Robichaud et al. (2000, Appendix B) identified protecting burned areas from grazing by livestock and native ungulates for several years as the most important factor favoring successful seeded species establishment and natural plant regeneration.

Ecological Consequences of Seeding

The practice of post-fire seeding has its critics (Keeley et al. 2006). The same qualities that make certain grass and forb species useful for watershed stabilization, namely fast growth and wide adaptability, also make them highly competitive with naturally-regenerating vegetation. Foresters have long been aware that successful grass establishment can interfere with tree seedling survival (Pearson 1942, Friedrich 1947a, Larson and Schubert 1969, Elliott

Fig. 2 Seeded grass seedling with roots mostly on top of the soil (arrow). Most of the seedlings on this site dried out and died (Photo credit: USDA Forest Service).

and White 1987, Ratliff and McDonald 1987). Sites in California, USA aerially-seeded after a fire with annual ryegrass had low pine (*Pinus*) seedling densities on plots with annual ryegrass cover greater than 40 percent (Griffin 1982, Conard et al. 1991); Barclay et al. (2004) found a similar relationship in New Mexico. In the southern Cascades, high mortality of planted sugar pine (*Pinus lambertiana*) seedlings occurred during the first year after fire on plots with high annual ryegrass cover (49 percent when seedlings were planted, 85 percent by mid-summer) (Amaranthus et al. 1993). Seedlings planted during the second year after a fire had greater survival on seeded plots, which by then contained only mulch from dead ryegrass, while pine seedlings in unseeded plots faced competition from native shrubs (Amaranthus et al. 1993). Once conifer seedlings are well established, grass competition is generally less detrimental to their growth than shrub competition (McDonald and Oliver 1984, McDonald 1986). Even cereals such as wheat can competitively suppress tree seedlings in the first year after fire when seeded grass density is high (Keeley 2004).

Successful establishment of seeded grasses frequently results in decreased growth of native herbaceous species as well (Beyers 2004). Considerable controversy arose over post-fire seeding in southern California chaparral ecosystems (Barro and Conard 1987) because a specialized annual flora takes advantage of the light, space and soil nutrients available immediately after fire (Sweeney 1956, Keeley et al. 1981). In addition, some dominant chaparral shrub species regenerate after fire only from seed (Sampson 1944, Keeley 1991; Fig. 3). Competition from seeded grasses could potentially have long-term impacts on plant community composition (Barro and Conard 1987). Because research demonstrated that seeded ryegrass reduced native plant cover, diversity, and shrub seedling density – often without increasing total plant cover – but seldom reduced erosion (Nadkarni and Odion 1986, Taskey et al. 1989, Beyers et al. 1994, 1998, Wohlgemuth et al. 1998), the U.S. Forest Service now does not seed chaparral slopes in southern California after most fires.

Fig. 3 Shrub seedlings (*Adenostoma fasciculatum, Ceanothus* spp.) and fire-following herbaceous plants take advantage of the high-light environment in post-fire chaparral (Photo credit: USDA Forest Service).

Seeded annual grasses that establish thick cover can become fuel for the next fire after they cure during summer. This increases the risk that a site will reburn before major tree and shrub species have matured enough to set seed (Crane et al. 1983, Zedler et al. 1983).

TO SEED OR NOT TO SEED?

Whether or not to seed after fire depends on the land manager's objectives, ecosystem, and ecological concerns. If immediate post-fire erosion protection is needed, mulch is much more effective at keeping soil on the slopes (see Chapter 13). With post-fire seeding, there is always the risk that the amount and timing of precipitation will not allow successful establishment. For rehabilitation of rangeland infested with cheatgrass or other invasive annuals, seeding with perennial grasses is the best strategy for establishing vegetation that will not be as likely to carry fire; the land manager's dilemma is whether to use proven non-native species or take a chance with more expensive native seed. Few studies have been conducted on the efficacy of post-fire grass seeding for suppression of invasive species in other vegetation types. Seeding remains the least expensive way to get some kind of ground cover on burned areas for site stabilization, unless expensive native species must be used, and it remains the only method applicable to very large areas.

The risk of woody plant regeneration failure with post-fire seeding in forests needs to be acknowledged. While 30 percent grass cover will begin to reduce erosion (Table 1), this is also the percentage grass cover that Schultz et al. (1955) found would increase shrub seedling mortality, and the 60 percent cover sufficient to virtually eliminate sediment movement is strikingly similar to the amount of grass cover that Schultz et al. (1955) would eliminate shrub seedlings altogether (55 percent). Tree seedlings probably respond similarly. The dilemma for the land manager is thus obvious: if seeding produces enough cover to effectively control erosion, it will also effectively suppress or eliminate woody plant seedlings in the seeded area. This could have repercussions for forest stand development and browse regeneration for wildlife.

For the land manager concerned primarily with erosion, Table 1 suggests that seeding may be a reasonable gamble for trying to increase plant cover during the first year after a fire. Seedling is likely to stabilize a site more quickly than natural regeneration. Where control of erosion for protection of life, property, or infrastructure is essential, however, seeding would not be a good choice – effective control of sediment movement is likely to be achieved at the end of the first year only one third of the time by seeding. More expensive but effective treatments such as straw mulch should be considered, especially where protection is needed from the first storms that occur after the fire.

Table 1 also expresses the risks to conifer and other woody seedling regeneration faced by the manager considering seeding. A greater proportion of seeded sites could experience woody regeneration failure due to competition with herbaceous plants than would occur naturally. Partial control of erosion with only partial suppression of woody regeneration is a more likely outcome, however. Land managers must weigh the potential erosion control benefit against the economic cost of seeding, the ecological cost of native plant suppression, and the possible economic cost to replant timber species. Rates of seeding or grass species which are likely to produce high levels of first year cover should probably be reserved for high value timberland that will be replanted and intensively managed.

In the western USA, where large high-intensity wildfires are a major public concern, post-fire seeding of burned areas has decreased dramatically since the 1980s (Robichaud et al. 2000). Efforts now focus on areas of high burn severity, where soil erosion is likely to be high and the potential for weed invasion is great, rather than on seeding every acre of every burn. The increasing use of post-fire seeding for suppression of invasive species in forest lands creates the same management dilemma as for erosion control: seeded grass establishment that will effectively suppress noxious weeds may also inhibit tree regeneration. The decision to seed after a fire must be made with these possible consequences in mind.

References

Allen, E.B. 1995. Restoration ecology: limits and possibilities in arid and semiarid lands. pp. 7-15. *In* B.A. Roundy, E.D. McArthur, J.S. Haley and D.K. Mann [compilers]. Proceedings: Wildland Shrub and Arid Land Restoration Symposium. General Technical Report INT-GTR-315. U.S. Department of Agriculture, Forest Service, Intermountain Research Station, Ogden, Utah, USA.

Amaranthus, M.P., J.M. Trappe, and D.A. Perry. 1993. Soil moisture, native revegetation, and *Pinus lambertiana* seedling survival, growth, and mycorrhiza formation following wildfire and grass seeding. *Restoration Ecology,* 1: 188-195.

Badía, D. and C. Martí. 2000. Seeding and mulching treatments as conservation measures of two burned soils in Central Ebro Vally, NE Spain. *Arid Soil Research and Rehabilitation,* 13: 219-232.

Barclay, A.D., J.L. Betancourt, and C.D. Allen. 2004. Effects of seeding ryegrass (*Lolium multiflorum*) on vegetation recovery following fire in a ponderosa pine (*Pinus ponderosa*) forest. *International Journal of Wildland Fire,* 13: 183-194.

Barro, S.C. and S.G. Conard. 1987. Use of ryegrass seeding as an emergency revegetation measure in chaparral ecosystems. General Technical Report PSW-102. U.S. Department of Agriculture, Forest Service, Pacific Southwest Forest and Range Experiment Station, Berkeley, California, USA.

Beyers, J.L., S.G. Conard, and C.D. Wakeman. 1994. Impacts of an introduced grass, seeded for erosion control, on post-fire community composition and species diversity in southern California chaparral. pp. 594-601. *In* Proceedings of the 12th International Conference on Fire and Forest Meteorology. Society of American Foresters, Bethesda, Maryland, USA.

Beyers, J.L., C.D. Wakeman, P.M. Wohlgemuth, and S.G. Conard. 1998. Effects of postfire grass seeding on native vegetation in southern California chaparral. pp. 52-64. *In* Proceedings, Nineteenth Annual Forest Vegetation Management Conference: Wildfire Rehabilitation. Available from: University of California, Division of Agriculture and Natural Resources, Cooperative Extension Program, 1851 Hartnell Avenue, Redding, California 96002-2217, USA.

Beyers, J.L. 2004. Postfire seeding for erosion control: Effectiveness and impacts on native plant communities. *Conservation Biology,* 18: 947-956.

Christ, J.H. 1934. Reseeding burned-over lands in northern Idaho. University of Idaho Agricultural Experiment Station Bulletin 201. University of Idaho, Moscow, Idaho, USA.

Conard, S.G., J.C. Regelbrugge, and R.D. Wills. 1991. Preliminary effects of ryegrass seeding on postfire establishment of natural vegetation in two California ecosystems. pp. 314-321. *In* P.L. Andrews and D.F Potts [eds.]. Proceedings of

the 11th Conference on Fire and Forest Meteorology. Society of American Foresters, Bethesda, Maryland, USA.
Crane, M.F., J.R. Habeck, and W.C. Fischer. 1983. Early postfire revegetation in a western Montana Douglas-fir forest. Research Paper INT-319. U.S. Department of Agriculture, Forest Service, Intermountain Forest and Range Experiment Station, Ogden, Utah, USA.
Dean, A.E. 2001. Evaluating effectiveness of watershed conservation treatments applied after the Cerro Grande Fire, Los Alamos, New Mexico. Masters thesis, University of Arizona, Tucson, Arizona, USA.
Densmore, R.V., G.P. Juday, and J.C. Zasada. 1999. Regeneration alternatives for upland white spruce after burning and logging in interior Alaska. *Canadian Journal of Forest Research,* 29: 413-423.
Elliot, K.J. and A.S. White. 1987. Competitive effects of various grasses and forbs on ponderosa pine seedlings. *Forest Science,* 33: 356-366.
Evanko, A.B. 1955. Species adaptability and seeding season for increased forage production in northeastern Washington. Research Paper INT-37. U.S. Department of Agriculture, Forest Service, Intermountain Forest and Range Experiment Station, Ogden, Utah, USA.
Evans, R.A. and J.A. Young. 1978. Effectiveness of rehabilitation practices following wildfire in a degraded big sagebrush-downy brome community. *Journal of Range Management,* 31: 185-188.
Floyd, M.L., D. Hanna, W.H. Romme, and T.E. Crews. 2006. Predicting and mitigating weed invasions to restore natural post-fire succession in Mesa Verde National Park, Colorado, USA. *International Journal of Wildland Fire,* 15: 247-259.
Forsling, C.L. 1931. Artificial reseeding on western mountain range lands. Circular No. 178. U.S. Department of Agriculture, Washington, D.C., USA.
Friedrich, C.A. 1947a. Seeding grass by airplane on western Montana's burned-over timberlands. Research Note No. 52. U.S. Department of Agriculture, Forest Service, Northern Rocky Mountain Forest and Range Experiment Station, Missoula, Montana, USA.
Friedrich, C.A. 1947b. A guide for range reseeding on and near the national forests of Montana. Station Paper No. 7. U.S. Department of Agriculture, Forest Service, Northern Rocky Mountain Forest and Range Experiment Station, Missoula, Montana, USA.
Goodrich, S. and D. Rooks. 1999. Control of weeds at a pinyon-juniper site by seeding grasses. pp. 403-407. *In* S.B. Monsen and R. Stevens [compilers] Proceedings: Ecology and Management of Pinyon-Juniper Communities within the Interior West. Proceeding RMRS-P-9. U.S. Department of Agriculture, Forest Service, Rocky Mountain Research Station, Ogden, Utah, USA.
Griffin, J.R. 1982. Pine seedlings, native ground cover, and *Lolium multiflorum* on the Marble-Cone burn, Santa Lucia Range, California. *Madroño,* 29: 177-188.
Hibbert, A.R., E.A. Davis, and D.G. Scholl. 1974. Chaparral conversion potential in Arizona. Part I: Water yield response and effects on other resources. Research Paper RM-126. U.S. Department of Agriculture, Forest Service, Rocky Mountain Forest and Range Experiment Station, Fort Collins, Colorado, USA.
Hull, A.C., Jr., and W.M. Johnson. 1955. Range seeding in the ponderosa pine zone in Colorado. Circular No. 953. U.S. Department of Agriculture, Washington, D.C., USA.

Hunter, M.E. and P.N. Omi. 2006. Seed supply of native and cultivated grasses in pine forests of the southwestern United States and the potential for vegetation recovery following wildfire. *Plant Ecology,* 183: 1-8.

Jones, T.A. 2003. The restoration gene pool concept: beyond the native versus non-native debate. *Restoration Ecology,* 11: 281-290.

Keeley, J.E. 1991. Seed germination and life history syndromes in the California chaparral. *The Botanical Review,* 57: 81-116.

Keeley, J.E. 2004. Ecological impacts of wheat seeding after a Sierra Nevada wildfire. *International Journal of Wildland Fire,* 13: 73-78.

Keeley, J.E., M. Carrington, and S. Trnka. 1995. Overview of management issues raised by the 1993 wildfires in southern California. pp. 83-89. *In* J.E. Keeley and T. Scott [eds.]. *Brushfires in California Wildlands: Ecology and Resource Management.* International Association of Wildland Fire, Fairfield, Washington, USA.

Keeley, S.C., J.E. Keeley, S.M. Hutchinson, and A.W. Johnson. 1981. Postfire succession of herbaceous flora in southern California chaparral. *Ecology,* 62: 1608-1621.

Klock, G.O., A.R. Tiedemann, and W. Lopushinsky. 1975. Seeding recommendations for disturbed mountain slopes in north central Washington. Research Note PNW-244. U.S. Department of Agriculture, Forest Service, Pacific Northwest Forest and Range Experiment Station, Portland, Oregon, USA.

Larson, M.M. and G.H. Schubert. 1969. Root competition between ponderosa pine seedlings and grass. Research Paper RM-54. U.S. Department of Agriculture, Forest Service, Rocky Mountain Forest and Range Experiment Station, Fort Collins, Colorado, USA.

McClure, N.R. 1956. Grass and legume seedings on burned-over forest lands in northern Idaho and adjacent Washington. Masters thesis, University of Idaho, Moscow, Idaho, USA.

McDonald, P.M. 1986. Grasses in young conifer plantations – hindrance and help. *Northwest Science,* 60: 271-278.

McDonald, P.M. and W.W. Oliver. 1984. Woody shrubs retard growth of ponderosa pine seedlings and saplings. pp. 65-89. *In* D. Coombes [ed.]. Proceedings, Fifth Annual Forest Vegetation Management Conference. Available from: University of California, Division of Agriculture and Natural Resources, Cooperative Extension Forestry Program, 1851 Hartnell Avenue, Redding, California 96002-2217, USA.

Miles, S.R., D.M. Haskins, and D.W. Ranken. 1989. Emergency burn rehabilitation: cost, risk, and effectiveness. pp. 97-102. *In* N.H. Berg [technical coordinator]. Proceedings of the Symposium on Fire and Watershed Management. General Technical Report PSW-GTR-109. U.S. Department of Agriculture, Forest Service, Pacific Southwest Forest and Range Experiment Station, Berkeley, California, USA.

Monsen, S.B. and E.D. McArthur. 1995. Implications of early intermountain range and watershed restoration practices. pp. 16-25. *In* B.A. Roundy, E.D. McArthur, J.S. Haley, and D.K. Mann [compilers]. Proceedings: Wildland Shrub and Arid Land Restoration Symposium. General Technical Report INT-GTR-315. U.S. Department of Agriculture, Forest Service, Intermountain Research Station, Ogden, Utah, USA.

Monsen, S.B., R. Stevens, and N.L. Shaw [compilers]. 2004. Restoring western ranges and wildlands. General Technical Report RMRS-GTR-136, vols. 1, 2. U.S. Department of Agriculture, Forest Service, Rocky Mountain Research Station, Fort Collins, Colorado, USA.

Nadkarni, N.M. and D.C. Odion. 1986. Effects of seeding an exotic grass *Lolium multiflorum* on native seedling regeneration following fire in a chaparral community. pp. 115-121. *In* J.J. DeVries [ed.]. Proceedings of the Chaparral Ecosystems Research Conference. Report No. 62. California Water Resources Center, University of California, Davis, California, USA.

Noble, E.L. 1965. Sediment reduction through watershed rehabilitation. pp. 114-123. *In* Proceedings of the Federal Inter-agency Sedimentation Conference 1963. Miscellaneous Publication 970. U.S. Department of Agriculture, Washington, D.C., USA.

Orr, H.K. 1970. Runoff and erosion control by seeded and native vegetation on a forest burn: Black Hills, South Dakota. Research Paper RM-60. U.S. Department of Agriculture, Forest Service, Rocky Mountain Forest and Range Experiment Station, Fort Collins, Colorado, USA.

Pearson, G.A. 1942. Herbaceous vegetation a factor in natural regeneration of ponderosa pine in the Southwest. *Ecological Monographs,* 12: 316-338.

Pellant, M. 1990. The cheatgrass-wildfire cycle – are there any solutions? pp. 11-18. *In* E.D. McArthur, E.M. Romney, S.D. Smith, and P.T. Tueller [compilers]. Proceedings – Symposium on Cheatgrass Invasion, Shrub Die-off, and Other Aspects of Shrub Biology and Management. General Technical Report INT-276. U.S. Department of Agriculture, Forest Service, Intermountain Research Station, Ogden, Utah, USA.

Pike, R.G. and J.G. Ussery. 2006. Key points to consider when pre-planning for post-wildfire rehabilitation. FORREX Series 19. FORREX Forest Research Extension Partnership, Kamloops, British Columbia, Canada. *Online at http:// www. forrex.org/publications/forrexseries/fs19.pdf* [accessed 11 Sep 2008]

Pinaya, I., B. Soto, M. Arias, and F. Díaz-Fierros. 2000. Rehabilitation of burnt areas: relative effectiveness of native and commercial seed mixtures. *Land Degradation & Development,* 11: 93-98.

Raftoyannis, Y. and I. Spanos. 2005. Evaluation of log and branch barriers as post-fire rehabilitation treatments in a Mediterranean pine forest in Greece. *International Journal of Wildland Fire,* 14: 183-188.

Ratliff, R.D. and P.M. McDonald. 1987. Postfire grass and legume seeding: what to seed and potential impacts on reforestation. pp. 111-123. *In* Proceedings ninth annual forest vegetation management conference. Available from: University of California, Division of Agriculture and Natural Resources, Cooperative Extension Forestry Program, 1851 Hartnell Avenue, Redding, California 96002-2217, USA.

Ratzlaff, T.D. and J.E. Anderson. 1995. Vegetal recovery following wildfire in seeded and unseeded sagebrush steppe. *Journal of Range Management,* 48: 386-391.

Rice, R.M., R.P. Crouse, and E.S. Corbett. 1965. Emergency measures to control erosion after a fire on the San Dimas Experimental Forest. pp. 123-130. *In* Proceedings of the Federal Inter-agency Sedimentation Conference 1963. Miscellaneous Publication 970. U.S. Department of Agriculture, Washington, D.C., USA.

Richards, R.T., J.C. Chambers, and C. Ross. 1998. Use of native plants on federal lands: policy and practice. *Journal of Range Management,* 51: 625-632.

Robichaud, P.R., J.L. Beyers, and D.G. Neary. 2000. Evaluating the effectiveness of postfire rehabilitation treatments. General Technical Report RMRS-GTR-63. U.S. Department of Agriculture, Forest Service, Rocky Mountain Research Station, Fort Collins, Colorado, USA.

Robichaud, P.R., T.R. Lillybridge, and J.W. Wagenbrenner. 2006. Effects of postfire seeding and fertilizing on hillslope erosion in north-central Washington, USA. *Catena,* 67: 56-67.

Sampson, A.W. 1944. Plant succession on burned chaparral lands in northern California. Bulletin 685. University of California, College of Agriculture, Berkeley, California, USA.

Schoennagel, T.L. and D.M. Waller. 1999. Understory responses to fire and artificial seeding in an eastern Cascades *Abies grandis* forest, U.S.A. *Canadian Journal of Forest Research,* 29: 1393-1401.

Schultz, A.M., J.L. Launchbaugh, and H.H. Biswell. 1955. Relationship between grass density and brush seedling survival. *Ecology,* 36: 226-238.

State of Victoria. 2006. Code of Practice for Fire Management on Public Lands. Victorian Government, Department of Sustainability and Environment, Melbourne, Australia. *Online at http://www.dse.vic.gov.au/fires* [accessed on 11 Sep 2008]

Sweeney, J.R. 1956. Responses of vegetation to fire: a study of the herbaceous vegetation following chaparral fires. *University of California Publications in Botany,* 28: 143-249.

Taskey, R.D., C.L. Curtis, and J. Stone. 1989. Wildfire, ryegrass, and watershed rehabilitation. pp. 149-161. *In* N.H. Berg [tech. coord.]. Proceedings of the Symposium on Fire and Watershed Management. General Technical Report PSW-GTR-109. U.S. Department of Agriculture, Forest Service, Pacific Southwest Forest and Range Experiment Station, Berkeley, California, USA.

USDA Forest Service. 2004. Emergency stabilization—burned-area emergency response (BAER). Section 2523 of Chapter 2520, Watershed protection and management. *Forest Service Manual FSM 2500, Watershed and Air Management.* U.S. Department of Agriculture, Forest Service, Washington, D.C., USA.

Wagenbrenner, J.W., L.H. MacDonald, and D. Rough. 2006. Effectiveness of three postfire rehabilitation treatments in the Colorado Front Range. *Hydrological Processes,* 20: 2989-3006.

Whisenant, S.G. 1990. Changing fire frequencies on Idaho's Snake River plains: ecological and management implications. pp. 4-10. *In* E.D. McArthur, E.M. Romney, S.D. Smith, and P.T. Tueller [compilers]. Proceedings – Symposium on Cheatgrass Invasion, Shrub Die-off, and Other Aspects of Shrub Biology and Management. General Technical Report INT-276. U.S. Department of Agriculture, Forest Service, Intermountain Research Station, Ogden, Utah, USA.

Wohlgemuth, P.M., J.L. Beyers, C.D. Wakeman, and S.G. Conard. 1998. Effects of fire and grass seeding on soil erosion in southern California chaparral. pp. 41-51. *In* Proceedings Nineteenth Annual Forest Vegetation Management Conference: Wildfire Rehabilitation. Available from: University of California, Division of Agriculture and Natural Resources, Cooperative Extension Forestry Program, 1851 Hartnell Avenue, Redding, California 96002-2217, USA.

Zagas, T., P. Ganatsas, T. Tsitsoni, and M. Tsakaldimi. 2004. Post-fire regeneration of *Pinus halepensis* Mill. stands in the Sithonia peninsula, northern Greece. *Plant Ecology*, 171: 91-99.

Zedler, P.H., C.R. Gautier, and G.S. McMaster. 1983. Vegetation change in response to extreme events: the effect of a short interval between fires in California chaparral and coastal scrub. *Ecology*, 64: 809-818.

12

Using Erosion Barriers for Post-fire Stabilization

Peter R. Robichaud

Abstract

Erosion barriers, made from a range of natural and engineered materials, have been used for the past 30 years to provide mechanical barriers to overland flow, promote infiltration, and trap sediment on burned hillsides through out the United States. Until recently, managers assumed that erosion barriers which trapped and held sediment on the burned hillslope were functioning as designed and reducing hillslope erosion. However, results from several studies where sediment from treated and untreated sites were measured and compared, have indicated that post-fire erosion barrier treatment effectiveness is less than expected. Immediately after installation, erosion barriers provide their best performance; however, the installation process can disturb and loosen soil making it easier to erode. Although erosion barriers can be effective for low to moderate intensity rainfall events, their effectiveness is greatly reduced for high-intensity rainfall events. Erosion barrier effectiveness is also dependent on the quality of the installation and decreases over time as the sediment storage areas above the logs become filled and the barrier can no longer trap mobilized sediment. The impact of these results has been a reduction in the use of erosion barriers for post-fire hillslope stabilization, especially when the potential storm size is considered.

INTRODUCTION

Erosion barriers, made from natural and engineered materials, have been used for post-wildfire runoff and erosion mitigation for decades. These structures, installed in tiers along hillslope contours, are designed to slow runoff, cause localized ponding, and store eroded sediment. When the erosion barriers function as designed, they can decrease the erosive energy of runoff, increase infiltration, and reduce downstream sedimentation (Robichaud et al. 2000).

U.S. Department of Agriculture, Forest Service, Rocky Mountain Research Station, 1221 S. Main St., Moscow, ID 83843 USA. Tel: 1 208 883 2349, Fax: 1 208 883 2318, e-mail: probichaud@fs.fed.us

Common post-wildfire hillslope erosion barriers include contour-felled logs (logs cut from burned trees; Fig. 1), straw wattles (0.25 m diameter, 4 to 6 m long nylon mesh tubes filled with straw; Fig. 2), contour trenches (hand or machine dug trenches), straw bales (blocks of straw bound with twine; Fig. 3), and constructed multi-log structures (Fig. 4). Since erosion barriers are installed immediately after a fire occurs, these structures provide protection in the first few post-fire years when erosion rates are likely to be the greatest.

Contour-felled log erosion barriers (LEBs) have always been the most widely used erosion barriers for post-fire stabilization and rehabilitation treatment, as most forest fires leave dead trees that can be felled and limbed for use as LEBs. In the 1960s and 1970s, land management agencies in the United States began installing LEBs on burned hillslopes to reduce post-fire runoff. It was assumed that reduced runoff would reduce erosion. Robichaud et al. (2000) surveyed land managers who had experience with post-fire stabilization treatments and found that 65 percent of the survey respondents self-reported that LEB installations provided 'good' or 'excellent' results. However, as with most post-fire stabilization and rehabilitation treatments,

Fig. 1 A contour-felled log erosion barrier with soil end berms to increase storage capacity.

Fig. 2 A straw wattle erosion barrier.

Fig. 3 Straw bale erosion barriers installed in a swale.

Fig. 4 Multi-log structure constructed for use as an erosion barrier. Effectiveness of this treatment is currently being measured using silt fence plots.

there was little quantitative information about LEB effectiveness prior to 2000, and these judgments were based on qualitative observations. Recent research efforts, where hillslope runoff and/or sediment have been measured, have provided insight as to the effectiveness and limitations of LEB post-fire treatments (Robichaud 2000, Dean 2001, Wohlgemuth et al. 2001, Gartner 2003, Spigel and Robichaud 2006, Wagenbrenner et al. 2006, Robichaud et al. 2008a, b).

POST-FIRE INSTALLATION OF EROSION BARRIERS

Each LEB is laid on the ground along the hillslope contour and staked in place to anchor the log and keep it from rolling downhill. To prevent underflow, a shallow basin is dug upslope from the LEB and the loose soil is packed against the LEB to seal the gaps between the LEB and soil surface (Robichaud 2000; Fig. 5). The shallow basin or depression on the upslope side of the LEB is where runoff may pond and entrained sediment may settle; consequently, the volume of this area is often referred to as the 'sediment storage capacity.' The storage capacity of a LEB is calculated as:

$$V = Ldw\,(1 - s)\,(1 - c) \tag{1}$$

where V is the LEB storage capacity (m^3); L is the maximum of either the LEB length minus 1 m (to account for log taper and reduced storage near the ends of the LEB) or the length of the hand-dug basin above the LEB (m); d is the mean depth of the storage space from three lengths measured from a horizontal line extended from the crest of the LEB to the upslope ground surface (m); w is the mean width of the storage space measured from the crest of the LEB to the upslope ground surface (m); s is the slope of the LEB ($m\ m^{-1}$); and c was the portion of the LEB that has poor ground contact ($m\ m^{-1}$; Fig. 6). The variables s and c reduce the storage capacity of the LEB

Fig. 5 Workers are digging the sediment storage basin upslope of a newly installation contour-felled log erosion barrier.

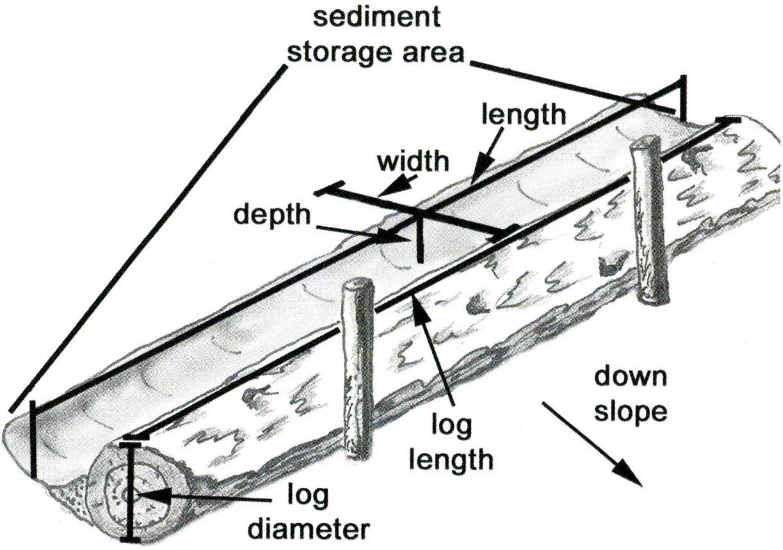

Fig. 6 Diagram of a contour-felled log with labeled dimensions used for calculating sediment storage capacity (from Robichaud et al. 2008a).

when the log is off-contour or has poor contact with the ground, respectively. Detailed discussion and measured values are provided in Robichaud et al. (2008a). Recent installations have included the use of additional loose soil mounded at the ends of the LEBs to form earthen end berms that angle upslope. These end berms increase the sediment storage capacity and inhibit concentrated runoff from flowing around the ends of the LEB. The addition of end berms on LEBs at research sites in Montana and Colorado increased the LEB sediment storage capacity by 16 and 10 percent, respectively (Robichaud et al. 2008b).

There is some evidence that the installation of erosion barriers can cause a short-term increase in erosion rates. In a paired watershed study to measure the effectiveness of LEBs, Robichaud et al. (2008b) reported that one of six sites had greater first post-fire year sediment yields in the treated watershed than in the control watershed; yet, in post-fire years two and three, the sediment yields generally were greater in the control than in the LEB treated watershed. In another two (out of six) sites, the first two sediment-producing rain events resulted in measurable sediment at the outlets of the treated watersheds but not in the untreated, control watersheds. Although differences in rainfall intensity between the treated and control watersheds were observed, these results suggest that LEB installation may cause enough soil disturbance to produce an increase in sediment yields, especially in the first few rain events following installation. This response to LEB installation was not observed at the two other sites where sediment-producing storms occurred in the first post-fire year.

The sediment-trapping ability of any LEB installation is dependent not only on the characteristics of the individual LEBs, but also on the density of LEBs over the landscape. LEB density measurements, such as total length of LEB per unit area (m ha^{-1}), number of LEB per unit area (LEBs ha^{-1}), or total LEB sediment storage capacity per unit area (m^3 ha^{-1} or Mg ha^{-1}), can be used to compare LEB installations. The layout of LEBs on a burned hillslope is designed to eliminate long, uninterrupted flow paths. The general pattern is staggered tiers, in which the center of each LEB is directly downslope from the gap between the two LEBs above it (Fig. 7). The total length of LEBs in each tier and the spacing between tiers is dependent on the number of trees available, the time, labor, and money available for installation, and the treatment plan established by the post-fire assessment team. As expected, the greater the LEB density, the greater is the potential sediment storage capacity of the installation. The total storage capacity of all LEBs at a site is determined by summing the individual LEB storage capacities (V; eq. 1).

Straw wattles provide a reasonable alternative to LEBs in burned areas where logs are scarce or poorly shaped. Straw wattles are permeable barriers that detain surface runoff long enough to reduce flow velocity and provide for some sediment storage (Fig. 2). Unlike LEBs, straw wattles are flexible and conform to the soil surface so that gaps rarely occur. Turning the ends of a

Fig. 7 Contour-felled logs installed in staggered tiers across a burned slope.

straw wattle upslope (forming the wattle into a broad 'smile' shape), like adding soil end berms to LEBs, can increase the straw wattle sediment-holding capacity. Similarly to LEBs, straw wattles can be laid out in staggered tiers on a hillslope (Fig. 8), but they are also used on road cuts or other highly erodible areas. The disadvantages of straw wattles over LEBs include the expense of manufacturing and shipping and the fact that the straw within the nylon web can be a source of non-native or invasive weed seed.

EROSION BARRIER EFFECTIVENESS

Early attempts to quantify LEB effectiveness were based on observing sediment trapped by the barrier (McCammon and Hughes 1980, Miles et al. 1989). Although these data provide insight into how LEBs function, the amount of sediment stored by LEBs is not a good measure of treatment effectiveness. The more important treatment effectiveness criteria is the reduction in runoff and/or sediment provided by the treatment, and this is generally determined by comparing runoff and/or sediment yields between equivalent treated and untreated (control) areas.

Recent studies have compared sediment yields (and, when possible, runoff) between treated and untreated sites:

- In a three-year study after the 2000 Bobcat Fire in Colorado, the calculated sediment storage capacity of the LEBs was greater than the sediment

Fig. 8 Straw wattles installed in staggered tiers across a burned hillslope of chaparral.

produced from an untreated site in an average year (Wagenbrenner et al. 2006). However, the authors emphasized that even with the ideal capacity to hold 100 percent of the eroded sediment from a hillslope, the actual LEB effectiveness was highly dependent on the magnitude and timing of the rain events affecting the hillslopes, as well as the size, density, and quality of the LEB installation. Although they did not measure runoff, Wagenbrenner et al. (2006) reported that LEBs increased infiltration, especially in the area upslope of the erosion barrier that was disturbed during installation.

- After the 2000 Valley Complex Fire in Montana, USA, two studies were done to compare three types of erosion barriers (LEBs, straw wattles, and hand-dug contour trenches) at the plot scale (Robichaud et al. 2008a). In a simulated rainfall (26 mm h^{-1} rainfall intensity) and inflow (48 L min^{-1} added concentrated flow) experiment, the LEBs and straw wattles reduced total runoff and all three erosion barrier treatments reduced peak runoff rates (Robichaud et al. 2008a). However, only the straw wattles reduced the sediment yields as compared to the controls due to less runoff going around the ends of the wattle. Using the same plots, a three-year natural rainfall erosion study (10 sediment-producing rainfall events) showed that event sediment yields were greater with increasing total rainfall and rainfall intensity. None of the treatments reduced sediment yields from the natural rainfall events in the 100 m^2 plots. Thus, the observed reduction in runoff and sediment yields with low intensity simulated rainfall was not apparent during high intensity summer storms typical of western Montana (Robichaud et al. 2008a).

- In a long-term effort to measure LEB treatment effectiveness at the small watershed scale, Robichaud et al. (2008b) compared runoff and sediment yields in six matched pairs of burned watersheds (1 to 13 ha) sites throughout the western United States. For four to six post-fire years, runoff and sediment yield were measured and correlated to rainfall properties. High intensity rainfall produced most of the measured runoff and sediment yields in five sites, while long duration rain events produced most of the runoff and erosion in the southern California site. Runoff, peak flows, and sediment yields were lower from the treated watersheds than from the control watersheds for small rain events (less than 2-yr return period for the 10 min duration), but there was no treatment effect for rain events with larger return periods (Fig. 9). Improper installation and degradation over time reduced the effectiveness of contour-felled log erosion barriers, and often resulted in scouring and the formation of rills.
- The scale at which the erosion barrier treatment is evaluated may affect the outcome. Gartner (2003) examined the effectiveness of LEB treatment at four spatial scales – plot (1 to 5 m^2), hillslope (~400 m^2), sub-catchment (1 to 5 ha), and catchment (~16)—after the 2000 Hi Meadows Fire in Colorado. During the study period, 107 mm of rainfall fell in 41 events, 2

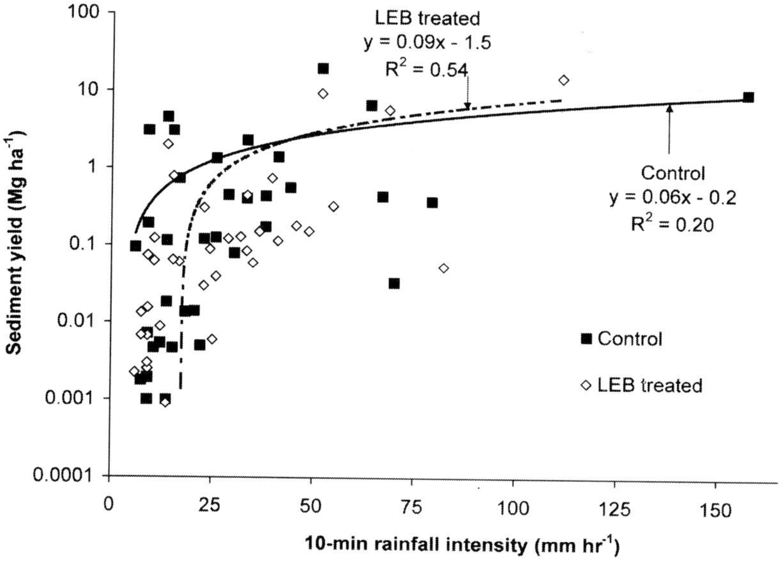

Fig. 9 Rainfall intensity versus sediment yield (log scale) from a paired watershed study (six sites in the western USA) of contour-felled log erosion barrier effectiveness. These data are from 61 sediment-producing rain events in post-fire years 0 to 6 (from Robichaud et al. 2008b).

of which had greater than 25 mm h^{-1} maximum 10-min intensity, and the LEB treatment reduced sediment yields at the hillslope and catchment scales, but not at the plot or sub-catchment scales. At the smallest scale (plots), where rainsplash and sheetwash erosion predominate, the effect of LEBs (which impact rill erosion, the largest source of hillslope sediment) could not be detected. Using replicated silt fence plots, the effectiveness of LEBs for the generally low intensity rain events at the hillslope scale was observed. Although the single paired sub-catchment and catchment sites produced contrary results, the author concluded that it was likely related to inexact pairing rather than scale or LEB treatment effects.

The consensus among these studies is that erosion barriers, and LEBs in particular, may reduce runoff and sediment yields for low intensity rain events, but they are unlikely to have a significant effect for high intensity rain events. Given that high intensity rain events produce the largest post-fire event sediment yields, the lack of treatment effectiveness for these storms is a serious consideration in treatment choice.

Efficiency and Performance Characteristics of Erosion Barrier Treatments

There are several ways to measure the efficiency of erosion barriers, but like all direct measurements of hillslope erosion, these data are difficult and labor-intensive to obtain. One method compares the amount of sediment held by the erosion barrier(s) to the total amount of sediment that was mobilized. This calculation is directly related to treatment effectiveness. Thus erosion barrier treatment efficiency, E_{TREAT} (%), would be

$$E_{TREAT} = \left(\frac{V_{EB}}{V_{EB} + V_{CS}} \right) 100 \tag{2}$$

where V_{EB} is the volume of sediment stored by the erosion barrier(s) (m^3), and V_{CS} is the volume of collected sediment below the erosion barrier treatment (m^3) (Robichaud et al. 2008a). If E_{TREAT} is calculated after several sediment-producing rain events, V_{EB} and V_{CS} are cumulative measurements. Once the erosion barriers reach a maximum amount of stored sediment (generally, ~67 percent of capacity), efficiency will decline with each subsequent sediment-producing event. As sediment is added, not only does the proportion of V_{CS} compared to V_{EB} increase, but also the LEBs deteriorate and move off contour, both of which decrease the overall efficiency. Robichaud et al. (2008a) calculated mean E_{TREAT} for each of the three types of erosion barriers following three sediment producing rain events during the first year of the natural rainfall experiment (2001). The mean E_{TREAT} for the first storm of 2001 was 87 percent for contour-felled logs, 83 percent for straw wattles, and 72 percent for contour trenches (Table 1). However, these barriers captured little additional

Table 1. The estimated sediment trapped, proportion of sediment storage capacity used, trap efficiency (%), and observations of erosion barrier performance for each erosion barrier over three consecutive rainfall events on 15, 21, and 30 July 2001 (from Robichaud et al. 2008a).

		Contour-felled logs				Straw wattles				Contour trenches			
		B	G	J	M	C	H	L	P	D	E	K	Q
Erosion barrier storage capacity (kg)		187	127	112	123	75	77	95	68	36	36	36	36
Estimated sediment trapped in erosion barrier[a b] (kg)	15 Jul	35 ◔	65 ◑	55 ◑	25 ◔	55 ◕	40 ◑	40 ◑	15 ◔	25 ◕	25 ◕	30 ◕	30 ◕
	21 Jul	55 ◔	125 ●	95 ◕	60 ◑	70 ●	45 ◑	40 ◑	35 ◑	30 ◕	30 ◕	n/e	n/e
	30 Jul	n/e	125 ●	95 ◕	95 ◕	n/e	55 ◕	55 ◑	40 ◑	p/f	30 ◕	p/f	p/f
Trap efficiency of erosion barrier[a] (%)	15 Jul	63	95	95	94	54	93	92	92	53	52	93	91
	21 Jul	21	52	32	46	15	7	51	56	8	6	–	–
	30 Jul	–	52	32	57	–	8	60	60	–	6	–	–
Observations of erosion barrier performance[c]	15 Jul	E		E U	E	E T U	T	E	E	T	E	E	E
	21 Jul	E	E T	E T	E T	E T U	E	T	E T	T	T		
	30 Jul		E T	E					T				

[a]n/e=not estimated; p/f=plot border failure; – = not calculable.
[b]Symbols for estimated proportion of storage filled: ◔ = 0-25%; ◑ = 26-50%; ◕ = 51-75%; and ● = 76-100%.
[c]Codes for erosion barrier performance: E=flowed around end(s); T=flowed over top; and U=flowed underneath.

sediment after that first storm, and their efficiency declined appreciably as additional rain events occurred. By the end of 2001, the mean E_{TREAT} had declined to less than 50 percent for contour-felled logs, 45 percent for straw wattles, and 10 percent for contour trenches.

Erosion barrier efficiency (E_{EB}) can also be determined by comparing the volume of sediment stored by the erosion barrier to the total sediment storage capacity of the barrier. By summation, this calculation can be applied over an erosion barrier installation, such that the total volume of sediment stored by all the erosion barriers, V_{EB} (m^3), is compared to the total sediment storage capacity (V_{TC}; m^3) of those barriers. Thus, erosion barrier trap efficiency, E_{EB} (%), would be

$$E_{EB} = \left(\frac{V_{EB}}{V_{TC}}\right)100 \qquad (3)$$

where and V_{TC} was the total site erosion barrier storage capacity as summed over all LEBs using Eq. 1. Although aggregation of individual erosion barrier efficiencies is not a measure of treatment effectiveness, it does describe the performance quality of the erosion barriers. In studies where the E_{EB} has been estimated, it rarely exceeds 67 percent (Robichaud et al. 2008b). As the volume of sediment stored in each erosion barrier is often estimated visually and large installations are usually sampled, reported data are often generalized – such as 'one-third full' or '50 percent efficient.'

Robichaud et al. (2008a) observed the performance characteristics of the three types of erosion barriers (LEBs, straw wattles, and contour trenches) during the first year of the natural rainfall experiment. In 13 of the 29 visual observations, runoff and sediment flowed over the top of the barrier; yet in only 3 of those observations were the barriers filled to capacity and 5 were at or below 50 percent full. The observations also document that sediment-laden runoff flowed around the ends (no end berms installed), and in many cases, over the tops of the erosion barriers. The flow around the ends of the structures was not related to the slope of the structures, since 3 of the 8 LEBs and straw wattle erosion barriers had slopes of 0 percent, and none of the structures had slopes greater than 4 percent. Thus, the flow around the ends as well as the flow over the tops of the structures was in response to runoff and sediment partially filling the structure storage capacity and blocking further input along the LEB. Also, 7 of the 8 contour-felled logs and straw wattles had 100 percent ground contact, yet 2 of these structures had flow underneath. Thus, the water that accumulated behind the log or straw wattle caused soil piping and undercut the loosened soil which had been used to fill the gap between the barrier and the ground surface. LEBs frequently intercept sediment-entrained concentrated flow, and the deposited sediment fills a small section uphill of the LEB leaving much of the total sediment storage capacity unused (Fig. 10).

Fig. 10 Sediment over-topped this partially filled contour-felled log without filling the unused portions of the sediment storage basin.

In Robichaud et al. (2008b), rills were often observed near the ends of LEBs (with and without end berms) that were installed off-contour and underneath LEBs that had gaps between the LEB and the soil surface. In these cases, the LEBs acted as runoff collectors, and the concentrated flow leaving the LEBs had greater flow velocity and sediment carrying capacity than less concentrated flow from above the LEB (Gartner 2003, Robichaud et al. 2008b). This would result in greater local erosion rates than if the LEBs had not been installed. About one-third of all the inspected LEBs showed evidence of flow beneath or around the end(s); yet, there was sufficient storage capacity in both the defective LEBs, which often were able to store some runoff and sediment, and the properly installed LEBs downslope of the defective LEBs to produce a net reduction in runoff and sediment yields for small, low-intensity rain events.

The erosion barrier performance observations reported in studies, as above, and in post-fire treatment monitoring reports indicate that erosion barrier trap efficiency, E_{EB} (%), can be improved by 1) adding soil berms to the ends (or turning the ends of the straw wattle upslope) of the erosion barriers; 2) increasing the erosion barrier density (erosion barrier length per unit area) on the hillslope; and 3) improving the quality control of installation to ensure that erosion barriers are placed on contour, securely anchored, and gaps between the erosion barrier and the ground are sealed. In post-fire field installations, with hundreds of barriers installed by crews of varying skill, attentiveness, and supervision, it is likely that some of the barriers will be poorly installed, compromising the potential storage capacity. In a study on contour-felled log erosion barriers installed by field crews in Colorado, an average of 32 percent of the barriers from 7 sites, and as many as 70 percent of the barriers from a single site, were either off-contour and/or had incomplete contact with the ground surface (Wagenbrenner et al. 2006). Improving the quality of erosion barrier installation may improve their performance, but it will also increase the time and labor costs for installation. In addition, increasing the sediment trapping efficiency (E_{EB}) will not likely increase the treatment efficiency (E_{TREAT}) by any appreciable amount. Thus, even with the improvements listed above, erosion barrier treatments have a short-lived capacity to retain hillslope erosion and, despite any improvements in efficiency, leave the hillslope vulnerable to high intensity rainfall events where their lack of effectiveness is well-documented.

MANAGEMENT IMPLICATIONS AND CONCLUSIONS

In the 1970s and 1980s, as erosion barriers (mostly LEBs) gained popularity for post-fire hillslope stabilization, the positive evaluations of effectiveness were often based on the qualitative observations of LEBs holding sediment on the burned hillslope (Robichaud et al. 2000). Managers assumed that without the LEBs in place, all the sediment held by the barriers would have been carried to the bottom of the slope and, consequently, the LEBs were functioning as designed. However, after seven years of study using quantitative measurements of erosion from both treated and untreated sites, the data show that LEBs do not lower the risk of post-fire hillslope erosion nearly as well as previously thought.

Erosion barriers are designed to provide immediate benefits after installation, in that they trap sediment during the first postfire year, which usually has the highest erosion rates (Robichaud and Brown 1999, revised 2000). Another advantage is that many erosion barriers, such as LEBs, are made from readily available local resources and do not introduce non-native plants to the burned area. However, the installation process may add to the erosion by disturbing the soil. Additionally, the concentration of intercepted runoff causes rill development below each barrier which also exacerbates erosion (Gartner 2003, Wagenbrenner et al. 2006, Robichaud et al. 2008b).

Recent studies show that erosion barriers are only effective for low to moderate intensity rainfall events, and that their effectiveness is greatly reduced for high-intensity events. This is a significant deficiency since greater than the 2-yr return interval 10-min rainfall intensity (high intensity) events can result in an order of magnitude increase in sediment as compared to low intensity events. In addition, these high intensity events drive the post-fire erosion process and produce the majority of sediment (Fig. 9; Robichaud 2005, Robichaud et al. 2008b). Thus, reducing the risk of damaging erosion usually requires reducing the risk of erosion for high intensity rain events.

The effectiveness of erosion barriers decreases over time as the sediment storage areas become filled and the barrier can no longer trap mobilized sediment (Robichaud et al. 2008a, Wagenbrenner et al. 2006). In some instances, LEBs have filled with sediment following the first several rain events, while others have taken one to two years to fill. The labor-intensive installation and the need for quality control make most erosion barrier treatments expensive for the limited erosion risk reduction obtained. Although erosion barriers are not as effective as needed for most post-fire hillslope treatments, they can be combined with other treatments, such as mulches and/or seeding, and contribute to the overall efficiency of the treatment (Dean 2001, deWolfe et al. 2008). Erosion barriers, particularly geotextile fabric barriers such as silt fences, can be effectively used after fires to protect small areas that contain high value resources (such as cultural sites) at risk for damage from increased erosion.

Post-fire rehabilitation treatment decisions involve balancing the need to reduce the post-fire risk of damage from increased runoff and erosion and the predicted effectiveness, availability, and installation costs of the treatments selected for use in the burned area. When potential storm size is included in the post-fire assessment, treatment decisions often do not favor the use of erosion barriers for hillslope erosion mitigation.

Acknowledgments

I gratefully acknowledge the U.S. Department of Interior and U.S. Department of Agriculture Forest Service Joint Fire Science Program (JFSP) for their generous support fire-related studies in the United States. Much of the work cited here and throughout this text are the results of JFSP supported projects.

References

Dean, A.E. 2001. Evaluating effectiveness of watershed conservation treatments applied after the Cerro Grande Fire, Los Alamos, New Mexico. Masters thesis, University of Arizona, Tucson, Arizona.

deWolfe, V.G., P.M. Santi, J. Ey, and J.E. Gartner. 2008. Effective mitigation of debris flows at Lemon Dam, La Plata County, Colorado. *Geomorphology*, 96: 366-377.

Gartner, J.D. 2003. Erosion after wildfire: the effectiveness of log erosion barrier mitigation. Masters thesis, University of Colorado, Fort Collins, Colorado, USA.

McCammon, B.P. and D. Hughes. 1980. Fire rehab in the Bend Municipal Watershed. pp. 252-259. *In* Proceedings of the 1980 Watershed Management Symposium, American Society of Civil Engineers, Reston, Virginia, USA.

Miles, S.R., D.M. Haskins, and D.W. Ranken. 1989. Emergency burn rehabilitation: cost, risk, and effectiveness. pp. 97-102. *In* N.H. Berg [tech. coord.]. Proceedings of the Symposium on Fire and Watershed Management. General Technical Report PSW-109. US Department of Agriculture, Forest Service, Pacific Southwest Forest and Range Experiment Station, Berkeley, California, USA.

Robichaud, P.R. 2000. Fire and erosion: evaluating the effectiveness of a post-fire rehabilitation treatment, contour-felled logs, 1-11. *In* M. Flug, D. Frevert, and D.W. Watkins Jr. [eds.]. Proceedings of the ASCE Watershed Management & Operations Management Conference, American Society of Civil Engineers, Reston, Virginia, USA.

Robichaud, P.R. 2005. Measurement of post-fire hillslope erosion to evaluate and model rehabilitation treatment effectiveness and recovery. *International Journal of Wildland Fire,* 14: 475-485.

Robichaud, P.R. and R.E. Brown. 1999, revised 2000. What happened after the smoke cleared: onsite erosion rates after a wildfire in eastern Oregon, 419-426. *In* D.S. Olsen and J.P. Potyondy [eds.]. Proceedings of the Wildland Hydrology Conference. American Water Resource Association, Herndon, Virginia, USA.

Robichaud P.R., J.L. Beyer and D.G. Neary. 2000. Evaluating the effectiveness of postfire rehabilitation treatments. General Technical Report RMRS-GTR-63. U.S. Department of Agriculture, Forest Service, Rocky Mountain Research Station, Fort Collins, Colorado, USA.

Robichaud, P.R., F.B. Pierson, R.E. Brown, and J.W. Wagenbrenner. 2008a. Comparing the postfire effectiveness of three hillslope erosion barrier treatments using simulated and natural rainfall. *Hydrological Processes,* 22: 159-170.

Robichaud, P.R., J.W. Wagenbrenner, R.E. Brown, P.M. Wohlgemuth, and J.L. Beyers. 2008b. Evaluating the effectiveness of contour-felled log erosion barriers as a post-fire runoff and erosion mitigation treatment in the western United States. *International Journal of Wildland Fire,* 17: 255-273.

Spigel, K.M. and P.R. Robichaud. 2006. First-year post-fire erosion rates in Bitterroot National Forest, Montana. *Hydrological Processes,* 21: 998-1005.

Wagenbrenner, J.W., L.H. MacDonald, and D. Rough. 2006. Effectiveness of three post-fire rehabilitation treatments in the Colorado Front Range. *Hydrological Processes,* 20: 2989-3006.

Wohlgemuth, P.M., K.R. Hubbert, and P.R. Robichaud. 2001. The effects of log erosion barriers on post-fire hydrologic response and sediment yield in small forested watersheds, southern California. *Hydrological Processes,* 15: 3053-3066.

13

Post-fire Mulching

Susana Bautista[1*], *Peter R. Robichaud*[2] and *Carme Bladé*[3]

Abstract

In post-fire rehabilitation, mulches are applied as emergency hillslope treatments intended to reduce the effects of rain impact and overland flow and keep the soil in place. Mulch immediately increases ground cover on the treated areas and is most effective during the critical first post-fire year when erosion is likely to be the greatest. A wide range of mulch materials (e.g., agricultural straw, wood chips, wood shreds) have been shown to be effective in reducing post-fire erosion. These advantages make mulching one of the most effective post-fire rehabilitation treatments available. However, mulch is a relatively expensive post-fire treatment, and this limits its use to burned areas with a high potential for soil erosion and important down stream values at risk.

The main disadvantage of mulching is the potential introduction of non-native plants (invasive weeds) with straw mulch. The use of certified 'weed-free' straw or straw from plants that will not grow in the burned area (e.g., rice straw applied on burned upland forest) can reduce this risk. Mulches made from shredded or manufactured forest materials generally do not contribute to the spread of invasive weeds. Thick mulches can suppress growth of vegetation. This is an advantage when mulches suppress invasive weeds and undesirable plants that spread after fires; however, it is a significant disadvantage when mulches inhibit natural recovery, endangered plant species, or seeded species.

Developments of new mulch materials and application methods (helimulching) have expanded the areas where mulch treatment successfully can be applied. As ecological restoration principles are integrated into post-fire rehabilitation, use of mulching treatments composed of local, site-specific

[1] *Department of Ecology, University of Alicante, Alicante, Spain.*

[2] *U.S. Department of Agriculture, Forest Service, Rocky Mountain Research Station, Moscow, Idaho, USA.*

[3] *CEAM Foundation, Paterna, Valencia, Spain.*

* *Corresponding author: Susana Bautista, Dpt. of Ecologia Apdo. 99, Universidad de Alicante, E-03080, Alicante, Spain. Tel: +34 96 590 34 00 ext. 3372, Fax: +34 96 590 98 32, e-mail: s.bautista@ua.es*

forest materials (wood chips, pine needles, shredded debris from post-fire logging, etc.) become more of a priority.

INTRODUCTION

Mulch is a cover placed over the soil to modify energy and water fluxes and to protect the soil in target areas. In agriculture and gardening, the main purposes of mulching are to modulate the soil temperature, control weed germination by reducing sunlight on the soil surface, preserve soil moisture, and control erosion. In post-fire rehabilitation, mulches are primarily intended to reduce rain impact and overland flow and keep the soil in place. Mulches are considered emergency hillslope treatments, which are regarded as a first line of defense against post-fire erosion and off-site impacts of sediments and floods.

The application of mulch treatments to burned lands uses techniques and materials from agriculture and civil engineering. Straw mulch is the most common material used to increase ground cover and reduce erosion rates on disturbed sites (Meyer et al. 1970, Kay 1983, Goldman et al. 1986), and is increasingly being applied as an emergency post-fire treatment in burned areas (Robichaud et al. 2000). It is often used in conjunction with grass seeding to provide ground cover in critical areas and to increase the success of seeding by improving moisture retention. The use of post-fire mulching dates back to the 1980s, when numerous rehabilitation measures were applied to severely burned forest watersheds after extensive wildfires in the western US (e.g., Gross et al. 1989, Miles et al. 1989). Since then, mulching use has increased and it is considered one of the most cost-effective emergency rehabilitation treatments. Mulching has been the primary treatment in some recent large post-fire rehabilitation efforts: the 2002 Hayman Fire in Colorado, USA, where more that 1200 ha were hydromulched and 3000 ha were straw mulched (Robichaud et al. 2003); the 2002 Rodeo-Chedeski Fire in Arizona, USA, where more than 8000 ha were treated with straw mulch (Richardson 2002); and the 2006 Tripod Fire in Washington, USA, where more than 14 000 ha were treated with straw mulch.

A major advantage of post-fire mulching is its immediate effectiveness after installation. However, treatment effectiveness does depend on the material used, application rate and technique, as well as site conditions. In addition, potential drawbacks to post-fire mulching have been acknowledged, such as introduction of invasive species into burned areas by mulch that contains these seeds. The decision to apply mulch in burned areas and the selection of mulch materials to be used must take into account the potential risks and the site-specific likelihood of treatment success. The following sections provide information about 1) post-fire mulch materials and application techniques; 2) advantages and disadvantages of post-fire mulch treatment and known treatment effectiveness; and 3) new approaches being used in post-fire mulching.

POST-FIRE MULCHING MATERIALS AND APPLICATION TECHNIQUES

Several natural and synthetic materials are applied as dry mulch or mixed with water as wet mulch (hydromulch). In general, organic mulches are more effective than inorganic materials at trapping soil particles and retaining soil moisture (Harding 1990). A number of organic materials, such as straw, jute, wood excelsior, wood chips, wood shreds, etc., have been used as mulch for post-fire soil conservation. Agricultural straw, particularly wheat, rice, and barley straw, are the most widely used mulch materials. Although woodchip mulching is less common, wood shreds and wood strands (thin wood strips manufactured from non-merchantable timber) are increasingly being developed and used for post-fire treatment. Woodchip mulching is often applied in the framework of post-fire logging operations, converting logging debris into woodchip biomass, which is spread on-site as mulch. Some recent innovations in post-fire mulching include the use of straw pellets and soil binding products. Compressed straw pellets greatly expand upon wetting, and can contain binding substances, such as soil polyacrylamide-family flocculants, that are released as the pellets expand. Other mulching materials, such as jute or wood excelsior, are rarely used in post-fire rehabilitation.

Depending on the type and size of the materials used, effectiveness and residence time of mulching treatments can vary greatly. Wood chips commonly provide a thicker mulch cover than other mulches, which may benefit soil conservation, but also may increase the risk of inhibiting seed germination and slowing post-fire vegetation recovery (Beyers et al. 2006). Groenier and Showers (2004) compared wood shavings, shredded wood, and excelsior as alternatives to wood chips for use as erosion control mulch. The authors concluded that shredded wood was the best alternative because shredding machines can handle the soil and rocks found in slash material, bark does not have to be removed from the slash, and the long fibers of shredded wood (from 3 to 4 mm thick, 6 to 50 mm wide, and 20 to 40 cm long) interlock, helping to keep the shredded wood in place. However, treatment effectiveness and ecological effects of these alternative wood mulches remain to be studied. In straw mulching, long fiber straw provides greater reduction in erosion than short fiber mulches, such as hydromulch fibers, as long fibers require greater shear force to displace them. Wood chips are decay-resistant mulches, while straw pellets and hydromulching fibers are designed to function for about a year and then rapidly decay. Rice straw is heavier and more durable than other types of straws.

In the past, seed germination from grain or hay mulch was considered to contribute to treatment effectiveness, as it added cover to the site. However, the use of straw may introduce non-native seed species that can persist and compete with the re-establishment of native vegetation. Land managers now seek 'weed-free' mulch, but it is not always available in the locations and quantities needed. In addition, straw and hay products may contain some species of invasive plants and still meet some weed-free standards. For instance, although certified 'weed-free' straw was used on the Hayman Fire, some of the straw brought used for rehabilitation treatment was contaminated with cheat grass (*Bromus tectorium*), an invasive grass difficult to control or eradicate once established (Robichaud et al. 2003). In the US, where each state maintains a listing of weed species, the mulches used for post-fire erosion control must be certified weed free for the specific state where it is applied. This can be a challenge when a wildfire occurs in more than one state and/or the mulch material is shipped from a different state than where it is applied. Rice straw is less likely to harbor noxious weeds, since rice is grown in moist habitats and the successful weeds from rice fields are unlikely to germinate or spread in dry forest environments. However, because weed seeds have wide ecological amplitude, this assumption should be tested by monitoring for invasive species over time after mulching treatments are applied (Kruse et al. 2004).

The effectiveness of straw mulch is related to the amount of ground cover it provides, as effective post-fire erosion mitigation requires at least 50 to 60 percent ground cover (Pannkuk and Robichaud 2003). Application rates of approximately 2.2 to 4.5 Mg ha^{-1} generally provide 70 to 80 percent cover and an average mulch depth of 2.5 to 5 cm (Napper 2006). Optimizing the thickness of post-fire mulch is a balance between soil protection and avoidance of adverse effects on plant recovery and/or seed germination of conjoint seeding treatments. The effectiveness of mulching also depends on the even application and consistent thickness of the mulch material. Strong winds can blow straw mulches off-site or pile it so deeply that seed germination is inhibited. These wind effects can be minimized by increasing the mulching rate (>3 Mg ha^{-1}) and/or by punching the mulch into the soil (Fig. 1), or by falling sub-merchantable trees and branches on top of the mulch to hold it down. Tackifiers/binders can be applied to stabilize mulch fibers on-site. Organic tackifiers, such as starch-derived materials, contain small amounts of natural nitrogen, which acts as a slow-release fertilizer, and readily absorb and retain water, which may increase vegetation recovery and improve seeding treatments.

Mulch can be spread from the ground and from helicopters (helimulching). Ground application is preferred for relatively small areas where 100 percent of the ground can be covered by thin, even straw mulch. Ground application is often done by hand (Fig. 2) and requires large crews to spread the mulch in a timely manner. This makes hand application expensive (about US $1000 to US $2500 per hectare; Napper 2006). Ground application can also be done with trailer- or truck-mounted straw blowers, which can

Fig. 1 Straw mulch was punched into the soil to resist movement by the wind.

Fig. 2 Straw mulch being applied by hand (photo from Napper 2006, p. 28).

spread the mulch much faster than hand application, but its use is limited to areas above and below roads or drivable fire lines. Mulch also can be spread on hillslopes in contour strips (strip mulching) placed 10 to 30 meters apart (strip mulching), such that approximately 50 percent of the burned slopes are covered. Compared to full-coverage mulching, strip mulching costs less and can be installed faster (Miles 2005); however, strip mulching is less effective (P.R. Robichaud, personal observation). Although straw mulch can be transported and distributed by helicopter, it is more economical to truck straw bales to staging areas located in close proximity to treatment areas. Within the staging area, straw bales can be prepared for application by helicopter, straw blowers, or hand spreading.

The use of helicopters to apply mulch for post-fire emergency rehabilitation first occurred in 2001, and has quickly become a popular application method as it allows large areas to be treated quickly and efficiently at a lower cost than hand application. Based on treatment cost tracking conducted in southwest US from 2000 to 2003, the reported cost of straw helimulching ranged from US \$500 to US \$2000 per hectare (Napper 2006). Helicopter application of straw mulch uses 100 to 1000 kg straw bales that are placed in cargo nets after which the baling twine is cut (Fig. 3). The cargo net is suspended 20 to 40 m beneath the helicopter. The helicopter flies over the target at about 60 knots and 60 m above the ground and releases all but one corner of the net. The straw bales break apart as they fall from the cargo net and spread further upon impact (Fig. 4). This application can provide a fairly even distribution of straw mulch over the ground surface; however, it has also been reported that aerial application may result in thick mounds and relatively uneven spreading patterns of the material (Santi et al. 2006). Some additional ground-based work may be needed to spread the mulch cover evenly and/or to apply on-site straw retention measures in areas where high winds are likely to occur.

Fig. 3 Staging area for helimulching operation (photo from Napper 2006, p. 28).

Fig. 4 Aerial straw mulching (Napper 2006, p. 25).

Wet mulches have been increasingly used for post-fire stabilization in recent years. There are numerous fiber mulches, soil stabilizers and combinations of materials (tackifier, polymers, seeds, etc.) that, when mixed with water and applied to the soil surface, form a matrix that help reduce erosion and foster plant growth. Numerous manufactures of hydromulch products have specific formulas for their mulches and others adapt their hydromulch mix to match the soil type and desired results. Hydromulch is most commonly applied on road cut and fill slopes, construction sites, and other disturbed areas with truck-mounted equipment. Truck-mounted sprayers can apply hydromulch along existing forest roads (Fig. 5). However, the effective range of truck-mounted hydromulch sprayers is about 50 m on either side of roads. Extension hoses can be added to truck-mounted systems to increase the distance from the roadway where hydromulch can be applied; however, this practice is of limited use in forest areas where large numbers of trees and/or standing tree trunks interfere with the use of extension hoses. Large-scale hydromulch application to burned hillslopes requires helicopters fitted with built-in or suspended slurry tanks that are opened while flying over a target treatment area (Fig. 6). In addition, aerial hydromulching requires large staging areas, access to water, and close proximity to treatment units (Fig. 7). Aerial application makes hydromulching expensive even when compared to other mulching treatments. For example, after the Hayman Fire,

Fig. 5 Ground-based hydromulching on high soil burn severity area near a road after the 2002 Hayman Fire in Colorado, USA.

Fig. 6 Aerial application of hydromulch on high soil burn severity area after the 2002 Hayman Fire in Colorado, USA.

Fig. 7 Staging area for aerial hydromulching (photo from Napper 2006, p. 13).

aerial hydromulching was applied to steep, inaccessible areas that drain directly to the South Platte River for an average cost of US $6000 per hectare (Robichaud et al. 2003).

TARGET AREAS FOR POST-FIRE MULCHING

Due to the cost and logistics of mulching, it is usually applied where there are downstream values at high risk for damage, such as above heavily used roads and critical stream reaches (Fig. 8). Given that helimulching is an efficient way to treat large land areas, post-fire mulching will likely increase in the future and may result in more areas being designated for mulch treatments.

Mulch is generally most effective on slopes up to 65 percent (Napper 2006) and in areas where high winds are not likely to occur. On steeper slopes, overland flow is more likely to wash the mulch downslope (Wagenbrenner et al. 2006). The post-fire straw mulching that was done after the Hayman Fire in Colorado (USA) occurred on slopes ranging from 20 to 60 percent, and with the exception of ridge tops, wind was not a significant issue. The mulch treatments effectively reduced hillslope runoff and erosion. In contrast, after the 2003 southern California (USA) fires, strong Santa Anna winds blew the straw mulch off many windward hillslopes that had been treated with wheat and rice straw. Numerous flooding events occurred the following December 25 (Christmas Day Storm), that could be partly attributed to the loss of effective straw cover in some treated watersheds (P.R. Robichaud, personal observation). The application of

Fig. 8 Due to high risks to downstream values, straw mulch was applied on some low soil burn severity slopes after the 2006 Brins Fire in Arizona, USA.

mulch on areas with some surface roughness, or the use of felled or limbed small trees to increase the site roughness may help to hold the mulch in place and avoid redistribution of the straw, and therefore improve mulch effectiveness.

After a wildfire, there is a mosaic of low, moderate, and high soil burn severity conditions within the burned area; however, the size and arrangement of the various soil conditions affects the potential erosion from a given area. Using the Water Erosion Prediction Project (WEPP) model, Robichaud and Monroe (1997) did a simulation study to determine how the spatial arrangement of hillslope soil burn severity conditions would effect erosion. They found that hillslopes with equal areas of high and low soil burn severity had greater predicted sediment yields when the high soil burn severity section(s) were above (i.e., higher on the hillslope) the low soil burn severity section(s). Some post-fire mulch treatments have been installed on lower hillslopes only, leaving the upper slopes untreated. However, results from Robichaud and Monroe (1997) and recent treatment monitoring (P.R. Robichaud, unpublished data) suggest that mulch treatments should include the top of the hillslope when high burn severity areas are upslope of low burn severity areas or when the hillslope is uniformly burned at high severity. Like most post-fire treatments, mulching is mostly applied in areas of high soil burn severity, as low soil burn severity areas generally produce low runoff and erosion (Benavides-Solorio and MacDonald 2001). Even in areas of moderate and high soil burn severity, erosion prediction technology can help determine where post-fire mulching is most needed and is likely to be most effective.

In conifer forests, low and moderate severity burned sites often have trees that are charred and partially consumed by fire, leaving dead needles in the canopy. These needles fall to the ground and provide a natural mulch ground cover. Pannkuk and Robichaud (2003) have shown a 60 to 80 percent reduction in interrill erosion and a 20 to 40 percent reduction in rill erosion with a 50 percent ground cover of dead needles. Thus, prudent use of post-fire rehabilitation treatments would exclude areas where needles are present to provide sufficient ground cover. Mulching should also be avoided in areas that contain sensitive or rare plants. Even careful mulching (use of weed-free mulching material and applications that result in minimum thickness) pose some potential risk of noxious weed invasion and/or suppression of natural vegetation recovery that would endanger these sensitive species.

HOW MULCH FUNCTIONS IN POST-FIRE REHABILITATION

The biological and physical role of mulches in the rehabilitation of burned soils remains largely unknown. Much of the available information comes from agricultural research, and only a few studies have examined the mechanisms responsible for mulching effects on burned areas. Mulch can influence many interactive processes that may affect soil conservation and ecosystem recovery after wildfires. First and foremost, mulching treatments increase ground cover and physically bind the loose surface soil together. The role of surface cover in reducing soil erosion is well established (Elwell and Stocking 1976), and is increasingly well documented for burned areas (e.g., Wagenbrenner et al. 2006). Mulches create surface obstructions that reduce runoff velocities, thereby reducing soil erosion by overland flow, and providing more opportunities for water infiltration. In addition, straw mulch often forms mini-debris dams with interlocking straw that store sediments and further reduce flow velocities. As a consequence of the reduction in raindrop impact, mulches also can reduce soil surface degradation, soil detachment, and soil transport by rainsplash. After a wildfire in a Mediterranean pine forest in southeastern Spain, Bautista et al. (1996, 1997) evaluated straw mulching effects on post-fire soil properties in paired mulched and control plots. One and two years after the treatment application, mulched plots showed significantly higher soil infiltration capacity and average size of soil aggregates, and lower soil surface compaction than untreated plots (Fig. 9). Furthermore, the mulch increased soil moisture content in treated plots over a 14-month study period (Fig. 10).

By changing microclimatic conditions and water availability in the surface soil, mulch can affect post-fire plant recovery and soil biological

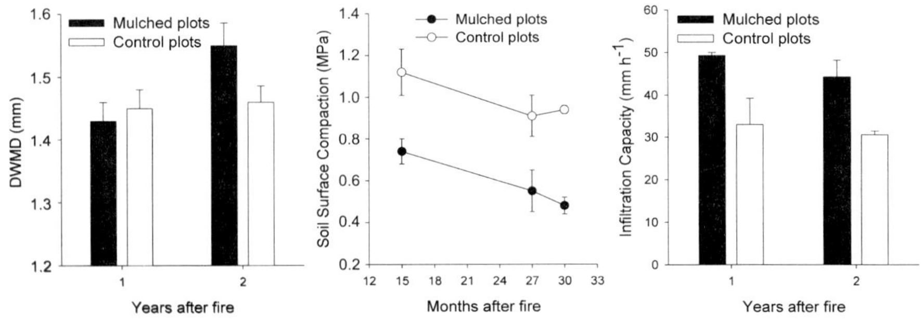

Fig. 9 Weighted mean size of surface soil (0 to 5 cm depth) aggregates (DWMD, mm), soil surface compaction (Mpa), and soil infiltration capacity (mm h^{-1}) for straw mulched and control (untreated) plots, one and two years after the 1993 Benidorm Fire in Southeast Spain.

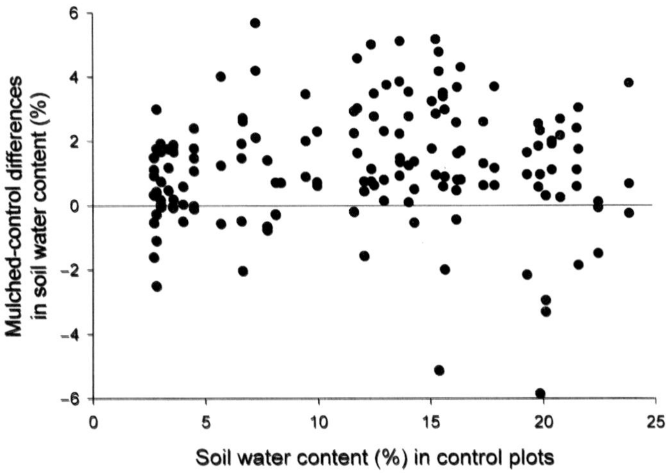

Fig. 10 Differences in soil water content (%) between straw mulched and control (untreated) plots. Data were recorded over a 14-month period after the 1993 Benidorm Fire, Southeast Spain.

activity, which in turn impacts soil runoff and erosion. Many of these changes may contribute to improve natural vegetation recovery and benefit seeded species. This is particularly true in arid areas and in the dry environments that result when vegetation is removed by wildfire. Thus, some studies have documented that post-fire mulching provided favorable conditions for plant recovery in severely burned areas (Bautista et al. 1996, Wagenbrenner et al. 2006). On the other hand, mulching as a weed control method is used in agriculture and agro-forestry (Gupta 1991); a thick layer of mulch can inhibit herb and shrub germination establishment by blocking light from reaching the soil surface and mechanically impeding seedling growth. This well-documented result of agricultural mulching, suggests that a potential side effect of post-fire mulching may adversely affect vegetation recovery and seeding effectiveness. Robichaud et al. (2000) reported that shrub seedlings were more abundant at the edge of mulch piles, where the mulch material was less than 2.5 cm deep. Thus, the thickness of the mulch layer is a key factor controlling positive and negative mulch effects on vegetation.

In post-fire rehabilitation research, the study of short- and long-term effects of mulch application on soil nutrient balance has received little attention. Decomposition of organic mulches release small amounts of nutrients and organic matter to the soil. However, organic mulches with a high carbon/nitrogen ratio will tie up nitrogen during the early stages of decomposition. The implications of these processes for post-fire ecosystem recovery are currently being studied (P.R. Robichaud, personal communication).

MULCH TREATMENT EFFECTIVENESS

Despite the increase in post-fire emergency rehabilitation activities in recent years, there are still few data to determine if these post-fire treatments are practical and effective. This is likely due to the high cost and effort required for direct measurement of runoff and erosion in remote areas where wildfires normally occur. Mulching is primarily intended to provide immediate and short-term erosion control. From the limited information available (MacDonald 1989, Robichaud et al. 2000), most data consistently show that mulches are highly effective in reducing post-fire erosion when a minimum ground cover of about 60 percent is applied. After a wildfire in southeastern Spain, straw mulch was applied at a rate of 2 Mg ha^{-1} to three 16-m^2 plots while 3 plots were left untreated. Over a 19-month period the mean sediment yield from the untreated plots was 1.1 Mg ha^{-1} versus 0.1 Mg ha^{-1} from the mulched plots (Bautista et al. 1996). After the Cerro Grande Fire in New Mexico, USA, the application of straw mulch with seed reduced sediment yields by 70 percent in the first year and 95 percent in the second year, although precipitation during the two study years was below normal (Dean 2001). Wagenbrenner et al. (2006) reported that straw mulch immediately increased the mean ground cover to nearly 80 percent and facilitated vegetative regrowth after the 2000 Bobcat Fire in Colorado, USA. Sediment yields from the mulched plots were less than 5 percent of the sediment yields from the control plots in 2001, 2002, and 2003. However, mulching did not reduce sediment yields in 2000 because a large amount of sediment was produced from a single 5 to 10-yr return interval storm. These results indicate that mulching may not be effective in reducing sediment yields from large, high-intensity storm events, and highlight the fact that effectiveness of any hillslope rehabilitation treatment will be dependent on the actual rainfall amounts and intensities, especially in the first years after the fire (Robichaud et al. 2000, Robichaud 2005). For example, after the 2003 Cedar Fire, in San Diego County, California, USA, aerial hydromulch reduced stream runoff and sediment yields during drier years, with less difference during the second post-fire year when numerous rain events occurred (Wohlgemuth et al. 2006).

Some mulch materials are more effective than others. On a burned site in Arizona, USA, Beyers et al. (2006) evaluated the effectiveness of wood chips, rice straw, and straw pellets containing flocculent, finding that wood chips provided the greatest total ground cover and greatest reduction in erosion over several post-fire years. Aerial hydromulching is relatively new in post-fire rehabilitation and effectiveness data are still scarce. However, monitoring of post-fire treatment effectiveness following recent wildfires in southwestern United States indicate that aerial hydromulch has limited effectiveness in reducing post-fire sediment yields. For example, after the aerial application of hydromulch at 2.5 Mg ha^{-1} on the 2002 Hayman Fire in Colorado, USA, measured erosion reduction was only 18 percent in the first post-fire year and 27 percent in the second post-fire year (P.R. Robichaud, unpublished data). Following the 2003 Cedar Fire in California, USA, hydromulch treatment was applied at two rates—100 percent coverage and 50 percent coverage in 30-m contour strips. Measured erosion reduction was low (maximum event sediment yield reduction of 53 percent), and monitoring results indicate that the amount of mulch coverage did not significantly affect runoff and sediment yields (Wohlgemuth et al. 2006). Hydromulch fibers form a smooth dense mat. However, the resistive force of the hydromulch materials against the shear force of concentrated runoff is low. Therefore, hydromulch is more effective on short slope lengths such as road cuts where concentrated flow is not as likely (Napper 2006). Residual canopy and standing trees may reduce treatment effectiveness by intercepting significant amounts of hydromulch during the application process, and reducing the amount of hydromulch on the soil surface.

Mulching effects on vegetation recovery are less known. The effect of hydromulch on native vegetation was monitored on the Cedar Fire in southern California, USA, and general vegetation recovery (percent cover) was not hindered by the hydromulch. However, it remains to be seen how species diversity and community structure will be affected in the long-term (Wohlgemuth et al. 2006). Beyers et al. (2006) reported that burned sites in Arizona and California (USA) treated with thick or decay-resistant mulches, such as rice straw and wood chips, had slower recovery of vegetation cover than sites treated with straw pellets and hydromulch. None of the mulches increased vegetation cover. Conversely, Bautista et al. (1996) and Wagenbrenner et al. (2006) reported that straw mulch slightly increased vegetation recovery.

Mulch is frequently applied to improve the germination of seeded grasses. Bautista et al. (1997, 2005) found that seeding without mulching did not significantly increase ground cover and protect the soil from erosion, while mulch combined with seeding produced the highest plant cover values. The mulch cover could enhance seeding success by increasing soil moisture content and by holding the seeds in place on the soil thereby avoiding the common loss of seeds through being washed down-slope. As the seeded species grow, the new vegetation helps hold the mulch onsite.

NEW APPROACHES TO POST-FIRE MULCHING

There is an increasing demand for new restoration and rehabilitation approaches that adhere to the principles of ecological restoration. This approach emphasizes application of treatments based on the use of native local species and materials, control of noxious weed invasions, and aims not only to protect the soil from erosion and degradation, but also to enhance ecosystem function and resilience. In post-fire rehabilitation, the ecological restoration prioritizes seeding treatments based on native-species seed mixes and mulching treatments based on local, site-specific forest materials (wood chips, pine needles, shredded debris from forest-clearing or post-fire logging, etc.). Following the ecological restoration approach, post-fire rehabilitation treatments (seeding, mulching, and seeding plus mulching) were applied on a Mediterranean pine forest that had been burned by a wildfire on November 2002 in Alicante, Spain. Mulching material was a heterogeneous mix of wood chips, small twigs, and pine needles made from shredded forest-clearing debris from nearby pine forests (Fig. 11). The seeding mix included perennial grasses, herbs, sub-shrubs and tall shrub native species. The treatments were

Fig. 11 Detailed view of the mulch forest materials (wood chips, small twigs and pine needles) mulch mix that was applied to treated plots after the 2002 Benifallim Fire in Alicante, Spain.

aimed to control soil degradation and enhance vegetation recovery. The effectiveness of the treatments as compared with control (untreated) areas was tested on 12 15-m^2 plots distributed on 3 different watersheds within the burned area and arranged following a randomized complete block design. Over a two year period, seeding-plus-mulching treatment enhanced plant cover; however, neither seeding nor mulch alone increased plant cover. The mulch layer greatly reduced post-fire soil loss (Fig. 12). During the first post-fire year, average soil loss on untreated plots was about 20 Mg ha^{-1}, whereas mulched sites showed negligible losses and even soil accumulation. Mulching also reduced soil surface compaction and enhanced water infiltration and soil microbiological activity. Seeding without mulching did not protect the soil surface from degradation. These results indicate that a mulch mixture of forest materials can effectively reduce erosion and protect soil functions while natural vegetation regenerates and/or seeded species establish. The mulch facilitated the establishment of the seeded species. The wide range of functional groups (grasses, herbs, shrubs) included in the seed mix is expected to contribute to the structural and functional recovery of the ecosystem.

Another innovation in post-fire mulching treatments is the use of synthetic products and technologies designed for stabilization and rehabilitation of disturbed areas, but not necessarily for post-fire environments. For example, during the last decade, polyacrylamide (PAM), a synthetic polymer that aids in aggregation of fine soil particles, has been widely used to reduce erosion and increase infiltration in irrigated agricultural soils and low-flow irrigation trenches (Sojka et al. 2007). More recently, PAM products have been introduced to hydromulch seed mixes to help bind soil particles. PAM has been used on road cuts and fills and on construction sites to stabilize soils and reduce erosion prior to seed

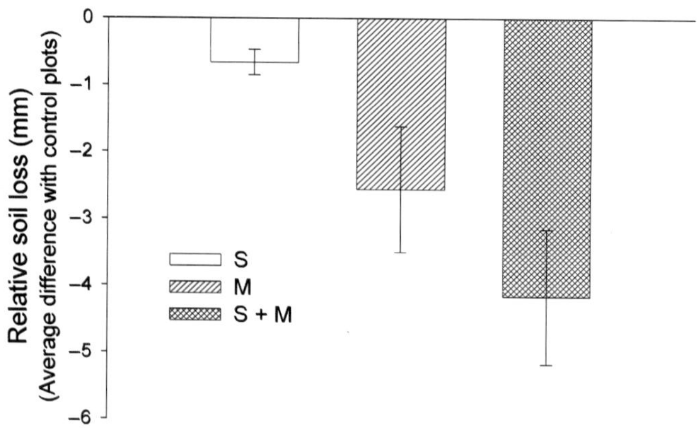

Fig. 12 Relative soil loss (mm) during the first year after treatment application as compared with soil loss in control (untreated) sites. S: seeding; M: mulch; S+M: seeding plus mulch. Mean values and standard error are indicated.

germination and plant establishment (Robichaud et al. 2003). The effectiveness of PAM as a postfire treatment is currently being evaluated. A single test using simulated rainfall on a severely burned plot in the northern Colorado Front Range found some initial erosion reduction that disappeared after a few rain events (L. MacDonald, personal communication).

CONCLUSIONS AND MANAGEMENT IMPLICATIONS

Post-fire treatment plans are generally a compromise between treatment costs and potential effectiveness and the potential damage to valued resources from unmitigated erosion. In some burned areas the 'no treatment' option may be the most appropriate response. This is particularly true for areas of low-burn severity, areas where natural mulch (needle cast) may provide the needed ground cover, and areas where rapid natural recovery is expected. However, once it is determined that post-fire treatments are necessary, mulching has several advantages over other treatment choices:

1. Mulch immediately increases ground cover on the target areas which makes mulch effective during the critical first post-fire year when erosion is likely to be the greatest;
2. Straw, wood shreds, and wood chip mulch have consistently been shown to be effective;
3. The risk of treatment failure is low and can be diminished by applying mulch on suitable areas and/or using measures that reduce potential mulch redistribution in windy sites;
4. Mulch materials are readily available in most areas; and
5. Mulch increases soil moisture retention which, in turn, improves biological and physical soil properties.

These advantages make dry mulching one of the most effective post-fire rehabilitation treatments available and it is increasingly being used as aerial application expands the potential treatment areas. Suitable sites for post-fire mulching include areas with downstream values at risk, high soil erosion potential, low to moderate slope angles, and unlikely to be exposed to strong winds. The limited success and high costs of aerial hydromulching makes it less cost-effective than aerial dry mulch.

The main disadvantage of mulching is the potential introduction of non-native plants (invasive weeds) with straw mulch. This risk is reduced by careful selection of the agricultural straw to be used for mulching, such as certified weed-free straw or a straw whose seed (and associated weed seeds) are not likely to thrive in the burned area (e.g., rice straw mulch used on burned upland forest hillslopes). The contribution of mulch to the spread of invasive weeds is nearly eliminated by using shredded or manufactured forest materials. Thick mulches can suppress natural recovery of fire-tolerant and fire-dependent plant species as well as the germination and the establishment of seeded species. This is an advantage when mulches suppress invasive weeds and undesirable plants that come in and spread after fires; however, it is a significant disadvantage when mulches inhibit natural recovery or endangered plant species. Mulch type and target thickness should be adapted to the particular site conditions and needs.

Like all post-fire treatments, mulch treatment performance should be monitored to determine the effectiveness of various mulches and application techniques in a range of ecosystems. Both short- and long-term ecological effects of post-fire mulching need to be evaluated and compared to the natural post-fire recovery process. These monitoring data provide the scientific information needed to improve future post-fire treatment decisions.

References

Bautista, S., J. Bellot, and V.R. Vallejo. 1996. Mulching treatment for post-fire soil conservation in a semiarid ecosystem. *Arid Soil Research and Rehabilitation,* 10: 235-242.

Bautista, S., N. Abad, J. Llovet, C. Bladé, A. Ferran, J.M. Ponce, R.N. Caturla, J.A. Alloza, J. Bellot, and V.R. Vallejo. 1997. Siembra de herbáceas y aplicación de mulch para la conservación de suelos afectados por incendios forestales (Spanish). pp. 395-434. *In* R. Vallejo [ed.]. La restauración de la cubierta vegetal en la Comunidad Valenciana. Fundación CEAM-Generalitat Valenciana, Valencia, Spain.

Bautista, S., C. Bladé, T. Gimeno, H. Kribeche, and V.R. Vallejo. 2005. Native species seeding and chopped-wood mulching as post-fire rehabilitation treatments in Mediterranean forests in SE Spain, p. 34. *In* Ecological Restoration: A Global Challenge, 17th Conference of the Society for Ecological Restoration International, Society for Ecological Restoration International, Tucson, Arizona, USA.

Benavides-Solorio, J.D. and L.H. MacDonald. 2001. Post-fire runoff and erosion from simulated rainfall on small plots, Colorado Front Range. *Hydrological Processes,* 15: 2931-2952. DOI: 10.1002/hyp.383

Benavides-Solorio, J.D. and L.H. MacDonald. 2002. Erratum for 'Post-fire runoff and erosion from simulated rainfall on small plots, Colorado Front Range'. *Hydrological Processes,* 16: 1131-1133. DOI: 10.1002/hyp.5017

Beyers, J., W. Christensen, P. Wohlgemuth, and K. Hubbert. 2006. Effects of postfire mulch treatments on vegetation recovery in the southwestern US. (poster abstract). *Presented at* The 91st annual meeting of the Ecological Society of America, Ecological Society of America, Washington, D.C., USA. *Online at* http://abstracts.co.allenpress.com/pweb/esa2006/document/62958 [accessed 11 Sep 2008].

Dean, A.E. 2001. Evaluating effectiveness of watershed conservation treatments applied after the Cerro Grande Fire, Los Alamos, New Mexico. M.S. Thesis, University of Arizona, Tuczon, Arizona, USA.

Elwell, H.A. and M.A. Stocking. 1976. Vegetal cover to estimate soil erosion hazard in Rhodesia. *Geoderma,* 15: 61-70.

Goldman, S.J., K. Jackson, and T.A. Bursztynsky. 1996. *Erosion and Sediment Control Handbook*. McGraw-Hill, San Francisco, California, USA.

Groenier, J.S. and C. Showers. 2004. Shredding small trees to create mulch for erosion control. Tech. Tip 0471-2335-MTDC. U.S. Department of Agriculture, Forest Service, Missoula Technology and Development Center, Missoula, Montana, USA.

Gross E., I. Steinblums, C. Ralston, and H. Jubas. 1989. Emergency Watershed treatments on burned land in southwest Oregon. pp. 109-114. *In* Proceedings of the symposium on Fire and Watershed Management. General Technical Report PSW-GTR-109. U.S. Department of Agriculture, Pacific Southwest Research Station, Berkeley, California, USA.

Gupta, G.N. 1991. Effects of mulching and fertilizer application on initial development of some tree species. *Forest Ecology and Management*, 44: 211-221.

Harding, M.V. 1990. Erosion control effectiveness: comparative studies of alternative mulching techniques. pp. 149-156. *In Environmental Restoration: Sciences and Strategies for Restoring the Earth*. Island Press, Covelo, California, USA.

Kay, B.L. 1983. Straw as an erosion control mulch. Agronomy Progress Report No. 140. U.S. Department of Agriculture, Agricultural Experiment Station, Cooperative Extension, University of California, Davis, California, USA.

Kruse, R., E. Bend, and P. Bierzychudek. 2004. Native plant regeneration and introduction of non-natives following post-fire rehabilitation with straw mulch and barley seeding. *Forest Ecology and Management*, 196: 299-310.

MacDonald L.H. 1989. Rehabilitation and recovery following wildfires: a synthesis. pp. 141-144. *In* Proceedings of the Symposium on Fire and Watershed Management. General Technical Report PSW-GTR-109. U.S. Department of Agriculture, Forest Service, Pacific Southwest Research Station, Berkeley, California, USA.

Meyer L.D., W.H. Wischmeier, and G.R. Foster. 1970. Mulch rates required for erosion control on steep slopes. *Soil Science Society of America Proceedings* 34: 928-931.

Miles, S.R. 2005. Listing of BAER Treatments: cost, risk, effectiveness, production rates and applications. Burned Area Emergency Response Lesson. U.S. Department of Agriculture, Forest Service, Burned Area Emergency Response Program. *Online at* http://www.fs.fed.us/r5/baer/4g.html, [accessed 11 Sep 2008]

Miles, S.R., D.M. Haskins, and D.W. Ranken. 1989. Emergency burn rehabilitation; cost, risk, and effectiveness. pp. 97-102. *In* N.H. Berg [tech. coord.]. Proceedings of the symposium on Fire and Watershed Management. General Technical Report PSW-109. U.S. Department of Agriculture, Forest Service, Pacific Southwest Research Station, Berkeley, California, USA.

Napper, C. 2006. The Burned Area Emergency Response Treatment Catalog (BAERCAT). Technical Report. 0625 1801-SDTDC. U.S. Department of Agriculture, Forest Service, National Technology and Development Program, Watershed, Soil, Air Management, Washington D.C. USA. *Online at* http://www.fs.fed.us/eng/pubs/pdf/BAERCAT/lo_res/06251801L.pdf [accessed 11 Sep 2008]

Pannkuk, C.D. and P.R. Robichaud. 2003. Effectiveness of needle cast at reducing erosion after forest fires. *Water Resources Research*, 39(12): 1333-1342. DOI: 10.1029/2003WR002318

Richardson, S. 2002. BAER Accomplishment Report as of 10/23/02. Rodeo-Chediski fire salvage project record. Vol. II. U.S. Department of Agriculture, Forest Service, Apache and Sitgreaves National Forests, Springerville, Arizona, USA. *Online at* http://www.fs.fed.us/r3/asnf/salvage/pdfs/023-200210230baer_accomp.pdf [accessed 11 Sep 2008].

Robichaud, P.R. 2005. Measurement of post-fire hillslope erosion to evaluate and model rehabilitation treatment effectiveness and recovery. *International Journal of Wildland Fire*, 14: 475-485. DOI: 10.1071/WF05031

Robichaud, P.R. and T.M. Monroe. 1997. Spatially-varied erosion modeling using WEPP for timber harvested and burned hillslopes, Paper No. 97-5015. *Presented at* The 1997 ASAE Annual International Meeting, American Society of Agricultural Engineers, St. Joseph, Michigan, USA.

Robichaud, P.R., J.L. Beyers, and D.G. Neary. 2000. Evaluating the effectiveness of post-fire rehabilitation treatments. General Technical Report RMRS-GTR-63. U.S. Department of Agriculture, Forest Service, Rocky Mountain Research Station, Fort Collins, Colorado, USA.

Robichaud, P.R., L.H. MacDonald, J. Freeouf, D. Neary, D. Martin, and L. Ashmun. 2003. Post-fire rehabilitation of the Hayman Fire. pp. 293-313. *In* R. Graham [tech. ed.]. Hayman Fire Case Study. General Technical Report RMRS-GTR-114. U.S. Department of Agriculture, Forest Service, Rocky Mountain Research Station, Fort Collins, Colorado, USA.

Sojka, R.E., D.L. Bjorneberg, J.A. Entry, R.D. Lentz, and W.J. Orts. 2007. Polyacrylamide in agriculture and environmental land management. *Advances in Agronomy*, 92: 75-162.

Santi, P., J. Higgins, S. Cannon, and J. DeGraff. 2006. Evaluation of post-wildfire debris flow mitigation methods and development of decision-support tools: Final Report to the Joint Fire Science Program (JFSP). U.S. Department of Agriculture, Forest Service and U.S. Department of Interior Joint Fire Science Program, Boise, Idaho, USA. *Online at* http://jfsp.nifc.gov/projects/03-1-4-14/03-1-4-14_final_report.pdf [accessed 11 Sep 2008].

Wagenbrenner, J.W., L.H. MacDonald, and D. Rough. 2006. Effectiveness of three post-fire rehabilitation treatments in the Colorado Front Range. *Hydrological Processes*, 20: 2989-3006. DOI:10.1002/hyp.6146

Wohlgemuth, P., P.R. Robichaud, and J.L. Beyers. 2006. The effects of aerial hydromulch as a post-fire erosion control treatment on the Capitan Grande Reservation. Unpublished report submitted to Viejas Band of Mission Indians. *Available from* P. Wohlgemuth, U.S. Department of Agriculture, Forest Service, Pacific Southwest Research Station, Riverside, California, USA.

14

Long-term Restoration Strategies and Techniques

V.R. Vallejo[1*], *I. Serrasolses*[2], *J.A. Alloza*[1], *M.J. Baeza*[1], *C. Bladé*[1], *E. Chirino*[1], *B. Duguy*[1], *D. Fuentes*[1], *J.G. Pausas*[1], *A. Valdecantos*[1] and *A. Vilagrosa*[1]

Abstract

Long-term post-fire restoration not only aims to restore the ecosystem structure and function, but also endeavors to recover ecosystem fire resilience and reduce future fire propogation potential. This generally requires restoration strategies that promote secondary succession towards more mature, more resilient plant communities at a landscape scale. Pre-fire planning is essential to prioritize vulnerable sites and develop plans for these areas. Fire behavior models are often used for this process.

Current restoration techniques (plant species selection, seeding of woody plants, development of quality nursery stock, site preparation, soil amendment and fertilization, etc.) typical of the semi-arid Mediterranean environment are described with recent study results providing examples. However, the usefulness of these techniques is proven in the field where the complex interaction of long-term climate, short-term weather events, introduced plants, soil properties, extant organisms, etc. make each restoration project unique. Given the uncertainties of environmental conditions and the myriad of interactions, adaptive management principles should be applied to long-term post-fire restoration.

INTRODUCTION

In general, long-term forest fire impacts requiring restoration actions are caused by: a) wildfires affecting fire-sensitive ecosystems in regions where natural fires are uncommon; b) unprecedented fire frequency or severity (i.e., altered fire regime) over fire-dependent ecosystems; c) unprecedented combination of fire regime and other disturbances over fire-dependent ecosystems. For example, Mediterranean ecosystems can be considered fire-

[1] *Centro de Estudios Ambientales del Mediterráneo (CEAM) Foundation, Paterna, Valencia, Spain.*
[2] *Unitat d'Ecologia, Facultad de Biosciences, Universitat Autònoma de Barcelona, Barcelona, Spain.*
* *Corresponding author: V.R. Vallejo, CEAM Foundation, Parque Tecnológico, Ch. Darwin 14, 46980 Paterna, Spain. Tel: + 34 96 131 8227, Fax: +34 96 131 8190, e-mail: vvallejo@ub.edu*

dependent/influenced as they have evolved under fire influence (Pausas and Verdú 2005, Pausas et al. 2006), and Mediterranean plants show a large array of adaptations to cope with fire impact (Pausas et al. 2004a). However, during the last decades, fire regimes have been deeply altered (Pausas and Vallejo 1999). This fact, in combination with other long-term anthropic disturbances, may cause further fire-induced degradation beyond the resilience domain of Mediterranean ecosystems.

The Mediterranean basin has been subjected to extensive and intensive exploitation for millennia (Vallejo et al. 2006). In many instances this exploitation has been excessive and resulted in land degradation. As far back as 2500 years BP, Plato complained about the degradation of upland forests and dramatic soil losses (Yassoglou 2000). As a consequence of this long-term human impact, most of the Mediterranean basin is now regarded as 'degraded' (TNC 2007), whereas most of the other Mediterranean-climate regions of the world have suffered less degradation. Therefore, fire impacts on ecosystems should be analyzed in terms of the interactions between direct fire-induced processes and previous human-induced degradation processes. And post-fire rehabilitation should include a long-term perspective on recuperating ecosystem integrity according to ecological restoration concepts (van Andel and Grootjans 2006). In addition, as fire hazards are inherent in the Mediterranean and other world ecosystems, fire prevention principles should be incorporated into post-fire rehabilitation strategies to reduce the number of future fire events.

Post-fire regeneration in fire-dependent ecosystems usually follows the autosuccession process, in which the same plant species composition and relative abundance regenerates after a fire (Trabaud 1994). However, this model does not always occur. There are several woody species that do not regenerate either after a single fire (Riera and Castell 1997, Retana et al. 2002) or after short fire intervals (e.g., *Pinus halepensis* and *P. pinaster*; Vallejo and Alloza 1998). In addition, post-fire weather conditions and/or seed bank exhaustion can drastically affect obligate seeder species regeneration (Faraco 1998, Baeza 2004).

This chapter will present the rationale for long-term post-fire restoration strategies and describe the techniques used. As most of the research in this area has been conducted in Mediterranean-climate and other dry regions of the world, and the Mediterranean basin in particular, these regions are the focus of the chapter.

POST-FIRE ECOSYSTEM RESILIENCE

Mature Mediterranean ecosystems are often dominated by shrub and tree species that have the ability to resprout after fire (and also after cutting and animal browsing). Resprouting plants quickly regenerate plant cover from below ground reserves, even in summer, quite independently of rainfall events (Vallejo and Alloza 1998). This trait of an ecosystem shows high resilience to wildfires (Ferran et al. 1992) as most of the pre-fire species reappear in similar density soon after a fire and soil protection is achieved rapidly – reducing the risk of increased runoff, soil erosion, and degradation (Abad et al. 1996).

Historically, Mediterranean ecosystems have been degraded by burning, crop abandonment, overgrazing, wood gathering, and charcoal production (which often involved uprooting the largest shrubs and trees), and these disturbances have been combined in multiple space and time sequences (Vallejo et al. 2006). However, in European Mediterranean countries, these practices have been strongly reduced in the last three to four decades through a generalized process of tertiarisation of rural economies. This is likely to occur in the near future in the Mediterranean countries of Africa. Thus, a generalized process of land abandonment has taken place since the 1960s in Europe, and is continuing under the Common Agricultural Policy.

Abandoned lands are colonized by opportunistic species, which in the early stages are mostly obligate seeders (Gallego et al. 2004). These species have short life cycles and many of them generate an abundant and persistent seed bank. Woody seeders are often strong fuel accumulators; they lead to high fuel load accumulation and thus to fire-prone shrublands (Baeza et al. 2006). In fact, the dramatic expansion of large wildfires in European Mediterranean countries has been partly attributed to the extensive land abandonment occurring in the region (Vallejo and Alloza 1998). Wildfires affecting fire-prone shrublands that have colonized old fields often enter into short-interval fire cycles that stop any further secondary succession towards more mature ecosystems. Even without fire, some early stages of secondary succession are stable and inhibit late-successional species colonization (Debussche et al. 1996). Many opportunistic shrubs have the ability to colonize both old fields and burned ecosystems (Baeza and Vallejo 2006). In the short-term, ecosystems dominated by obligate seeders regenerate slowly after fire, thus leaving bare soil exposed to wind and water erosion for relatively long periods of time (Vallejo 1999). This may result in irreversible soil degradation/loss at the ecological scale and enhance the long-term, whole-ecosystem degradation that had started prior to the fire. Therefore, land abandonment promotes short-term fire cycles that result in ecosystem degradation loops. Recovering ecosystem resilience would thus require breaking these loops and promoting secondary succession towards more mature, more resilient plant communities (Vallejo and Alloza 1998).

Woody resprouters often produce big seeds, with dispersion mediated by animals (Pausas et al. 2004a). Some of these seeds show high water requirements for germination (Montoya 1993), and require the presence of dispersors (Alcantara et al. 1997). Their often fleshy and highly nutritional fruits are very attractive to predators (Waller 1993). All these circumstances place serious constraints on the ability of these species to colonize degraded sites (Laguna and Reyna 1990). Therefore, where natural colonization of late-successional woody resprouters is not sufficient, artificial introduction through seeding or plantation may be required to improve ecosystem resilience (Vallejo et al. 2006).

PLANNING POST-FIRE RESTORATION

Identification of Vulnerable Ecosystems

For fire-prone areas, preparedness must be incorporated into forest management planning. Post-fire rehabilitation and restoration measures require pre-fire planning to prioritize vulnerable sites and timely post-fire implementation of restoration actions. Assuming that the first objective in post-fire rehabilitation is the mitigation of runoff, flash floods, and soil erosion, early post-fire interventions have to be concentrated at the most vulnerable sites. These can be identified in a given area by using erosion models, basic cartography, and GIS (see Alloza and Vallejo 2006). When a fire occurs, emergency seeding and other techniques can be applied (see Chapters 10, 11, 12 and 13 in this book). Although early interventions may not be part of a long-term perspective on post-fire restoration, it is important to avoid early interventions that may work against the long-term plan. For example, if early interventions introduce alien herbaceous species, this might hinder the normal progression of recovery through secondary succession (see examples in Robichaud et al. 2000).

Longer-term restoration is appropriate when the existing vegetation shows low resilience to forest fires, loss of key forest species has occurred, and/or regeneration of fire-prone formations is likely to occur (Fig. 1). These vulnerable plant formations can be identified from vegetation cartography and/or forest inventories (Alloza and Vallejo 2006).

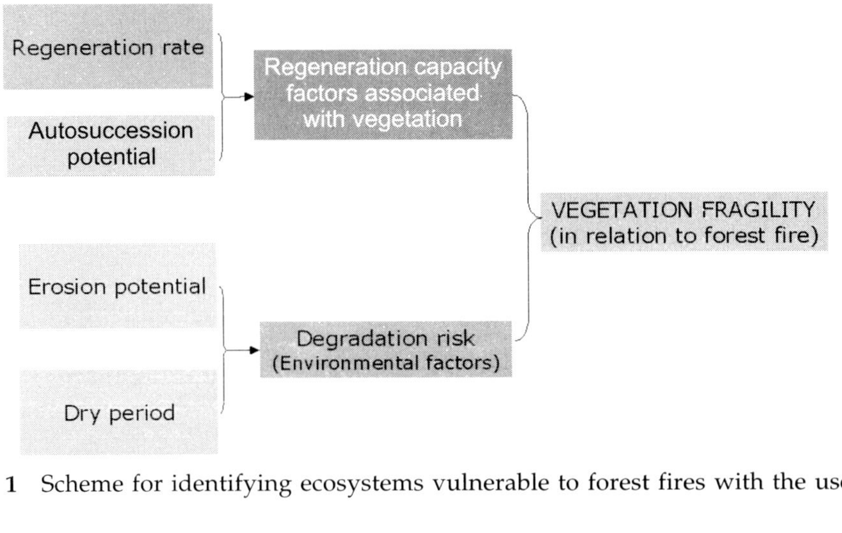

Fig. 1 Scheme for identifying ecosystems vulnerable to forest fires with the use of GIS.

Planning Restoration at the Landscape Scale

Planning restoration in fire-prone regions at the landscape scale should aim at reducing landscape combustibility. Traditionally, it is believed that disturbances are more likely to spread across a homogeneous area (Wiens et al. 1985), but the opposite also occurs (Turner 1987). It has been proposed that in highly fragmented landscapes disturbances require a higher boundary-crossing frequency and a more convoluted route and, therefore, spread less easily (Turner and Romme 1994, Forman 1995). In the case of fires, it is generally accepted that greater landscape heterogeneity retards fire propagation (Minnich 1983, Wiens et al. 1985, Knight 1987), although landscape pattern may have little influence on crown fire behavior when burning conditions are extreme (Turner et al. 1994, Keeley et al. 1999). No universal correlation has been found between fire propagation rate and landscape heterogeneity (Morvan et al. 1995). Landscape-scale fire patterns are the result of complex interactions among topography, weather and vegetation (fuel type, moisture, quantity, and spatial distribution) (Turner and Romme 1994, Hargrove et al. 2000). The topographic and physiographic features of the landscape influence the local probabilities of initial ignition and burning patterns, while the spatial arrangement of fuel categories also influences initial ignition as well as fire growth and behavior.

Large increases in fire occurrence were experienced in many of the Mediterranean areas in the 1970s due to land-use changes. Since the beginning of the 20th century, intensive land abandonment and decrease in grazing activities have generally resulted in increased fuel loads and expansion of large, interconnected non-wooded patches (Duguy 2003) throughout the ecosystem. The landscapes became highly fire-prone and the risk of large fires increased. The large number of fires that did occur generally caused further homogenization of these landscapes (Debussche et al. 1987, Vos 1993, Vázquez and Moreno 1998).

To reduce both fire occurrence and fire spread while promoting the expansion of a forest in the landscape, Forman and Collinge (1996) proposed three main approaches which focused on landscape pattern:

1) minimize the sites that are especially susceptible to fire ignition;
2) increase landscape spatial heterogeneity; and
3) increase barriers or filters that inhibit fire spread.

Using the FARSITE model (Finney 1998) for fire simulation, we determine fuel model distributions and fire-break networks that would reduce fire risk at the landscape level, and hence provided guidance for forest restoration aimed at fire prevention (Fig. 2; Duguy et al. 2005). FARSITE simulations showed that large interconnected patches of heavy surface fuels (mature dense shrublands) favored fast and intense fires. The fragmentation of this highly fire-prone matrix through the introduction of dense woodlands (i.e., the creation of a more fine-grained landscape *sensu*; Forman 1995) was very

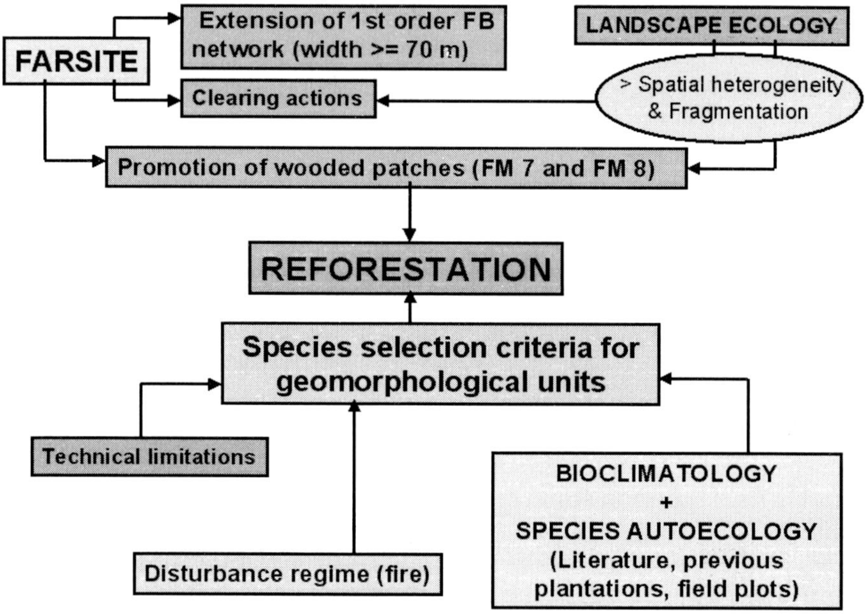

Fig. 2 Outline of the planning procedure for managing and restoring the landscape for fire prevention and landscape functional quality enhancement: fuel breaks (FB) design, fuel clearing treatments in forest to promote fuel models (FM) 8 and 9 (low combustibility) and reforestation.

effective in reducing the fire size and, in most cases, in reducing burning conditions (rate of spread, fireline intensity). Other effective landscape-level fuel alterations were the introduction of forest corridors between woodlands and the promotion of complex patches (high perimeter/area ratios or high fractal dimensions) among wooded patches. As these latter patches are potential sources for colonization processes, all actions that increase the edge length between wooded patches and non-wooded patches favor forest expansion. Surface fuel reduction actions applied over large areas (e.g., extensive clearing actions) were also an effective way of controlling fire spread, limiting fireline intensity, and lowering potential fire-caused damages (Byram 1959, Ryan and Noste 1985). Fuel reduction on fire-prone shrublands dominated by seeder species can be conducted in conjunction with plantations of woody resprouters (Baeza et al. 2005) to achieve the double benefit of reducing fire hazard and improving ecosystem resilience and diversity (see the section, *Plant Species Selection,* below). Our results also showed that similar degrees of fragmentation might lead to different fire sizes and fireline intensities, depending on the precise spatial arrangement of the various woodland successional stages.

It appears that a certain degree of heterogeneity and fragmentation of the vegetation structural diversity provides resistance to fire spread (Agee et al.

2000). It also provides a wider range of environmental resources and conditions, thereby promoting higher biodiversity in the landscape. Nevertheless, further research is still needed to identify the relationships between fuel models (used to determine fire growth and behavior and associated critical values of target landscape structures) and sustainable landscape management strategies. Coupling firebreak networks with appropriate landscape-level fuel treatments also seems to be a good strategy for limiting the occurrence of large, high-intensity fires, and thereby, reducing the associated negative effects on the ecosystems. However, fire management at the landscape scale may be very expensive, and a cost-benefit analysis would be needed, especially in areas where strong winds may reduce their effectiveness.

PLANT SPECIES SELECTION

In the last decades, species selection criteria in reforestation plans have been immersed in the native/exotic and conifer/hardwood discussion. The use of native flora is a priority in conservation-based reforestation (FAO 1989), but it has frequently been hampered by the limited success of seedling establishment. The success of exotic species may be related in many cases to the absence of specific pests, biogeographical isolation and, especially, to their early-successional features, although there is often a risk of either lack of adaptation or its opposite: extreme aggressiveness. Some of the native trees proposed for reforestation are late-successional, which makes them more sensitive to biotic and abiotic conditions (Hughes and Styles 1987, Zobel et al. 1987). This is the case of native tall shrubs and broad leaved resprouters. It is generally assumed that late successional species have low survival possibilities when introduced in open or degraded lands, although this is not always so (Ashby 1987). In the case of Mediterranean forests, attempts to introduce *Quercus* spp. seedlings commonly faced high mortality rates, making this alternative very expensive (Mesón and Montoya 1993). In addition, for historical and biological reasons, the techniques for introducing late-successional native species are poorly developed (Zobel et al. 1987), and up to a few years ago the scarcity of native plant material available from nurseries was a barrier to diversifying restoration practices. The forestry tradition uses conifers as pioneers to restore degraded lands, and after some years of silvicultural treatments, hardwoods are introduced under the pine canopy in an improved soil and microclimate conditions (see, for example, Montero and Alcanda 1993). Nevertheless, young pine plantations are very vulnerable to fire and the use of pines alone in reforestations is especially risky in wildfire 'hot spots'. Recent advances in the ecophysiology of hardwoods offer much-improved seedling plantation results on open degraded lands in the Mediterranean (Baeza et al. 2005).

In the context of ecological restoration, plant species selection for post-fire recovery of the ecosystem (i.e., structure and function), reduction of future fire risk, and improving fire resilience involves the following considerations (Fig. 3; Vallejo and Alloza 2004):

- The first step is to determine what native species are suitable for restoring the habitat. This is not straightforward in extremely degraded ecosystems. Often, remnants of the (supposedly) original vegetation are used as a reference after phytosociology investigations. This should be regarded, however, as a broad indication only. The presence of a species close to the site to be restored, under similar physiographic conditions, is a more reasonable indicator of species compatibility with the habitat since soil degradation may have made the habitat unsuitable for the reference species. More direct information is provided by auto-ecological studies, but these are still scarce for many native species of potential interest. This

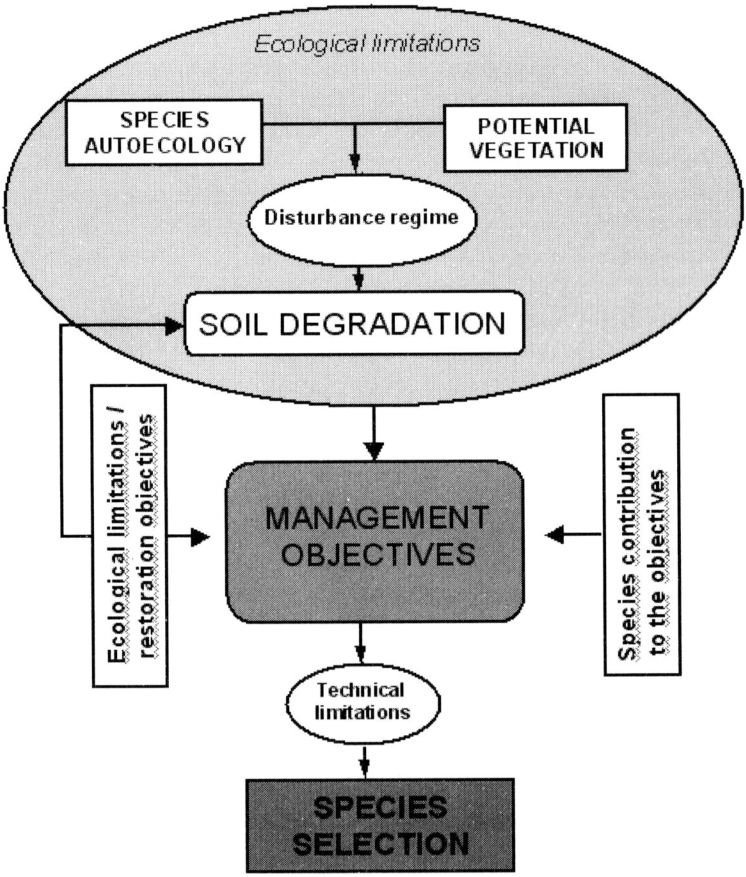

Fig. 3 Species selection constraints and criteria for afforestation.

area deserves further research. The plant species selected have to be adapted not only to the climate (including extreme events) and soil conditions, but also to the prevailing perturbation regime (e.g., wildfires and pests). For example, pine species were routinely used in afforestation actions in the past, both in the Mediterranean and in many other regions of the world. Most pines do not survive fire intervals that are shorter than the time period required to produce enough seeds, which is around 15 to 20 years in the Mediterranean (Pausas et al. 2004b).

- From the set of species found to be suitable for restoring a given habitat, the species that best fit the management objectives should be selected. In the case of post-fire restoration we would select woody resprouters according to the above-stated objectives of increasing fire ecosystem resilience and reducing fire risk. Resistance to fire, defined at the individual species level, should be related to species flammability, which is determined by plant structure (fuel density and size), necromass proportion, moisture content, and the presence of components that enhance or diminish flammability (volatile organics, resins). At the community level, resistance should be related to the combustibility of the ecosystem, including species composition, structure of the stand and characteristics of the litter bed. For example, in the Mediterranean, *Ulex parviflorus* is considered highly flammable, especially in mature stands that accumulate a lot of standing dead fuel; *Quercus coccifera, Erica multiflora, Rhamnus lycioides* and *Juniperus oxycedrus* are considered to show medium flammability; and *Pistacia lentiscus* and *Rhamnus alaternus* show low flammability (Elvira and Hernando 1989). Considering fuel loading, especially fine and dead fuel, and surface/volume ratio, Papió and Trabaud (1991) found that *Pistacia lentiscus* presented low fire hazard, whereas *Genista scorpius* (with a similar structure as *Ulex parviflorus*) presented high fire hazard. Trabaud (1976) emphasized the role of the litter layer in the combustibility of Aleppo pine (*Pinus halepensis*) forests, more than the flammability of the pine species itself. Other objectives might be considered, such as improving soil fertility through introducing N-fixers (Binkley and Giardina 1998) or enhancing carbon sequestration (Lal 1999).
- Restoration usually consists of introducing one or several keystone species. These species, typically trees or tall shrubs, are supposed to play a critical role in determining ecosystem structure and functioning, acting as 'ecosystem engineers' (Jones et al. 1994) that modify the habitat. It is assumed that these species will improve soil properties, create a forest floor habitat, improve the microclimate, and indirectly facilitate the importation of seeds by birds. Finally, the introduction of a woody species could not be enough for its complete establishment if symbionts, pollinators, or dispersers are lacking (Hobbs and Norton 1996). Mycorrhiza and/or rhizobateria inoculation in the nursery is a way to

ensure efficient symbiosis for seedlings to be introduced (Barea and Honrubia 2004).
- Technical constraints may impede the introduction of a specific species in a restoration project. Adequate technical knowledge of species cultivation requirements and plantation techniques are essential for successful introduction. Species growing in the same habitat may show contrasting growth and physiological strategies (Vilagrosa et al. 2003a, Vilagrosa et al. 2005), and hence may require different cultivation techniques in the nursery. In addition, this basic ecophysiological knowledge is very limited for many of the most promising species for restoration worldwide and especially in tropical regions. The main environmental limitation for a successful introduction of plants on degraded Mediterranean sites is water stress, and this is, of course, also applicable to other arid regions of the world. In Mediterranean regions, the most critical situations are located in the transition between semi-arid and dry sub humid climates, where high water stress is combined with high disturbances, especially fire.
- Finally, cost constraints always limit the practice of restoration and its innovation.

SEEDING WOODY SPECIES

For afforestation, seeding may offer many advantages over planting, especially in time and cost savings. It has also been suggested that seeding is easier to mechanize and reduces the risk of root deformation. Seedlings developed directly on site are expected to acclimatize better to the site conditions from the early plant development phases. However, the unreliability of this direct seeding method, which yields inconsistent seedling emergence and survival and growth rates is the main reason for its limited use (Winsa and Bergstern 1994).

In general, seeding techniques include: a) row seeding (sowing seeds in strips across an area); b) spot seeding (dropping a number of seeds on a small spot to ensure the emergence of at least one seedling in each spot); and c) broadcast seeding (scattering seeds over the entire reforestation area; Barnett and Baker 1991). The first two techniques include soil preparation and seed covering with a thin soil layer that facilitates seed germination by keeping soil moisture around the seed. In addition, covering the seeds with soil may reduce seed predation because it makes them undetectable for visually-searching seed-predators and may also limit detectability for predators that use olfaction to find seeds (Nilson and Kjältén 2003). Broadcast seeding offers fewer guarantees for germination, unless a mulch layer is applied after seeding; it does represent, however, the most rapid and inexpensive seeding technique. Other specific techniques have been developed including micro-site preparation (Bergstern 1988, Wennström et al. 1999), which consists of sowing seeds in small inverted pyramidal indentations that increase seed moisture by improving soil-seed contact and reducing evaporation.

Aerial seeding, a type of broadcast seeding, has been used since the 1950s and involves the dropping of seeds from helicopters or fixed wing aircraft. The most important advantage of aerial seeding is its potential to seed remote areas with limited access and to treat large areas in a short time and at a low cost. On a calm day under optimum conditions, a helicopter can cover up to 1200 ha; however, the usual daily average is about 600 to 800 ha (Barnett and Baker 1991). Aerial seeding may be appropriate in circumstances where the previous vegetation has been removed, such as after a fire, extensive logging, or in reclamation of a mine site. It has been widely used for emergency seeding after fire (see Chapter 11 on non-native and native seeding in this book) in western USA, and as a supplement to natural regeneration after logging in northern Europe, USA, and Canada. Some attempts at pine forest restoration after fire using aerial pine seeding have been made in eastern Spain (Peman and Navarro 1998); however, only the results of a single experiment conducted in the Sierra del Garraf (Barcelona) have been published to date (Castell and Castelló 1996). In this study, 2 kg ha^{-1} of *Pinus halepensis* seeds mixed with 18 kg ha^{-1} of inert wheat (added to ensure better pine seed distribution and to provide predator satiation) were sowed. A relatively successful average pine germination of 5 percent, representing an overall density of 6000 seedlings per hectare, was reported. Nevertheless, the results were highly variable depending on the terrain (from 12,000 seedlings per hectare in old fields to seven-times-lower densities on hillslopes and at the bottom of valleys). These reasonably good results were probably related to mild temperatures and with abundant rainfall that occurred just after seeding.

Seed predation is one of the major causes of direct seeding failure. After a recent fire in the Valencia region, seed predation was assessed in a *Pinus halepensis* aerial seeding project completed in late November. In four of the six monitored plots, more than 80 percent of the pine seeds were predated in two months, and at the end of spring no germinated pine was found (Pausas et al. 2004c). Therefore seeding success may depend, at least in part, on reducing the seed predation.

Correct timing of seeding and appropriate use of seed pre-treatments to overcome seed dormancy may be critical for taking advantage of the most favorable conditions for germination and thus ensuring rapid germination. In a study to determine the effects of seed priming on subsequent *Pinus halepensis* seed germination in the field, seeds were primed for 6 days at 20°C in sand moistened with a 10^{-3} M gibberelline solution. The primed seeds germinated earlier than the control seeds, and in some cases, the prime seeds had higher germination rates (Fig. 4). The effectiveness of seed priming may vary with climatic conditions. In this experiment, seed priming was most effective in suboptimal temperature conditions (i.e., late autumn). Seed pre-treatment may be useful to ensure rapid germination when seeding is conducted in autumn.

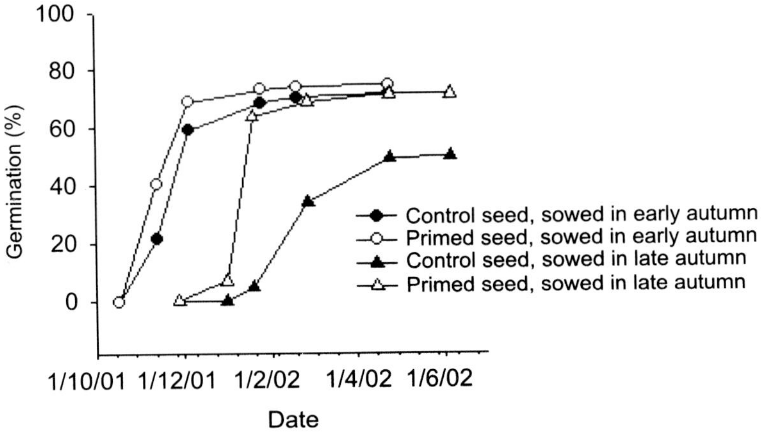

Fig. 4 Cumulative percent germination of primed and control seeds of *Pinus halepensis* sowed in early or late autumn. Experimental seeding was conducted in Alicante (eastern Spain), using pots placed outdoors with moderate watering.

PLANTATIONS

Plant Quality: Nursery Cultivation

Seedling plantations on drylands and degraded soils are often discouraging because of high mortality rates and poor growth. In general, climatic conditions after planting are one of the major limiting factors for seedling establishment. Suitable restoration techniques may help the seedlings to get through the transplant shock and first summer drought, and establish successfully. These include several nursery techniques that take into account the morpho-functional characteristics of seedlings to promote their resistance to drought conditions and increase their acclimation to the reforestation site. The main technical elements in the nursery culture are:

- Substrates or growing media.
- Containers
- Drought preconditioning
- Fertilization.

Substrates or growing media

The characteristics of the growing media are important for good root development—a key step in the success of a plantation (Peñuelas and Ocaña 1996). The recommended growing media includes standard components, such as peat moss or other organic materials (coconut fiber, composted sawdust, bark, or composted sewage sludge) in combination with aeration materials (e.g., perlite, sand, vermiculite, tuff, or polystyrene; Landis et al. 1990). Several decades ago, foresters thought that the use of raw substrates based on topsoil produced better rustic plants that were well adapted to harsh field conditions. Natural topsoil is difficult to standardize; not only is it very heavy, which hinders planting operations, but also it often comes from excavations for constructions and has poor fertility. Our experiments in eastern Spain using different types of growing media showed that those based on topsoil produced poor results in terms of survival and growth (Fig. 5). A mixture with small amounts of hydrogels or some clays (sepiolite) can increase the water holding capacity of the plug, thus providing the seedlings with high water availability for a longer period of time (Fig. 6). This can be especially important in semi-arid climates with high rainfall variability.

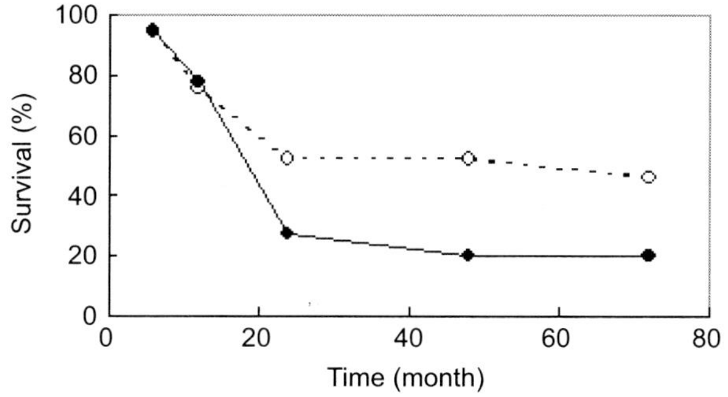

Fig. 5 Field survival dynamics of *Pinus halepensis* seedlings cultivated in two different growing media: compacted low quality topsoil with sand (black circles), and high quality topsoil with peat (white circles).

Containers and root systems

Several studies have related the planting stock quality of the seedlings to the type of container used (Landis et al. 1990, Peñuelas 1995, Vilagrosa et al. 1997, Dominguez et al. 1999). An appropriate container should have a shape and dimension that allow the seedling to develop correctly, especially its root system. In past decades, seedlings were grown in pots and polyethylene bags that often produced deformations in the root system, like taproot spiralling and/or reduced lateral root growth (Peñuelas and Ocaña 1996, Vilagrosa et al. 1997). Recently, producers tend to use pipe-shaped containers suspended in the air with channels or ribs inside them. This type of container prevents taproot spiralling by facilitating aerial root pruning, which in turn favors the development of secondary roots. Moreover, the interior channels or ribs promote the downward growth of roots and avoid spiralling. Nevertheless, it is difficult to find a universally acceptable type of container because the container must be adapted to many factors, such as species, nursery management, and planting needs. In general, high-volume containers (300 cm^3 or more) are recommended for reforestations in dry and semi-arid climates and for species with high root-to-shoot ratio, because they allow for the critical root system development during the first stages after planting. In our experience, long containers are preferred for species that develop a tap root,

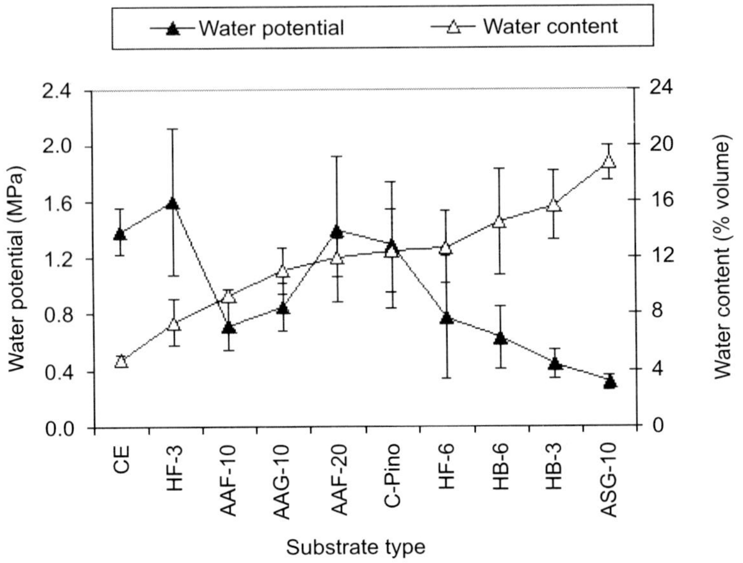

Fig. 6 Comparison of different substrate type during a drought period (Chirino and Vilagrosa, unpublished data). After 7 days of drought, substrate mixed with (ASG-10) maintained high water content and high predawn water potential compared to the control substrate (CE). The mixture with hydrogels (HB-3, HB-6, and HF-6) improved the water-holding capacity of the substrate when compared with the control substrate. Substrate type: CE substrate mixed with composted pine bark 25% (C-Pino); hydrogel Bures at 3 and 6% (HB-3 and HB-6); hydrogel Stockosorb at 3 and 6% (HF-3 and HF-6); coarse clay Sepiolite at 10% (ASG-10); fine clay Attapulgite 20/70 at 10 and 20% (AAF-10 and AAF-20); and coarse Attapulgite 4/20 at 10% (AAG-10).

like *Quercus* sp., while wider containers are recommended for species that show important secondary root development.

Drought preconditioning

There is a great deal of evidence indicating that a major obstacle to plantation success is transplant shock, that is, the intense short-term stress experienced by seedlings when they are transferred from favorable nursery conditions to the more adverse field environment. Drought preconditioning, the induction of drought resistance mechanisms, is one of the main techniques used to prepare seedlings for drought stress (Landis et al. 1998). However, a characteristic of arid region plant species is their ontogenetically high resistance to stress conditions. The most commonly used drought-preconditioning techniques (i.e., short-term preconditioning), designed for plant species characteristics in humid or subhumid climates, are of little benefit when applied to dry-climate species (Fonseca 1999, Vilagrosa et al. 2003b). For dryland species, long-term drought preconditioning in the nursery promotes greater benefits to plant morpho-functional characteristics than short-term preconditioning (Rubio et al. 2001, Chirino et al. 2003). On the other hand, the response to drought preconditioning seems to depend on the plant species. For example, species like *Pistacia lentiscus* are very responsive to preconditioning while species like *Quercus coccifera* are not. The kind of response is probably related to the drought strategy developed by each species (Vilagrosa et al. 2003b).

The main responses obtained in drought preconditioning experiments are: higher root-shoot ratio in the nursery (Chirino et al. 2003), changes in allocation patterns (i.e., higher fine root colonization in the plantation hole and lower above-ground development; Fonseca 1999, Rubio et al. 2001, Chirino et al. 2003), higher tolerance to drought conditions by means of higher elasticity of the cell membrane (Rubio et al. 2001) or better photochemical efficiency (Vilagrosa et al. 2003b), and drought-avoidance mechanisms such as higher root hydraulic conductivity for supplying water to leaves, higher leaf capacitance to water, and lower transpiration rates (Villar-Salvador et al. 1999, Vilagrosa et al. 2003b). In general, drought preconditioning does not improve survival, but it produces healthier seedling in field conditions (Rubio et al. 2001).

Fertilization

Given that nutritional status affects basic morphological and physiological plant processes, fertilization influences seedling growth and development. In the last decades, forest seedling fertilization practices have moved from using the lowest fertilization rates possible to maximize hardening of the seedling, to the current strategy of increasing fertilization to produce a seedling that can resist stress with sufficient photosynthetic capacity and carbohydrate reserves to initiate vigorous growth in the field. Recent studies indicate that larger, well-fertilized seedlings respond to field conditions better than smaller, less-fertilized seedlings (Villar-Salvador et al. 2000, Puértolas et al. 2003). Similarly, a positive relationship has been observed between survival or growth and nitrogen content in leaves (Oliet et al. 1997, Puértolas et al. 2003). However, under more limiting environments (semi-arid climate, irregular rainfalls) these results might not apply. Trubat et al. (2004), analyzing a wide range of species in semi-arid climates, observed that in years with a scarcity of rainfall, the bigger and better fertilized seedlings showed higher mortality rates than the smaller, less fertilized seedlings. Root growth potential was not promoted by higher fertilization, and seedlings were observed to develop root biomass according to their initial size (Trubat et al. 2004).

Field Testing

Recent findings on seedling quality have stressed the importance of promoting morpho-functional characteristics acclimated to a target ecosystem.

Doing so will reduce variability in seedling success. However, acclimatizing to a target ecosystem means that seedling quality cannot be determined at the nursery alone; it must to be tested in the field.

Site Preparation

Site preparation for reforestation generates a certain degree of soil disturbance, which may temporarily increase the risk of soil erosion (Shakesby et al. 1994). Thus, it is recommended that soil preparation work for plantations be applied at least two years after a fire, when the soil is less vulnerable and the regenerated plant cover provides a minimum protective threshold. The objective of site preparation is to increase the effective soil volume for root growth, to improve the capture of runoff, and to increase the soil water-holding capacity, in order to enhance seedling survival in the short-term. Due to its suitability for steep slopes, pit planting is a commonly used spot-treatment in soils with abundant rock outcrops, or in degraded areas where the existing vegetation can play an important role both in the recovery process and in soil conservation. When no machinery is employed, its effectiveness in increasing the soil water-holding capacity is low due to the small volume of soil affected by this technique. However, the small disturbance is a positive feature in terms of reducing the risk of soil erosion (Alloza 2003), preserving the specific richness and woody seedlings density, and reducing possible damage to the natural standing vegetation. Linear subsoiling is one of the most widely used soil preparation techniques, and it generally yields higher seedling growth and survival than spot treatments (Espelta et al. 2003, Bocio et al. 2004). This method provides a higher volume of effective soil for root growth, and a higher water-holding capacity. On the other hand, it may increase soil erosion and negatively affect the visual impact on the landscape, especially in rocky soils.

Water availability is the main factor hampering ecosystem restoration in dry or semi-arid areas (Vallejo et al. 2000). Current techniques that increase the amount of water available in the planting hole are: the application of different inorganic (hydrogels; Hüttermann et al. 1999) or organic amendments (composted or uncomposted refuses; Querejeta et al. 2000) or the construction of small water-harvesting structures associated with the planting holes (micro-catchments; De Simón 1990, Fuentes et al. 2004). The micro-cachment technique involves dividing the slope into several units that reduce its length and, as a consequence, the erosive strength of the runoff water. This soil preparation includes the excavation of shallow furrows to collect the runoff water in the plantation hole, and the excavation of a bench with a ridge to retain water. An inaccurate procedure or the occurrence of extreme rainfall events may generate the breakdown of the structure, leading to concentrated runoff and rill erosion.

Soil Amendments

Shallow soils or soils with poor structure may need high nutrient pools to maintain an acceptable seedling performance, and fertilization may compensate for these physical drawbacks. Planting holes may benefit from the application of biosolids, which act as a slow-release fertilizer and can provide longer-lasting effects than inorganic fertilizers. Additionally, biosolids promote microbial activity and increase the soil water-holding capacity and infiltration rates, resulting in higher water availability for the target seedlings. The negative effects of biosolid application are related to increased salinity and, if using semi-liquid sludges (slurry), changes in soil physical properties that occur as the sludge dries. Determining the optimum application rate is the key to this technique. Some studies suggest that doses of 15 to 30 Mg (dry weight) ha^{-1} are best for a *Pinus halepensis* plantation under dry-sub-humid Mediterranean conditions (Valdecantos et al. 2004). Using composted biosolids as mulch in the restoration of semi-arid open shrublands has proven to be effective in increasing the soil microbial functional diversity based on the local microflora.

Hydrophilic gels are synthetic products with the ability to absorb and retain high amounts of water in relation to the volume they occupy, thereby increasing the soil water-holding capacity (Hüttermann et al. 1999). The hydrogels retain water at a high matric potential, so it is easily available to the root. Once applied to the soil, the moisture in the rhizosphere lasts longer. In addition, hydrogels may incorporate some fertilizer properties or even fungal inoculum (Mikkelsen 1994). Applications of these products change the soil structure by modifying the size of soil aggregates and porosity, which implies an improvement in water storage capacity, soil aeration, and drainage. Hence, hydrogels can reduce transplant shock during the short time when the seedling roots are within the zone of the hydrogel influence. Nevertheless, in areas where the water deficit is extremely high, applying these polymers to the soil may result in negative consequences for the target seedling due to the high affinity of the hydrogel for the small amount of available water. In loamy or finer textured soils, the hydrogel moisture is subjected to suction by the clays under drying processes, thus reducing its chances of being used by the roots. Therefore, the positive effects of hydrogels are likely to be more relevant in sandy soils than in finer textured soils.

Tree Shelters

High radiation levels and high evaporative demand characterize dry environments. Under these conditions, seedling survival is usually higher under the protection of a canopy than in open areas (Espelta 1996, Vilagrosa et al. 1997, Vallejo et al. 2006), but exceptions are not uncommon (Vilagrosa et al. 2001, Pérez-Devesa et al. 2004). The use of tree shelters may ameliorate harsh conditions and improve species survival and growth. These positive effects have been attributed to the fact that tree shelters modify the plant environment by creating a greenhouse microclimate with increased temperature, relative humidity, and carbon dioxide levels (Burger et al. 1992).

Most tested species, including some growing under Mediterranean-humid conditions, showed a positive response to tree shelters (Costello et al. 1996). Nevertheless, several experiments in dry Mediterranean conditions showed that tree shelters did not improve the overall survival, even though positive interactions were found among tree shelter treatments, site conditions, and species (Vilagrosa 2001). A detailed analysis revealed that, in terms of survival, the effect of tree shelters was more important in the driest regions. The implementation of treeshelters is especially recommended in restorations that involve introduction of pre-germinated acorns that may be predated in high percentage by small rodents (Seva et al. 2004).

In relation to growth, tree shelters mainly improve stem elongation (Rey Benayas 1998, Domínguez et al. 1999, Cortina et al. 2004, Seva et al. 2004). The main effects reported from the use of tree shelters involve reductions in both the water deficit and the incoming radiation (Kjelgren and Rupp 1997, Kjelgren et al. 1997). These conditions would favor the development of morpho-functional traits of shade-tolerant plants: stem elongation, larger leaves with lower specific leaf weight, higher chlorophyll content, higher shoot weight ratio, etc. (Kozlowski et al. 1991). Despite the fact that these traits may seem negative for the survival of introduced seedlings, the lower radiation and higher relative humidity inside the tree shelter may favor more efficient photosynthetic machinery and lower transpiration rates, thus increasing water-use efficiency.

One of the main problems described for unventilated tree shelters (i.e., those with no lateral holes) was the increased temperature inside, which may be deleterious for seedling growth and survival (Burger et al. 1992, Bergez and Dupraz 1997). However, in ventilated tree shelters, temperature changes were minimal (Seva and Cortina 1999).

CONCLUSION

At present, strategies and techniques are available to address the long-term ecological restoration of degraded ecosystems/landscapes after wildfires. However, the restoration process is subject to many uncertainties that cannot be foreseen and will undoubtedly affect the success of restoration. Of course, most restoration efforts involve vegetation enhancements that are very dependent on both long term climate (wet-dry cycles) as well as short term weather events (adequate rainfall for germination), both of which are uncertain. In addition, detailed knowledge of the ecosystem to be restored and the potential interactions between introduced plants, soil properties, extant organisms, etc. may also be incomplete and add to the uncertainties. Therefore, restoration projects should follow adaptive management principles

(Whisenant 1999), including monitoring and project modification as circumstances change over time. Although adaptive management will lead to greater restoration success, this dynamic approach requires more time and longer-term funding than usual projects that do not include monitoring or adaptation actions.

Long-term post-fire restoration is an expensive process that should be clearly justified in terms of improving landscape and ecosystems quality (biodiversity, resilience, structure, function, etc.) and reducing wildfire propagation. Therefore, quality control and evaluation should be incorporated in the design and budget of a restoration project (Vallauri et al. 2005, www.ceam.es/reaction). Unfortunately, from a public perception point of view, the results of restoration actions are not immediately apparent making it difficult to justify the expense to the public. Long-term demonstration projects may be used to showcase the value of long-term restoration activities.

Acknowledgements

This research was funded by the Generalitat Valenciana, Bancaja Foundation and EC Research (Energy, Environment and Sustainable Development, REACTION project: EVK2-CT2002-80025, REDMED: ENV4-CT97-0682, GEORANGE: EVK2-2000-22089, SPREAD: EVGI-2001-00027 y CREOAK: QLRT-2001-01594).

References

Abad, N., R.N. Caturla, J. Baeza, C. Bladé, F. Vieira, E. Carbó, A. Valdecantos, A. Bonet, I. Serrasolsas, R. Guardia, J. Raventós, J.A. Alloza, A. Escarré, J. Bellot, and V.R. Vallejo. 1996. Regeneración de los montes quemados. pp. 51-148. *In* V. R. Vallejo [ed.]. La restauración de la cubierta vegetal en la Comunidad Valenciana. CEAM, Valencia, Spain.

Agee, J.K., B. Bahro, M.A. Finney, P.N. Omi, D.B. Sapsis, C.N. Skinner, J.W. van Wagtendonk, and C.P. Weatherspoon. 2000. The use of shaded fuelbreaks in landscape fire management. *Forest Ecology & Management*, 127: 55-66.

Alcantara, J.M., P.J. Rey, F. Valera, A.M. Sánchez-lafuente, and J.E. Gutierrez. 1997. Habitat alteration and plant intra-specific competition for seed dispersers. An example with *Olea europaea* var. *sylvestris*. *Oikos*, 79: 291-300.

Alloza, J.A. 2003. Análisis de repoblaciones forestales en la Comunidad Valenciana. Desarrollo de criterios y procedimientos de evaluación. Ph.D. thesis, Universidad Politécnica de Valencia. Departamento de Producción Vegetal. Valencia, Spain.

Alloza, J.A. and V.R. Vallejo. 2006. Restoration of burned areas in forest management plans. pp. 475–488. *In* W.G. Kepner, J.L. Rubio, D.A. Mouat, and F. Pedrazzini [eds.]. *Desertification in the Mediterranean Region: a Security Issue*, NATO Security through Science Series, Number 3 Springer-Verlang, New York, New York, USA.

Ashby, W.C. 1987. Forests. pp. 89–108. *In* W.R. Jordan III, M.E. Gilpin, and J.D. Aber [eds.]. *Restoration Ecology*. Cambridge University Press, Cambridge, UK.
Baeza, M.J. 2004. El manejo del matorral en la prevención de incendios forestales. pp. 65-92. *In* V.R. Vallejo and J.A. Alloza [eds.]. *Avances en el estudio de la gestión del monte mediterráneo*. Fundación CEAM, Valencia, Spain.
Baeza, M.J. and V.R. Vallejo. 2006. Ecological mechanisms involved in dormancy breakage in *Ulex parviflorus* seeds. *Plant Ecology*, 183: 191-205.
Baeza, M.J., A. Valdecantos, and V.R. Vallejo. 2005. Management of Mediterranean shrubands for forest fire prevention. pp. 37–60. *In* A.R. Burk [ed.]. *New Research on Forest Ecosystems*. Nova Science Publisher, New York, New York, USA.
Baeza, M.J., J. Raventós, A. Escarré, and V.R. Vallejo. 2006. Fire risk and vegetation structural dynamics in Mediterranean shrubland. *Plant Ecology*, 187: 189-201.
Barea, J.M. and M. Honrubia. 2004. La micorrización dirigida de la planta forestal. pp. 215-260. *In* V.R. Vallejo and J.A. Alloza [eds.]. *Avances en el estudio de la gestión del monte mediterráneo*. Fundación CEAM, Valencia, Spain.
Barnett, J.P. and J.B. Baker. 1991. Regeneration methods. pp. 35–50. *In* M.L. Duryea and P.M. Dougherty [eds.]. *Forest Regeneration Manual*. Kluwer Academic Publishers, Dordrecht, The Netherlands.
Bergez, J.E. and C. Dupraz· 1997. Transpiration rate of *Prunus avium* L. Seedlings inside an unventilated treeshelter. *Forest Ecology and Management*, 97: 255-264.
Bergstern, U. 1988. Pyramidal indentations as a microsite preparation for direct seeding of *Pinus sylvestris*. *Scandinavian Journal of Forest Research*, 3: 493-503.
Binkley, D. and C. Giardina. 1998. Why tree species affect soil? The warp and woof of tree-soil interaction. *Biogeochemistry*, 42: 89-106.
Bocio, I., F.B. Navarro, M.A. Ripoll, M.N. Jiménez, and E. De Simón. 2004. Holm oak (*Quercus rotundifolia* Lam.) and Aleppo pine (*Pinus halepensis* Mill.) response to different soil preparation techniques applied to forestation in abandoned farmland. *Annals of Forest Science*, 61: 171-178.
Burger, D.W., P. Svhra, and H. Harris. 1992. Treeshelters use in producing container-grown trees. *HortScience*, 27: 30-32.
Byram, G.M. 1959. Combustion of forest fuels. pp. 113-126. *In* K.P. Davis [ed.]. *Forest Fire: Control and Use*. McGraw-Hill, New York, New York, USA.
Castell, C. and J.I. Castelló. 1996. Metodología y resultados de la siembra aérea efectuada en el parque natural del Garraf. *Montes*, 46: 51-57.
Chirino, E., A. Vilagrosa, and E. Rubio. 2003. Efectos de la reducción del riego y la fertilización en las características morfométricas de *Quercus suber*. *Cuadernos de la Sociedad Española de Ciencias Forestales*, 17: 51-56.
Cortina, J., J. Bellot, A. Vilagrosa, R.N. Caturla, F. Maestre, E. Rubio, J.M. Ortíz De Urbina, and A. Bonet. 2004. Restauración en semiárido. pp. 345–406. *In* V.R. Vallejo and J.A. Alloza [eds.]. *Avances en el Estudio de la Gestión del Monte Mediterráneo*. CEAM, Valencia, Spain.
Costello, L.R., A. Peters, and G.A. Giusti.· 1996. An evaluation of treeshelters effects on plant survival and growth in a Mediterranean climate. *Journal of Arboriculture*, 22: 1-9.
De Simón, E. 1990. Restauración de la vegetación de las cuencas mediterráneas. Repoblaciones en zonas áridas. *Ecología. Fuera de serie n°1*, 401-427.
Debussche, M., S. Rambal, and J. Lepart. 1987. Les changements de l'occupation des terres en région méditerranéenne humide: évaluation des conséquences hydrologiques. *Acta Oecologica Oecologia Appliata*, 8: 317-332.

Debussche, M., J. Escarre, J. Lepart, C. Houssard, and S. Lavorel. 1996. Changes in Mediterranean plant succession: Old-fields revisited. *Journal of Vegetation Science,* 7: 519-526.

Domínguez, S., P. Villar, J.L. Peñuelas, N. Herrero, and J.L. Nicolás· 1999. Técnicas para cultivar encinas en suelos agrícolas. *Quercus,* 166: 22-25.

Duguy, B. 2003. Interacción de la historia de usos del suelo y el fuego en condiciones Mediterráneas. Respuesta de los ecosistemas y estructura del paisaje. PhD Thesis. Departamento de Ecología, Universidad de Alicante, Alicante, Spain.

Duguy, B., J.A. Alloza, V.R. Vallejo, A. Röder, and J. Hill. 2005. Use of the FARSITE fire growth model to design strategies for fire prevention and promotion of landscape diversity and resilience in a fire-prone area (Ayora, eastern Spain). pp. 91-95. *In* J. de la Riva, F. Pérez-Cabello, and E. Chuvieco [eds.]. Proceedings of the 5th EARSeL International Workshop on Remote Sensing and GIS. Applications to Forest Fire Management: Fire Effects Assessment, ISBN: 84-96214-52-4, Universidad de Zaragoza, Zaragoza, Spain.

Elvira, L.M. and C. Hernando. 1989. *Inflamabilidad y energía de las especies de sotobosque.* Instituto Nacional de Investìgación y Tecnología Agreria y Alimentaria, Madrid, Spain.

Espelta, J.M. 1996. La regeneració de boscos d'alzina (*Quercus ilex* L.) i pi blanc (*Pinus halepensis* Mill.): Estudi experimental de la resposta de les plàntules a la intensitat de llum i a la disponibilitat d'aigua. PhD thesis. Autonomous University of Barcelona, Bellaterra, Spain.

Espelta, J.M., J. Retana, and A. Habrouk. 2003. An economic and ecological multi-criteria evaluation of reforestation methods to recover burned *Pinus nigra* forests in NE Spain. *Forest Ecology and Management,* 180: 185-198.

FAO. 1989. *Arid Zone Forestry. A guide for field technicians.* Food and Agricultural Organization of the United Nations, Rome, Italy.

Faraco, A. 1998. Gravedad del fuego y patrones espaciales y temporales post-incendio de las plantas de un escobonal de la Sierra de Gredos. Ph.D. thesis. Universidad Complutense de Madrid, Madrid, Spain.

Ferran, A., I. Serrasolsas, and V.R. Vallejo. 1992. Soil evolution after fire in *Quercus ilex* and *Pinus halepensis* forests. pp. 397-405. *In* A. Teller, P. Mathy and J.N.R. Jeffers [eds.]. *Responses of Forest Ecosystems to Environmental Changes.* Elsevier, London, UK.

Finney, M.A. 1998. FARSITE: Fire area simulator-model development and evaluation. Research Paper RMRS-RP-4. U.S. Department of Agriculture, Forest Service, Rocky Mountain Research Station, Ogden, Utah, USA.

Fonseca, D. 1999. Manipulacion de las características morfo-estructurales de plantones de especies forestales mediterráneas producidas en vivero. Master-thesis. CIHEAM-IAMZ, Zaragoza, Spain.

Forman, R.T.T. 1995. Some general principles of landscape and regional ecology. *Landscape Ecology,* 10: 133-142.

Forman, R.T.T. and S.K. Collinge. 1996. The "spatial solution" to conserving biodiversity in landscapes and regions. pp. 537–568. *In* R.M. DeGraaf and R.I. Miller [eds.]. *Conservation of Faunal Diversity in Forested Landscapes.* Chapman and Hail, London, UK.

Fuentes, D., A. Valdecantos, and V.R. Vallejo. 2004. Plantación de *Pinus halepensis* Mill. y *Quercus ilex* subsp. *ballota* (Desf) Samp. en condiciones mediterráneas secas utilizando microcuencas. *Cuadernos de la Sociedad Española de Ciencias Forestales,* 17: 157-161.

Gallego Fernández, J.B., M.R. García Mora, and F. García Novo. 2004. Vegetation dynamics of mediterranean shrubland in former cultural landscape at Grazalema mountains in South Spain. *Plant Ecology,* 172: 83-94.

Hargrove, W.W., R.H. Gardner, M.G. Turner, W.H. Romme, and D.G. Despain. 2000. Simulating patterns in heterogeneous landscapes. *Ecological Modelling,* 135: 243-263.

Hobbs, R.J. and D.A. Norton. 1996. Towards a conceptual framework for restoration ecology. *Restoration Ecology,* 4: 93-110.

Hughes, C.E. and B.T. Styles. 1987. The benefits and potential risks of woody legume introductions. *The International Tree Crops Journal,* 4: 209-248.

Hüttermann A., M. Zommorodi, and K. Reise. 1999. Addition of hydrogels to soil for prolonging the survival of *Pinus halepensis* seedlings subjected to drought. *Soil & Tillage Research,* 50: 295-304.

Jones C.G., J.H. Lawton, and M. Shachak. 1994. Organisms as ecosystem engineers. *Oikos,* 69: 373-386.

Keeley, J.E., C.J. Fotheringham, and M. Morais. 1999. Reexamining fire suppression impacts on brushland fire regimes. *Science,* 284: 1829-1832.

Kjelgren, R. and L.A. Rupp. 1997. Establishment in treeshelters I: Shelters reduce growth, water use, and hardiness, but not drought avoidance. *HortScience,* 32: 1281-1283.

Kjelgren, R., D.T. Montague, and L.A. Rupp. 1997. Establishment in treeshelters II: Effect of shelter color on gas exchange and hardiness. *HortScience,* 32: 1284-1287.

Knight, D.H. 1987. Parasites, lightning and the vegetation mosaic in wilderness landscapes. pp. 59-84. *In* M.G. Turner [ed.]. *Landscape Heterogeneity and Disturbance*. Ecological Studies, 64. Springer-Verlag, New York, New York, USA.

Kozlowski T.T., P.J. Kramer, and S.G. Pallardy. 1991. *The Physiological Ecology of Woody Plants.* Academic Press, Toronto, Ontario, Canada.

Laguna, E. and S. Reyna. 1990. Diferencias entre los óptimos natural y forestal de las vegetaciones valencianas y alternativas futuras de gestión, *Ecología, Fuera de Serie,* 1: 321-330.

Lal, R. 1999. Soil management and restoration for C sequestration to mitigate the accelerated greenhouse effect. *Progress in Environmental Science,* 1: 307-326.

Landis, T.D., R.W. Tinus, S.E. Mcdonald, and J.P. Barnett. 1990. Containers and growing media. Vol. 2 of The Container Tree Nursery Manual, Agricultural Handbook 674. U.S. Department of Agriculture, Forest Service, Washington, DC, USA.

Landis, T.D., R.W. Tinus, and J.P. Barnett. 1998. Seedling propagation. Vol. 6 of The Container Tree Nursery Manual, Agricultural Handbook 674. U.S. Department of Agriculture, Forest Service, Washington, DC, USA.

Mesón, M. and M. Montoya. 1993. *Selvicultura Mediterránea.* Mundiprensa, Madrid, Spain.

Mikkelsen, R. 1994. Using hydrophilic polymers to control nutrient release. *Fertilizer Research,* 38: 53-59.

Minnich, R.A. 1983. Fire mosaics in southern California, and northern Baja California. *Science,* 219: 1287-1294.

Montero, J.L. and P. Alcanda. 1993. Reforestación y biodiversidad. *Montes,* 33: 57-76.

Montoya, J.M. 1993. *Encinas y encinares,* Mundi-Prensa, Madrid, Spain.

Morvan, N., F. Burel, J. Baudry, P. Tréhen, A. Bellido, Y.R. Delettre, and D. Cluzeau. 1995. Landscape and fire in Brittany heathlands. *Landscape and Urban Planning*, 31: 81-88.

Nilson, M.E. and J. Kjältén. 2003. Covering pine seeds immediately alter seeding: effects on seedling emergence and mortality through seed predation. *Forest Ecology and Management*, 176: 449-457.

Oliet J., R. Planelles, M. López-Arias, and F. Artero. 1997. Efecto de la fertilización en vivero sobre la supervivencia en plantación de *Pinus halepensis*. pp. 69-79. *In* Actas de la Reunión de Madrid. Sociedad Española de Ciencias Forestales, Madrid, Spain.

Papió, C. and L. Trabaud. 1991. Comparative study of the aerial structure of five shrubs of Mediterranean shrublands. *Forest Science*, 37: 146-159.

Pausas, J.G. and V.R. Vallejo. 1999. The role of fire in European Mediterranean ecosystems. pp. 3-16. *In* E. Chuvieco [ed.]. *Remote Sensing of Large Wildfires*. Springer-Verlag, Berlin, Germany.

Pausas J.G. and M. Verdú. 2005. Plant persistence traits in fire-prone ecosystems of the Mediterranean Basin: A phylogenetic approach. *Oikos*, 109: 196-202.

Pausas, J.G., R.A. Bradstock, D.A. Keith, J.E. Keeley, and GCTE Fire Network. 2004a. Plant functional traits in relation to fire in crown-fire ecosystems. *Ecology*, 85: 1085-1100.

Pausas, J.G., C. Bladé, A. Valdecantos, J.P. Seva, D. Fuentes, J.A. Alloza, A. Vilagrosa, S. Bautista, J. Cortina, and V.R. Vallejo. 2004b. Pines and oaks in the restoration of Mediterranean landscapes of Spain: New perspectives for an old practice – a review. *Plant Ecology*, 209: 209-220.

Pausas, J.G., E. Ribeiro, and V.R. Vallejo. 2004c. Post-fire regeneration variability of *Pinus halepensis* in the eastern Iberian Peninsula. *Forest Ecology and Management*, 203: 251-259.

Pausas J.G., J.E. Keeley, and M. Verdú. 2006. Inferring differential evolutionary processes of plant persistence traits in Northern Hemisphere Mediterranean fire-prone ecosystems. *Journal of Ecology*, 94: 31-39.

Peman, J. and R. Navarro. 1998. Repoblaciones Forestales Centra. Colección eines, Universitat de Lleida, Lleida, Spain.

Peñuelas, J.L. 1995. Medios de producción: substratos y contenedores. pp. 101-122. *In* J.F. Ballester-Olmos [ed.]. *Producción de plantas forestales*. Universidad Politécnica de Valencia, Valencia, Spain.

Peñuelas, J.L and L. Ocaña. 1996. *Cultivo de plantas forestales en contenedor, principios y fundamentos*. MAPA, Mundi-Prensa, Madrid, Spain.

Pérez-Devesa, M., J. Cortina, and A. Vilagrosa. 2004. Factors affecting *Quercus suber* establishment in dense shrublands. p. 132. *In* The proceedings of LINKECOL. Linking community and ecosystem ecology: recent advances and future challenges, European Science Foundation Conference, imedea, Univeristat de les Illes Balears, Palma de Mallorca, Spain. *Online at* http://www.imedea.uib.es/natural/terrestrial_ecology/Book%20of%20abstracts%20LINKECOL%20Palma%202004.pdf [accessed 28 Aug 2008]

Puértolas J., L. Gil, and J.A. Pardos. 2003. Effects of nutritional status and seedling size on field performance of *Pinus halepensis* planted on former arable land in the Mediterranean basin. *Forestry*, 76: 159-168.

Querejeta, J.I., A. Roldán, J. Albaladejo, and V. Castillo. 2000. Soil physical properties and moisture content affected by site preparation in the afforestation of a semiarid rangeland. *Soil Science Society of America Journal*, 64: 2087-2096.

Retana, J., J.M. Espelta, A. Habrouk, J.L. Ordóñez, and F. Solà-Morales. 2002. Regeneration patterns of three Mediterranean pines and forest changes after a large wildfire in NE Spain. *Ecoscience*, 9: 89-97.

Rey Benayas, J.M. 1998. Growth and survival in *Q. ilex* seedlings after irrigation and artificial shading in Mediterranean set-aside agricultural land. *Annales Sciences Forestiers*, 55: 801-807.

Riera, J. and C. Castell. 1997. Efectes dels incendis forestals recurrents sobre la distribució de dues espècies del Parc Natural del Garraf: el pi blanc (*Pinus halepensis*) i la savina (*Juniperus phoenicea*). *Butlleti de l'Institut Català d'Història Natural*, 65: 105-116.

Robichaud, P.R., J.L. Bayers, and D.G. Neary. 2000. Evaluating the Effectiveness of Postfire Rehabilitation Treatments. General Technical Report RMRS-GTR-63, U.S. Department of Agriculture, Forest Service, Rocky Mountain Research Station, Fort Collins, Colorado, USA.

Rubio, E., A. Vilagrosa, J. Cortina, and J. Bellot· 2001. Modificaciones morfofisiológicas en plantones de *Pistacia lentiscus* y *Quercus rotundifolia* como consecuencia del endurecimiento hídrico en vivero. Efectos sobre supervivencia y crecimiento en campo, pp. 527-538. *In* Proceedings of III Congreso Forestal Español. Junta de Andalucía, Consejería de Medio Ambiente, Granada, Spain.

Ryan, K.C. and V.N. Noste. 1985. Evaluating prescribed fires. pp. 230-238. *In* J. Lotan, B. Kilgore, W. Fischer, and R. Mutch [tech. coords.]. Proceedings, symposium and workshop on wilderness fire, General Technical Report INT-182. U.S. Department of Agriculture, Forest Service, Intermountain Research Station, Ogden, Utah, USA.

Seva, J.P. and J. Cortina. 1999. Diferentes técnicas de plantación. *In* Proyecto I+D en relación con la Restauración de la Cubierta vegetal en la Comunidad Valenciana. CEAM, Valencia, Spain.

Seva, J.P., A. Valdecantos, J. Cortina, and V.R. Vallejo. 2004. Different techniques for afforestation with *Quercus ilex* ssp. *ballota* (Desf.) Samp. in degraded lands (Comunidad Valenciana). *Cuadernos de la Sociedad Española de Ciencias Forestales*, 17: 233-238.

Shakesby, R.A., D.J. Boakes, C.O.A. Coelho, A.J.B. Gonçalves, and R.P.D. Walsh. 1994. Limiting soil loss after forest fire in Portugal: the influence of different post-fire timber clearance practices. Vol. 11 D-39, pp. 1161-1170. *In* Proceedings of 2nd International Conference of Forest Fire Research, Coimbra, Portugal.

TNC (The Nature Conservancy). 2007. Fire, Ecosystems, and People: Threats and Strategies for Global Biodiversity Conservation. The Nature Conservancy, Global Fire Initiative, Tallahassee, Florida, USA. *Online at* http://www.nature.org/initiatives/fire/files/fire_ecosystems_and_people.pdf [accessed 15 Sep 2008].

Trabaud, L. 1976. Inflammabilité et combustibilité des principales espèces des garrigues de la région méditerranéenne. *Oecologia Plantarum*, 11(2): 117–136.

Trabaud, L. 1994. Post-fire plant community dynamics in the Mediterranean Basin, pp. 1–15 *In* J.M. Moreno and W.C. Oechel [eds.]. *The Role of Fire in Mediterranean-type Ecosystems*. Ecological Studies 107. Springer-Verlag, Berlin, Germany.

Trubat, R., J. Cortina, and A. Vilagrosa. 2004. Estado nutricional y establecimiento de especies leñosas en ambiente semiárido. *Caudernos de la Sociedad Española de Ciencias Forestales,* 17: 245-251.

Turner, M.G. 1987. *Landscape Heterogeneity and Disturbance.* Ecological Studies 64. Springer-Verlag, New York, New York, USA.

Turner, M.G. and W.H. Romme. 1994. Landscape dynamics in crown fire ecosystems. *Landscape Ecology,* 9(1): 59-77.

Turner, M.G., W.W. Hargrove, R.H. Gardner, and W.H. Romme. 1994. Effects of fire on landscape heterogeneity in Yellowstone National Park, Wyoming. *Journal of Vegetation Science,* 5: 731–742.

Valdecantos, A., D. Fuentes, and J. Cortina. 2004. Utilización de biosólidos en la restauración forestal. pp. 313–344. *In* V.R. Vallejo and J.A. Alloza [eds.]. *Avances en el estudio de la gestión del monte Mediterráneo.* Fundación CEAM, Valencia, Spain.

Vallauri, D., J. Aronson, N. Dudley, and R. Vallejo. 2005. Monitoring and evaluating forest restoration success. pp. 150–156. *In* S. Mansourian, D. Vallauri, and N. Dudley [eds.]. *Forest Restoration in Landscapes. Beyond Planting Trees.* Springer-Verlag, New York, New York, USA.

Vallejo, V.R. 1999. Post-fire restoration in Mediterranean ecosystems. pp. 199-208. *In* G. Eftichidis, P. Balabanis, and A. Ghazi [eds.]. Wildfire Management. Proceedings of the Advanced Study Course. Algosystems SA & European Commission DGXII, Athens, Greece.

Vallejo, V.R. and J.A. Alloza. 1998. The restoration of burned lands: the case of eastern Spain. pp. 91–108. *In* J.M. Moreno [ed.]. *Large Forest Fires.* Backhuys Publishers, Leiden, The Netherlands.

Vallejo, V.R., S. Bautista, and J. Cortina. 2000. Restoration for soil protection after disturbances. pp. 301–343. *In* L. Trabaud [ed.]. *Life and Environment in the Mediterranean.* WIT Press, Boston, Massachusetts, USA.

Vallejo, V.R. and J.A. Alloza. 2004. La selección de especie en restauración forestal. pp. 195-214. *In* V.R. Vallejo and J.A. Alloza [eds.]. *Avances en el Estudio de la Gestión del Monte Mediterráneo.* Fundación CEAM, Valencia, Spain.

Vallejo, R., J. Aronson, J.G. Pausas, and J. Cortina. 2006. Restoration of Mediterranean woodlands. pp. 193–207. *In* J. van Andel and J. Aronson [eds.]. *Restoration Ecology: The New Frontier.* Blackwell Publishing, Malden, Massachusetts USA.

Van Andel, J. and A.P. Grootjans. 2006. Concepts in restoration ecology. pp. 16-28. *In* J. van Andel and J. Aronson [eds.]. *Restoration Ecology. The New Frontier.* Blackwell Publishing, Malden, Massachusetts USA.

Vázquez, A. and J.M. Moreno. 1998. Patterns of lightning- and people-caused fires in peninsular Spain. *International Journal of Wildland Fire,* 8(2): 103-115.

Vilagrosa, A. 2001. Field techniques for the reduction of transplant shock: treeshelters and hydrophilic gels. pp. 85-101. *In* Restoration of degraded ecosystems in Mediterranean region (REDMED project, EC ENV4-CT97-0682). Cap. 8. Final report. Fundación CEAM, Valencia, Spain.

Vilagrosa, A., J.P. Seva, A. Valdecantos, J. Cortina, J.A. Alloza, I. Serrasolsas, V. Diego, M. Abril, A. Ferran, J. Bellot, and V.R. Vallejo. 1997. Plantaciones para la restauración forestal en la Comunidad Valenciana. pp. 435-548. *In* V. R. Vallejo [ed.]. 1997. *La Restauración de la Cubierta Vegetal en la Comunidad Valenciana.* Fundación CEAM, Valencia, Spain.

Vilagrosa, A., R.N. Caturla, N. Hernández, J. Cortina, J. Bellot, and V.R. Vallejo. 2001. Reforestaciones en ambiente semiárido del sureste peninsular. Resultados de las investigaciones desarrolladas para optimizar la supervivencia y el crecimiento de especies autóctonas. pp. 213-219. [*Online* http://www.juntadeandalucia.es/servicios/publicaciones/detalle/45857.html] *In* The Proceedings of the III Congreso Forestal Español. Montes para la sociedad del nuevo milenio. Junta de Andalucía, Conserjeria de Medio Ambient, Sevilla, Spain. [accessed 2 Sep 2008]

Vilagrosa, A., J. Bellot, V.R. Vallejo, and E. Gil. 2003a. Cavitation, stomatal conductance and leaf dieback in seedlings of two co-occurring Mediterranean shrubs during an intense drought. *Journal of Experimental Botany*, 54: 2015-2024.

Vilagrosa, A., J. Cortina, E. Gil-Pelegrín, and J. Bellot. 2003b. Suitability of drought preconditioning techniques in Mediterranean land restoration. *Restoration Ecology*, 11: 208-216.

Vilagrosa, A., J. Cortina, E. Rubio, R. Trubat, E. Chirino, E. Gil-Pelegrín, and V.R. Vallejo. 2005. El papel de la ecofisiología en la restauración forestal de ecosistemas mediterráneos. *Investigación agraria. Sistemas y recursos forestales*, 14(3): 446-461.

Villar-Salvador, P., L. Ocaña, J. L. Peñuelas, and I. Carrasco. 1999. Effect of water stress conditioning on the water relations, root growth capacity, and the nitrogen and non-structural carbohydrate concentration of *Pinus halepensis* Mill. (Aleppo pine) seedlings. *Annals of Forest Science*, 56: 459-465.

Villar-Salvador, P., S. Domínguez, J.L. Peñuelas, L. Carrasco, N. Herrero, J.L. Nicolás, and L. Ocaña. 2000. Plantas grandes y mejor nutridas de *P. pinea* tienen mejor desarrollo en campo. Volume I, pp. 219–228. *In* The Proceedings of the 1er Simposio del Pino Piñonero (*Pinus pinea* L.). Junta de Castilla y León, INIA y Cose, Valladolid, Spain.

Vos, W. 1993. Recent landscape transformation in the Tuscan Apennines caused by changing land use. *Landscape and Urban Planning*, 24: 63-68.

Waller, D.M. 1993. How does mast-fruiting get started? *Tree*, 8: 122-123.

Wennström, U., U. Bergsten, and J.E. Nilsson. 1999. Mechanized microsite preparation and direct seeding of *Pinus sylvestris* in boreal forests – a way to create desired spacing at low cost. *New Forests*, 18: 179-198.

Whisenant, S. 1999. *Repairing Damaged Wildlands*. Cambridge University Press, Cambridge, UK.

Wiens, J.A., C.S. Crawford, and J.R. Gosz. 1985. Boundary dynamics: a conceptual framework for studying landscape ecosystems. *Oikos*, 45: 421-427.

Winsa, H. and U. Bergstern. 1994. Direct seeding of *Pinus sylvestris* using microsite preparation and invigorated seed lots of different quality: 2-year results. *Canadian Journal of Forest Research*, 24: 77-86.

Yassoglou N. 2000. Land, desertification vulnerability and management in Mediterranean landscapes. Volume 1, pp. 87-113. *In* P. Balabanis, D. Peter, A. Ghazi, and M. Tsogas [eds.]. *Mediterranean Desertification*. European Commission, Brussels, Belguim.

Zobel, B.J., G. Van Wyk, and P. Stahl. 1987. *Growing Exotic Forests*. J. Wiley & Sons, New York, New York, USA.

Section III
Regional Strategies

15

The Portuguese Experience in Managing Fire Effects

António J.D. Ferreira[1,2*], *Joaquim Sande Silva*[1], *Celeste Coelho*[2], *Anne-Karine Boulet*[2], and *Jan Jacob Keizer*[2]

Abstract

Portugal has a warm temperate Mediterranean climate, with marked dry Mediterranean conditions in the south to more wet Atlantic conditions in the Northwest. Nevertheless all the regions are characterized by hot, dry summers and cooler, wet winters. Since 1980, the mean annual area burned in Portugal is 1070 km^2*, which slightly exceeds one percent of the total area of the country, and has the highest fire incidence in Europe. After forest fires, the largest amounts of sediment and solute loss occur during the first four months. Important nutrient losses occur during the first rainfall events that are large enough to saturate the ash layer and promote overland flow. Our review of the recent literature suggests that prescribed fire effects on soil and water conservation seldom reach the magnitude of wildfires; therefore, one of the main solutions for the forest fire problem in Portugal is to reintroduce prescribed fire. Prescribe fire can control shrubs and the forest understory, allowing a higher landscape diversity, which creates natural breaks to the forest fires progression and, therefore, eases their suppression. However, forest fires are going to continue to occur, therefore it is necessary to identify areas for intervention immediately after a fire and implement cost-effective mitigation strategies.*

INTRODUCTION

Wildfire in Portugal

In the past century, the Mediterranean countries of Europe have experienced large-scale land transformations. Areas of marginally productive agriculture have been converted to forest plantations or abandoned to the natural process

[1] *CERNAS, Escola Superior Agrária de Coimbra, Bencanta, Coimbra, Portugal.*
[2] *CESAM, Departamento de Ambíente e Ordenamento, Universidade de Aveiro, Aveiro, Portugal.*
* *Corresponding author: António J.D. Ferreira, CERNAS, Escola Superior Agrária de Coimbra, Bencanta, P-3040-316 Coimbra, Portugal. Tel: +351 938 487479, e-mail: aferreira@esac.pt*

of ecological succession creating shrublands and woodlands similar to other regions of southern Europe (Moreno 1999). Increased wildfires have accompanied this land abandonment (Margaris et al. 1996), a pattern that has been attributed to biomass accumulation in forest and shrub areas in wet Mediterranean regions (Ferreira et al. 2005a).

Portugal has a warm temperate climate, in transition between the Mediterranean and the Atlantic types, with a gradient between the dryer climates in the south to wetter climates in the Northwest. All of them are nevertheless characterized by hot, dry summers and cool, wet winters. Areas of rugged terrain are common, and the natural vegetation is typically evergreen, resistant to drought, and pyrophytic (Nunes et al. 2005). Wildfire is considered the most important agent of land cover change in Portugal and is a matter of concern in the forests and shrublands of northern and central regions, where, over the last decade, increasingly catastrophic wildfires are occurring (Pereira and Santos 2003). Wildfire is a major threat to the economic viability of commercial forestry and to the ecological health of diverse ecosystems affected by high fire incidence (Bunting and Rego 1988, Nogueira 1990, Silva 1990, Moreira et al. 2001, Vasconcelos et al. 2001).

Since 1980, the mean annual area burned in Portugal was 1070 km^2, which slightly exceeds one percent of the total area of the country and, as a result, Portugal has the highest fire incidence in Europe (Fig. 1; Nunes et al. 2005). The magnitude of Portugal's burned area is large even when compared to other Southern European Union countries, which share the same Mediterranean climate pattern (Fig. 2). In recent years, the forest fire problem has become overwhelming. Very long, hot, dry periods, resulting from Sahara winds, represent a climate change for which Portuguese forests are not well-adapted. The abnormal hot winds that occurred for long periods of the year

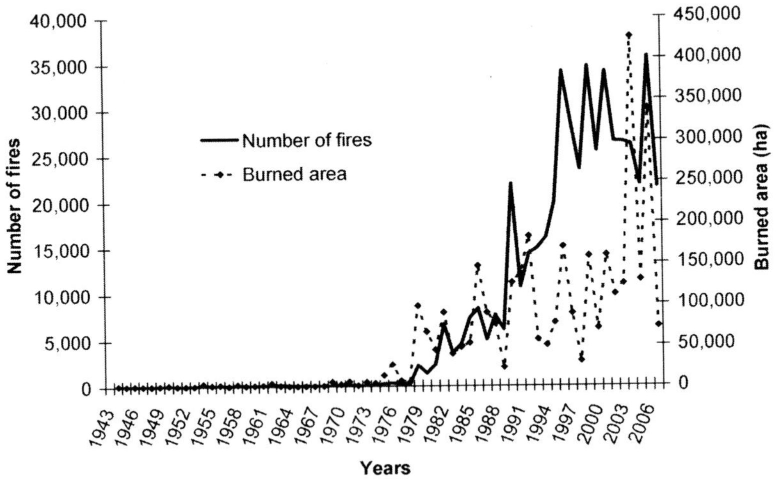

Fig. 1 Number of fires and burned area in Portugal from 1943 through 2006.

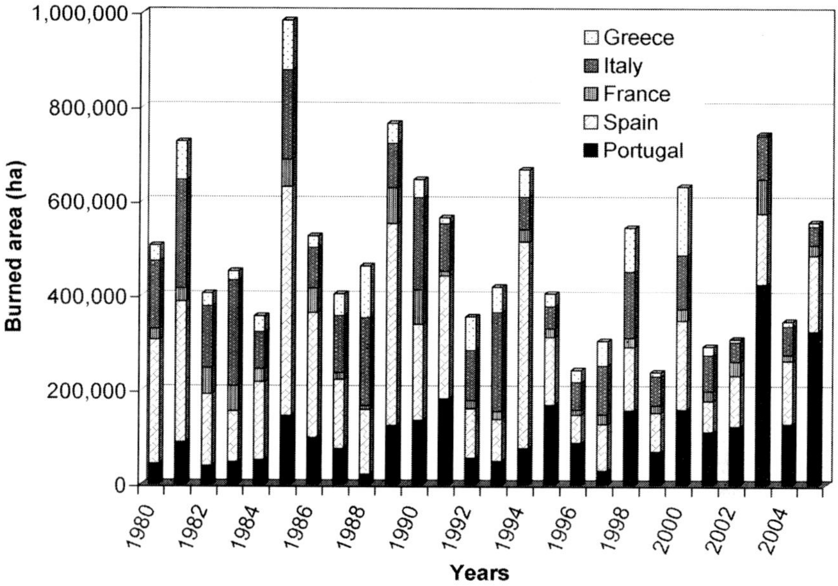

Fig. 2 Burned areas from the southern European Union countries from 1980 to 2005.

were responsible for the calamitous fire seasons of 2003 and 2005, when Portugal accounted more than 50 percent of the total area burned in the Southern European Union countries.

FIRE EFFECTS IN PORTUGAL

Post-fire Hydrology

Infiltration and soil water repellency

Post-fire changes in vegetation and top soil are known to have significant impacts on the hydrological regime (Cerdà and Doerr 2005). In Portugal these impacts have been measured at both at the plot (Walsh et al. 1994, Ferreira 1997, Ferreira et al. 1997, Thomas et al. 1999, 2000a, b) and catchment scales (Ferreira et al. 1997, Ferreira et al. 2005b). Fire-affected soil properties, such as infiltration rate, porosity, conductivity, and water storage capacity can result in deteriorated hydrologic functioning and quickly lead to decreases in ecosystem sustainability (Neary et al. 1999).

Forest fires can create or enhance soil water repellency immediately below the ashes (DeBano 1981, Giovannini 1987, Giovannini et al. 1988). This effect has been found after high-intensity fires with certain litter types, such as *Eucalyptus globules* and *Pinus pinaster* litter typical of Portuguese forests (Ferreira et al. 2000, 2005a, 2005b, Coelho et al. 2004). Volatized soil organic components move into the soil, cool, and coalesce on mineral particles (DeBano et al. 1970, Giovannini and Lucchesi 1984, Giovannini 1994). This results in a discreet layer of water repellent soil just below and parallel to the surface (DeBano 1981). The severity of fire-induced soil water repellency depends on a number of soil characteristics including moisture content, texture, and pre-fire organic matter quantity and composition (Botelho et al. 1994, Giovannini 1994). Burning can also destroy and/or displace the natural soil surface water repellency and drive the water repellent layer deeper into the soil such that a highly wettable soil layer overlays a highly water repellent layer (Doerr et al. 2006). Soil water repellency can be developed when soil temperatures rise above 176°C and destroyed at temperatures greater than 288°C (DeBano et al. 1976, DeBano 1981).

In Portugal, burned *Eucalyptus globules* forests have presented a two-fold response to fire-induced soil water repellency and, hence, to infiltration. In areas where strong and homogeneous soil water repellency has been developed, there is a sharp decrease in infiltration capacity to values below 2 mm h^{-1} (Ferreira et al. 2000). In the areas where extensive root systems were burned completely (very common in the case of rotten root systems from previous thinning forest management practices), the temperatures within the root macropores may be high enough to destroy water repellency and converted the roots to ash. This promotes the formation of concentric hydrophilic areas around the macropores where a mean infiltration capacity of 249 mm h^{-1} in areas of burned out root macropores was measured (Table 1). This may account for the relatively low overland flow (11.6 percent of rainfall) measured 4 months after a wildfire despite the lack of vegetation and litter layer, and the existence of water repellent soils (Ferreira et al. 2005 a, b).

Table 1 Soil hydraulic properties and processes (after Ferreira et al. 2005a).

Land use	Soil density (g cm^{-3})	Organic cover (%)	Average infiltration capacity (mm h^{-1})	Overland flow (% rainfall)
Mature pine	0.49	99.0	47.8	0.09
Mature eucalyptus	0.70	82.8	45.9	1.8
Burned pine	0.8	0	Water repellent[a] 1.77 Macropore[b] 249.0	11.6

[a]Homogeneous water repellent areas.
[b]Concentric hydrophilic areas around incinerated root systems (resulting from the complete combustion of rotten stumps left after forest thinning practices).

A rain simulation experiment (rain intensity 50.5 mm h^{-1}, applied for 45 to 60 min on 1 m^2 plots) was done immediately following a wildfire in Caramulo and Lousa mountains of central Portugal. Soil moisture content, overland flow, and nutrient concentrations were measured (Ferreira et al.

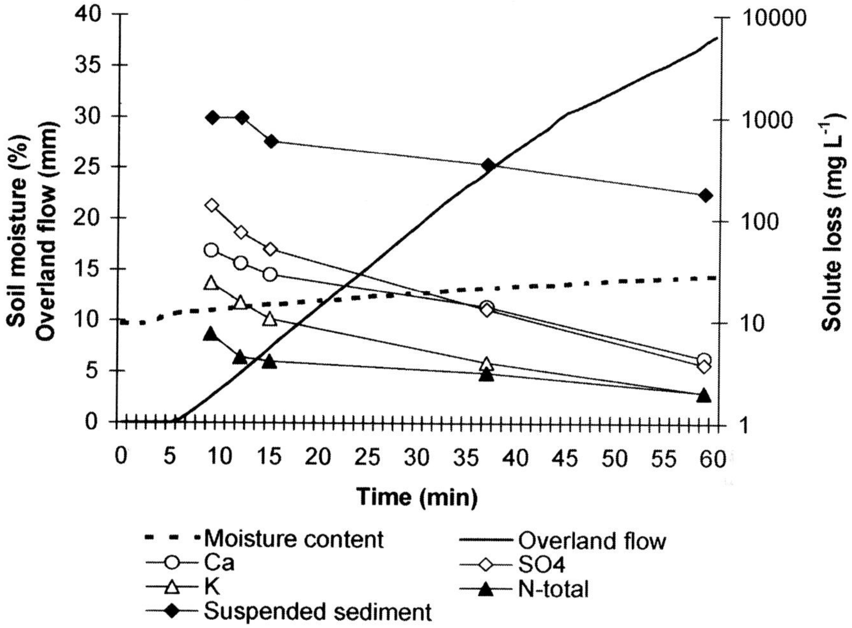

Fig. 3 Overland flow, suspended sediment and solute loss during rainfall simulation in a recently burned area (after Ferreira et al. 2005b).

2005b). The results indicated that in the burned area a water repellent soil layer had reduced infiltration capacity compared to an unburned mature pine forest (Fig. 3). The only rainfall sinks were the macropores that allowed water to bypass the soil water repellent layer.

Runoff

In burned forest environments overland flow responses can be enhanced by reduced infiltration capacities and the development or enhanced effectiveness of a water repellent layer (Sevink et al. 1989, Imeson et al. 1992, Doerr et al. 1996, Cerdà 1998, Ferreira et al. 2005b). At catchment scale, several authors have demonstrated an increase of runoff following forest fires (Scott and Van Wyk 1990, Scott and Schulze 1992, Scott 1993, Lavabre et al. 1993, Ferreira et al. 1997, 2005b, Ubeda and Sala 2001, Cosandey et al. 2005). In the rain simulation experiment described above, Ferreira et al. (2005b) reported enhanced overland flows of 37 mm in 1 hour, which represents almost 75 percent of the incident rainfall over that period—not unexpected given the reduced infiltration (Fig. 3).

In a study that compared runoff from burned and unburned areas in the litoral mountain range of Central Portugal, for the first 14 months after a forest fire, runoff at the catchment scale is considerably higher in the burned catchment than in the unburned control catchment (Fig. 4a; Ferreira et al.

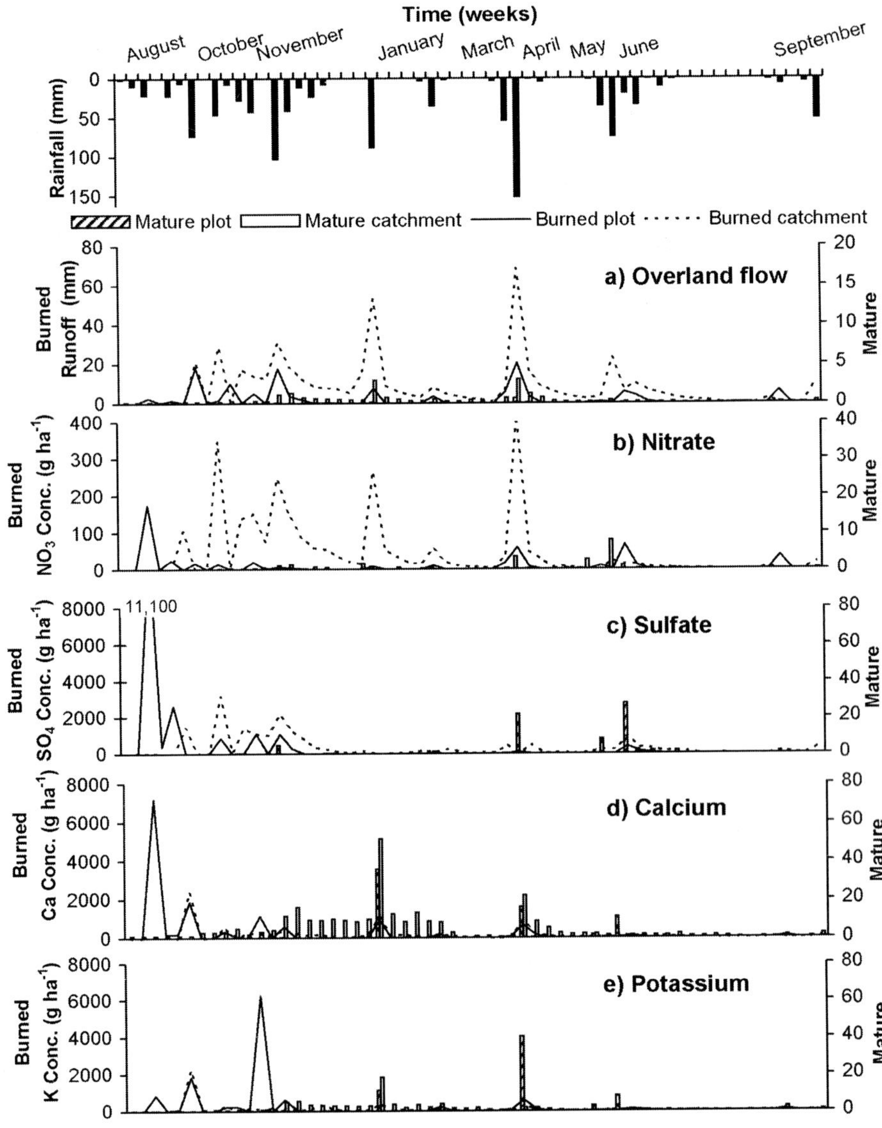

Fig. 4 Comparison of overland flow and nutrient loss from burned and mature (unburned) plots and catchments under natural rainfall. Note that data from burned plots and catchments are shown as lines using the scale on the left Y-axis, and the data from the mature plots and catchments are shown as bars using the scale from the right Y-axis. Also, the left Y-axis scale (burned) is up to two orders of magnitude greater than the right Y-axis scale (mature/unburned).

2005a). Given the runoff amount and response time, it is clear that most of the runoff from the unburned mature forested area came from the soil matrix. In contrast, the burned catchment had large runoff magnitude (50 percent of rainfall) and rapid (within 2 hours on average) peak runoff response. The catchment runoff per unit of area values were greater than the overland flow yields observed at plot scale. In general, there is a slight decrease in the magnitude of runoff (catchments) and overland flow (plots) over time for the same class of weekly rainfall amounts. The first rain events after the fire produce sharp responses with runoff equal to 50 percent of rainfall. One year later, the runoff generation threshold had increased, and only rain events greater than 25 mm produce runoff that averaged less than 10 percent of rainfall (Fig. 4a).

Erosion

The increased erosion that follows wildfires in Portugal is a logical response to the generally mountainous terrain of the forest lands, wet winter seasons, bare mineral soil exposed to rain drop impact, fire-induced soil water repellency, decreased infiltration, and increased runoff that have been discussed above. Coelho et al. (1995b) found erosion rates of 2 Mg $ha^{-1}yr^{-1}$ immediately following a forest fire, significantly lower when compared with the 50 Mg $ha^{-1}yr^{-1}$ found in the same region for rip-ploughed forest management practices but still greater than unburned mature pine (Shakesby et al. 1993).

Post-fire Soil Nutrients

Wet Mediterranean regions are characterized by ecosystems with dense vegetation covers, which allow the fast spread of fire. Forest fires often burn the entire litter layer and all small bushes, leading to a complete change in vegetation and surface soil structure. The fires in northern and central Portugal generally consume the L, F and, where present, H organic layers as well as most of the understory vegetation. Thus, large amounts of nutrients are mineralized as ground cover and litter are consumed and the ashes on the soil surface represent a substantial part of the nutrient pool.

After a forest fire (generally occurring during the dry summer months), the first autumn rainfall events and subsequent overland flow increases soil erosion and nutrient loss (Ferreira et al. 1997, 2005a). Post-fire nutrient losses have a significant impact in wet Mediterranean regions, since despite their luxurious vegetation, these mountainous areas often have weakly developed soil that are poor in nutrients. The mountain areas of central and northern Portugal are characterized by poorly developed *Humic cambisols* where the only nutrient pool is located at the L, F and H organic layers, and these layers are often completely consumed in forest fires.

Solutes and suspended sediment decrease logarithmically during a rainfall simulation experiment with a rainfall intensity of 50.5 mm h^{-1} (Fig. 3). The highest amounts of sediment and solute loss occur within the first 15 min after the rainfall simulation start, which coincides with the export of the highest amounts of ashes. Solute loss reflects the availability of the different nutrients in the ashes.

In the second and third years after burning, nutrient loss due to runoff (nutrients in solution), erosion (nutrients adsorbed to sediments), and ash (nutrients in high concentrations on the surface) were found to be substantially higher on burned terrain than on mature forest stands (Thomas et al. 1999, 2000a, b). Recent evidence has shown that even greater loss of nutrients occurs within the first six months after the forest fire (Cerdà and Lasanta 2005, Ferreira et al. 2005a).

The loss of solutes both from the plots and catchments for the studied period is considerably higher for burned areas when compared with the mature forest stands (Table 2; Ferreira et al. 2005a). Losses at plot scale range from 29 times higher for nitrates to 500 times more for magnesium. At catchment scale nitrate losses are considerably more (250 times higher), and sulphates had a comparatively higher loss (4400 times more). At this stage, there is an apparent decrease in the amounts of cations per unit of area; nevertheless losses are at least 10 times higher than for the mature forest catchment.

Table 2 Solute loss in overland flow and runoff at burned and mature control plots and catchments (after Ferreira et al. 2005a).

Catchment/ Plot/ Solute	Mature pine plot ($kg\ ha^{-1}\ yr^{-1}$)	Bouça (mature vegetation catchment) ($kg\ ha^{-1}\ yr^{-1}$)	Burned pine plot ($kg\ ha^{-1}\ yr^{-1}$)	Lourizela (burned catchment) ($kg\ ha^{-1}\ yr^{-1}$)
NO_3	0.017	0.01	0.49	2.5
SO_4	0.065	0.003	18.1	13.2
Cl	n.d.	2.09	n.d.	39.9
Ca	0.068	0.27	13.8	6.5
Mg	0.027	0.67	13.9	8.7
K	0.067	0.08	5.9	3.1
Na	n.d.	2.73	n.d.	30.1

This is partially caused by the increase in runoff amounts, which in turn increases the less reactive ions, such as chloride and sodium. At this stage, potassium also has the highest percentage loss increase. Being a limiting nutrient in the studied areas, there is a sharp control over potassium, which explains the very low amounts lost in mature forest catchments, only 0.08 $kg\ ha^{-1}yr^{-1}$.

These results suggest that there is an important exportation of nutrients lost in the soluble form from the ash layer in the first four months. After this initial period, peak losses of solutes are restricted to extreme events which are able to mobilize the remaining ashes. Nevertheless, during the second and third year, Thomas et al. (1999, 2000a, b) found that nutrient losses, either as solutes or adsorbed by the eroded sediments were significantly higher than the control plots.

The increase in nutrient loss is attributed to: 1) increased erosion as a result of bare soil being exposed to raindrop impact and overland flow, and 2) the easily erodible surface layer of non-cohesive ash containing high concentrations of nutrients (Thomas et al. 1999). Additional factors influencing the exportation of nutrients are related to hillslope hydrological processes, due to the changes in surface roughness as a result of consumption of the forest floor and understory vegetation. Additionally the formation of water repellent soil conditions reduces infiltration. These hydrological factors have a direct impact on overland flow at plot, hillslope, and catchment scale and thus contribute to solute loss (Ferreira 1997, Ferreira et al. 1997, Ferreira et al. 2000, Shakesby et al. 2000, Coelho et al. 2002a, b).

Change in rate of nutrient loss over time

According to Thomas et al. (1999), there are three likely reasons for the decline in the nutrient loss over time after a forest fire. First, a finite supply of nutrient-rich ash exists, which is depleted by erosion and forms a progressively smaller component of eroded material. Second, ash in the sediments are in a soluble form and vulnerable to leaching, thus the concentrations decline with time. Third, progressive coarsening of the eroded sediment leads to reduced nutrient concentrations. Since forest fires lead to increased overland flow, accelerated soil and nutrient loss occurs for two to three years after fire (Thomas et al. 2000a). Nevertheless, there is some controversy related to the details of nutrient loss. Gimeno-Garcia et al. (2004) found a decrease in organic matter, total nitrogen (N), nitrates (NO_3), and calcium (Ca) in the soil after fire and an increase in the levels of ammonia (NH_4), available phosphorus (P), sodium (Na), potassium (K), and magnesium (Mg). These results are not consistent with Thomas et al. (1999) where the exportation of Mg and K was high, which should cause a loss of those nutrients in the soil.

As opposed to N losses to the atmosphere, soil N can be more available following low-intensity burning by non-biological and biological means, converting organic forms to inorganic ammonium nitrogen (NH_4–N) and nitrate nitrogen (NO_3–N) (Pickett and White 1985, Hungerford et al. 1995). High-intensity fires can cause large losses of N directly through consumption or NH_4 volatilization. Furthermore, the NO_3 that was converted following the fire can be lost through denitrification, leaching, or overland flow (Neary et al. 1999).

In a recent study, NO_3 presented a distinct loss pattern (Fig. 4b) with enormous losses being measured following the first rainfall events. In the third week after the fire, rainfall exceeded 20 mm, saturating the ash layer and producing overland flow (Ferreira 2005b). Sharp decreases in NO_3 concentrations occurred in the following weeks. Subsequent peak concentrations were lower and were restricted to extreme rainfall events. At the catchment scale, the response was dependent on rainfall that was sufficient to produce runoff, and there was no flow during the dry periods. Increases in NO_3 concentration were dependent on rainfall amount, generally from October onwards, which was probably related to the movement of ash to the catchment stream. NO_3 seemed to be exhausted after the highest peaks occur during an extreme event in April.

Significant sulphate (S) loss occurred at the plot level during the first four months when sufficient rainfall saturates the ash and overland flow was able to transport it in either a dissolved or adsorbed form (Fig. 4c). Concentrations at catchment scale are significantly lower, but also almost disappear after 4 months, with the exception of the weeks when rainfall exceeds 50 mm, which triggered the sediment transport processes. Calcium (Ca) and the other solutes had significant losses in the third week after the fire with lower subsequent peak concentrations occurring when rainfall events exceed 25 mm of rainfall (Fig. 4d). The highest concentrations of Ca export at the catchment scale first occurred during the seventh week when runoff occurred. After this period, both plot and catchment losses occurred during rainy weeks when overland flow and runoff responses were high.

Potassium (K) losses at plot scale occur mainly during the first four months after the fire. The solute loss was dependent on the mobilized ash, which depends on rainfall amount and intensity. Throughout the first four months, source depletion was observed as small potassium responses following increasing wetter weeks (Fig. 4e). Catchment response was delayed when compared to the plot response. These results are probably explained by the smaller contributing area of the plots which were located at the top of the catchments near the watershed divide where the lack of upslope inputs lead to faster source exhaustion. Thus, ash in the catchments may undergo successive cycles of mobilization and deposition before it reaches the stream channel. As with other solutes, the control mature (unburned) plot and catchment had negligible losses.

Losses through overland flow and runoff removal are not the only nutrient loss processes; volatilization also plays an important role. Direct loss of nutrients to the atmosphere is temperature dependent. N is the element most prone to this type of loss as it starts to volatilize at 200°C, and at temperatures >500°C over half the N in organic matter can be volatilized. Higher temperatures are needed to vaporize potassium (K) >760°C, phosphorus (P) >774°C, sulfur (S) >800°C, sodium (Na) >880°C, magnesium (Mg) >1107°C, and calcium (Ca) >1240°C (Weast 1988).

Fire Severity Impacts on Soil and Water Conservation

One possible solution to reduce impacts of wildfire on soil and water conservation is to use prescribed fire to create organic matter discontinuities during the wet season, through the selective burning of shrublands . However, the question arises as to whether or not prescribed fires have impacts of the same magnitude as wildfires. Coelho et al. (2004) found that prescribed, experimental (moderate burn severity when burned at the end of May with drier conditions), and wildfire study sites differ significantly in their degree of soil water repellency. In the case of the wildfire sites, soil water repellency assessed using the MED method (Doerr et al. 1996, Coelho et al. 2004) revealed a strong to extreme water repellency, with some of the measurements attaining the maximum repellency (36 percent ethanol). All the individual tests done in two wildfire burned areas had greater than 13 percent ethanol readings (Fig. 5). All the measurements for the prescribed burn site (Cadafaz), and the experimental fire (Gestosa), show weaker water repellency with greater spatial variability and some non-water repellent areas. The prescribed burn had less soil water repellency than the average of either the experimental or the wildfire. The patterns along the sampled transects suggest a more contiguous spatial distribution of water repellency for the wildfires, when compared with prescribed and experimental fires. These fire-induced changes to the soil may increase overland flow and erosion rates due to reduced infiltration. These factors and the consumption of the forest floor reduce the number of obstacles to overland flow, which permits more efficient transport of sediments and solutes.

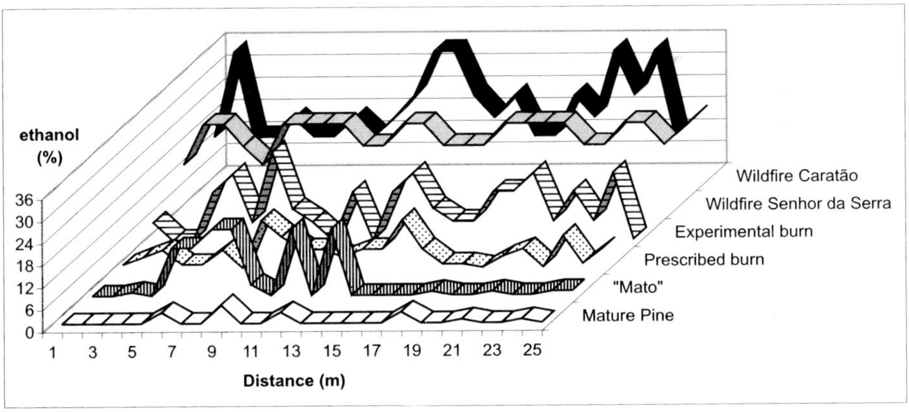

Fig. 5 Soil water repellency degree and spatial variability expressed as percent ethanol for the unburned and various burn severities (after Ferreira et al. 2005b).

However, the literature is not consistent as to the significance of soil water repellency on increased overland flow and other hydrological responses at larger scales. Vacca et al. (2000) found no significant increases in overland flow or erosion following a wildfire, although they started the measurements nine months after the burn. Scott and Van Wyk (1990), Scott and Schulze (1992), Lavabre et al. (1993), Scott (1993), and Ferreira (1996) observed significant increases of catchment runoff following forest fires. Coelho et al. (2004) and Ferreira et al. (2005b) conclude that the hydrological impact was due to the burn severity of the wildfire and the soil water repellency.

Prescribed fires result in an increase in overland flow when compared to the unburned control areas, such as mature pine or shrubland ('mato'). The total amount of overland flow is, however, significantly lower than overland flow rates produced on areas burned by wildfires, which can be as much as three to four times higher (Table 3). The experimental fire of Gestosa occurred in May 2002 when the soils were still wet, but an intermediate runoff response occurred mainly because the organic matter accumulation and the temperatures attained were somewhat greater than those for the prescribed fires. Although the initial soil moisture content was comparable for the experimental and prescribed fires, the prescribed fires had less overland flow. This difference could be related to differences in fuel quantity; the prescribed fires are seldom performed in areas where shrubs are more than 1 m high, and the Gestosa experimental fire was performed on shrubs more than 2 m high.

Erosion rates and nutrient loss depend principally on transport (i.e., hydrological processes) and on material available for transport. The wildfire sites show the highest erosion rates. This may be due to the high organic matter content prior to the burning, because mature *Pinus pinaster* stands accumulate a thick litter layer and some undergrowth. Additionally, overland flow plays an important role as the mass of eroded material is almost three times higher than observed from the prescribed fire site. The experimental fire of Gestosa had surprisingly lower erosion rates than those measured at the prescribed fire, 3.8 g m^{-2} h^{-1} on average for rainfall simulations with 50.5 mm per hour rainfall intensity.

Lost nutrients are mainly in the ash resulting from the combustion of vegetation and litter layer and, to a lesser extent, the mineral soil particles. Since the ash layer is highly erodible, it is easily transported by the first rainfall events of the rainy season. The concentration of solutes is higher in the wildfire burned area than in the experimental fire burned area, with the exception of potassium, which may be ascribed to differences in vegetation and litter. Considering overland flow amounts, the losses are twice as high for Mg, four times for sulphates and six times for Ca. The similar K loss for both land uses indicates a limitation in the availability of this nutrient.

At plot scale (16 m^2), overland flow, erosion rates, and solute loss for two unburned control plots (mature pine and shrub 'mato'), prescribed fires, and wildfires are listed in Table 4. It is expected that the intensity and spatial distribution of soil water repellency will have an increased effect on overland flow generation, erosion yield, and solute loss because of the larger surface area (16 m^2). Hydrological processes are controlled by rainfall intensities which seldom reach the 50.5 mm hr^{-1} produced by the rainfall simulator.

Table 3 Overland flow, erosion yields, and nutrient loss for different fire intensities measured during rainfall simulation experiments (after Ferreira et al. 2005b).

	Mature pine (Caratão)	"Mato" (shrubs) (Aigra Nova)	Prescribed fire (Cadafaz)	Experimental fire (Gestosa)	Wildfire (Caratão)	Wildfire (Senhor da Serra)
Overland flow (mm)	2.8	< 0.5	7.3	11.7	32.9	27.7
Erosion rates ($g\ m^{-2}\ h^{-1}$)	0.4	< detection limit	6.5	3.8	15.6	16.6
Solute loss (average of samples during 1-hr rainfall simulation, from a 0.24 m^2 plot).						
Ca ($mg\ l^{-1}$)	—	—	—	19.5	—	49.4
Mg ($mg\ l^{-1}$)	—	—	—	11.1	—	11.7
K ($mg\ l^{-1}$)	—	—	—	49.9	—	22.3
SO_4 ($mg\ l^{-1}$)	—	—	—	64.2	—	107.9

Thus, overland flow generation thresholds might not be achieved as frequently in the natural environment as compared to the rainfall simulations. For example, the control land uses had negligible overland flow and erosion, which were significantly lower than those measured with the rainfall simulator. The lower rainfall intensity and the larger area where processes of accumulation and transport interact may explain this general decrease, especially since these burns had some areas of non-water repellent soil where some infiltration occurred.

Table 4 Overland flow, erosion yields, and nutrient loss for different fire intensities as measured on 16 m^2 plots (after Ferreira et al. 2005b).

	Mature pine	"Mato" (Shrub)	Prescribed fire	Wildfire
Overland flow (% rainfall)	0.09	0.05	0.08	11.6
Erosion rates ($ton\ ha^{-1}\ yr^{-1}$)	0.02	0.08	0.2	2.2
Solute loss				
Ca ($mg\ l^{-1}$)	6.1	1.14	1.03	36.8
Mg ($mg\ l^{-1}$)	2.5	0.47	0.35	36.6
K ($mg\ l^{-1}$)	5.6	1.45	1.35	17.3
SO_4 ($mg\ l^{-1}$)	7.1	2.53	1.75	105.3

Prescribed fire solute losses were not significantly different from those at the control plots, although the wildfire plot recorded lower overland flow and erosion yields but similar concentrations of solute loss in comparison with the rainfall simulations.

In spite of varying temperatures and fire intensities, prescribed fires, experimental fires, and wildfires usually burn the entire litter layer and most vegetation to an extent that the remaining wood debris has little effect on hydrological or erosion processes. In some intense fires, the root systems are also burned, producing macropores that act as sinks for overland flow, sediment and solute yield (Doerr et al. 2003, Ferreira et al. 2000, 2003, 2005a). A critical result of varying burn severity is the degree and spatial distribution pattern of soil water repellency and its effect on hydrological and erosion processes. There is a direct relationship between soil burn severity and the degree of soil water repellency, and its contiguous spatial distribution. Wildfires tend to produce stronger and more contiguous distribution of soil water repellency. Therefore, it is expected that more intense wildfires will generate higher amounts of overland flow and erosion.

Areas burned with more intense fires produce greater overland flow, both at the micro-plot (rainfall simulation plot) and 16 m^2 plot scale. The soil water repellent layer formed (Giovanini 1994, Ferreira et al. 1997, 2005b) sharply reduces infiltration capacity (Coelho et al. 2002a, Ferreira et al. 2005b). The only sinks are the macropores and non-water repellent soil patches that allow the water to bypass the repellent layer. The significant differences in the degree and spatial distribution of soil water repellency result primarily from the distinct conditions that occur before and during the fires, in particular fire intensity and – to a lesser extent – soil and litter moisture contents (Botelho et al. 1994, Giovannini 1994). In addition, site-specific factors such as pre-fire organic matter content and composition could also play a role.

The spatial distribution of soil water repellency may have far-reaching land degradation implications. In situations where a strong degree of soil water repellency is restricted to small and hydrologically isolated areas, overland flow produced on water repellent soils could well infiltrate at hydrophilic areas down slope. This may explain the sharp decreases in overland flow amounts and sediment and solute transport from the micro-plot scale to the 16 m^2 plot scale.

In situations where large and/or inter-connected slope areas are strongly water repellent, even minor amounts of overland flow generated at the head of the slope can be enhanced down-slope, thereby increasing erosion rates and solute loss. The role of the spatial distribution of soil water repellency on hydrological processes at slope scale is also suggested by Ferreira et al. (2000, 2003) for unburned *Eucalyptus globulus* forest stands.

Overland flow and erosion rates are significantly lower at the 16 m^2 plot scale than at the micro-plot scale. This difference can be partially ascribed to the lower rainfall intensity and the greater measuring period, where soil characteristics become normalized, and to the hydrological processes, predominately deposition and infiltration through macropores, within the plot. They may play an important role in the water and sediment reduction observed in upscaling. The only significant increase is the sediment yield, which nevertheless was an order of magnitude less than that measured at the wildfire burned plot. Ferreira et al. (2005b) suggest that wildfire burned areas have a major influence on runoff at catchment scale immediately after a fire, and that this influence decreases significantly during the first year after the fire.

Lines of Action

Use of prescribed fire

From the studies presented, the use of prescribed fire to reduce organic matter accumulation and increase landscape diversity may be useful in reducing the risk of forest fire. Prescribed fire was originally used by shepherds to improve pastures before the mid-20th century reforestation effort. In the last two decades, prescribed fire has been slowly and steadily re-introduced as a forest and shrubland management practice (see Fernandes and Botelho 2003).

This tendency also seems to be occurring in the United States. Neary et al. (1999) states that, in contrast to the fire suppression paradigm of the past 50 years, more land managers are now reintroducing fire as a natural process into these ecosystems. This practice is being widely used in many areas, particularly in wildland-urban interfaces.

Preventive reforestation

An effort is underway to introduce new ideas in the reforestation process. These are described in the *Afforestation Manual to Prevent Forest Fires* (Silva and Pascoa 2002). One of the main themes is land planning. Several forest plans were developed recently, at various scales, serving as guidelines for the concerted management of the so-called 'Zonas de Intervenção Florestal.' These plans allow the ranking and identification of fire risk and promote the creation of discontinuity to reduce fire progression (see Silva 2002b).

The reforestation process is also guided by fire prevention philosophies that translate into different management practices for various geomorphologic landscape units, climatic regions, and vegetation types. For example, the choice of tree species for use in restoration will vary by the broad factors listed above as well as microclimatic conditions (Gomes and Silva 2002). Road and firebreak location can also be optimized to reduce the risk of fire progression by providing a discontinuity in the organic matter (Silva and Lima 2002).

In addition to the use of prescribed fires, other management practices exist to control the understory to reduce fire risk. This includes the use of herbicides, pasture, understory vegetation clearing, and soil plowing. These management practices can be applied around tree trunks, inside forest stands, or used to create a buffer zone around forested areas (Silva and Lopes 2002). Special care must be taken while working or walking on forested areas (Silva 2002c).

Approaches to reducing fire effects

The previous solutions are designed to prevent forest fires from occurring and, therefore, limit soil degradation by reduction of the frequency of burning and its severity. Nevertheless, forest fires have always occurred and will continue to occur in wet Mediterranean regions, and a plan of action is needed to promote soil and water conservation on burned soils. The research performed so far shows the delicacy of this task. Burned areas are frequently enormous, as shown in Figure 1, and to be effective, the intervention has to be made in the first few months after the fire. This poses overwhelming logistic and financial difficulties. It is unrealistic to treat all the landscape burned by a major forest fire. Therefore, new approaches have to be designed. Some ideas were developed and presented recently to the Portuguese Government. Although urgent and of utmost importance, the implementation of methodologies to reduce soil degradation processes after forest fires must be cost-effective and applied to selected locations where their implementation will render the biggest conservation impacts.

Since forest fires generally impact wide areas, it is impossible to intervene in the entire area to reduce nutrient loss, which happens as a result of ash removal in the first months after the fire. The nature and speed at which degradation processes occur make it impossible from a logistic and cost perspective. Therefore, a conceptual framework must identify the points where interventions are more effective. Geographical information systems can play an important role in locating areas for intervention to reduce soil and nutrient flush at the lowest cost possible.

The complexity and different scale of action for various mitigation techniques poses severe problems to the design and an implementation of experiments to be able to fully assess the efficiency of the selected mitigations techniques. Therefore a complex, multiple methodology approach must be developed including several scales of analyses.

CONCLUSION

We showed that most of the solute loss occurs during the first four months after the fire. In fact, important nutrient losses occur during the first rainfall events that are large enough to saturate the ash layer and promote overland flow. Solute loss depends on the availability of the easily mobilized ash layer. After the first four months, ash scarcity leads to solute loss only under extreme rainfall events. This presents an overwhelming problem for soil and water sustainability in Portugal, where since 1980 more than 115,000 ha are burned on an annual average. The enormous area burned each year and the rapid soil degradation (i.e., nutrient loss) following fires makes any attempt to implement mitigation measures a logistic nightmare and not cost effective. This rational supports the policy of forest fire prevention and is seen as the only efficient way to conserve forest productivity, soil, and water, and has resulted in a manual (Silva and Páscoa 2002) for forest managers, describing best management practices to reduce forest fire risk. This manual covers several forest management dimensions, including improvement of forest planning in accord with the geomorphologic and climatic conditions (which also drives species selection), optimization of firebreak and road locations, and practical advice on how to conduct forest management practices.

One of the main factors that has caused an increase in forest fire risk is the afforestation practices that plant exotic species in continuous commercial crop stands with little or no diversity. One of the main solutions for the forest fire problem in Portugal is a clear policy to increase landscape and economic structural diversity.

The reintroduction of prescribed fire has increasingly been considered a useful technique to control shrubs and the forest understory, allowing for greater landscape diversity, which creates breaks to the forest fires progression and, therefore, eases their suppression. The results presented show that prescribed fire effects on soil and water conservation seldom reach those of wildfires. The use of post-fire mitigation treatments is just beginning. At this early stage, the overwhelming problem is obtaining agreement among professionals and researchers involved in forest fire management as to how to identify areas for intervention, and with over 115,000 ha of burned area annually, how to mitigate post-fire effects in less than four months in a cost-effective manner. Several strategies have been developed and will be tested in the coming years.

Acknowledgments

This work was financed by several research projects, namely Contract 037046 (GOCE) DESIRE DESertification mItigation and REmediation of land a global approach for local solutions, financed by the European Union, and several research projects financed by the Portuguese Government, namely: POCTI/MGS/49210/2002 'SILVAQUA Assessment of climatic change impact on water resources and CO2 fixation in fast growing forest stands in Portugal.' POCI/AGR/60354/2004 EROSFIRE – 'A model-based, decision-support tool for soil erosion hazard assessment following forest wildfires.' POCI/AGR/58896/2004 PHOENIX – Forest conversion in burned areas. Projecto 2004 09 002629 7 do Fundo Florestal Permanente – 'Recuperação de Areas Ardidas'.

References

Botelho, H., J. Vega, P. Fernandes, and F. Rego. 1994. Prescribed fire behavior and fine fuel consumption in Northern Portugal and Galiza maritime pine stands. pp. 343-353. *In* D.X. Viegas [ed.]. The Proceedings of the 2nd International Conference on Forest Fire Research, University of Coimbra, Coimbra, Portugal.

Bunting, S. and F. Rego. 1988. Human impact on Portugal's vegetation. *Rangelands*, 10(6): 251-255.

Cerdà, A. 1998. Changes in overland flow and infiltration after a rangeland fire in a Mediterranean scrubland. *Hydrological Processes*, 12: 1031-1042.

Cerdà, A. and S.H. Doerr. 2005. The influence of vegetation recovery on soil hydrology and erodibility following fire: an eleven-year research. *International Journal of Wildland Fire*, 14(4): 339-342.

Cerdà, A. and T. Lasanta. 2005. Long-term erosional responses after fire in the Central Spanish Pyrenees. 1. Water and sediment yield. *Catena*, 60: 59-80.

Coelho C.O.A., R.A. Shakesby, M. Gonzalez del Tanago, J.L. Ternan, R.P.D. Walsh, and A.G. Williams. 1995b. Land management practice and erosion limitation in contrasting wildfire and gullied locations in the Iberian Peninsula. Final (unpublished) Report to the EU for Project EV5V-0041 'Land Management Practice and Erosion Limitation in Contrasting Wildfire and Gullied Locations in the Iberian Peninsula'.

Coelho, C.O.A., A.J.D. Ferreira, M. Baake, and J.J. Keizer. 2002a. Impacts of prescribed shrubland fire and forest wildfire on overland flow and soil erosion generating processes. pp. 1485-1496. *In* J.L. Rubio, R.P.C. Morgan, S. Asins, and V. Andreu [eds.]. The Proceedings of the 3rd International Congress, Man and Soil at the Third Millennium, Geoforma Ediciones, Logroño, Spain.

Coelho, C.O.A., A.J.D. Ferreira, A.K. Boulet, and J.J. Keizer. 2002b. Overland flow generation processes, erosion yields and nutrient loss under fires with different intensities: Lessons learned from analysis at different scales. pp. 197. *In* D.X. Viegas [ed.]. *Forest Fire Research and Wildland Fire Safety*. Millpress, Rotterdam, The Netherlands.

Coelho, C.O.A., A.J.D. Ferreira, A.K. Boulet, and J.J. Keizer. 2004. Overland flow generation processes, erosion yields and solute loss following different intensity fires. *Quarterly Journal of Engineering Geology and Hydrogeology*, 37(3): 233-240.

Cosandey, C., V. Andrèassian, C. Martin, J.F. Didon-Lescot, J. Lavabre, N. Folton, N. Mathys, and D. Richard. 2005. The hydrological impact of the Mediterranean forest: a review of French research. *Journal of Hydrology*, 301: 235-249.

DeBano, L.F. 1981. Water repellent soils: A state-of-the-art. General Technical Report, PSW-46. U.S. Department of Agriculture, Forestry Service, Pacific Southwest Research Station, Berkley, California, USA.

DeBano, L.F., L.D. Mann, and A. Hamilton. 1970. Translocation of hydrophobic substances into soil by burning organic litter. *Soil Science Society of America Proceedings*, 34: 130-133.

Doerr, S.H., R.A. Shakesby, and R.P.D. Walsh. 1996. Soil hydrophobicity variations with depth and particle size fraction in burned and unburnt *Eucalyptus globulus* and *Pinus pinaster* forest terrain in the Águeda basin, Portugal. *Catena*, 27: 25-47.

Doerr S.H., A.J.D. Ferreira, R.P.D. Walsh, R.A. Shakesby, G. Leighton-Boyce, and C.O.A. Coelho. 2003. Soil water repellency as a potential parameter in rainfall-runoff modelling: experimental evidence at point to catchment scales from Portugal. *Hydrological Processes*, 17: 363-377.

Fernandes, P.M. and H.S. Botelho. 2003. A review of prescribed burning effectiveness in fire hazard reduction. *International Journal of Wildland Fire*, 12(2): 117-128.

Ferreira, A.J.D. 1996. Processos hidrológicos e hidro-químicos em povoamentos de *Eucalyptus globulus* Labill. *E Pinus pinaster* Aiton. Ph.D. thesis, Aveiro University, Aveira, Portugal.

Ferreira, A.J.D. 1997. Soil erosion in a burned plot: Evaluation based on single event record. pp. 373-380. *In* Forest fire risk and Management Proceedings of the European School of Climatology and Natural Hazards Course, European Commission DG XII, Brussels, Belgium.

Ferreira, A.J.D., C.O.A. Coelho, R.A. Shakesby, and R.P.D. Walsh. 1997. Sediment and solute yield in forest ecosystems affected by forest fire and rip-ploughing techniques, central Portugal: a plot and catchment analysis approach. *Physics and Chemistry of the Earth*, 22: 309-314.

Ferreira, A.J.D., C.O.A. Coelho, A.J.B. Gonçalves, R.A. Shakesby, and R.P.D. Walsh. 1998. Impact of climatic change on slope and catchment hydrology in forest areas, Central Portugal. *Geookodynamic*, 19: 165-178.

Ferreira, A.J.D., C.O.A. Coelho, R.P.D. Walsh, R.A. Shakesby, A. Ceballos, and S.H. Doerr. 2000. Hydrological implications of soil water repellency in *Eucalyptus globulus* forests, north-central Portugal. *Journal of Hydrology*, 231/232: 165-177.

Ferreira, A.J.D., C.O.A. Coelho, R.P.D. Walsh, R.A. Shakesby, A. Ceballos and S.H. Doerr. 2003. The impact of water-repellency on overland flow and runoff in Portugal. pp. 167-178. *In* C.J. Ritsema and L.W. Dekker [eds.]. *Soil Water Repellency – Occurrence, Consequences, and Amelioration*. Elsevier Science Publishers, Amsterdam, The Netherlands.

Ferreira, A.J.D., C.O.A. Coelho, A.K. Boulet, and F.P. Lopes. 2005a. Temporal patterns of solute loss following wildfires in Central Portugal. *International Journal of Wildland Fire*, 14: 401-412.

Ferreira, A.J.D., C.O.A. Coelho, A.K. Boulet, G. Leighton-Boyce, J.J. Keizer, and C.J. Ritsema. 2005b. Influence of burning intensity on water repellence and hydrological processes at forest sites in Portugal. *Australian Journal of Soil Research*, 43: 327-336.

Gimeno-García, E., V. Andreu, and J.L. Rubio. 2004. Spatial patterns of soil temperatures during experimental fires. *Geoderma*, 118: 17-38.

Giovannini, G. 1987. Effect of fire and associate heating wave on the physicochemical parameters related to the soil potential erodibility. *Ecologia Mediterranea*, 13: 111-117.

Giovannini, G. 1994. The effect of fire on soil quality. pp. 15-27. *In* M. Sala and J.L. Rubio [eds.]. *Soil Erosion and Degradation as a Consequence of Forest Fires*. Geoforma Ediciones, Logroño, Spain.

Giovannini, G. and S. Lucchesi. 1984. DTA and IR investigation on soil hydrophobic substances. *Soil Science*, 137: 457-463.

Giovannini, G., S. Lucchesi, and M. Giachetti. 1988. Effect of heating on some physical and chemical parameters related to soil aggregation and erodibility. *Soil Science,* 146: 255-261.

Gomes, F. and J.S. Silva. 2002. Normas preventivas no planeamento das arborizações. *In* J.S. Silva and F. Páscoa [eds.]. Manual de Silvicultura para a Prevenção de Incêndios. VIII, 1-8, Direcção Geral das Florestas, Lisboa, Portugal.

Hungerford, R.D., M.G. Harrington, W.H. Frandsen, K.C. Ryan, and G.J. Niehoff. 1995. The influence of fire on factors that affect site productivity. pp. 32-50. *In* A.C. Harvey and L.F. Nuenschwander [eds.]. The Proceedings of the Management and Productivity of Western-Montane Forest Soils. General Technical Report INT-280. U.S. Department of Agriculture, Forest Service, Intermountain Research Station, Ogden, Utah, USA.

Imeson, A.C., J.M. Verstraten, E.J. Van Mullingen, and J. Sevink. 1992. The effects of fire and water repellency on infiltration and runoff under Mediterranean type forests. *Catena,* 19: 345-361.

Lavabre, J., D.S. Torres, and F. Cernesson. 1993. Changes in the hydrological response of a small Mediterranean basin a year after a wildfire. *Journal of Hydrology,* 142: 273-299.

Margaris, N.S., E. Koutsidou, and C. Giourga. 1996. Changes in traditional Mediterranean land-use systems. pp. 29-42. *In* C.J. Brandt and J. Thornes [eds.]. *Mediterranean Desertification and Land-Use.* John Wiley & Sons, Chichester, UK.

Moreira, F., F.C., Rego, and P.G. Ferreira, 2001. Temporal (1958-1959) pattern of change in a cultural landscape of northwestern Portugal: implications for fire occurence. *Landscape Ecology,* 16: 557-567.

Moreno, J.M. 1999. Forest fires: trends and implications in desertification prone areas of Southern Europe. pp. 115-150. *In* P. Balabanis, D. Peter, A. Ghazi and M. Tsogas [eds.]. Mediterranean Desertification: Research Results and Policy Implications 1. DG Research, European Commission, Bussels, Belgium.

Neary, D.G., C.C. Klopatek, L.F. DeBano, and P.F. Ffolliott. 1999. Fire effects on belowground sustainability: a review and synthesis. *Forest Ecology and Management,* 122: 51-71.

Nogueira, C.D.S. 1990. A floresta Portuguesa, DGF. *Informação,* 2: 18-28.

Nunes, M.C.S., M.J. Vasconcelos, J.M.C. Pereira, N. Dasgupta, R.J. Alldredge, and F.C. Rego. 2005. Land cover type and fire in Portugal: Do fires burn land cover selectively? *Landscape Ecology,* 20: 661-673.

Pereira, J.M.C. and T.N. Santos. 2003. Fire risk and burned area mapping in Portugal. Direcção Geral das Florestas, Lisboa, Portugal.

Pickett, S. and P.S. White. 1985. *The Ecology of Natural Disturbance and Patch Dynamics.* Academic Press, San Diego, California, USA.

Scott, D.F. 1993. The hydrological effects of fire in South African mountain catchments. *Journal of Hydrology,* 150: 409-432.

Scott, D.F. and D.B. Van Wyk. 1990. The effects of wildfire on soil wettability and hydrological behavior of an afforested catchment. *Journal of Hydrology,* 121: 239-256.

Scott, D.F. and R.E. Schulze. 1992. The hydrological effects of a wildfire in a eucalypt afforested catchment. *South African Forestry Journal,* 160: 67-74.

Sevink, J., A.C. Imeson, and J.M. Verstraten. 1989. Humus form development and hillslope runoff, and the effects of fire and management, under Mediterranean forest in NE Spain. *Catena*, 16: 461-475.

Shakesby, R.A., C.O.A. Coelho, A.J.D. Ferreira, J.P. Terry, and R.P.D. Walsh. 1993. Wildfire impacts on soil erosion and hydrology in wet Mediterranean forest, Portugal. *International Journal of Wildland Fire*, 3: 95-110.

Shakesby, R.A., S.H. Doerr, and R.P.D. Walsh. 2000. Problems and prospects in determining the hydrological and erosional significance of soil hydrophobicity. *Journal of Hydrology*, 231-232: 178-191.

Silva, J.M. 1990. La gestion forestie´ re et la silviculture de prevention des espaces forestiers menaces par les incendies au Portugal. *Revue Forestière Française*, 40 (spécial): 337-345.

Silva J.S. 2002a. As espécies florestais e a propagação do fogo. *In* J.S. Silva and F. Páscoa [eds.]. Manual de Silvicultura para a Prevenção de Incêndios. III, 1-8, Direcção Geral das Florestas, Lisboa, Portugal.

Silva, J.S. 2002b. O ordenamento do território. *In* J.S. Silva and F. Páscoa [eds.]. Manual de Silvicultura para a Prevenção de Incêndios. VII, 1-8, Direcção Geral das Florestas, Lisboa, Portugal.

Silva, J.S. 2002c. O fogo controlado. *In* J.S. Silva and F. Páscoa [eds.]. Manual de Silvicultura para a Prevenção de Incêndios. XI, 1-8, Direcção Geral das Florestas, Lisboa, Portugal.

Silva J.S. and J.R. Lopes 2002. A prevenção de incêndios através da intervenção no subcoberto. *In* J.S. Silva and F. Páscoa [eds.]. Manual de Silvicultura para a Prevenção de Incêndios. X, 1-8, Direcção Geral das Florestas, Lisboa, Portugal.

Silva, J.S. and I.P. Lima. 2002. Rede viária e divisional. *In* J.S. Silva and F. Páscoa [eds.]. Manual de Silvicultura para a Prevenção de Incêndios. IX, 1-8, Direcção Geral das Florestas, Lisboa, Portugal.

Silva, J.S. and F. Páscoa [eds.]. 2002. Manual de Silvicultura para a Prevenção de Incêndios. VIII, 1-8, Direcção Geral das Florestas, Lisboa, Portugal.

Thomas, A.D., R.P.D. Walsh, and R.A. Shakesby. 1999. Nutrient losses in eroded sediment after fire in eucalyptus and pine forests in the wet Mediterranean environment of northern Portugal. *Catena*, 36: 283-302.

Thomas, A.D., R.P.D. Walsh, and R.A. Shakesby. 2000a. Post-fire forestry management and nutrient losses in eucalyptus and pine plantations, northern Portugal. *Land Degradation and Development*, 11: 257-271.

Thomas, A.D., R.P.D. Walsh, and R.A. Shakesby. 2000b. Solutes in overland flow following fire in eucalyptus and pine forests, northern Portugal. *Hydrological Processes*, 14: 971-985.

Úbeda, X. and M. Sala. 2001. Chemical concentrations in overland flow from different forested areas in a Mediterranean Environment: burned forest at different fire intensity and unpaved road. *Zeitschrift fur Geomorphologie*, 45: 225-238.

Vacca, A., S. Loddo, G. Ollesch, R. Puddu, G. Serra, D. Tomasi, and A. Aru. 2000. Measurement of runoff and soil erosion in three areas under different land use in Sardinia (Italy). *Catena*, 40: 69-92.

Vasconcelos, M.J.P., S. Silva, M. Tomé, M. Alvim, and J.M.C. Pereira. 2001. Spatial prediction of fire ignition probabilities: comparing logistic regression and neural networks. *Photogrammetric Engineering and Remote Sensing*, 67(1): 73-81.

Walsh, R.P.D., D.J. Boakes, C.O.A. Coelho, A.J.B. Gonçalves, R.A. Shakesby, and A.D. Thomas. 1994. Impact of fire-induced hydrophobicity and post-fire forest litter on overland flow in northern and central Portugal. pp. 1149-1159. *In* Proceedings of the 2nd International Conference on Forest Fire Research. University of Coimbra, Coimbra, Spain.

Weast, R.C. 1988. *Handbook of Chemistry and Physics*. CRC Press, Boca Raton, Florida, USA.

16

Effects of Forest Fires and Post-Fire Rehabilitation: A Colorado, USA Case Study

Lee H. MacDonald[1]* and *Isaac J. Larsen*[1]

Abstract

Anthropogenic activities have increased the number of large, high-burn severity wildfires in the lower and mid-elevation coniferous forests in Colorado as well as much of the western US. Forests provide most of the water for cities and agriculture, and the increased runoff and erosion after wildfires is a major concern because of the potential adverse effects on flooding, water quality, and other aquatic resources. Areas burned at high severity are of primary concern because rainfall intensities of only 8 to 10 mm h^{-1} can generate substantial amounts of runoff and surface erosion. Typical post-fire erosion rates from areas burned at high severity are 5 to 10 Mg ha^{-1} yr^{-1} for the first 2 to 3 yr after burning, and this is about 5 to 80 times the values measured from areas burned at moderate or low severity. Post-fire sediment yields are most closely associated with the amount of surface cover and rainfall erosivity. Three to five years are typically required before hillslope-scale sediment yields decline to near-background levels.

Studies on multiple fires indicate that the most effective post-fire rehabilitation treatments are those that immediately provide surface cover, such as straw mulching. Seeding and seeding combined with scarification did not increase the rate of vegetative regrowth and therefore did not reduce post-fire sediment yields. Hydromulching varied in its effectiveness, and this was attributed to the differences in the mixtures applied to different sites. Contour-felled log erosion barriers were effective only for small and moderate-sized storms, and the effectiveness of this treatment is easily negated by poor installation. The application of a polyacrylamide also failed to significantly reduce post-fire sediment yields. Mulching is the most cost-effective treatment at US$50 to US$150 per megagram reduction in sediment yields.

---

[1] *Department of Forest, Rangeland, and Watershed Stewardship, Colorado State University, Fort Collins, Colorado, USA.*

* *Corresponding author: Lee H. MacDonald, Department of Forest, Rangeland, and Watershed Stewardship, Colorado State University, Fort Collins, Colorado 80523-1472, USA. Tel: 970 491 6109, e-mail: leemac@cnr.colostate.edu*

Post-fire sediment yields from Colorado are within the range of values reported from the western US and other countries. The results of this case study can provide useful guidance to land managers and researchers in other areas, as the basic principles and processes identified in this chapter are more broadly applicable.

INTRODUCTION

Over the last several decades there has been an increase in the number of large wildfires in the western US (Westerling et al. 2006). The increase in wildfires is a major concern for the public and resource managers because of the potentially large increases in runoff and erosion, and the resulting adverse effects on life, property, and aquatic resources. Flooding after the 1996 Buffalo Creek Fire southwest of Denver, Colorado caused two fatalities and repeatedly washed out a state highway, and the increased sediment load reduced the storage capacity in Strontia Springs Reservoir by approximately one-third (Agnew et al. 1997). Debris flows after the 2002 Coal Seam and Missionary Ridge Fires in western Colorado damaged homes, roads, and railways (Cannon et al. 2003). The high sediment and ash loads after high severity fires greatly increase water treatment costs and reduce macro-invertebrate and fish populations (Rinne 1996, Rieman and Clayton 1997, Minshall et al. 2001, Kershner et al. 2003).

The hydrologic and geomorphic effects of high severity wildfires are of particular concern in Colorado because most of the state's water supply is derived from forested areas and water-related resources are an important economic asset (MacDonald and Stednick 2003). There also has been a large increase in the number of people living in the wildland-urban interface, and this has increased the potential loss of life and property from high-severity wildfires and the subsequent flooding and erosion.

Land managers commonly apply rehabilitation treatments after high severity wildfires in order to reduce the potential increases in runoff and erosion. Mitigation treatments include seeding, scarification, mulching, hydromulching, and the application of soil binding agents such as polyacrylamides. The application of such treatments over large areas is quite costly, as evidenced by the US$25 million spent after the 2002 Hayman Fire by the U.S. Department of Agriculture, Forest Service (USFS) and the Denver Water Board (Robichaud et al. 2003, Wiley, personal communication 2005), and the approximately US$100 million spent after the Cerro Grande Fire in northern New Mexico (Morton et al. 2003). The problem is that there have been very few studies in the central Rocky Mountains on post-fire erosion rates or the effectiveness of mitigation treatments in reducing post-fire sediment yields. There also is an urgent need to better understand the underlying processes that cause the observed increases in post-fire runoff and erosion rates, as this is crucial to predicting post-fire effects and the application of cost-effective rehabilitation treatments.

The objectives of this chapter are to: 1) provide a basic understanding of the historic fire frequency and severity in the major forest types in Colorado; 2) summarize our current understanding of post-fire erosion processes; 3) quantify the effects of wild and prescribed fires on soil and aquatic resources at both the hillslope and small catchment scales; and 4) summarize our data on the effectiveness of post-fire rehabilitation treatments. The data presented in this case study are derived from intensive, multi-year studies on how wild and prescribed fires affect soil properties, vegetative cover, runoff, and erosion rates. The fortuitous collection of hillslope and catchment-scale data prior to the Hayman Fire allows us to directly compare pre- and post-fire conditions. For three wildfires data also have been collected on the effectiveness of different post-fire rehabilitation treatments. The combined dataset includes nearly 600 plot-yr of data at the hillslope scale, and catchment-scale runoff, cross-section change, and sediment yield data from three wildfires (Moody and Martin 2001a, Eccleston and MacDonald 2006, Kunze and Stednick 2006, Eccelston 2008). Rainfall simulations and process-based studies provide more detailed insights into the causes of the observed increases in runoff and erosion after wild and prescribed fires, and help explain why different post-fire rehabilitation treatments vary in their effectiveness.

This combination of studies provides a unique, in-depth understanding of the effects of forest fires on runoff and erosion at different spatial scales, and the effectiveness of post-fire rehabilitation treatments. The resulting information should be of considerable use for researchers and land managers in other areas, as the underlying processes will vary in rates and magnitude but are generally applicable to other burned areas.

FOREST TYPES AND FIRE REGIMES

Colorado contains a variety of forest types (Fig. 1), and the type of forest is largely controlled by the amount of precipitation in relation to potential evapotranspiration (PET). In general, a rise in elevation increases precipitation while decreasing temperatures and PET. This moisture gradient means that higher-elevation forests provide most of the runoff for both municipal water supply and agriculture. Conversely, fire risk and fire frequency generally decline with increasing elevation. The moisture and temperature gradients largely control the presence of the different forest types in Colorado, and these can be broadly classified into the lower montane, montane, and subalpine zones. Each zone has a different moisture regime, species composition, fuel density, and historic fire regime (Romme et al. 2003a).

At the dry end of the moisture gradient is the lower montane zone (~1675 to 2000 m), and this is dominated by ponderosa pine (*Pinus ponderosa*) (Romme

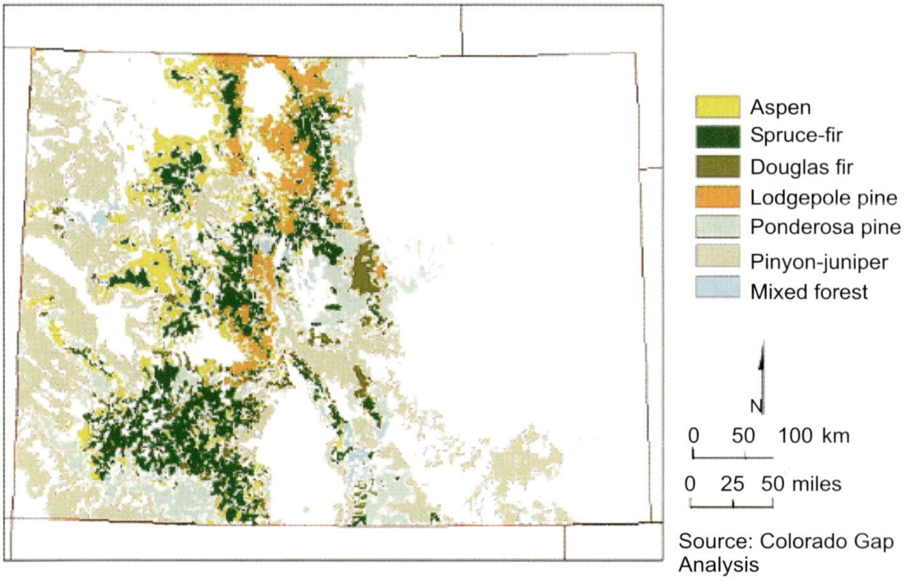

Fig. 1 Map of the major forest types in Colorado.

et al. 2003a). The natural fire regime is characterized by low-severity surface fires with a recurrence interval of 5 to 40 yr (Veblen 2000, Veblen et al. 2000, Grissino-Mayer et al. 2004). The frequent fires tended to maintain an open, park-like forest (Veblen 2000, Romme et al. 2003b).

The predominant forest type in the montane zone (~2000 to 2600 m) is intermixed ponderosa pine and Douglas-fir (*Pseudotsuga menziesii*), and in more mesic areas these grow in dense, closed-canopy stands (Romme et al. 2003a). The natural fire regime is mixed severity, with both frequent, low severity fires and infrequent, high severity fires (Brown et al. 1999, Ehle and Baker 2003, Romme et al. 2003a). Under the natural fire regime individual stands burned at intervals ranging from every 10 to 100 yr, and the larger, high severity fires tended to occur during severe droughts after a wetter period that allowed more fuels to accumulate. The larger fires could be up to thousands of hectares in size, but within the fire perimeter there would be a heterogeneous patchwork of high severity, low severity, and unburned areas (Romme et al. 2003a).

The forests in the subalpine zone (~2600 to 3400 m) are composed of relatively dense stands dominated by lodgepole pine (*Pinus contorta*), Engelmann spruce (*Picea engelmannii*), and subalpine fir (*Abies lasiocarpa*) (Romme et al. 2003a). The cooler, moister environment and shorter summers means that most fires were naturally extinguished before they spread, but there were infrequent, high severity, crown fires (e.g., Romme et al. 2003a, Buechling and Baker 2004, Sibold et al. 2006). On average, individual stands burned only once every 100 to 500 yr because of the infrequent congruence of the drought conditions necessary for fire spread and a natural ignition source, such as lightning (Veblen 2000). In contrast to Engelmann spruce and subalpine fir, lodgepole pine is highly dependent on these infrequent, stand-replacing fires because it is not shade tolerant and requires bare mineral soil for establishment.

Native Americans used fire for hunting and manipulating vegetative cover, but they are not believed to have greatly affected the natural fire regime. European settlement in the second half of the 19th century had a much greater effect on fuel loads and the fire regime of Colorado forests. In many areas there was an initial reduction in forest density due to timber harvest, grazing, and clearing for pasture. Since the early 1900s there has been a decrease in fire frequency and an increase in forest density due to fire suppression, the cessation of widespread burning by settlers, and reductions in grazing and logging.

The changes in fuel loading and fire frequency have been most pronounced in the lower montane and montane forests (Romme et al. 2003a). In the lower montane forests there has been an estimated 2 to 14 fold decrease in fire frequency since about 1920 (Veblen 2000). The increased forest densities have increased the vulnerability of these forests to large, high severity fires during severe droughts (Keane et al. 2002, Romme et al. 2003b). Similarly, the montane forests have denser, even-aged stands following logging and fires in the 19th century and the relatively wet conditions early in the 20th century (Romme et al. 2003b). As in the lower montane zone, the increased density is believed to have increased the risk of large, high-severity fires (Romme et al. 2003b).

In the subalpine forests, European settlement has had a much smaller effect on the fire regime. Fire suppression has been in effect for less than 100 yr, while large portions of the spruce-fir forests have not been affected by fire for 400 yr. This means that the period of fire suppression is still too short to have greatly altered the natural fire regime (Buechling and Baker 2004, Sibold et al. 2006). Timber harvest, grazing, and other uses have altered the stand structure and species composition in some areas, but most stands are still within their natural range of variability in terms of forest density and fuel loadings (Romme et al. 2003b).

The number, size, and severity of wildfires since 1996 provides strong empirical evidence for an altered fire regime in the lower montane and montane zones in Colorado. Major fires in these zones include the 1996 Buffalo Creek fire, which burned 48 km^2 in the South Platte River Watershed southwest of Denver; the June 2000 High Meadows and Bobcat Fires, which each burned more than 40 km^2; and the record 2002 fire season, which included the 557 km^2 Hayman Fire southwest of Denver, the 295 km^2 Missionary Ridge Fire in southwestern Colorado, and the 49 km^2 Coal Seam Fire in western Colorado (Cannon et al. 2003, Graham 2003). The Hayman Fire was unprecedented in terms of both the size of the fire and homogeneity of high severity burns (Romme et al. 2003a). The resulting increases in runoff, flooding and erosion, together with the degradation of downstream aquatic resources, stimulated much of the research that is summarized in this case study.

EFFECTS OF FIRES ON SURFACE COVER, SOIL WATER REPELLENCY, RUNOFF AND EROSION

Surface Cover and Soil Water Repellency

Under unburned conditions the lower montane and montane forests typically have greater than 85 percent surface cover (Libohova 2004) and infiltration rates in excess of 100 mm h^{-1} (Martin and Moody 2001). After a high-severity fire 85 to 95 percent of the surface is either bare mineral soil or bare soil covered with ash (Fig. 2) (Libohova 2004, Pietraszek 2006). In moderate severity fires the litter layer is completely consumed, but there is no alteration of the underlying mineral soil. Low severity fires are defined by the incomplete combustion of the surface litter (Wells et al. 1979). The consumption of the protective litter cover in high severity fires greatly increases the amount of rainsplash, propensity for overland flow, and wind erosion (e.g., Terry and Shakesby 1993, Prosser and Williams 1998, Whicker et al. 2006).

Burning also alters the strength, persistence, and depth of soil water repellency (see Chapter 7). In coniferous forests in Colorado, as in many other

Fig. 2 Photo from summer 2003 showing a pair of sediment fences, the piles of sediment excavated from the fences, and the relatively bare hillslopes one year after the June 2002 Hayman Fire.

areas, the soil surface is often strongly water repellent under unburned conditions (Huffman et al. 2001). Burning at high and moderate severity vaporizes a variety of hydrophobic compounds in ponderosa and lodgepole pine forests, and the condensation of these compounds induces strong soil water repellency from the soil surface to a depth of approximately 6 cm (Fig. 3) (Huffman et al. 2001, Rough and MacDonald 2005). Data from several fires suggests that post-fire soil water repellency is slightly stronger and deeper after prescribed fires than wildfires, and this may be attributed to higher fuel loadings and greater heating due to the slower rate of fire spread (Huffman et al. 2001). Overall, the strength of soil water repellency rose with both increasing burn severity and sand content, and decreased with increasing soil moisture (Huffman et al., 2001). These trends are consistent with other studies (e.g., DeBano 1981, Chapter 7), but the high spatial and temporal variability means that these three variables explained only 30 to 41 percent of the observed variability (Huffman et al. 2001). The large spatial and temporal variability in soil water repellency within fires and severity classes appears to

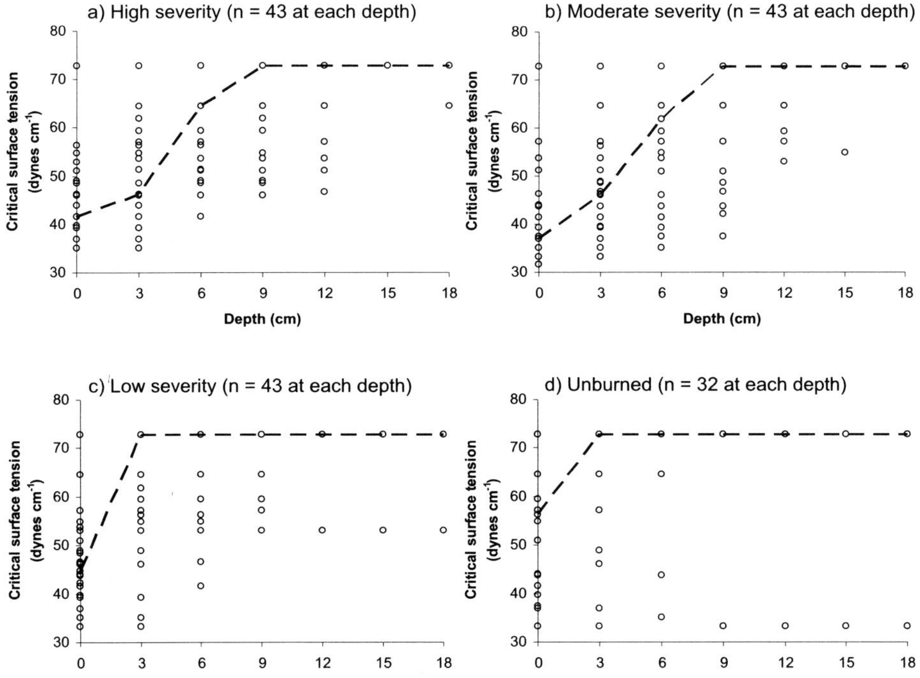

Fig. 3 Soil water repellency versus soil depth for: a) high burn severity, b) moderate burn severity, c) low burn severity, and d) unburned sites (from Huffman et al. 2001). The burned sites represent data from two wild and three prescribed fires the Colorado Front Range. Higher values indicate weaker water repellency and the dashed lines indicate the median values.

be characteristic of wildfires in Colorado and elsewhere (Hubbert et al. 2006, Woods et al. 2007).

Other researchers have identified a soil moisture threshold, which is when soils shift from being water repellent to hydrophilic (e.g., Doerr and Thomas 2000). Data from the Bobcat Fire indicate that the soil moisture threshold increases with fire severity, as the soil moisture threshold was only 10 percent in unburned sites, 13 percent in sites burned at low severity, and at least 26 to 28 percent in sites burned at moderate and high severity (MacDonald and Huffman 2004). The presence of a soil moisture threshold probably helps explain why burning has little effect on winter runoff and erosion rates as discussed below.

Both repeated measurements on the same fire and comparisons from fires of different ages indicate that post-fire soil water repellency is relatively short-lived in the lower montane and montane forests in Colorado. At the Bobcat Fire, the soil water repellency was much weaker three months after burning and was statistically non-detectable 12 months after burning (MacDonald and Huffman 2004). At the Hayman Fire the post-fire soil water repellency broke down most rapidly at the soil surface, and was statistically undetectable at all depths within two years after burning (Fig. 4) (MacDonald et al. 2005). The more rapid decay at the soil surface was attributed to the preferential erosion of the finer-grained water repellent particles, chemical breakdown due to solar radiation, the physical disturbance induced by repeated freezing and thawing, and the greater biological activity at the soil surface. As discussed later, this rapid decay means that soil water repellency is unlikely to be the

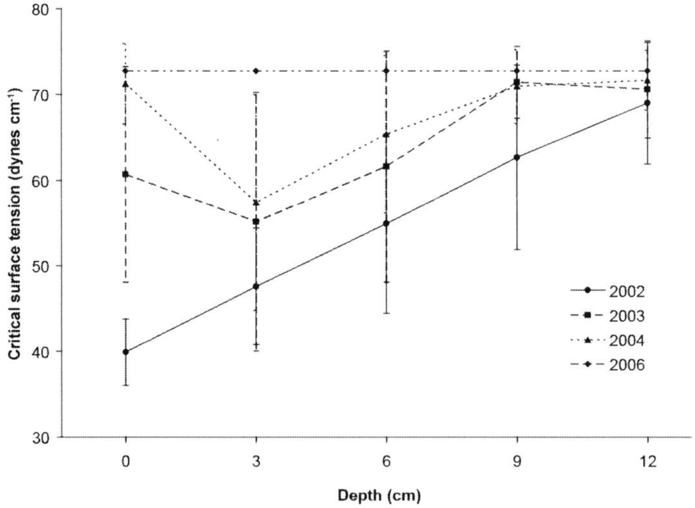

Fig. 4 Mean soil water repellency over time from the June 2002 Hayman Fire. Higher values indicate weaker soil water repellency and the bars indicate one standard deviation. There was no water repellency in 2006.

primary cause of the observed increases in runoff and surface erosion after burning (MacDonald et al. 2005).

Runoff

In unburned forests infiltration rates typically are much greater than rainfall intensities, and this means that infiltration-excess (Horton) overland flow is very rare (MacDonald and Stednick 2003). In areas adjacent to the Hayman fire with coarse granitic soils, rainfall intensities of 60 to 65 mm h^{-1} did not induce any overland flow (Libohova, 2004). After a high-severity wildfire overland flow is much more prevalent, and data from wildfires in Colorado, New Mexico, and western South Dakota indicate that storms with a maximum 30 min (I_{30}) rainfall intensity of only 7 to 10 mm h^{-1} can induce Horton overland flow (Cannon et al. 2001a, Moody and Martin 2001b, Benavides-Solorio 2003, Pietraszek 2006, Kunze and Stednick 2006, Wagenbrenner et al. 2006). The dramatic change from subsurface to surface runoff can increase the size of peak flows by one to two orders of magnitude (Bolin and Ward 1987, Moody and Martin 2001a, Gottfried et al. 2003), and readily explains the observed flooding, scour in low-order channels, and increase in debris flows in steep, headwater basins (Cannon and Reneau 2000, Cannon et al. 2001b).

The problem is that we cannot yet quantify the relative importance of the various processes that are believed to contribute to the observed decrease in infiltration. In addition to the post-fire increase in soil water repellency, burning consumes the surface organic layer and this decreases interception. In high severity fires the consumption of the organic matter at the soil surface effectively disaggregates the soil particles (Giovannini and Lucchesi 1983), and this increases the potential for soil sealing (Neary et al. 1999). The loss of the protective litter cover reduces surface roughness and thereby increases overland flow velocities and the size of peak flows. The combined effect on runoff rates are well documented for different areas (e.g., Helvey 1980, Prosser and Williams 1998, Kunze and Stednick 2006), but more detailed, process-based experiments are needed to determine the role of each factor under different conditions.

Post-fire Sediment Yields in Lower Montane and Montane Forests

Post-fire erosion processes and sediment yields

The same set of processes that increase post-fire runoff rates play a major role in increasing post-fire erosion rates. The loss of surface cover decreases interception, increases rainsplash erosion, and increases runoff velocities. The disaggregation of soil particles increases soil erodibility (Moody et al. 2005) and the susceptibility to soil sealing. The increase in soil erodibility and surface runoff increases sheetwash, rilling, and channel erosion.

In the lower montane and montane forests in Colorado, a series of studies have shown that high severity wildfires increase hillslope- and catchment-scale sediment yields by several orders of magnitude (Morris and Moses 1987, Moody and Martin 2001a, Benavides-Solorio and MacDonald 2005, Pietraszek 2006). Plots established before the Hayman Fire generated no sediment or overland flow in the year prior to burning (Libohova 2004). After burning at high severity, individual storms with rainfall intensities of 8 to 40 mm hr^{-1} generated up to 15 Mg ha^{-1} of sediment from the same plots. On average, the plots burned at high severity in the Hayman Fire generated 7 to 11 Mg ha^{-1} of sediment during each of the first three years after burning (Fig. 2).

Controls on post-fire sediment yields

Data from six different fires in the Colorado Front Range show that over 90 percent of the post-fire sediment is generated by high intensity summer thunderstorms (Benavides-Solorio and MacDonald 2005). Little sediment is generated by snowmelt because the soils are not water repellent at higher soil moisture contents and snowmelt rates generally do not exceed the infiltration capacity. The spatial and temporal variability in summer thunderstorms causes a corresponding variability in post-fire sediment yields, and this limits our ability to deterministically predict post-fire sediment yields at different spatial scales (Larsen and MacDonald 2007).

In general, the areas burned at high severity are of greatest concern because in the first year after burning these areas produce about 5 to 40 times more sediment than the plots burned at moderate severity (Fig. 5; Benavides-

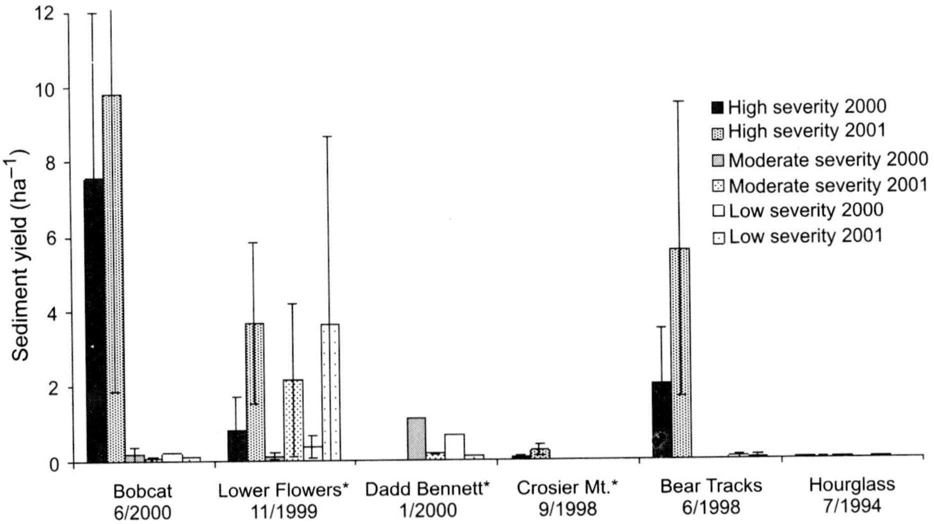

Fig. 5 Sediment yields by burn severity for six Colorado fires for June-October 2000 and June-October 2001. The bars indicate one standard deviation and an asterisk denotes a prescribed fire. Month and year of burning are listed under each fire. Not all severities were monitored in each fire.

Solorio and MacDonald 2005). The low burn severity plots generally produced only about half as much sediment as moderate burn severity plots (Fig. 5), but the validity of these relative values are constrained by the much smaller number of low and moderate burn severity plots (Benavides-Solorio and MacDonald 2005).

For plots burned at high severity, the hillslope-scale sediment yields from wildfires were substantially greater than the sediment yields from prescribed fires. The lower sediment yields from prescribed fires can be attributed to more needlefall, the patchier distribution of burn severity, and the resultant potential for downslope areas to capture some of runoff and sediment coming from the more severely burned areas (Benavides-Solorio and MacDonald 2005).

Data from the Bobcat fire show that convergent hillslopes produced 3 to 4 times more sediment per unit area than planar hillslopes, and this difference is attributed to the rilling observed in the convergent hillslopes (Benavides-Solorio and MacDonald 2005). Subsequent measurements have shown that rill incision in convergent hillslopes accounts for about 60 to 80 percent of the hillslope-scale sediment yields from the Hayman and Schoonover Fires (Pietraszek 2006). Hillslope erosion and channel incision measurements after the 1996 Buffalo Creek Fire also indicate that channel incision generated about 80 percent of the estimated sediment yield from a 27 km^2 basin (Moody and Martin 2001a). Our current conceptual model is that most of the surface runoff is being generated from the hillslopes, but most of the sediment is being generated by concentrated flow and incision in the convergent rills and lower order channels. In the most extreme storms we posit that the more planar sideslopes generate and deliver a greater proportion of the sediment through the development of a dense rill network.

Both univariate and multivariate analyses show that percent surface cover is the predominant control on post-fire sediment yields, as this explains approximately 61 percent of the variability in post-fire sediment yields (Fig. 6; Benavides-Solorio and MacDonald 2005, Pietraszek 2006). If cover is held constant, rainfall erosivity becomes the most important control on post-fire sediment yields (Benavides-Solorio and MacDonald 2005, Pietraszek 2006). A plot of the annual sediment yields from 72 hillslopes that burned at high severity in nine different fires shows that median sediment yields are highest in the second year after burning (Fig. 7), and this is due to the greater summer rainfall and slow rate of regrowth. By the fourth summer after burning the median sediment yield drops from nearly 10 Mg ha^{-1} to only 0.1 Mg ha^{-1}, but there is tremendous variability due to the variations in summer rainfall and rate of vegetative regrowth (Fig. 7). Sediment yields drop to near background levels once the percent bare soil drops below 30 percent (Fig. 6), and this typically requires about four years for plots burned at high severity, two years for plots burned at moderate severity, and less than one year for plots burned at low severity. Plots with coarser soils generally have slower regrowth rates due to their poorer water holding capacity (Benavides-Solorio and MacDonald

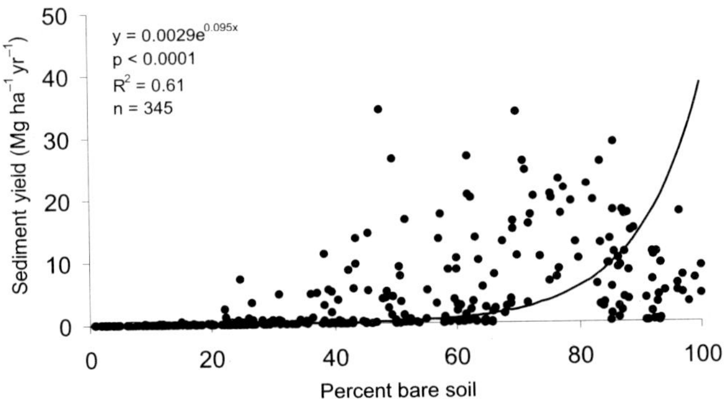

Fig. 6 Relationship between percent bare soil and annual sediment yield. Data were collected from seven wildfires and three prescribed fires in the Colorado Front Range.

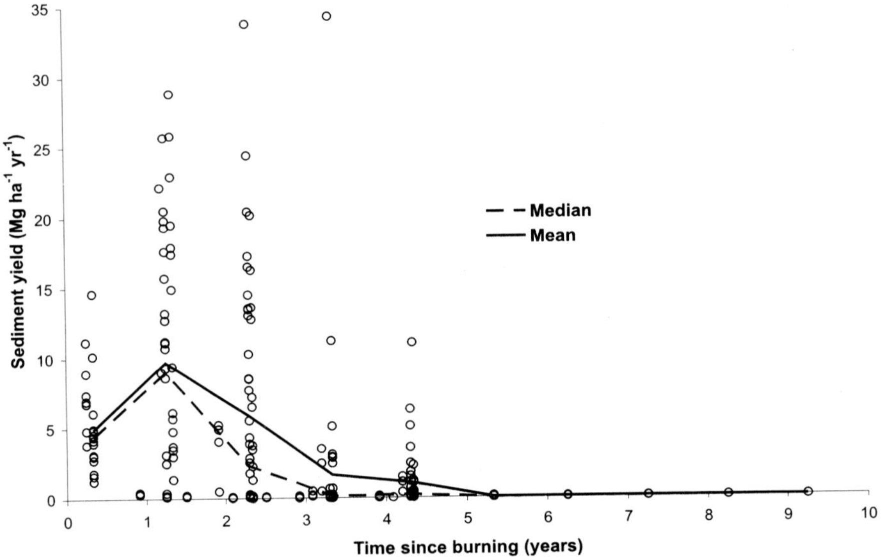

Fig. 7 Annual sediment yields versus time since burning for plots burned at high burn severity in six wildfires and three prescribed fires in the Colorado Front Range.

2005). In the case of the Hayman Fire, which has very coarse-textured soils, sediment yields were still elevated in the summer of 2006, which is four full years after burning.

Comparison of post-fire sediment yields from Colorado against other regions

Post-fire sediment yields for the Colorado Front Range are within the range of values reported in the western US and other countries (Fig. 8). First-year sediment yields from Colorado are generally similar to values from southern

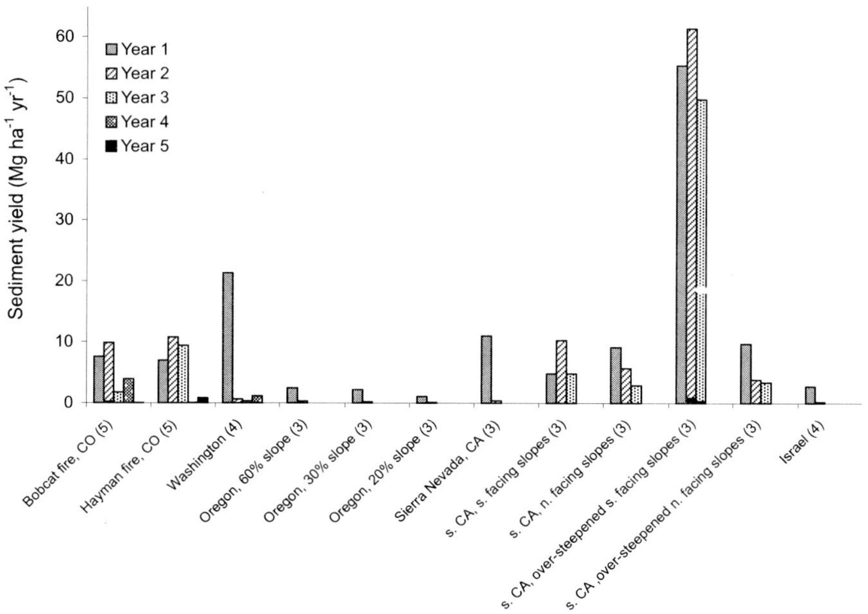

Fig. 8 Annual post-fire sediment yields over time for different locations. Year 1 refers to the year burned, and the number of years of data for each location is in parentheses. The sediment yield data are taken from Robichaud et al. (2006) for Washington, Robichaud and Brown (1999) for Oregon, MacDonald et al. (2004) and Chase (2006) for the Sierra Nevada, Krammes (1965) for southern California (all in USA), and Inbar et al. (1997) for Israel. South and north are abbreviated by s. and n. respectively.

California (Krammes 1965) and the Sierra Nevada (MacDonald et al. 2004, Chase 2006), but are only about 15 percent of the values from burned south-facing, over-steepened hillslopes in southern California and about 35 percent of the values from north-central Washington (Krammes 1965, Robichaud et al. 2006; Fig. 8). First-year sediment yields in Colorado are about three to seven times greater than values from Oregon and Israel (Inbar et al. 1997, Robichaud and Brown 1999).

The time needed for post-fire erosion rates to return to near-background levels can be longer for Colorado than most other areas (Fig. 8). In relatively wet areas, such as Washington, Oregon and California's Sierra Nevada, post-fire erosion rates decline to near-background levels by the third year after burning. In the Colorado Front Range the median sediment yield from sites burned at high severity is only 0.13 Mg ha^{-1} for the fourth summer after burning, but the mean value is 1.6 Mg ha^{-1} because the maximum value was 34 Mg ha^{-1} (Fig. 7). The slower recovery rates in Colorado can be attributed to the dry, cold climate and relatively poor soils, and the longest recovery rates are usually in areas with particularly coarse-textured soils because these have the poorest growing conditions and slowest rates of vegetative regrowth.

Effects fires on channels

The predominance of rill and channel erosion means that much of the sediment generated after fires is delivered to streams. On hillslopes and in the steeper headwater channels the predominant post-fire response is rill and channel incision, but further downstream the predominant post-fire response is aggradation (Fig. 9). The shift from incision to aggradation is attributed to the lower transport capacity associated with decrease in channel gradient, and the decrease in runoff with increasing catchment size due to the small size of the convective thunderstorms that generate most of the surface runoff and erosion (Eccleston 2008).

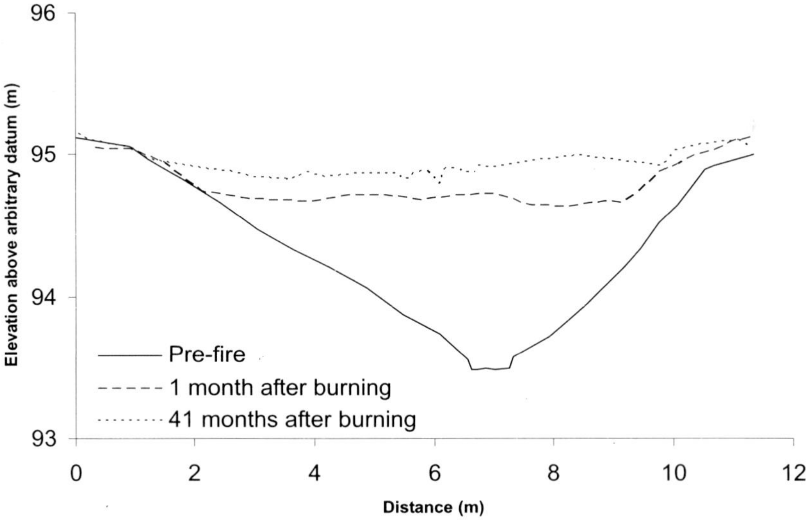

Fig. 9 Channel cross-section in a 3.4 km^2 watershed that was burned by the Hayman Fire. The cross-section was surveyed in the year before the fire, shortly after the second post-fire runoff event, and 41 months after burning.

Measurements of channel cross-sections and sediment transport rates indicate that the recovery rate for downstream, aggraded channels is likely to be at least an order of magnitude slower than the recovery rate for hillslope erosion. In most cases the median hillslope erosion rate is close to zero by the third summer after burning, and by the fifth year after burning all of our hillslope study sites produce little or no sediment (Fig. 7). The drop in sediment production indicates that infiltration and surface runoff rates also have recovered to near-background levels. The reduction in surface runoff will decrease downstream runoff and sediment transport capacities, and the decline in high flows will limit the rate at which the downstream channels can export the accumulated sediment (Fig. 9; Eccleston 2008). The estimated residence time of the post-fire sediment stored in channels after the Buffalo Creek Fire is 300 yr (Moody and Martin 2001a), and an even longer residence time is expected for the channel in Fig. 9 because all of the discharge is currently subsurface (i.e., within the coarse aggraded material). In the absence of surface runoff, the aggraded sediment will not be transported to the channels further downstream that have perennial flow and a greater capacity to transport sediment. The implication is that the altered fire regime in the lower montane and montane forests could have long-term effects on channel morphology and other aquatic resources.

Fire Effects in Subalpine Forests

Few post-fire runoff and sediment yield data are available for the subalpine zone, but these areas are of lesser concern for several reasons. First, wildfires are much less frequent and humans have not yet greatly altered the natural fire regime. Second, the lower population density means a lower risk for life and property. Third, model simulations using Disturbed WEPP (Elliot 2004) indicate that subalpine forests have a much lower risk for post-fire flooding and erosion than the lower montane and montane forests Simulations were done for 14 different climate stations assuming a 100 m-long hillslope with a 30 percent slope that had burned at high severity. The mean predicted sediment yields for the sites above 2400 m was 4.3 Mg ha^{-1} yr^{-1}, or just 31 percent of the mean value for the seven sites below 2400 m.

To the best of our knowledge, only one study has documented the effects of a wildfire on runoff and erosion rates in the subalpine zone. The 1967 Comanche Fire burned 190 ha of lodgepole pine and spruce-fir forest, and field measurements indicated higher infiltration rates in the burned areas than unburned areas and no evidence of soil water repellency (Striffler and Mogren, 1971). Soil tracer studies indicated that the maximum particle displacement on a 62 percent slope was only about 8 m. The limited erosion can be at least partly attributed to the lack of intense rainfall, as the post-fire rainstorms had maximum 30 min intensities of only 5 to 10 mm h^{-1} (Striffler and Mogren 1971). The limited data from this study are consistent with the Disturbed WEPP simulations and help confirm that post-fire erosion risks are substantially lower in the subalpine zone.

EFFECTIVENESS OF POST-FIRE REHABILITATION TREATMENTS

Bobcat Fire

Rehabilitation treatments are commonly applied after forest fires in order to minimize the increases in runoff and erosion, but very few studies have documented the effectiveness of these treatments (Robichaud et al. 2000). In Colorado treatment effectiveness has been evaluated by comparisons of hillslope-scale sediment production rates on three different wildfires – the June 2000 Bobcat Fire near Fort Collins and the 2002 Hayman and Schoonover Fires southwest of Denver.

At the Bobcat Fire three sets of replicated plots were set up to compare the mean surface cover and annual sediment yields for three treatments – seeding, straw mulching, and contour-felling – against the mean values from untreated control plots (Wagenbrenner et al. 2006). All of the treatments were applied to hillslopes burned at high severity by the USFS or following USFS protocols. The seeding treatment included slender wheatgrass (*Elymus trachycaulus*), mountain brome (*Bromus marginatus*), and a commercial mix of sterile grass seed applied at a target rate of 34 kg ha^{-1} or 430 seeds m^{-2}; two plots were seeded by air and two plots were seeded by hand. In the mulch treatment wheat straw was applied to three plots at a rate of 2.2 Mg ha^{-1}. In the contour-felled log erosion barrier (LEB) treatment the burned trees were cut down, de-limbed, and placed on the contour to act as sediment traps. Earthen berms were constructed on the uphill side of each log to prevent underflow, and the target density was 300 to 450 m of logs per ha. Log density, sediment storage capacity, and log failure rates were assessed on two sites in the Bobcat Fire and two sites in each of two other fires (Wagenbrenner et al. 2006). For each treated and control plot vegetative recovery was assessed by classifying the surface cover at a minimum of 100 points in late spring and early fall, and hillslope-scale sediment yields were monitored with sediment fences (Fig. 2) (Robichaud and Brown 2002, http://www.fs.fed.us/institute/middle_east/platte_pics/silt_fence.htm).

In the case of the Bobcat Fire, a storm with an I_{30} of 48 mm h^{-1} and an estimated recurrence interval of 5 to 10 years occurred 2 months after burning (Wagenbrenner et al. 2006). The sediment generated by this storm caused all of the sediment fences to fill and overtop except for three of the mulched plots and one of the LEB plots. This meant that the measured sediment yields were primarily a function of the total storage capacity, and none of the treatments had significantly lower sediment yields than the controls for the first year after burning (Wagenbrenner et al. 2006). Following this storm, three new mulched plots and seven new LEB plots were established.

Seed densities in the seeded plots were 25 to 50 percent lower than the target density, and field observations indicated that much of the seed was washed downslope during the first rainstorm. Seeding did not significantly increase the amount of surface cover in either the aerial- or hand-seeded plots at any point during the study, and in the absence of any difference in surface cover there were no significant differences in sediment yields (Fig. 10; Wagenbrenner et al. 2006).

After mulching there was only 26 percent bare soil as compared to the mean value of 67 percent on the control plots, and this difference was highly significant. Vegetative regrowth was significantly higher on the three old mulched plots than the control plots, and the combination of mulching and vegetative regrowth resulted in significantly more surface cover on the old and new mulched plots than the control plots for each of the first three years after burning. In the second to fourth years after burning the mean sediment yields from the mulched plots were only about 5 percent of the mean value from the corresponding control plots (Fig. 10; Wagenbrenner et al. 2006).

The sediment storage capacity in the first set of LEB plots was completely

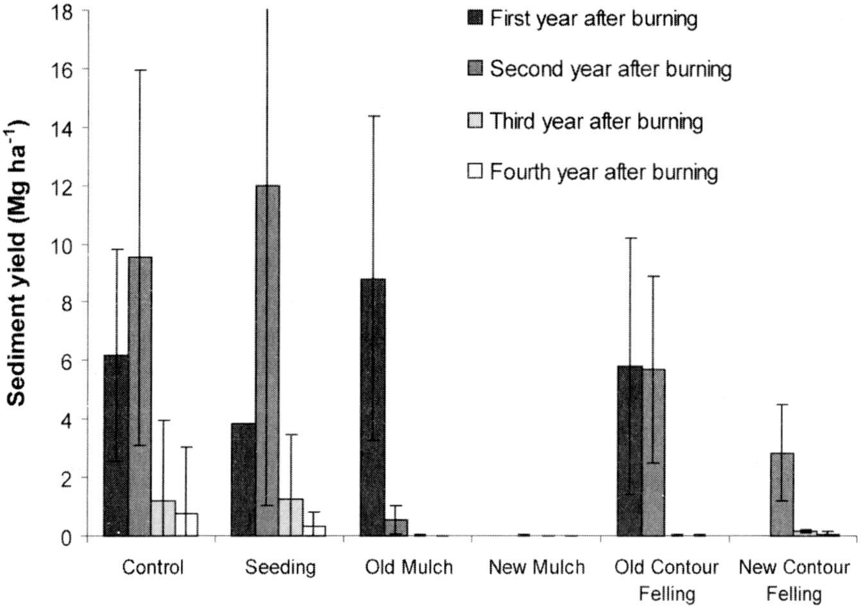

Fig. 10 Mean annual sediment yields by year since burning from the control plots and the different rehabilitation treatments at the Bobcat Fire. The old mulch and old contour-felled log erosion barrier treatments were applied before the very large storm that occurred two months after the fire. The new mulch and new contour-felled log treatments were applied after this storm. The bars indicate one standard deviation.

filled as a result of the large storm in August 2000, and there was no difference in sediment yields between these plots and the corresponding controls because all but one of the sediment fences had overtopped (Fig. 10). After this storm 7 new LEB plots were established and sediment yields in the new LEB plots were reduced by 71 to 90 percent relative to the controls (Fig. 10), but the high variability within treatments meant that this difference was only significant for the second year after burning (i.e., the summer after installation) (Wagenbrenner et al. 2006). As might be expected, the LEB treatment had no significant effect on the total amount of surface cover or the rate of revegetation. The survey of 210 contour felled logs at 6 sites showed that 32 percent of the logs were ineffective in trapping sediment because they were installed off-contour, had poor ground contact, or both (Wagenbrenner et al. 2006). The mean sediment storage capacity was 16 m^3 ha^{-1}, but both the failure rate and the estimated sediment storage capacity varied widely among the 6 sites.

Hayman and Schoonover Fires

At the Hayman and Schoonover Fires a similar approach was used to evaluate the effectiveness of five different post-fire rehabilitation treatments – seeding combined with scarification, straw mulch with grass seeding, hydromulch applied by ground spraying, hydromulch applied by helicopter, and a polyacrylamide applied in both a wet and a dry formulation (Rough et al. 2004). Since most of the treatments were applied relatively late in the summer and there were only 1 to 3 small storm events after these treatments had been installed, treatment effectiveness is only evaluated for the second through fifth years after burning.

The scarification treatment was done by hand using a heavy metal rake with long tines (McLeod), and the subsequent seeding used a barley (*Hordeum vulgare*) and triticale (*Triticosecale rimpaui*) mixture with a target density of 80 kg ha^{-1} or 280 seeds m^{-2}. The mean scarification depth was only 1.6 cm, which was too shallow to break up the observed water repellent soil layer (Fig. 3). Similar to the Bobcat Fire, the scarification and seeding treatment did not significantly increase the amount of surface cover, nor did it have any significant effect on sediment yields (Fig. 11a). The data suggest that the scarification treatment increased sediment yields in the first few storms after

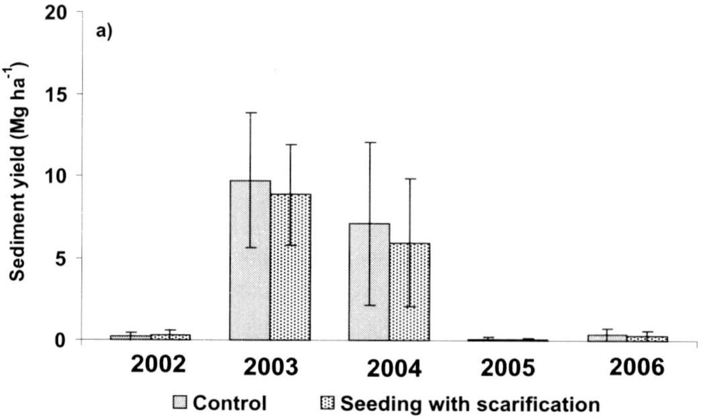

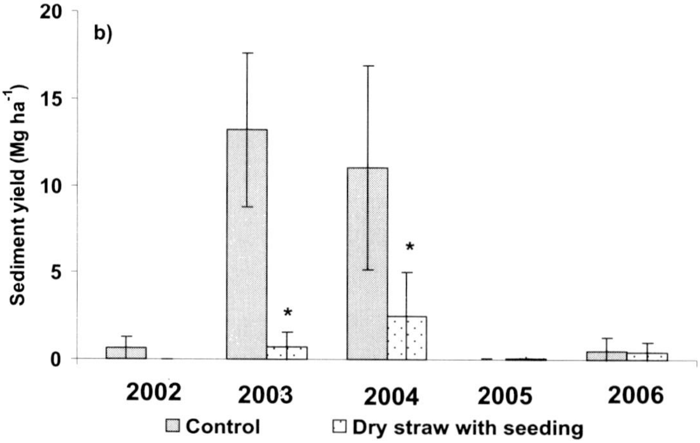

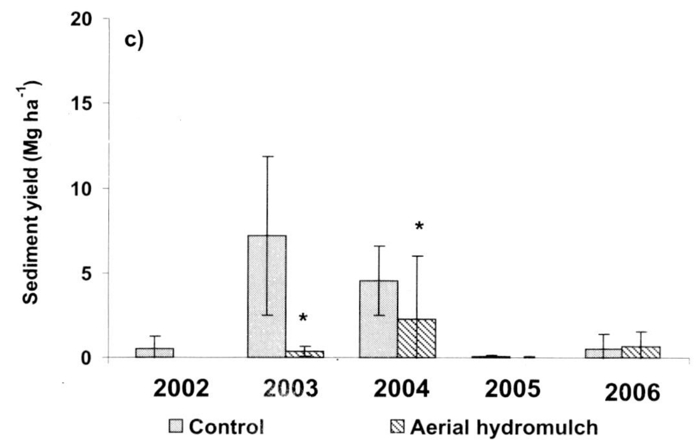

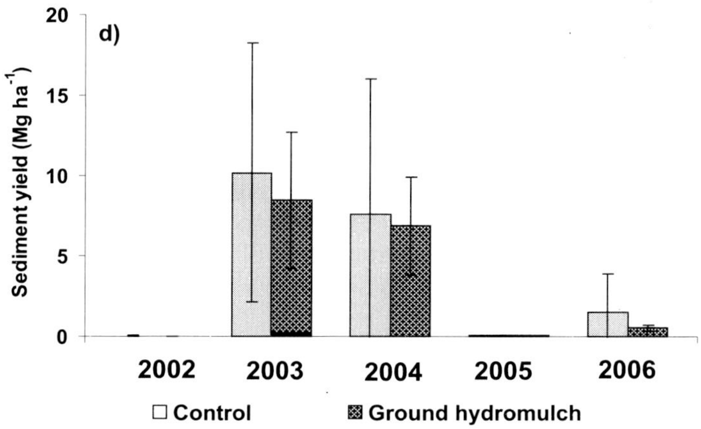

Fig. 11 Mean annual sediment yields from replicated control and treated plots in the Hayman Fire from 2002 to 2006. The treatments include: a) seeding with scarification; b) dry mulch with seeding; c) aerially-applied hydromulch; and d) ground-applied hydromulch. Bars represent one standard deviation. The Hayman Fire burned in June 2002, and the summers in 2002 and 2005 were exceptionally dry. An asterisk indicates a significant treatment effect (p = 0.05).

treatment by disturbing the soil surface and increasing the soil erodibility, but the 45 percent increase was not statistically significant relative to the controls.

The straw mulch and seeding treatment significantly increased the amount of surface cover relative to the controls for the first 2 yr after burning. In the second year after burning mean sediment yields for the mulched plots were only 5 percent of the mean value for the control plots, and in the third year after burning the mean sediment yield for the mulched plots was still only 23 percent of the mean value from the controls (Rough and MacDonald 2005). Both of these differences in sediment yields were significant (Fig. 11b). In the fourth and fifth years after burning the mulch and seeding treatment ceased to be effective in reducing sediment yields relative to the untreated controls, and this is attributed to the progressive increase in ground cover on the untreated controls.

The results for the hydromulch treatments were mixed. Both the aerially- and ground-applied hydromulch treatments had about 65 percent surface cover in the second year after burning, and this was significantly higher than the mean value from the corresponding control plots. The aerially-applied hydromulch reduced sediment yields by more than 90 percent in the second year after burning, and by about 50 percent in the third year after burning, and these differences were significant. The ground-applied hydromulch did not significantly reduce sediment yields relative to the controls (Fig. 11d) despite providing a similar amount of surface cover as the aerially-applied hydromulch, and the lack of a significant effect is attributed primarily to the differences in the formulation of the aerial- and ground-applied hydromulch mixtures (Rough, 2007). Neither of the hydromulch treatments significantly reduced sediment yields in the fourth and fifth year after burning (Fig. 11c).

The polyacrylamide (PAM) applied in a wet formulation appeared to significantly reduce sediment yields in the first year after application while the dry formulation had no significant effect. The wet formulation was then applied to the 3 plots that had been treated with the dry PAM, but in contrast to the first year results, the new wet PAM treatment had no significant effect on sediment yields. Subsequent laboratory tests showed that the PAM tended to bind with the residual ash, and the potential for PAM treatments to reduce post-fire erosion is not clear due to the complications of soil type, amount of ash present on the soil surface, the application rate, the high variability among sites, and the limited number of experimental plots studied to date (n = 3 for each treatment in our study; Rough 2007).

DISCUSSION OF REHABILITATION EFFECTIVENESS IN COLORADO

Factors Contributing to Treatment Effectiveness

The data from the untreated plots and the different post-fire rehabilitation treatments indicate that percent surface cover is the most important control on sediment yields. Very similar results have been obtained from a series of small (~5 ha) treated and untreated catchments set up after the 2002 Hayman Fire (P.R. Robichaud, personal communication 2006). The resulting principle is that any treatment that immediately increases the amount of surface cover is most likely to reduce post-fire sediment yields. The straw mulch and aerially-applied hydromulch were the most effective treatments because they protected the soil from raindrop impact and soil sealing, and this helped sustain high infiltration rates. The straw mulch was more effective than the hydromulch in terms of increasing the surface roughness, and the increased roughness will help slow overland flow, increase infiltration, and reduce particle entrainment. Studies from other types of disturbed areas also indicate that straw mulch increases seed germination and plant growth by increasing soil moisture and reducing surface temperatures (Goldman et al. 1986), but our data generally do not show significantly more live vegetation in the mulched plots than untreated control plots (Wagenbrenner et al. 2006, Rough 2007). The disadvantage of straw mulch is that it is more susceptible to redistribution by wind and overland flow than a well-formulated hydromulch, and it can contribute to the introduction of noxious weeds.

The ground-applied hydromulch was ineffective despite immediately increasing surface cover. The ground- and aerially-applied hydromulch used different mixtures, and the binding agent in the aerially-applied hydromulch was specifically selected for the coarse-grained soils at the Hayman Fire (P.R. Robichaud, personal communication 2006). Field observations indicate that the ground-applied hydromulch did not bind to the soil surface and was readily broken up or displaced by overland flow (P.R. Robichaud, personal communication 2006, Rough 2007). The results suggest that the exact formulation of the hydromulch can greatly affect its ability to bind with the soil and its effectiveness in reducing post-fire erosion.

Seeding treatments, including scarification and seeding, were not effective because they had no significant effect on the amount of surface cover, rate of vegetative regrowth, or hillslope-scale sediment yields (Wagenbrenner et al. 2006, Rough 2007). The failure of post-fire seeding to significantly reduce sediment yields is consistent with most other studies, and the lack of effectiveness is attributed to the fact that much of the erosion occurs before a dense plant cover can be established (Robichaud et al. 2000). Seeding should be most effective when a fire is followed by a well spaced series of gentle storms, but this sequence would facilitate natural regrowth and rarely occurs. In Colorado the effectiveness of seeding also is limited by the tendency for the seeds to be washed downslope in small- or moderate-sized storms and the relatively poor growing conditions (e.g., limited summer precipitation, coarse-textured soils, and low fertility).

The effectiveness of LEB treatments depends primarily on the sediment storage capacity relative to the post-fire erosion rates. The estimated mean sediment storage capacity of 16 m^3 ha^{-1} is about equal to the total mass of sediment captured in the control plots, and this would suggest that the contour-felled logs should, on average, be able to capture most of the sediment generated by a high-severity wildfire. The LEB plots installed after the large storm were on planar hillslopes (Wagenbrenner et al. 2006), but post-fire sediment is derived primarily from convergent rills and channel incision. The problem is that contour-felled logs are designed to trap sediment on planar hillslopes, and they cannot be easily placed to reduce erosion or trap the sediment from central rills in convergent topography or small headwater channels.

Some proponents have claimed that LEB treatments can reduce the amount of surface runoff and hence the amount of rill and channel erosion by enhancing infiltration and trapping overland flow. Infiltration tests after the Bobcat Fire did show a significantly higher permeability in the trenches upslope of the logs relative to the hillslopes (Wagenbrenner et al. 2006). The potential reduction in surface runoff was calculated from the increase in infiltration and the potential runoff storage capacity. The results showed that the amount of runoff from a 10 mm storm would be reduced by about 26 percent, but this value would be progressively smaller for larger storms (Wagenbrenner et al. 2006). Subsequent measurements indicated that the potential for LEB treatments to reduce runoff would rapidly diminish as the deposition of fine sediment reduced the infiltration rates in the trenches, infiltration rates increased on the untreated hillslopes, and the capacity for storing overland flow was reduced by the accumulation of sediment behind the logs (Wagenbrenner et al. 2006). We conclude that LEB treatments can reduce the amount of runoff only from the first and smaller storms after installation; potentially reduce sediment yields primarily through the storage of sediment rather than runoff; and will be more effective on planar rather than convergent hillslopes.

Treatment Effectiveness in Relation to Storm Size and Time Since Burning

The sediment yield data from the different storms on the Bobcat Fire suggest that treatment effectiveness declines with increasing storm size. None of the treatments was effective in reducing sediment yields when subjected to a 5 to 10 yr storm event, but both the old and new mulch treatments and the new LEB treatment significantly reduced sediment yields in the following summer when the storm events were less severe. For the Hayman and Schoonover Fires there is no evidence of a decrease in effectiveness with increasing storm size, but there were no storm events with a recurrence interval greater than 2 yr.

The effectiveness of each of the rehabilitation treatments will decline over time, and there are several reasons for this. Both straw mulch and hydromulch break down over time, but the data presented here indicate that these treatments were effective in reducing sediment yields for as long as the third summer after burning (Figs. 10 and 11). In the case of contour-felled log erosion barriers, the effectiveness will decline as the sediment storage capacity fills up. Installation of the logs off contour or leaks beneath the logs will tend to concentrate flow and initiate rill erosion, and these problems are likely to increase over time. The absolute effectiveness of any treatment also will decline over time because of the natural decline in sediment production rates from untreated hillslopes (Figs. 10 and 11). It also should be recognized that our ability to detect treatment effectiveness is limited by the high variability in sediment production rates within replicated treatments as well as the spatial variability in rainfall.

Treatment Cost-effectiveness

As noted earlier, large amounts of public and occasionally private funds are spent on post-fire rehabilitation treatments. Seeding has long been the most commonly-applied treatment in forested areas because it costs only US$45 per hectare and is easily applied by airplanes over rough, unroaded terrain. Scarification requires much more labor and this increases the cost per hectare by a factor of about 13 (Robichaud et al. 2003). Straw mulching costs about US$1000 to US$1600 per hectare for ground application by machine and US$1850 to 3000 for hand application (http://www.fs.fed.us/r5/baer/index.html). For logistical reasons straw mulching has been limited to areas with road access, but after the Hayman Fire straw mulch was successfully applied by helicopters with bales of hay in cargo nets suspended beneath the helicopter. The cost of mulching from air is roughly US$1800 per hectare, which is similar to the cost of ground mulching because the reduced labor costs compensate for the high cost of helicopter time. Hydromulching is generally the most expensive treatment as ground-applied hydromulch after the Hayman Fire cost US$2350 per hectare (Robichaud et al. 2003). Aerial hydromulching is about three times the cost of ground-based hydromulching (Robichaud et al. 2003).

These cost data can be combined with our measured reductions in sediment yields to estimate the cost effectiveness of the different treatments (Table 1). The results show that ground-applied dry mulch is the most cost-effective at approximately US$50 to US$150 per megagram reduction in sediment yields. Hydromulching is roughly 5 to 15 times as expensive as mulching, and there was not a large difference between ground- and aerially-applied hydromulching because the ground-based hydromulch treatment was less expensive and less effective in reducing sediment yields. Ground-based hydromulching could be more cost-effective than aerial hydromulching if one assumes a similar hydromulch formulation and a similar effectiveness, but hydromulching is still much more expensive than straw mulching. The cost-effectiveness of seeding with scarification was calculated for Table 1, but this calculation assumes that the statistically insignificant reduction in sediment yield is a real value.

Both the public and land managers need to recognize that the sheer size of the 2002 wildfires in Colorado far exceeded the resources available for post-fire rehabilitation treatments. Table 1 clearly indicates which treatments are most cost-effective at the hillslope scale, but there are no data to indicate these treatments would be effective if they were applied across catchments larger than a few hectares. Since treatment effectiveness also declines with increasing storm size, both financial and physical constraints will limit our ability to reduce larger-scale runoff and sediment yields after large, high severity wildfires.

Table 1 The cost of post-fire rehabilitation treatments applied after the 2002 Hayman Fire, the mean reduction in sediment yields, and the calculated treatment cost per Mg reduction in sediment yields at the hillslope scale. All treatment costs except ground-based dry mulching are from Robichaud et al. (2003). Ground-based dry mulch costs are from the U.S. Department of Agriculture, Forest Service Region 5 Burned Area Emergency Response website (http://www.fs.fed.us/r5/baer/index.html). The cost-effectiveness of machine and aerially-applied dry mulch assumes that both treatments are as effective in reducing sediment yields as mulch applied by hand. The cost-effectiveness of the seeding with scarification and the ground-applied hydromulch assume that the statistically insignificant reductions in sediment yield are real.

Treatment	Cost	Mean sediment yield reduction from 2002–2006	Cost per unit sediment yield reduction
	US$ ha^{-1}	Mg ha^{-1}	US$ Mg^{-1}
Seeding with scarification	640	2.1	305
Ground-applied dry mulch (machine application)	990–1600	21.0	47-76
Ground-applied dry mulch (hand application)	1830-2970	21.0	86
Aerially-applied dry mulch	1800	21.0	86
Ground-applied hydromulch	2350	3.5	673
Aerially-applied hydromulch	7410	8.9	828

CONCLUSIONS

Approximately one-third of Colorado is covered by forests, and the changes in precipitation and temperature with increasing elevation allow the forested areas to be classified into three distinct zones with widely varying forest densities, fuel loadings, and natural fire regimes. The lower montane and montane forests are of greatest concern because: 1) the drier conditions result in more frequent fires; 2) human activities have increased the risk of large, high-severity fires; 3) these forests have higher potential post-fire runoff and erosion rates; and 4) there are a large number of human and natural resources at risk.

Detailed studies on a series of wildfire and prescribed fires have resulted in an extensive and unique dataset for assessing the effects of wildfires, the effectiveness of different post-fire rehabilitation treatments, and post-fire recovery rates. Areas burned at high severity are of greatest concern because runoff and sediment yields increase by several orders of magnitude, and summer rainfall intensities of 8 to 10 mm h^{-1} can generate substantial amounts of overland flow and surface erosion. Percent surface cover and rainfall erosivity are the most important controls on post-fire sediment yields; three to five years are generally required for hillslope sediment yields to decline to near-background levels. The persistence of elevated sediment yields is longer than in most other areas, and this is attributed to the relatively poor conditions for post-fire regrowth.

The most effective post-fire rehabilitation treatments are those that immediately provide surface cover, such as straw mulching or hydromulching. Seeding, or seeding combined with scarification, did not significantly affect vegetative regrowth or sediment yields. Contour-felled log erosion barriers were only effective for small and moderate-sized storms because of the limited sediment storage capacity, and the effectiveness of this treatment can be negated by poor installation. The application of a polyacrylamide did not consistently reduce sediment yields. Mulching is by far the most cost-effective post-fire rehabilitation treatment at US$50 to US$150 per megagram reduction in sediment yields.

Climate projections indicate an increased likelihood of extreme fire weather and high-severity wildfires in the Rocky Mountains (Baker 2003), and this translates to a greater need for predicting the effects of future fires and post-fire rehabilitation treatments. The results presented here can provide useful guidance to land managers and researchers in other areas, as the basic principles and processes identified in this chapter are more broadly applicable.

Acknowledgments

A large number of organizations and individuals have contributed to the data and analyses presented in this chapter. We would like to particularly thank the numerous graduate students who have worked on various aspects of these issues, and in chronological order these include Tedd Huffman, Matt Kunze, Joe Wagenbrenner, Zamir Libohova, Juan de Dios Benavides-Solorio, Daniella Rough, Jay Pietraszek, Darren Hughes, Duncan Eccleston, Ethan Brown, Matt Welsh, and Keelin Schaffrath. Support for this work has come from the U.S. Department of Agriculture Arapaho-Roosevelt National Forest, Pike-San Isabel National Forest, Rocky Mountain Research Station, Pacific Southwest Forest and Range Experiment Station, and the U.S. Department of Agriculture and Department of Interior Joint Fire Science Program. Individuals too numerous to list have helped refine our ideas and provided additional information, but we would like to specifically acknowledge Pete Robichaud, Carl Chambers, Deb Entwistle, Bill Romme, and Bill Elliot.

References

Agnew, W., R.E. Labn, and M.V. Harding. 1997. Buffalo Creek, Colorado, fire and flood of 1996. *Land and Water*, 41: 27-29.

Baker, W.L. 2003. Fires and climate in forested landscapes of the U.S. Rocky Mountains. pp. 120-157. *In* T.T. Veblen, W.L. Baker, G. Montenegro, and T.W. Swetnam [eds.]. *Fire and Climatic Change in Temperate Ecosystems of the Western Americas. Ecological Studies 160*, Springer-Verlag, New York, New York, USA.

Benavides-Solorio, J.D. 2003. Post-fire runoff and erosion at the plot and hillslope scale, Colorado Front Range. Ph.D. dissertation, Colorado State University, Fort Collins, Colorado, USA.

Benavides-Solorio, J. and L.H. MacDonald. 2001. Post-fire runoff and erosion from simulated rainfall on small plots, Colorado Front Range. *Hydrological Processes*, 15: 2931-2952.

Benavides-Solorio, J.D. and L.H. MacDonald. 2005. Measurement and prediction of post-fire erosion at the hillslope scale, Colorado Front Range. *International Journal of Wildland Fire*, 14: 457-474.

Bolin, S.B. and T.J. Ward. 1987. Recovery of a New Mexico drainage basin from a forest fire. pp. 191-198. *In* R.H. Swanson, P.Y. Bernier, and P.D. Woodward [eds.]. 1992. Forest Hydrology and Watershed Management. IAHS Publication no 167, International Association of Hydrological Sciences Press, Wallingford, Oxfordshire, UK.

Brown, P.M., M.R. Kaufmann, and W.D. Shepperd. 1999. Long-term, landscape patterns of past fire events in a montane ponderosa pine forest of central Colorado. *Landscape Ecology*, 14: 513-532.

Buechling, A. and W.L. Baker. 2004. A fire history from tree rings in a high-elevation forest of Rocky Mountain National Park. *Canadian Journal of Forest Research*, 34: 1259-1273.

Cannon, S.H. and S.L. Reneau. 2000. Conditions for generation of fire-related debris flows, Capulin Canyon, New Mexico. *Earth Surface Processes and Landforms*, 25: 1103-1121.

Cannon, S.H., E.R. Bigio, and E. Mine. 2001a. A process for fire-related debris flow initiation, Cerro Grande fire, New Mexico. *Hydrological Processes*, 15: 3011-3023.

Cannon, S.H., R.M. Kirkham, and M. Parise. 2001b. Wildfire-related debris-flow initiation processes, Storm King Mountain, Colorado. *Geomorphology*, 39: 171-188.

Cannon, S.H., J.E. Gartner, A. Holland-Sears, B.M. Thurston, and J.A. Gleason. 2003. Debris flow response of basins burned by the 2002 Coal Seam and Missionary Ridge fires, Colorado. *In* D.D. Boyer, P.M. Santi, and W.P. Rogers [eds.]. Engineering Geology in Colorado: Contributions, Trends, and Case Histories. Association of Engineering Geologists Special Publication no. 15 and Colorado Geological Survey Special Publication 55, on CD-ROM, Denver, Colorado, USA.

Chase, E.H. 2006. Effects of wildfire and salvage logging on site conditions and hillslope sediment production: Placer County, California. M.S. thesis, Colorado State University, Fort Collins, Colorado, USA.

DeBano, L.F. 1981. Water repellent soils: a state-of-the art. U.S. Department of Agriculture, Forest Service General Technical Report PSW-46, U.S. Department of Agriculture, Forest Service, Pacific Southwest Forest and Range Experiment Station, Berkeley, California, USA.

Doerr, S.H. and A.D. Thomas. 2000. The role of soil moisture in controlling water repellency: new evidence from forest soils in Portugal. *Journal of Hydrology*, 231-232: 134-147.

Eccleston, D. 2008. Post-fire channel change in two small watersheds in the Colorado Front Range. M.S. thesis, Colorado State University, Fort Collins, Colorado, USA.

Eccleston, D. and L. MacDonald. 2006. Post-fire channel change in small mountainous catchments. Proceedings, AGU Hydrology Days, Colorado State University, Fort Collins, Colorado, USA.

Ehle, D.S. and W.L. Baker. 2003. Disturbance and stand dynamics in ponderosa pine forests in Rocky Mountain National Park, USA. *Ecological Monographs*, 73: 543-566.

Elliot, W.J. 2004. WEPP internet interfaces for forest erosion prediction. *Journal of the American Water Resources Association*, 40: 299-309.

Giovannini, G. and S. Lucchesi. 1983. Effect of fire on hydrophobic and cementing substances of soil aggregates. *Soil Science*, 136: 231-236.

Goldman, S.J., K. Jackson, and T.A. Bursztynsky. 1986. *Erosion and Sediment Control Handbook*. McGraw-Hill Book Co., New York, New York, USA.

Gottfried, G.J., D.G. Neary, M.B. Baker Jr., and P.F. Ffolliott. 2003. Impacts of wildfire on hydrologic processes in forest ecosystems: Two case studies. *In* Proceedings of the First Interagency Conference on Research in Watersheds. U.S. Department of Agriculture, Agricultural Research Service, Tucson, Arizona, USA. *Online at* http://www.tucson.ars.ag.gov/icrw/proceedings.htm [accessed 15 Sep 2008].

Graham, R.T. [ed.]. 2003. Hayman fire case study. General Technical Report RMRS-GTR-114. U.S. Department of Agriculture, Forest Service, Rocky Mountain Research Station, Fort Collins, Colorado, USA.

Grissino-Mayer, H.D., W.H. Romme, M.L. Floyd, and D.D. Hanna. 2004. Climatic and human influences on fire regimes of the southern San Juan Mountains, Colorado, USA. *Ecology*, 85: 1708-1724.

Helvey, J.D. 1980. Effects of a north central Washington wildfire on runoff and sediment production. *Water Resources Bulletin*, 16: 627-634.

Hubbert, K.R., H.K. Preisler, P.M. Wohlgemuth, R.C. Graham, and M.G. Narog. 2006. Prescribed burning effects on soil physical properties and soil water repellency in a steep chaparral watershed, southern California, USA. *Geoderma*, 130: 284-298.

Huffman, E.L., L.H. MacDonald, and J.D. Stednick. 2001. Strength and persistence of fire-induced soil hydrophobicity under ponderosa and lodgepole pine, Colorado Front Range. *Hydrological Processes*, 15: 2877-2892.

Inbar, M., L. Wittenberg, and M. Tamir. 1997. Soil erosion and forestry management after wildfire in a Mediterranean woodland, Mt. Carmel, Israel. *International Journal of Wildland Fire*, 7: 285-294.

Keane, R.E., K.C. Ryan, T.T. Veblen, C.D. Allen, J. Logan, and B. Hawkes. 2002. Cascading effects of fire exclusion in Rocky Mountain ecosystems: A literature review, General Technical Report RMRS-GTR-91. U.S. Department of Agriculture, Forest Service, Rocky Mountain Research Station, Fort Collins, Colorado, USA.

Kershner, J.L., L. MacDonald, L.M. Decker, D. Winters, and Z. Libohova. 2003. Fire induced changes to aquatic ecosystems. pp. 232-243. *In* R.T. Graham [ed.]. Hayman Fire Case Study. General Technical Report RMRS-GTR-114. U.S. Department of Agriculture, Forest Service, Rocky Mountain Research Station, Fort Collins, Colorado, USA.

Krammes, J.S. 1965. Seasonal debris movement from steep mountainside slopes in southern California. pp. 85–88. *In* Proceedings of the Federal Inter-Agency Sedimentation Conference, Miscellaneous Publication 970. U.S. Department of Agriculture, Washington, D.C., USA.

Kunze, M.D. and J.D. Stednick. 2006. Streamflow and suspended sediment yield following the 2000 Bobcat fire, Colorado. *Hydrological Processes*, 20: 1661-1681.

Larsen, I.J. and L.H. MacDonald. 2007. Predicting post-fire sediment yields at the hillslope scale: Testing RUSLE and Disturbed WEPP. *Water Resources Research*, 43: W11412. doi: 10. 1029/2006WR005560.

Libohova, Z. 2004. Effects of thinning and a wildfire on sediment production rates, channel morphology, and water quality in the Upper South Platte River watershed. M.S. thesis, Colorado State University, Fort Collins, Colorado, USA.

MacDonald, L.H. and J.D. Stednick. 2003. Forests and water: A state-of-the-art review for Colorado. Colorado Water Resources Research Institute Report no. 196. Colorado State University, Fort Collins, Colorado, USA.

MacDonald, L.H. and E.L. Huffman. 2004. Post-fire soil water repellency: Persistence and soil moisture thresholds. *Soil Science Society of America Journal*, 68: 1729-1734.

MacDonald, L.H., D.B. Coe, and S.E. Litschert. 2004. Assessing cumulative watershed effects in the Central Sierra Nevada: hillslope measurements and catchment-scale modeling. pp. 149-157. *In* D.D. Murphy and P. A. Stine [eds.]. Proceedings of the Sierra Nevada Science Symposium. General Technical Report PSW-GTR-193. U.S. Department of Agriculture, Forest Service, Pacific Southwest Research Station, Albany, California, USA.

MacDonald, L., D. Rough, and Z. Libohova. 2005. Effects of forest fires on the strength and persistence of soil water repellency in the Colorado Front Range. *Geophysical Research Abstracts*, Vol. 7, 08613, SRef-ID: 1607-7962/gra/EGU05-A-08613.

Martin, D.A. and J.A. Moody. 2001. Comparison of soil infiltration rates in burned and unburned mountainous watersheds. *Hydrological Processes*, 15: 2893-2903.

Minshall, G.W., J.T. Brock, D.A. Andrews, and C.T. Robinson. 2001. Water quality, substratum and biotic responses of five central Idaho (USA) streams during the first year following the Mortar Creek fire. *International Journal of Wildland Fire*, 10: 185-199.

Moody, J.A. and D.A. Martin. 2001a. Initial hydrologic and geomorphic response following a wildfire in the Colorado Front Range. *Earth Surface Processes and Landforms*, 26: 1049-1070.

Moody, J.A. and D.A. Martin. 2001b. Post-fire, rainfall intensity—peak discharge relations for three mountainous watersheds in the western USA. *Hydrological Processes*, 15: 2981-2993.

Moody, J.A., J.D. Smith, and B.W. Ragan. 2005. Critical shear stress for erosion of cohesive soils subjected to temperatures typical of wildfires. *Journal of Geophysical Research*, 110: F01004.

Morris, S.E. and T.A. Moses. 1987. Forest fire and the natural soil erosion regime in the Colorado Front Range. *Annals of the Association of American Geographers*, 77: 245-254.

Morton, D.C., M.E. Roessing, A.E. Camp, and M.L. Tyrrell. 2003. Assessing the environmental, social, and economic impacts of wildfire. Yale University Global Institute of Sustainable Forestry, Research Paper 001, New Haven, Connecticut, USA.

Neary, D.G., C.C. Klopatek, L.F. DeBano, and P.F. Ffolliott. 1999. Fire effects on belowground sustainability: A review and synthesis. *Forest Ecology and Management*, 122: 51-71.

Pietrazsek, J.H. 2006. Controls on post-fire erosion at the hillslope-scale, Colorado Front Range. M.S. thesis, Colorado State University, Fort Collins, Colorado, USA.

Prosser, I.P. and L. Williams. 1998. The effect of wildfire on runoff and erosion in native Eucalyptus forest. *Hydrological Processes*, 12: 251-265.

Rieman, B. and J. Clayton. 1997. Wildfire and native fish: Issues of forest health and conservation of sensitive species. *Fisheries*, 22: 6-15.

Rinne, J.N. 1996. Short-term effects of wildfire on fishes and aquatic macroinvertebrates in the southwestern United States. *North American Journal of Fisheries Management*, 16: 653-658.

Robichaud, P.R. and R.E. Brown. 1999 (revised 2000). What happened after the smoke cleared: onsite erosion rates after a wildfire in eastern Oregon. pp. 419-426. *In* D. Olsen and J.P. Potyondy [eds.]. Proceedings of the American Water Resources Association Specialty Conference on Wildland Hydrology. American Water Resources Association, Herdon, Virginia, USA.

Robichaud, P.R. and R.E. Brown. 2002. Silt fences: an economical technique for measuring hillslope soil erosion. General Technical Report RMRS-GTR-94. U.S. Department of Agriculture, Forest Service, Rocky Mountain Research Station, Fort Collins, Colorado, USA.

Robichaud, P.R., J.L. Beyers, and D.G. Neary. 2000. Evaluating the effectiveness of postfire rehabilitation treatments, General Technical Report RMRS-GTR-63. U.S. Department of Agriculture, Forest Service, Rocky Mountain Research Station, Fort Collins, Colorado, USA.

Robichaud, P.R., L. MacDonald, J. Freeouf, D. Neary, D. Martin, and L. Ashmun. 2003. Post-fire rehabilitation of the Hayman fire. pp. 293-313. *In* R.T. Graham [ed.]. Hayman Fire Case Study. General Technical Report RMRS-GTR-114. U.S. Department of Agriculture, Forest Service, Rocky Mountain Research Station, Fort Collins, Colorado, USA.

Robichaud, P.R., T.R. Lillybridge, and J.W. Wagenbrenner. 2006. Effects of postfire seeding and fertilizing on hillslope erosion in north-central Washington, USA. *Catena*, 67: 56-67.

Romme, W.H., T.T. Veblen, M.R. Kaufmann, R. Sherriff, and C.M. Regan. 2003a. Historical (pre-1860) and current (1860-2002) fire regimes. pp. 181-195. *In* R.T. Graham [ed.]. Hayman Fire Case Study. General Technical Report RMRS-GTR-114. U.S. Department of Agriculture, Forest Service, Rocky Mountain Research Station, Fort Collins, Colorado, USA.

Romme, W.H., M.R. Kaufmann, T.T. Veblen, R. Sherriff, and C.M. Regan. 2003b. Historical (pre-1860) and current (1860-2002) forest and landscape structure. pp. 196-203. *In* R.T. Graham [ed.]. Hayman Fire Case Study. General Technical Report RMRS-GTR-114. U.S. Department of Agriculture, Forest Service, Rocky Mountain Research Station, Fort Collins, Colorado, USA.

Rough, D. 2007. Effectiveness of burned-area emergency rehabilitation treatments in reducing post-fire erosion after the Hayman and Schoonover fires in the Colorado Front Range. M.S. thesis, Colorado State University, Fort Collins, Colorado, USA.

Rough, D. and MacDonald, L.H. 2005. Effectiveness of BAER treatments in reducing post-fire erosion after the Hayman Fire, Colorado Front Range. Proceedings, AGU Hydrology Days, Colorado State University, Fort Collins, Colorado, USA.

Rough, D., L.H. MacDonald, and J. Wagenbrenner. 2004. Effectiveness of post-fire rehabilitation treatments in reducing post-fire sediment yields. *Geophysical Research Abstracts*, Vol. 6, 06016, SRef-ID: 1607-7962/gra/EGU04-A-06016.

Sibold, J.S., T.T. Veblen, and M.E. González. 2006. Spatial and temporal variation in historic fire regimes in subalpine forests across the Colorado Front Range in Rocky Mountain National Park, Colorado, USA. *Journal of Biogeography*, 32: 631-647.

Striffler, W.D. and E.W. Mogren. 1971. Erosion, soil properties, and revegetation following a severe burn in the Colorado Rockies. pp. 25-36. *In* C.W. Slaughter, R.J. Barney, and G.M. Hansen [eds.]. Proceedings—Fire in the Northern Environment—A Symposium. U.S. Department of Agriculture, Forest Service Pacific Northwest Forest and Range Experiment Station, Portland, Oregon, USA.

Terry, J.P. and R.A. Shakesby. 1993. Soil hydrophobicity effects on rainsplash: simulated rainfall and photographic evidence. *Earth Surface Processes and Landforms*, 18: 519-525.

Veblen, T.T. 2000. Disturbance patterns in southern Rocky Mountain forests. pp. 31-54. *In* R.L. Knight, F.W. Smith, S.W. Buskirk, W.H. Romme, and W.L. Baker [eds.]. *Forest Fragmentation in the Southern Rocky Mountains*. University Press of Colorado, Boulder, Colorado, USA.

Veblen, T.T., T. Kitzberger, and J. Donnegan. 2000. Climatic and human influences on fire regimes in ponderosa pine forests in the Colorado Front Range. *Ecological Applications*, 10: 1178-1195.

Wagenbrenner, J.W., L.H. MacDonald, and D. Rough. 2006. Effectiveness of three post-fire rehabilitation treatments in the Colorado Front Range. *Hydrological Processes*, 20: 2989-3006.

Wells, C.G., R. E. Campbell, L.F. DeBano, C.E. Lewis, R.L. Fredriksen, E.C. Franklin, R.C. Froelich, and P.H. Dunn. 1979. Effects of fire on soil, General Technical Report WO-7. U.S. Department of Agriculture, Forest Service, Washington, D.C., USA.

Westerling, A.L., H.G. Hidalgo, D.R. Cayan, and T.W. Swetnam. 2006. Warming and earlier spring increase western U.S. forest wildfire activity. *Science*, 313: 940-943.

Whicker, J.J., J.E. Pinder III, and D.D. Breshears. 2006. Increased wind erosion from forest wildfire: Implications for contaminant-related risks. *Journal of Environmental Quality*, 35: 468-478.

Woods, S.W., A. Birkas, and R. Ahl. 2007. Spatial variability of soil hydrophobicity after wildfires in Montana and Colorado. *Geomorphology*, 86: 465-479.

17

Fire Landscapes in Canada: How to Restore or Prevent Them

Michael P. Curran[1]* and *David F. Scott*[2]

Abstract

Canada contains a wide range of fire landscapes, with the steeper and considerably higher mountainous areas of British Columbia and Alberta, in the west, representing the greatest risks for post-wildfire erosion. The full potential of this problem was recently recognized following the 2003 wildfire season. Traditionally, 'rehabilitation' following wildfire has focused primarily on the access structures (firelines, etc.) constructed to fight the fire, along with some broadcast seeding or reforestation. Opportunities exist to capitalize on ecosystem restoration which attempt to restore a 'natural disturbance regime' to forests, particularly in urban interface areas where wildfire risks are of serious concern. This is because reducing the risk of severe wildfires will also reduce the risk of subsequent soil erosion. In recognition of the greater potential for soil erosion after wildfires, we are planning for increasing problems as variable climate, due to global change, impacts Canadian forests. This climate variation has already created increased areas of dead timber (fuels) due to pests like mountain pine beetle, and increased potential for more severe wildfires. Policy revision is ongoing in some provinces, with risk assessment procedures being drafted and tested, and provisions being developed for targeted hillslope restoration of burned slopes, which may become standard practice.

INTRODUCTION TO THE CANADIAN FOREST ENVIRONMENT

Canada is comprised of a wide variety of forest landscapes, and a correspondingly wide range of fire regimes. Forest regions (or landscapes) of

[1] *British Columbia Ministry of Forests and Range, Kootenay Lake Forestry Centre, Nelson, British Columbia, Canada and Agroecology (Forest Soils), University of British Columbia, Vancouver, British Columbia, Canada.*

[2] *Watershed Management, University of British Columbia Okanagan, Kelowna, British Columbia, Canada.*

* *Corresponding author: Michael P. Curran, British Columbia Ministry of Forests and Range, British Columbia Forest Service, Kootenay Lake Forestry Centre, 1907 Ridgewood Rd, Nelson, British Columbia, Canada V1L 6K1, Tel: 250 825 1100, Fax: 250 825 9657, e-mail: mike.curran@gov.bc.ca*

Canada have been described by Rowe (1972), and this classification is still relevant today in considering fire regimes. Locally, forest regions are further divided based on climate and resulting vegetation into ecozones or ecoregions (e.g., Strong and Leggat 1981 for Alberta), or biogeoclimatic zones in British Columbia (BC) (Meidinger and Pojar 1991). Inherent in each of these classifications is an understanding of the typical stand development history and resulting climax forest condition. Forests with similar stand development histories can be grouped according to how natural disturbances either initiate or maintain that stand condition. As most of Canada, particularly the remaining forested areas, has a fairly recent development history, ecologists have been able to describe and map what are considered natural disturbance regimes for the various landscapes.

The most common natural disturbance is wildfire and corresponding 'natural disturbance regimes' have been described for various areas as being stand maintaining or stand initiating and categorized as rare, infrequent, or frequent events (BC Ministries of Forests and Environment 1995). In the humid coastal and some interior temperate rainforest areas of western Canada, stand initiating fires are rare and may occur only once every several hundred years. However, in the spruce dominated forests of both the western Cordillera and circumpolar boreal areas of Canada, stand-initiating fires can be very common, resulting in stands of jack or lodgepole pine. At the drier extreme, in western Canada, we have grassland and open forests (e.g., open growing Ponderosa pine) where naturally frequent stand-maintaining fires would occur, without fire suppression efforts, at intervals of less than 10 yr.

Fire plays a number of roles in ecosystems. As evident in the naming of the natural disturbance regimes, fire can either maintain or initiate a resulting habitat condition. This is normally to the benefit or detriment of various organisms, both above and below ground. In the Canadian context, much of the forest and range lands are publicly owned and under the stewardship of the Provincial governments. The current aim of public forest and rangeland management is generally to mimic natural disturbance regimes while managing a range of values, including the production of timber and range resources (e.g., grazing for livestock production). Historically, fire has been excluded from forest and range areas to protect life and property, and to ensure provision of ecosystem and watershed services such as fresh water supplies for fisheries and domestic use, and other social values such as recreation and visual aesthetics. Beyond the immediate concerns for life, property, and valuable forest commodities like timber, fire has often been considered un-aesthetic as well, visually scarring the landscape. It is because of these values that, like other parts of the world, the exclusion of fire from its normal role in Canadian ecosystems has occurred for over 100 yr in some, and at least 50 yr, in many other locations.

As in many areas of the world, wildfires are usually started by lightning, and can burn millions of hectares in the less inhabited northern boreal forests. Most of Canada was glaciated in the last ice-age and since this time (as little as 10,000 yr ago in some areas), fire is considered to have played a role of shaping the landscape through accelerated erosion in the post-fire environment. However, while wildfire erosion may be expected on sandy soils in the boreal forests of Central and Eastern Canada (Chanasyk et al. 2003), the primary observations have been of increased phosphorous levels in streams and lakes due to erosion of exposed organic materials rather than mineral soil erosion (Prepas et al. 2003, Burke et al. 2005), presumably due to enough forest floor remaining to protect the underlying erodible mineral soil. This is consistent with the findings of Carcaillet et al. (2006) who, based on sedimentary records in eastern Canada, concluded that fire played little role in erosion by water action (because litter layers were seldom consumed), but that in drier areas with sandy soils, wind erosion was initiated by wildfire. In addition, an analysis of climatic records back to the previous hypsothermal period, some 6,000 yr ago, suggests that fire may decrease in some parts of eastern Canada as the global temperature becomes warmer (Flannigan et al. 2001). Therefore, in regards to post-wildfire erosion, it is primarily the fires that occur in the mountainous areas of western interior Canada (British Columbia (BC) Interior, western Alberta, and southern Yukon) that are of concern.

Wildfire erosion occurs when the right combination of soil and weather conditions exist. Evidence of fires and fire frequency, in the form of charcoal, is often seen in alluvial and colluvial deposits. While this is not necessarily considered a direct correlation with immediate post-wildfire erosion, it is certainly likely in many areas based on our observations of recent events in southern BC (Curran et al. 2006). For example, in east-central BC, Sanborn et al. (2006) studied charcoal recovered from colluvial and alluvial fan deposits and concluded that major runoff events coincided with wildfire events every 800 to 1200 yr and were even broadly related to periods of higher fire frequency in southwestern BC.

On a larger scale, fire can contribute to large-scale erosion events, which can range from flooding to debris floods and flows, and overall slope instability in hilly or mountainous terrain (Curran et al. 2006). Fire effects on the soil, underlying physiographic characteristics, and probable precipitation patterns in the area must be considered in planning post-fire restoration (rehabilitation). To reduce overall erosion risk, areas of high risk and soil conditions that can be ameliorated (or prevented) must be identified. In mountainous areas, the greatest hazard is often found on slopes burned at high severity in steep catchment areas, which are prone to erosion given the right precipitation event (e.g., thunderstorms following a period of summer drought[1]). The consequences of post-wildfire soil erosion due to high intensity storms are often include damage or obliteration of settlements on alluvial/colluvial fans in the valley bottoms and associated transportation networks.

[1] *In some cases, in areas where snowmelt is the primary source of annual runoff, the increased 'equivalent clearcut area' due to a wildfire (and hence decreased interception of snow by forest vegetation), can lead to a larger snowpack that will melt faster and could create a peak runoff event that triggers erosion or flooding during the spring runoff. However, the authors are not aware of any examples of this possibility occurring to date.*

MANAGING OR PREVENTING THE RISK OF WILDFIRE EROSION

Regarding the factors described above, it is clear that fire restoration efforts may succeed in preventing wildfire erosion if we either recognize and manage risk due to fire-related events or prevent wildfires in the first place. This can be accomplished through three existing approaches, which have been undertaken to varying degrees and for various reasons in Canada:

1) **Prevention of wildfire problems,** through the pro-active re-introduction of fire into the ecosystem (although this has generally been done for ecosystem restoration more than for wildfire or erosion risk reduction, these reasons should also be considered);
2) **Post-fire rehabilitation** within the burned area of roads, constructed fire-breaks, and other disturbances made during fire-fighting efforts, and/or downslope or downstream drainage catchment areas at risk; and/or
3) **Protection of downslope or downstream developments and settlements** through the construction of large sediment catchbasins, channel protection, or actual relocation of the object at risk.

The focus of this chapter is on post-fire rehabilitation, item 2 above. However, there has been a growing trend towards ecosystem maintenance burning (item 1) in a number of regions of Canada, which is discussed below. Also, protection of downslope values (item 3) has been around as long as the values have existed in fire prone areas, such as the trans-continental railway constructed through the mountainous areas of western Canada starting in the 1880s.

Prevention of Wildfire Problems

Current situation

Fires at the interface between forests and settlements (so-called urban interface fires) have long been a concern in most of Canada and have been the topic of much emergency preparedness planning and more recently fire proofing of areas surrounding buildings and cultural icons, to reduce the risk (e.g., the 'FireSmart' programs disseminated by various Canadian provincial agencies). In the recent past, ecosystem maintenance burning has become common practice in a number of mountainous areas in western Canada, including the Rocky Mountain Trench in southeastern BC (Rocky Mountain Trench Society 2006) and the Canadian National Park System (Van Wagner and Methven 1980, Rogeau 2007). Efforts to return a site to a 'natural fire regime' are termed 'restoration' and are applied in areas where fire exclusion has altered the balance of ecosystems and habitats represented across the landscape.

Future expectation and opportunities

With predicted climate change and population increases, there is an increased concern about urban interface fires and the need for forest ecosystems to change in response to the predicted shifting rainfall patterns. Increased climatic variation has already increased the areas of dead timber (fuels) due to proliferation of pests like mountain pine beetle, and also increased the potential for more severe wildfires. Climate predictions for the short-, medium- and long-term future, suggest that dry areas will become drier, some wet areas may become wetter, and the weather will likely become more variable and extreme on both ends of the spectrum. For example, more violent thunderstorms in summer and rain-on-snow events in winter are expected. This pattern will be more like the types of storms currently observed in the western USA, which would result in similar erosion patterns shifting further north into Canada. Fuel management and ecosystem restoration are ways to reduce the hazard associated with wildfire erosion by reducing the frequency and/or severity of wildfires by reducing fuel loads. As a number of the urban interface forests are expected to become too dry for some of the current vegetation, proactive measures are desirable or even required in some areas, to prevent high burn severity wildfires and reduce the potential for catastrophic post-wildfire erosion.

Post-fire Rehabilitation

Current situation

In Canada, restoration efforts after forest fires have historically been focussed on rehabilitating access structures (e.g., roads and trails) developed during fire fighting efforts, and broadcast seeding and/or replanting of trees to hasten the re-growth of commercial forests. Little work has been done on actual hillslope restoration to prevent erosion or associated problems.

On forest land, the restoration has focussed on both the access structure network and to some extent the hillslope. In Canada, most forest land is publicly owned, either in areas managed for timber production, domestic water supply (with or without timber extraction), or in parks or other nature reserves. Most public natural resource management agencies have firefighting capabilities and employ aggressive firefighting and mop-up techniques to prevent the spread, and reduce the extent, of wildfires. As part of the cleanup efforts, access structures (roads, trails, landings) and fire guards (constructed fuel breaks, typically cleared by bulldozers or excavators depending on hillslope conditions) are rehabilitated to reduce erosion, restore natural drainage pathways, and promote re-vegetation. This typically involves re-contouring of excavated roads and scarification of other disturbances such as overland trails. Re-vegetation may be in the form of grass seeding, with either agronomic or natural species, or in the support of natural regeneration, such as placement of clumps of shrubs, during soil rehabilitation.

Hillslope restoration is not done the same everywhere. For example, in BC there used to be a policy of broadcast seeding and reforesting after all wildfires. This policy was changed in the recent past, but some reforestation continues to occur. To our knowledge, apart from the broadcast seeding which has often been ineffective, especially during the first post-fire year (see Chapter 11), no physical hillslope restoration or erosion control measures were normally taken after wildfires.

Future expectation and opportunities

The severe 2003 wildfire season in BC brought policies and approaches into question (Filmon 2003), including the need to address erosion and runoff problems that were observed following these fires (Curran et al. 2006). Large-scale erosion and flooding and the formation of water-repellent soils following wildfires had been reported for many years in the literature from the western United States, including California, Oregon, Washington, Idaho, and Montana. However, this phenomenon was not widely recognized in western Canada before 2003 because there were fewer urban – forest interface wildfires and fires were less frequent and not as widespread as in the United States. The 2003 wildfire season was different, since a much larger area burned at the more populated urban interface and soils and fuels were drier than average. These conditions led to more severe soil heating and created conditions, like soil water repellency, that left sites more vulnerable to damage from high-intensity rainfall. Also, more post-wildfire, high-intensity precipitation events may have occurred in some areas (Curran et al. 2006).

In direct response to 2003 post-wildfire erosion events in BC, a number of initiatives began, including: 1) government policy reviews; 2) development of assessment procedures that were first used during the 2006 fire season (British Columbia Forest Service (BCFS) Southern Interior Forest Region 2006); 3) installation of erosion control trials (Scott et al. 2007); and 4) testing of burn severity mapping (BCFS, unpublished data). It is expected that policy and procedures will continue to evolve as part of the risk assessment procedure. Although the Canadian risk assessment procedure is not as comprehensive as the Burned Area Emergency Response (BAER) process used by federal public land agencies in the USA (USDA FS 2007a), it is similar in structure. The objectives of the BC program include a post-fire assessment of fire effects, identification of potential risks, and communication of these findings to all affected jurisdictions and responsible agencies. It is possible that some preventative treatments may be undertaken where justified for the public good. The assessment procedures, burn severity mapping, and hillslope erosion control treatments, as well as the trial applications of this new policy, have been modelled after the U.S. Department of Agriculture, Forest Service (USDA-FS; Robichaud et al. 2000, USDA-FS 2007b), in collaboration with USDA FS colleagues, and will evolve over time based on further local adaptation and testing.

Protection of Downslope Developments and Settlements

Current situation

Since the development of transportation corridors in mountainous areas, there has been a need to protect these from natural processes, including wildfire erosion. However, it is difficult to provide definite examples of works that were upgraded or constructed in direct response to fires. This is due to the fact that the mountainous areas of Canada have non-fire related erosion and runoff issues on many alluvial and colluvial fans, and even hillslope drainage channels, that are crossed by roads and railways. Strategies from railway and highway engineering for mitigation of runoff and stability problems continue to evolve as new materials and methods become available. Typically, a peak runoff event causes infrastructure damage, which is repaired with larger capacity culverts or bridges, and often will have a trash rack installed to catch and hold debris being carried downslope. In one instance, a catchbasin was planned after a catastrophic incident of wildfire erosion, but it was not built. This is probably because the first major runoff event carrying a large pulse of sediment had already occurred, and the risk for future events of equal magnitude decreased rapidly – even below the risk for unburned catchments in some areas.

Future expectation and opportunities

If more post-fire erosion mitigation treatments are employed, it would lessen the impact and need for works further downstream (presuming that the hillslope efforts are successful). In BC, two efforts are being brought together under evolving government policies that address natural hazards mitigation and post-wildfire hillslope erosion.

RESTORATION EFFORTS THAT ARE CONTEMPLATED IN THE FUTURE

Future expectations for all three kinds of restoration activities (prevention of the problem in the first place, within-fire restoration, and downslope works) will likely occur in the future. We anticipate that more fuel management and ecosystem maintenance burning will occur in response to urban interface wildfire concerns, natural disturbance concepts, and climate change. The removal of invasive plants in some areas will also help to lessen wildfire risk because some of these are highly flammable (e.g., scotch broom). Below is a sample scenario of wildfire risk assessment and mitigation/restoration in the future, with photo and data examples.

Immediate Wildfire Control and Public Safety

Government agencies and emergency preparedness staff aggressively fight the fire and ensure public safety through assignment of resources relative to values at risk. The approach is described for BC by the BCFS Protection Branch (2007). During the fire fighting efforts, earth science specialists are called in to provide advice on sensitive areas (e.g., regarding sediment control on fire guards and access trails during the planning or construction).

Wildfire Erosion Risk Analysis

Widfires are screened by the BCFS to assess the increased threat to public safety, property, and infrastructure posed by the potential for post-wildfire landslides, erosion, and flooding events. Where screening identifies a candidate fire, specialists in soil science, hydrology, geomorphology, and geotechnical engineering (and others as required) review all available information, following basic principles of natural hazard risk assessments, as summarized for Southern Interior British Columbia (BCFS Southern Interior Forest Region 2006) or more generally discussed in Pike and Ussery (2006). When possible, satellite imagery or aerial photographs are used to generate a burn severity map, following procedures similar to USDA-FS (2007b), to help guide field assessments of risk.

Notification of affected parties and decisions on preventative works

In the interest of public safety and protection of infrastructure, ongoing communication with all affected government and transportation agencies occurs during and after the wildfire. The level of effort depends on the degree of threat and elements at risk (e.g., infrastructure or actual settlements) with public safety being the paramount concern. The people or resources at risk may be removed from the hazardous area, protected, or the risk may be mitigated through preventative works (e.g., hillslope restoration) if funding is approved as warranted. It is possible that rainfall monitoring may be done as part of a flood warning system, if considered practical and warranted. If there is an imminent threat to public safety the Provincial Emergency Program will be immediately notified. If the risk to identified elements is moderate or high, based on the results of the field assessments, the results of the risk analysis are communicated to appropriate stakeholders.

Preventative works

As these have not been actively employed before, an adaptive management process will be followed whereby more than one approach may be tried on a given fire and some follow-up monitoring will occur (the amount depends on available research funding). Once again, works are modeled after USDA-FS experience as documented by Robichaud et al. (2000), but also modified based on local experience over time.

Experience is currently being gained on sites burned in the 2003 fires where various mulching techniques are currently being tested and evaluated (Scott et al. 2007). Five research sites in the Southern Interior of British

Columbia were selected as representative of landscapes burned in the fire season of 2003. The study sites were placed on slopes that were at a high risk for post-fire erosion (i.e., severely burned, steep and relatively even slopes) that offered sufficient space and local uniformity to allow 15 plots to be established at each location. Treatments of straw mulch, bark mulch, hand seeding (50 g 100 m^{-2}) and needle fall were tested against an untreated control, and replicated three times at each of the five fire site locations. Silt fences were installed to capture the eroded sediment (Fig. 1). Sediment was collected periodically and weighed.

Fig. 1 Soils eroded off a bark mulch experimental plot, that were captured in a silt fence so the mass and water content can be measured, as part of the rehabilitation trial described in the text.

In year 1, the bark mulch, straw mulch, and needle fall significantly reduced erosion, with needles and bark mulch being the most effective (Fig. 2). Seeding alone did not reduce erosion relative to the untreated plots. Erosion rates were significantly different between fire site locations. In year 2, erosion was an order of magnitude lower than in the first post-fire year (Fig. 3), and the treatment effects were no longer significant.

Our experience to date confirms USDA-FS findings that ground cover provides effective protection against accelerated erosion, but seeding does not (see Chapters 11 and 13). Early application of treatments after a fire is important. If foliage is not consumed in the fire, natural needle fall will offer effective erosion control with no intervention.

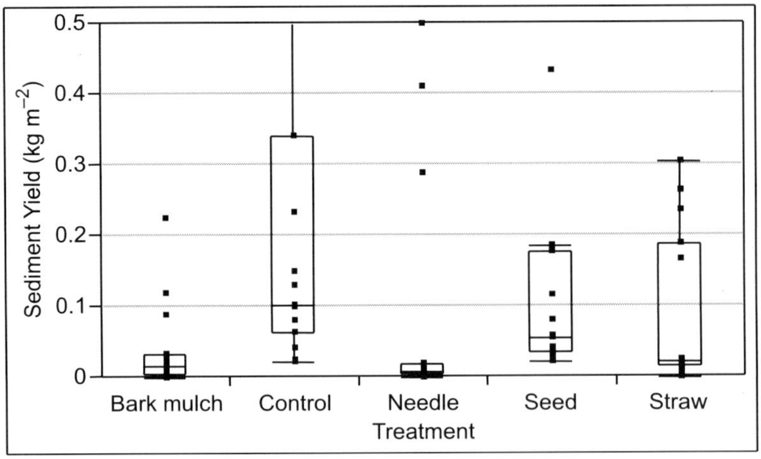

Fig. 2 Sediment yields in the **first year** after fire, by treatment for all locations, showing the broad range of variation, and the high sediment yields associated with the control and seeded plots. The box indicates the 25th, 50th and 75th percentiles, and the whiskers show the 2nd and 98th percentiles.

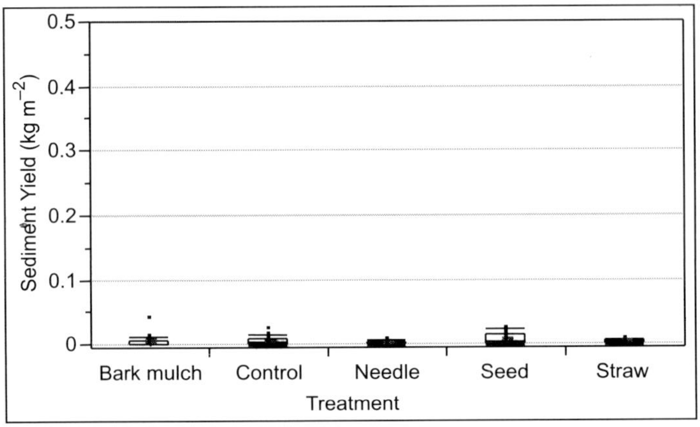

Fig. 3 The sediment yields in the second year, averaged across all plots within treatments plotted with the same vertical scale as the first year results to show differences in erosion during the second post-fire season.

During the risk assessment process, the areas of concern, in terms of risks to life, infrastructure, and personal property, are determined to select and pinpoint placement of treatments because blanket treatment of the entire burned area is not feasible. The size of, and accessibility to, an area being treated may determine the treatment applied. The relative costs of treatments over small (10's to 100's of square meters) and large (10's to 100's of hectares) will also be a deciding factor.

CONCLUSIONS

Within the mountainous areas of western Canada, the risks of post-wildfire erosion have been recognized. Policies are being revised presently, and risk assessment procedures are being drafted and tested. Provisions for targeted hillslope restoration of burned slopes may become standard practice. We are planning for potential problems that may result from a more variable climate. Changing climate has already created increased areas of dead timber (fuels) from the proliferation and spread of the mountain pine beetle, increasing the potential for more severe wildfires.

With climate change, erosion following wildfire is expected to become a bigger issue, not just in the mountainous areas, but elsewhere in Canada. For example, much of the Canadian Shield has coarse textured soils, which are considered more prone to developing water repellent conditions following severe wildfires that consume the forest floor. In addition, the many organic soils that exist throughout Canada may be more prone to erosion following destruction of the protective live vegetation and fibrous organic material near the surface during severe burns. In areas of permafrost, burning may cause subsequent soil warming leading to solifluction and other related forms of erosion. Lastly, in grassland environments, wildfire can make the site more susceptible to wind erosion. Although these phenomena currently are not common in Canada, the observed and predicted trends make them likely in the near future.

Acknowledgement

The authors appreciated the review comments from Dr. Artemi Cerdà, Ken Soneff, BC Ministry of Forests and Range in Kamloops, British Columbia and Mark Heathcott, Parks Canada, Calgary, Alberta for their comments which have improved the final manuscript.

References

B.C. Ministries of Forests and Environment. 1995. Biodiversity guidebook. Forest Practices Code. *Online at* http://www.for.gov.bc.ca/tasb/legsregs/fpc/fpcguide/biodiv/biotoc.htm [accessed 15 Sep 2008].

BCFS Protection Branch. 2007. How we fight fires. *Online at* http://bcwildfire.ca/ [accessed 15 Sep 2008].

BCFS Southern Interior Forest Region 2006. Interim standard operating procedures and risk assessment protocols for wildfire erosion. BC Ministry of Forests and Range, Southern Interior Regional Office, Kamloops, British Columbia, Canada.

Burke, J.M., E.E. Prepas, and S. Pinder. 2005. Runoff and phosphorus export patterns in large forested watersheds on the western Canadian Boreal Plain before and for 4 years after wildfire. *Journal of Environmental Engineering Science*, 4(5): 319-325.

Carcaillet, C., P.J.H. Richard, H. Asnong, L. Capece, and Y. Bergeron. 2006. Fire and soil erosion history in East Canadian boreal and temperate forests. *Quaternary Science Reviews,* 25: 1489-1500.

Chanasyk, D.S., I.R. Whitson, E. Mapfumo, J.M. Burke, and E.E. Prepas. 2003. The impacts of forest harvest and wildfire on soils and hydrology in temperate forests: A baseline to develop hyphotheses for the Boreal Plan. *Journal of Environmental Engineering Science,* 2: S51-S62.

Curran, M.P., B. Chapman, G.D. Hope, and D. Scott. 2006. Large-scale erosion and flooding after wildfires: understanding the soil conditions, Technical Report 030. B.C. Ministry of Forests and Range, Research Branch, Victoria, British Columbia, Canada.

Filmon, G. 2003. Firstorm 2003. Provincial Review. *Online at* http://www.2003 firestorm.gov.bc.ca/firestormreport/FirestormReport.pdf [accessed 16 Sep 2008].

Flannigan, M., I. Campbell, M. Wotton, C. Carcaillet, P. Richard, and Y. Bergeron. 2001. Future fire in Canada's boreal forest: paleoecology results and general circulation model – regional climate model simulations. *Canadian Journal of Forestry Research,* 31: 854-864.

Meidinger, D.V. and J. Pojar [eds.]. 1991. Ecosystems of British Columbia. Special Report Series 06, Ministry of Forests, Research Branch, Victoria, British Columbia, Canada.

Pike, R.G. and J.G. Ussery. 2006. Key points to consider when pre-planning for post-wildfire rehabilitation. FORREX Series 19. FORREX Forest Research Extension Partnership, Kamloops, British Columbia, Canada. *Online at* http://www.forrex.org/publications/FORREXSeries/FS19.pdf [accessed 16 Sep 2008].

Prepas, E.E., J.M. Burke, D.S. Chanayk, D.W. Smith, G. Putz, S. Gabos, W. Chen, D. Millions, and M. Serediak. 2003. Impact of wildfire on discharge and phosphorus export from the Sakwatamu watershed in the Swan Hills, Alberta, during the first two years. *Journal of Environmental Engineering Science,* 2: S63-S72.

Robichaud, P.R., J.L. Beyers, and D.G. Neary. 2000. Evaluating the effectiveness of post-fire rehabilitation treatments. General Technical Report RMRS-GTR-63. U.S. Department of Agriculture, Forest Service, Rocky Mountain Research Station, Ft. Collins, Colorado, USA.

Rocky Mountain Trench Society. 2006. Fire-maintained ecosystem restoration in BC's Rocky Mountain Trench: Principles, Strategy, Progress. Blueprint for Action. *Online at* http://www.trenchsociety.com/setup/content/Blueprint_for_Action_2006.pdf [accessed 16 Sep 2008].

Rogeau, M.P. 2007. Synopsis of fire regime work: National Parks of Western Canada. Prepared for Dan Perrakis, Fire Ecologist, Parks, Canada, Western and Northern Service Centre, Winnipeg, Manitoba, Canada.

Rowe, J.S. 1972. Forest Regions of Canada. Information Canada, Ottawa, Quebec, Canada.

Sanborn, P., M. Geertsema, A.J.T. Jull, and B. Hawkes. 2006. Soil and sedimentary charcoal evidence for Holocene forest fires in an inland temperate rainforest, east-central British Columbia, Canada. *The Holocene,* 16(3): 415-427.

Scott, D.F., K. Forsyth, K. Geist, and S.R. Thomson. 2007. Evaluation of fire site rehabilitation methods in terms of controlling erosion and sedimentation. Technical Report, Forest Investment Account (FIA) Project YO73327. University of British Columbia-Okanagan, Kelowna, British Columbia, Canada. *Online at* http://www.for.gov.bc.ca/hfd/library/FIA/HTML/FIA2007MR409.htm [accessed 3 Oct 2008].

Strong, W.L. and K.R. Leggat. 1981. Ecoregions of Alberta. Alberta Energy and Natural Resources – Resource Evaluation and Planning Division, Edmonton, Alberta, Canada.

USDA Forest Service. 2007a. Burned Area Emergency Response, BAER program. U.S. Department of Agriculture, Forest Service, Washington, D.C. *Online at* http://www.fs.fed.us/biology/watershed/burnareas/background.html [accessed 16 Sep 2008].

USDA Forest Service. 2007b. Burned Area Emergency Response (BAER) Imagery Support. U.S. Department of Agriculture, Forest Service, Remote Sensing Applications Center, Salt Lake City, Utah, USA, *Online at* http://www.fs.fed.us/eng/rsac/baer/[accessed 16 Sep 2008].

Van Wagner, C.E. and I.R. Methven. 1980. Fire in the management of Canada's National Parks: philosophy and strategy. Occasional Paper 1. Parks Canada, Hull, Quebec, Canada.

18

Fire, Catchment Runoff and Erosion Processes, and Post-fire Rehabilitation Programs: Recent Australian Experience

David Dunkerley[1*], *Natalie Martin*[2], *Shaun Berg*[1] and *Rob Ferguson*[3]

Abstract

Fire is an ever-present influence on much of the Australian landscape, and has been a driver of ecosystem change through the Tertiary and Quaternary periods. The arrival of aboriginal people at perhaps 50 to 60 ka BP, and subsequently European settlement in the last 200 years, has resulted in major changes in fire regimes. Much of the Australian forest vegetation is fire-prone as well as fire-adapted, and major fires periodically consume more than one million ha in a single fire event. Enormous financial costs are incurred in fire control, as well as in the loss of assets and primary production. This chapter describes the impacts of the most recent disastrous forest fires that swept across large areas of New South Wales and Victoria in the early summer months of 2003, following a season of very dry conditions. These fires were ignited by lightning, and burned for about 60 days.

Under the emergency conditions of the fire suppression and control program, which involved more than 15,000 firefighters, about 9,000 km of fire control lines were constructed. Some of these were made by heavy machinery, and were up to 60 m in width. Rehabilitation of these cleared lines began immediately following the declaration that the fires were safe, and involved covering with earth and vegetation residues, as well as cross-draining and some replanting. The protection of waterways was a high priority in this work.

[1] *School of Geography and Environmental Science, Clayton Campus, Monash University, Victoria, Australia.*
[2] *North East Catchment Management Authority, Wodonga, Victoria, Australia.*
[3] *Melbourne Water, Melbourne, Victoria, Australia.*
* *Corresponding author: David Dunkerley, School of Geography and Environmental Science, Building 11, Clayton Campus, Monash University, Victoria 3800, Australia. Tel: +61 3 99052914, Fax: +61 3 990052948, e-mail: david.dunkerley@arts.monash.edu.au*

In areas of alpine vegetation, ground cover was so reduced that surface runoff was generated on areas where it was previously not present. This was accompanied by the widespread erosion of alpine humus soils. Highly significant areas of moss and peat were burned, and rehabilitation works targeted the retention of seepage water in these ecosystems, in order to promote regrowth of the mosses and associated wetland species. In lower montane forest environments, remarkably high catchment runoff efficiency resulted from the complete loss of canopy and ground cover. Intense thunderstorm rains followed the fires in many areas, and exceptionally large loads of coarse bedload, as well as ash and finer sediments, were carried into streams and rivers. Rehabilitation works have sought to limit the coarse sediment flux arriving at major regional streams and rivers, but sediment control works have been required along many roads and culverts in fire-affected areas. Hillslope sediment movement has remained very active for the four years post-fire, and this is accompanied by ongoing readjustments in stream channel form. It is clear that the fires of 2003 were a major driver of landscape change, and it seems likely that in fire-prone Australian environments, the geomorphic and hydrologic changes wrought by such events vastly overshadow the incremental changes that occur in the years between catastrophic fire seasons.

INTRODUCTION – FIRE IN THE AUSTRALIAN ENVIRONMENT

Both natural and anthropogenic fires have long histories in the Australian environment, but neither history is well known. The presence of natural fire in Australia is known to extend far back into the Tertiary period. The changing abundance of charcoal particles preserved in dated sedimentary deposits in the Murray Basin, a large sedimentary basin primarily lying beneath New South Wales (NSW) and Victoria, and containing about 500 m of sediment infill, suggests that fire occurred infrequently during the Eocene epoch, somewhat more frequently in the Oligocene, and abundantly in the Late Miocene (Kershaw et al. 2002). This evident increase in fire occurrence as the Tertiary progressed is considered to reflect a progressive drying of the Australian environment, associated with which were incremental pedologic and vegetation changes, including the development of sclerophyllous vegetation and of the tall open-forest communities (Kershaw et al. 2002, Barlow 2004). The process of aridification was caused by cooling of the surrounding oceans, which curtailed evaporative transfer of moisture to the atmosphere, but bolstered by feedback effects including the rising albedo of the landmass as paler dryland ecosystems developed. The aridification process was linked to global tectonic changes, including the northward drift of the Australian landmass (the bulk of which took place during the Tertiary period), and the opening of the Southern Ocean that separates Australia and Antarctica. The progressive loss of the warm and moist continental climate of the early Tertiary saw the gradual disappearance of an early Tertiary rainforest cover, and its replacement by sclerophyllous taxa adapted to fire and to nutrient scarcity. According to pollen data from sedimentary sources, it is certain that by the start of the Quaternary period, at a little more than 2 Ma BP, fire-prone sclerophyllous vegetation was widely established in south-eastern Australia (Kershaw 1994). From Late Quaternary and especially Holocene deposits in lakes, river billabongs, and in the marine sedimentary record, there is an extensive and increasingly complete record of the presence of fire. The most detailed information on charcoal abundance is contained in records from Holocene times (the last 10 ka). Only rare terrestrial records, including those from crater lakes in North Queensland, and from Lake George in the Australian Capital Territory (ACT) near Canberra, extend beyond 100 ka BP (Kershaw et al. 2006).

Finding the means to ascertain when natural fires began to be augmented by anthropogenic burning is a contemporary challenge (e.g., Kershaw et al. 2006). Intervals in the late Quaternary that show a rise in charcoal abundance have been interpreted to signal anthropogenic burning by early populations of Australian aboriginal people. This event has been placed as early as 130 ka BP (e.g., Singh et al. 1981) but support for this interpretation has faded, especially in the absence of corroborating archaeological evidence of human presence, which only extends to about 60 ka BP in northern Australia (Roberts et al. 1994) and for southern Australia, only to about 50 ka BP (Turney et al. 2001). Currently, it is argued on various lines of evidence that human populations probably only had a significant impact on fire in the Australian environment from about 45 ka BP (Kershaw et al. 2006). Nevertheless, 45 millennia of fire manipulation by people is sufficient time for pervasive fire effects on ecosystems to have developed, as is suggested below.

In parts of Australia, contemporary land management involves an attempt to sustain the kind of fire regime generated by indigenous people through this long interval of time (e.g., Kakadu Board of Management 1999). While aboriginal burning for purposes related to food gathering would have been managed according to a regime of customary practices, it seems likely that hunting or cooking fires would periodically have escaped and resulted in additional, unplanned fires (Dodson 2004). Lightning strike across Australia can be presumed to have caused fires in past millennia as it does over large areas today.

This brief review suggests that fire has been present as an influence on the Australian environment through an extensive interval of geological time, with the onset of anthropogenic burning on a significant scale from perhaps 45 ka BP. Entire fire regimes seem likely to have been altered after this date, with changed seasonal incidence of fires, as well as altered patterns of intensity and spatial extent. Additional changes were wrought in the last 200 years, following European settlement, including additional sources of ignition, as well as attempts to control or eliminate fire in some environments.

The Nature and Frequency of Fires in Pre-European Times

In the tall open-forests of southern Australia, the presence of charcoal at depth in the soil is one of the indicators of major bushfires in the past (Ashton and Attiwill 2004), the charcoal perhaps being emplaced as roots were burned, or by falling into the pits created by tree-fall following fires. Various estimates of the frequency of forest fire have been made, using the age-distribution of long-lived stands, and these have suggested a mean interval between stand-killing fires of 50 to 100 years (McCarthy et al. 1999, Gill and Catling 2002). The distinctive occurrence of extensive even-aged stands of taxa such as the Mountain Ash *Eucalyptus regnans* (which regenerates from seed shed from the canopy following fires) have been taken to indicate that extensive major bushfires, sufficiently intense to destroy forest stands, occurred prior to European settlement (Ashton 1976). However, Lindenmayer et al. (2000) have presented evidence from Victorian montane forests suggesting an important role for low-intensity, non-stand-replacing fires in these forests. More generally, though the palaeo-fire record referred to earlier sheds useful light on the relative frequency of fire in various time periods, frustratingly, it is not capable at present of shedding any light on fire intensity and its variation through time or across the varying vegetation communities. A common view is that small forest fires were probably more common in aboriginal times, with large damaging fires becoming more evident following European settlement (Ashton and Attiwill 2004).

There are undoubtedly marked regional differences in particular post-European fire histories. In alpine and subalpine areas of southeasthern Australia, for example, fire was frequently applied to benefit the conditions of fodder availability for summer cattle grazing, so that the fire frequency thought to have characterised aboriginal times rose by a factor of two to five in the European period (Williams and Costin 1994, Pickering and Barry 2005). Large wildfires such as those of 1939 and 2003 have swept across the alpine environments, consuming grass, shrub heath, and snowgum communities. In the more arid inland, the displacement of indigenous populations from tribal lands probably in some cases resulted in altered fire regimes, though ignition by lightning remains common.

Fire in the Contemporary Australian Environment

Fire is one of the greatest forms of environmental disturbance in Australia. In one 12-month period (1999 to 2000) the total area burned nationally was >700,000 km^2, mostly in the northern tropics and the arid interior (Environment Australia 2001). Burning, much of it deliberate, is a widespread, annual event across much of the northern tropics. For the period 1997 to 1999, it has been estimated that an average of >300,000 km^2 of the northern savannas were burned or fire affected annually (Russell-Smith et al. 2003). In Victoria, in southern Australia, an average of 620 wildfires burn about 110,000 ha each summer season (Department of Sustainability and Environment (DSE), Victoria, 2006).

In forested areas of Australia, unplanned fire behavior can be extremely severe. This circumstance relates to several factors, including the climatic environment, the characteristics of the vegetation, and fire management practices. Instances of severe wildfires in Victorian forests include the infamous 'Black Friday' wildfires of 13 January 1939 which burned $>1.38 \times 10^6$ ha, the 'Ash Wednesday' fires of 1983, which consumed 87×10^3 ha (Rawson et al. 1983, Attiwill 1994), and the fires of January and February 2003 which, in Victoria alone, consumed $>1.1 \times 10^6$ ha (Department of Sustainability and Environment, 2006), with additional large areas burned in adjoining parts of New South Wales.

The economic costs of fire are substantial. For 'disaster' category fires, defined as those whose insurance costs exceed A$10M, the annual average cost is A$77 million (Australian Institute of Criminology 2004). The 1939 'Black Friday' fires cost A$750 million in current terms; the 1983 Ash Wednesday fires cost >A$400 million; while the 2003 fires in Canberra cost >A$300 million (Australian Institute of Criminology 2004).

In the arid continental interior, hundreds of fires occur annually, many ignited by lightning. These burn through ecosystems of all kinds, including fire-prone tussock grasses such as Spinifex (*Triodia* spp.), and maintain mosaics of fire scars of differing ages, with a fire return period in the Great Victoria Desert in Western Australia estimated to be approximately 20 years (Haydon et al. 2000). The fires tend to be small, and only about two to five percent of the burnable landscape of the Great Victoria Desert is actually burned in any year (Haydon et al. 2000). In these remote areas, little is known about the hydrologic and geomorphic consequences of fire, although almost certainly burning facilitates dust entrainment by vortex winds generated on hot soil surfaces in summer.

The role of the climatic environment

Large areas of Australia may experience extreme conditions during summer (December through February), with temperatures often >35°C, accompanied by relative humidity <20 percent and winds gusting to >80 km^{-1} h (Ashton and Attiwill 2004). For example, during the severe Ash Wednesday fires of February 1983, Adelaide recorded a temperature of 42°C and a relative humidity of <10 percent (Keeves and Douglas 1983); in Victoria, Melbourne reached 43°C with a relative humidity as low as five percent (Rawson et al. (1983). Fine fuel loads dry quickly under these conditions, reaching as little as a few percent moisture by weight, and the risk of fire is consequently extreme. When ignited, the fires can be of high intensity (>100,000 kw m^{-1}) and sweep across large areas; they are virtually uncontrollable without favorable changes in weather conditions. Repeatedly during the European period, such fires have destroyed areas covering $>10^6$ ha in the course of a few weeks.

Flammability of sclerophyllous vegetation

As noted above, the aridity of Australia and the nutrient-poor soils have combined to result in widespread sclerophylly, whose attributes includes small and hardy leaves. Many of these contain flammable resins or volatile oils, which increase the vigor of fire (Attiwill 1994). Moreover, many of the Eucalypt species are highly productive of litter, including abundant fine twig materials that form highly combustible fine fuel loads on the forest floor. Indeed, because leaf materials decompose more rapidly than woody detritus, woody fine fuels become enriched on the forest floor in comparison with the original litterfall. In some Australian forests, the litter fuel loadings can reach 35 t ha^{-1} (Gill and Catling 2002). Other features of the vegetation are relevant to the spread of fire, including the release of burning strands of bark which can be carried as firebrands and result in fire spotting well ahead of the main fire front.

Fire management

As in the USA, there has been a customary practice of suppressing major fires, primarily for the protection of landscapes and infrastructure (Rieman et al. 2003, Hatten et al. 2005). Fire suppression, in combination with other drivers, such as grazing pressure and climate change, have resulted in various ecosystem changes, including a decline in grasses and a rise in woody shrubs across the rangelands (Neave and Abrahams 2002). Fire management in many kinds of landscapes, but especially in forests, has become a complex issue that involves perception of what is of value in natural or cultural landscape attributes, of sustainability, biodiversity, economic activity, recreation, primary production, human health and safety, water supply, river health, weed and pest management, and an array of other issues (e.g., Gill 2005). Superimposed on the contemporary challenge of fire management is the ever-growing issue of global and regional environmental change, and its implications for fire occurrence and behavior (Williams et al. 2001, Cary 2002), which do nothing but increase the challenge that will have to be faced by fire managers in coming decades. In southeastern Australia especially, drought is a recurrent feature of the climate system, related to the operation of the El Niño – Southern Oscillation. Seasons when rainfall is very much below the long-term average often presage very active bushfire seasons during the summer months.

FIRE AND LANDSCAPE PROCESSES IN AUSTRALIA

In the remainder of this chapter, we leave aside fire history, and the complex social and economic considerations relating to fire management in the fire-prone Australian environment, and turn instead to the effects of fire on hydro-geomorphic processes. Much of what is discussed relates to the effects that are seen immediately post-fire, and through a relatively short interval of years subsequently, during which the landscape undergoes relaxation, moving from high levels of post-fire disturbance toward a condition of greater stability. We illustrate where appropriate the variety of approaches to rehabilitation of landscapes that are employed in Australia.

It is important to recall that fire has acted on Australian environments for vast periods of time, including the systematic manipulation of fire regimes by Australian indigenous people through more than 40 millennia. Aridity has been present to varying degrees for many millennia, and nutrient scarcity in Australian soils, relating to the enormously long periods of subaerial weathering, and an absence of glacial scour or tectonism capable of exposing unweathered basement rock, has been a long-term influence on vegetation. While there was minor glacial activity in south-eastern Australia during the Quaternary period, the elevation of the bulk of the continent is too low to have allowed even periglacial activity. The poor soil resource has been a factor driving the evolution of the highly flammable Australian sclerophyllous vegetation. High loads of fine fuel, in combination with the climatic context already described, have ensured that severe fire is a recurring event (Ashton and Attiwill 2004). Following severe fires, the kinds of hillslope and channel events described later in this chapter, including accelerated soil loss and the intensified export of organic matter and nutrients from drainage basins, further deplete the resource-base available to the vegetation. Thus, as expressed by Ashton and Attiwill (2004), the outcome for ecosystem productivity of repeated fire and fire-related ecosystem losses is a 'downward spiral' of resources degradation and the reinforcement of the conditions leading to fire-prone vegetation communities. Large areas of forest in Australia appear to be governed by this kind of fire-driven regime.

In considering such long-term outcomes of repeated severe fires, it is helpful to envision a time-lapse view of the affected landscape. In montane forests, for example, severe fire triggers processes beyond simple accelerated hillslope erosion. Owing to the mechanical weakening caused by fire, frequent tree-falls are commonly observed, and the root network often brings to the surface attached clasts of weathered bedrock and attached soil material from the lowest soil horizons; at the same time, the tilted root mass leaves a pit in the soil surface into which water, sediment, surface soil materials, and organic matter can be delivered by overland flow. These materials arrive at subsoil depths within the tree-fall pits, and become buried by arriving hillslope sediments. These processes, repeated incessantly, amount to an ongoing process of soil overturning or mixing, occurring in conjunction with the export of soil materials flushed away by post-fire erosion (including export of nutrients dissolved in hillslope runoff).

It seems reasonable to envision that mechanisms such as this could steer the course of soil development, given the vast available time. Understanding such fundamental, low-level ecosystem and landscape roles for fire is one of the tasks facing researchers. One possible outcome of the fire-driven overturning mechanism is the loss, by wind, raindrop impact, and overland flow, of fines from the upper, overturning part of the soil column, and its differentiation from the less disturbed lower layers. It has recently been argued that the important class of texture-contrast soils in Australia may indeed be the product of systematic burning through the millennia of aboriginal occupation (McIntosh et al. 2005). These authors have drawn attention to the less marked development of similar soils in New Zealand, where the Maori people have been present and using fire for only about two percent of the time through which Australian ecosystems have been exposed to anthropogenic burning.

This conceptual framework provides a necessary basis for interpreting contemporary post-fire landscape change. The occurrence of fire today, and the effects of these fires on landscape processes, is undoubtedly conditioned to a significant extent by the less immediately striking, but nonetheless pervasive, effects of fire through prehistory and geologic time that have set the predisposing soil and vegetation parameters which today support the ongoing role for fires in many environments. In particular, this perspective is of fundamental relevance to assessing the appropriateness of measures to rehabilitate landscapes disturbed by fire in the modern era.

THE EFFECTS OF FIRE ON HYDROLOGIC AND GEOMORPHIC PROCESSES

We have adopted a simple classification of the kinds of hydro-geomorphic process changes that may arise following major fires. These changes fall into three classes:

Hillslope and Channel Process Changes

This category includes process changes that arise more or less immediately following a fire of sufficient intensity and areal extent. These include such outcomes as:

- the *intensification* of processes that were operating pre-fire, such as soil dislodgment by splash, or bedload transport in streams;
- the *occurrence of new processes* that were not active pre-fire, such as Hortonian surface runoff, or mud and debris flows, on hillslopes;
- changes in the *spatial locations* of erosion and deposition of soil and sediment on hillslopes or of sediment in streams;
- changes in hillslope properties triggering *consequent* effects not directly related to the fire event, and which may arise after the lapse of a significant time. An example is the intensification of soil splash as tree canopies redevelop following a crown fire. The canopies then release gravity drops onto the soil surface where a protective litter cover has not yet become replenished.

- changes in *channel form and geometry,* including changes in the riparian zone, and in the locations of scour, transport and deposition of sediments. Altered catchment runoff ratios, for example, may lead to widespread bank retreat as increased boundary shear stress results in an increase in stream channel capacity.
- changes related to *channel network behavior,* including effects arising at stream junctions. These include slackwater deposition at stream junctions, as well as the progradation of gravel deltas from the bedload delivered by steep tributary streams.

Complex Response Chains

This category includes effects that were referred to as complex response by Schumm (1979) and others. Complex response chains are cascades of events that may be widely separated in space and/or time from an initial trigger event. Complex response chains arise post-fire in the following contexts:

- Hillslope complex response chains. These may arise from declining sediment availability, the slow armoring of soil surfaces as coarse erosional lag deposits accumulate, and the changing rates of recovery of canopy and ground cover. Thus, in the example noted previously, splash dislodgment of soils may not peak immediately post-fire, but rather several years later, when tree crowns have developed sufficiently to intercept rain and transform small incident raindrops into larger gravity drops, which may strike the ground repeatedly beneath persistent drip-points. This effect would presumably be eliminated as understory and contact cover regenerated. But the fire-adapted Eucalypt trees in many cases regenerate foliage quickly, through the mechanism of epicormic budding, and before sufficient litter has accumulated at the soil surface to provide protection from drop impacts.
- Channel complex response chains. An example is provided by the initial aggradation of channels in response to the large fluxes of eroded hillslope materials. However, exhaustion of supply on the hillslopes, or the recovery of vegetation, inevitably reduces the sediment flux. Thus, incision may begin again some years post-fire, after a lag determined by the recovery of hillslope sediment delivery processes.
- Channel network complex response chains. The functional reintegration of stream channel networks is dependent on the readjustment of all of the tributary sub-catchments. This undoubtedly proceeds at different rates for channels of different stream orders, owing to related differences in stream gradients, in the volume of aggraded sediments emplaced post-fire, the extent of post-fire collapse in undercut stream banks, and the differing impacts of storm events on small and large catchments. Thus, network readjustments will follow a complex temporal sequence influenced strongly by local hydrologic and geomorphic conditions.

Longer-term Incremental Changes

These are not dealt with further here, but include potentially persistent changes in soil moisture status and weathering processes related to a loss of soil depth or permeability; changes in the relative magnitudes of baseflow and stormflow, and of solute and particulate loads; and weathering changes related to progressive floristic change and alteration to processes of bioturbation, changed litter fluxes and decomposition rates, etc. Some effects of this kind probably have relaxation times in the decades-to-centuries range, though little is known of the slow processes involved. Some of the altered landscape processes responsible for long-term incremental change are very probably triggered when fire-induced hydro-geomorphic process changes become essentially intransitive (Chappell 1983). An example would be the local stripping of soil by intense post-fire erosion, sufficient to expose bedrock or impermeable soil horizons. Recovery from perturbations such as this would be extremely slow.

Instances of all of the categories of post-fire geomorphic change listed here will be provided in the following sections of this chapter.

CASE STUDIES

We will consider three instances of wildfire in Victorian landscapes. These provide representative illustrations of the range and severity of impacts on geomorphic and hydrologic processes. The examples selected also illustrate a range of approaches to post-fire rehabilitation. The case studies are: 1) fire effects on hillslope runoff and erosion in steep montane forests of the eastern uplands of Victoria. 2) fire effects on hydrology and erosion in alpine terrain, from the Bogong High Plains. This is a part of the major Alpine National Park, and is an area of largely treeless vegetation where grasses, shrub heath, and *Sphagnum* moss beds are widespread. 3) fire effects on channel sediment delivery and channel form in rivers and streams draining the eastern uplands of Victoria

In all cases, we draw on recent studies carried out following the major bushfire season in the summer of 2003. These fires are described first to provide the required background for the geomorphic and hydrologic studies.

The Southeast Australian Fires of Early 2003

After very dry antecedent conditions, fires were ignited by lightning on 7 January 2003 (approximately the middle of the summer season). Multiple lightning ignition points resulted in fires that burned for about 60 days, consuming >1.3×10^6 ha in Victoria, and a further 800,000 ha in adjoining parts of the New South Wales (NSW) Alps (Fig. 1). In Victoria, >15,000 firefighters were involved in attempts at fire control. Government funding for environmental and agriculture rehabilitation alone amounted to >A$70 million

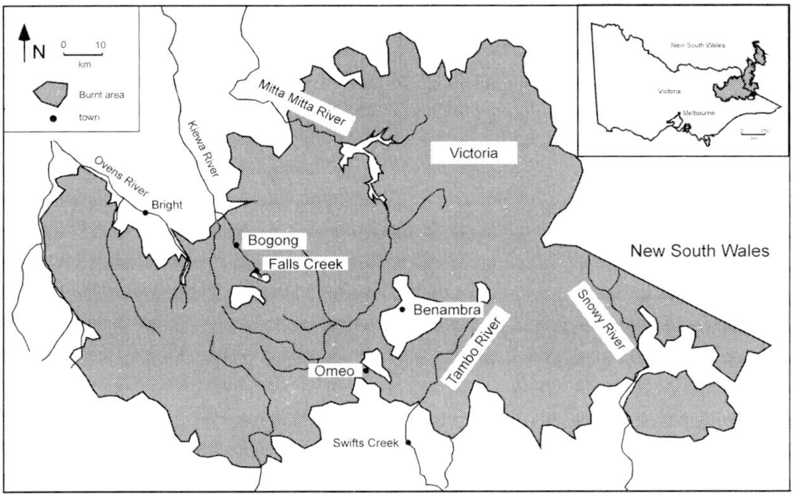

Fig. 1 Location of the areas of eastern Victoria affected by the fires of early 2003. The inset shows the extension of the burned area into adjoining parts of New South Wales. All of the rivers shown drain outwards toward the edges of the map. The field study areas discussed in this chapter lie within a triangular area bounded by the towns of Omeo, Benambra, and Falls Creek.

(Department of Sustainability and Environment, Victoria; DSE 2005). The fires burned through extensive tracts of steep and rugged forest terrain. Fire fighting and asset protection (e.g., of alpine resort townships) required the construction of emergency fire control lines. In some cases these were cleared by hand, but elsewhere heavy earthmoving machinery was used to clear breaks up to 60 m wide (DSE 2005). In total, >9,000 km of fire control lines were established. For reasons of erosion control (including silt runoff into waterways), visual amenity, and the exclusion of non-essential vehicle traffic, rehabilitation of these control lines was a high priority. An objective set very early following the fires was to stabilize disturbed sites with unprotected soils prior to the arrival of potentially erosive rain events (Jacobs and Walker 2003). This work was considered urgent because of the potential for major delivery of sediment into waterways. However, a rapid commencement of rehabilitation work was also seen as desirable because some fire suppression machinery was still in place, and could be employed speedily for rehabilitation works (Jacobs and Walker 2003). Government works rehabilitated >6,000 km of these control lines, largely within three months of the fires (DSE 2005). Rehabilitation included respreading of soil and the use of fallen trees and shrubs to provide soil protection and shelter for regenerating plants. A study of the impacts of a forest control line in Victoria has been presented by Ham (2003). The work is costly; covering bare ground with salvaged soil and vegetation, and cross-draining of tracks was estimated to cost up to A$5,500 per km. In the alpine environments, where the conditions necessitated more care with drainage, and sometimes replanting by hand, the costs reached A$10,000 to A$15,000 per km of single-blade-width fire control line (Jacobs and Walker 2003).

Case Study 1. Hillslope Process in Montane Forests of the Eastern Uplands of Victoria

Much of the area burned in the 2003 fires in eastern Victorian mountain terrain was forested public land. The studies reported here were made in the Omeo – Benambra area, in the catchment of the Mitta Mitta River (Fig. 1). Much of the terrain is within the Alpine National Park, but some is state forest, and some is privately owned agricultural land. The fires here were severe, burning tree crowns and consuming understory vegetation and litter over large areas.

Regional bedrock in the Omeo-Benambra study area includes Devonian granites and Ordovician sedimentary and metamorphic rocks. The granitic terrain in particular became vulnerable to hillslope erosion following the fires, as the sandy grus was readily splashed and carried by intense post-fire hillslope runoff. Field inspections following the fires indicated that the very widespread stripping of hillslope soils and grus was occurring largely as a result of unchannellized slope runoff; rills and gullies were noted but their occurrence was not common.

Hobgen (2005) carried out an investigation of hillslope sediment delivery in a typical burned catchment of 1.9 km^2 lying within this area and draining into the Mitta Mitta River. She installed 12 geotextile silt fences, 3 m wide, in footslope locations, within a few metres of the 5.26 km-long main stream channel. Hillslope gradients immediately above the silt fences were typically 20 to 30°. Eroded sediment trapped by the fences would thus certainly have left the slopes and entered the channel, but may have entered channel storage lower in the channel network. The trapped sediment was collected, weighed, and sieved on 5 occasions in the period September 2004 to July 2005 (20 to 30 months post-fire). The silt fences collected substantial volumes of soil and grus washed from the hillslopes until cessation of systematic monitoring (end of 2006).

The fires were followed by some days of intense thunderstorm rain, often quite localized, which generated high loads of hillslope sediment. Sediment trapped in the silt fences declined with time, but in December 2004, when further heavy rain occurred (monthly rainfall was 93.5 mm; I_{30} was 30 mm h^{-1}), reached nearly 1300 g per mm rain depth, or >120 kg at a single silt fence (see Fig. 2). The silt fence data revealed a strong dependence of hillslope sediment loss on slope aspect, northwest-facing slopes (hotter and drier due to exposure to the afternoon sun) shedding an estimated 97.5 Mg of sediment to the stream during the study. In contrast, the estimated load from southeast-facing slopes was only 27.6 Mg. Over the 194 ha catchment, these sediment fluxes are the equivalent of about 1 Mg ha^{-1} yr^{-1}. However, there is extensive bare granite outcrop in the upper catchment, so that on soil-covered areas, soil

Fig. 2 Upper: Overflowing silt fence in the landscape north of Omeo, from Hobgen (2005). Lower: silt fence in the same area (photo taken September 2006), showing the absence of ground cover more than three years post-fire.

loss rates would exceed this value. Trapped sediment was dominantly sand and coarse sand; granules and pebbles were commonly trapped, but silts and clays comprised <10 percent of the sediment by weight.

Rates of hillslope sediment delivery immediately post-fire are inferred to have been much greater than the rates just described. Reconnaissance field inspections established that hillslope sediment delivery was more efficient than channel conveyance of the sediment. Channels throughout the fire-affected region rapidly became choked with sand, gravel, cobbles and boulders (Fig. 3). The flushing of these materials from the stream channels has now supplanted stream aggradation. However, it is not clear whether this change has been the result of vegetation recovery and a consequent decline in the rate of sediment supply from the catchment hillslopes, or of an exhaustion of available sediment supply on the slopes. Observations in the study area suggest that this progression of events is now, nearly four years post-fire, triggering a new wave of bank collapse and gully erosion running upslope from the riparian zone along many over-deepened stream channels. This is an instance of a *complex response chain* triggered by the 2003 fires, since new kinds of geomorphic process (gullying, bank collapse), attributable to the 2003 fires, are being initiated after a lag of some years. It is not yet possible to determine when stability will return to these landscapes. It is conceivable that another severe fire will occur beforehand, again providing an impetus driving landscape disturbance. To develop a conceptual model of this kind of landscape change, we need evidence of the longer-term occurrence of fires like those of 2003. This information might be derived by identifying and dating ancient channel deposits indicative of post-fire erosional processes such as seen today. However, such information is currently not available.

Fig. 3 Aggraded channel of a typical small tributary to the Mitta Mitta River. Photo location about 25 km north of Omeo.

The large fluxes of hillslope sediment reaching stream channels and causing channel aggradation suggest highly efficient hillslope runoff, even in storms of only moderate rainfall intensity. A rainfall intensity-duration relationship for the field area was derived from continuous tipping bucket rainfall records spanning the period of silt-fence data collection. It revealed that the maximum 1-hour intensity was only 19 mm h^{-1}. The maximum 12-hour intensity was only 4.3 mm h^{-1}, while for 24 hours, it was just 2.3 mm h^{-1} (i.e., about 55 mm in one day). Forest soils ordinarily would be capable of infiltrating rain at these intensities without the production of surface runoff, except locally by saturation overland flow. Highly efficient surface runoff and hillslope sediment transport may be driven by one or more of the following processes:

- runoff from existing bedrock outcrops
- runoff from local areas of thin soil generated by post-fire erosion
- water repellent soils, perhaps generated locally by high fuel loads, burning logs, etc.
- low infiltration capacities in the post-fire soil surface, which has lost the uppermost organic layers, including the litter.

In all cases, the lack of litter and understory vegetation would contribute to a relatively open path for hillslope runoff.

In order to investigate soil infiltration properties, cylinder infiltrometer tests were carried out on two hillslope transects in the same catchment as the silt fences. Results show that soil properties are highly non-uniform. Infiltration rates in areas where the soil surface has been scoured by splash and runoff erosion exhibit very low infiltration rates, in some cases <5 mm h^{-1}. In such sites, all litter together with the fine root mat have been removed either by the fires or by post-fire erosion, leaving relatively dense subsurface soil exposed. Elsewhere, soil has lodged against fallen trees, exposed roots, or where the slope locally flattens. Such sites exhibit high infiltration rates, of 30 to 60 mm h^{-1} (the maximum measured rate was 130 mm h^{-1}). Thus, given the rainfall intensities noted above, impermeable patches would shed runoff in even low-intensity rain events in this area, while the permeable patches should be able to absorb even the maximum 30-minute intensity (30 mm h^{-1}). In addition, though it has not been investigated, it seems likely that slope aspect is linked to systematic variations in soil infiltration rates. Thus, it can be concluded that post-fire soil properties in this intensely-burned forest environment exhibit high spatial variability, and that efficient hillslope runoff (and hence, sediment delivery) probably depend on linkages between impermeable patches, or on impermeable patches located fairly close to drainage lines.

The rarity of rills and gullies suggests that wide-area particle dislodgment was dominant post-fire. Splash appears to be a primary agent of soil dislodgment in this area, as well as in the monsoonal tropics of Australia (Fig. 4). This provides another example of a *complex response chain* in the

Fig. 4 Splash pedestals resulting from post-fire soil dislodgment. Upper: splash pedestal in the Omeo area. Coin is 23 mm in diameter. Middle and lower: splash pedestals from the monsoonal tropics of Australia. The lower image shows a single pedestal capped by a large stone. Lens cap is 49 mm in diameter. The pedestal is approximately 50 cm in height, to the base of the cap stone. Intense splash dislodgment in northern Australia results from annual burning just before the start of the monsoon rain season.

forested Victorian uplands. Hobgen (2005) found the mean of nearly 300 splash pedestal heights in her 1.9 km^2 catchment to be 24 mm. On the presumption that the pedestals reflect area-wide stripping of soils, this equates to a loss of 370 Mg ha^{-1}. However, observations of a smaller sample of splash pedestals made in August 2006 showed that these were preferentially developed where tree canopies were regenerating, and were thus probably produced by large gravity drops released from overhead foliage (the lower parts of the tree canopy in the study area stand at about 8 m above the soil surface: sufficient for terminal velocity to be closely approached by gravity drops falling from the leaves and branches). The 2006 pedestal heights averaged 39.4 mm (n=23, standard deviation 13.1 mm), and pedestals are clearly most strongly developed beneath regenerating tree crowns. One splash pedestal was 70 mm in height. Thus, as the forest recovers, it appears that local splash dislodgment of soil is becoming more evident, nearly four years post-fire. It seems probable that this process will decline only when ground cover, including litter, recovers to pre-fire levels. Once again, this evidence shows that post-fire hydrologic and geomorphic process adjustments involve complex response, rather than a simpler, monotonic trend toward pre-fire conditions.

Case Study 2: Hillslope Hydrologic and Erosional Processes in Victorian Alpine Environments

Extent and hydrological importance of Australia's alpine areas

Australia's alpine region includes the highest parts of the Eastern Highlands and associated uplands in the southeast of the continent (including regions in Tasmania), comprising an area of approximately 25,000 km^2 (Costin 1989). These uplands are the areas of highest precipitation and have the highest water yield per unit area of any part of the river basins which drain the uplands (Downes 1962). In the Upper Murray River Basin, for example, land above 1,050 m comprises only 23 percent of the catchment area yet produces 43 percent of the total yield (Department of Water Resources Victoria 1989).

Though a common event at lower elevations, bushfire is much less frequent in the Australian Alps, because the combination of factors required – drought and severe fire weather – only occur several times per century (Commonwealth Scientific and Industrial Research Organisation (CSIRO) 2003). However, the fires of 2003 swept through 60 percent of the Alpine National Park (Department of Sustainability and Environment 2005). This fire was only one of three landscape-scale fires in the Victorian Alps that have affected treeless vegetation in the last 100 years; the other fires being in 1939 and 1985 (Wahren et al. 2001).

Post-fire hydrology and erosion studies have been made on the Bogong High Plains in northeastern Victoria since the 2003 fires (Berg 2006). The High Plains are a series of alpine and subalpine plateaus that range from approximately 1500 m to 1900 m (Williams and Ashton 1987). Many areas are either above the treeline or treeless because of the effects of cold air drainage. The 2003 fires in the alpine region of the Bogong High Plains affected several distinct plant communities, including *Eucalyptus pauciflora* (Snow Gum) on ridges and knolls, grasslands, heathlands, and many *Sphagnum* bogs and other wetlands. In a small (<2 km^2), severely fire-affected (>90 percent burned) alpine catchment in the Rocky Knobs region near the ski village of Falls Creek (S 36° 53', E 147° 17'), post-fire groundcover surveys revealed pronounced effects on vegetation cover. Approximately one year post-fire, 34 percent bare ground was recorded on steep (>10°) hillslopes which once supported dense shrubs, including *Orites lancifolia*, *Phebalium squamulosum*, *Hovea longifolia*, and *Bossiaea foliosa*, and over 20 percent bare ground was recorded in flatter areas dominated by *Poa hiemata* and *Kunzea muelleri*. Two years later, bare ground constituted approximately 5 percent of total groundcover on the steep slopes, due primarily to the coloniser *Poa hothamensis*, while the percentage of bare ground on flatter terrain was reduced by just 5 percent (through the spread of existing *Poa* species).

Post-fire hillslope runoff data were collected from six runoff plots, each covering 1 m^2, equipped with tipping-bucket flow recorders and event data loggers. Runoff was recorded during natural rain events larger than 2.5 mm, and the data revealed mean runoff fractions ranging from 8.0 percent (generated from a fully enclosed plot for which water arriving from upslope was excluded by the top boundary wall of the plot) and 19.0 percent (an unbound plot, with no upslope wall and hence open to water arriving from upslope). Under pre-fire vegetation conditions, it is reasoned that overland flow would have been a rare occurrence, owing to the dense foliar and surface litter covers, except during times of snowmelt where saturation-excess may dominate. Pre-fire, there was very little bare ground (less than 1 to 2 percent; Williams and Costin 1994).

To assist in interpretation of the runoff plot results, infiltration rates in the catchment were measured at many locations using a cylinder infiltrometer. In conjunction with high resolution rainfall data collected by a tipping-bucket gauge located close to the runoff plots, the infiltration test results indicated that infiltration excess was not likely to occur regularly, with rainfall intensities only exceeding measured infiltration rates over very short time periods (less than five minutes). Likewise, sediment transport on steep slopes is also presumed to be minimal in the absence of fire, given the absence of surface runoff. However, under the post-fire conditions, soil erosion data from runoff plots and silt fences revealed maximum values of 178 g m^{-2} mo^{-1} and 135 g m^{-2} mo^{-1} respectively. These are significant values in light of the shallowness of the skeletal soils in the steeper terrain (these soils are generally less than 50 cm deep), and the slow rates of soil development in the low temperatures of the alpine environment.

Installations of erosion pins have also provided some insight into soil movement on burned hillslopes in the High Plains study catchment, with mean soil denudation per erosion pin over a 15-month post-fire period being nearly 10 mm; again, a considerable value given the depth of existing soil and the relatively short time period following fire. This depth of soil loss is lower than the 24 mm deduced from natural splash pedestals in the Omeo area, but in that case, tree canopies were present to generate erosive gravity drops.

The field investigations made in the Bogong High Plains area were conducted in support of a broader study of the hydrology of the alpine environment. The study will employ catchment modelling as a tool to further investigate the role of different parts of the landscape in generating runoff, including post-fire runoff. Water yield from the experimental catchment was monitored continuously at a V-notch weir installed near the basin outlet. The weir was equipped with a capacitive water level recorder, and a solar powered data logger. Post-fire storm hydrographs derived from the 8 largest runoff events in the study catchment during the 3 years following fire revealed reasonably flashy responses, with time of rise values between 5.5 and 14 h, ratio of direct runoff to baseflow values of up to 1.32:1, and base times as little as 14.5 h. While no pre-fire streamflow records were available, storm runoff and daily streamflow on the Bogong High Plains has been shown to increase immediately post fire (Lawrence 1999). Thus, in summary, unequivocal evidence of fire effects on surface runoff, sediment transport, and catchment hydrology exists in the alpine region of the Bogong High Plains. Recovery times are expected to be long, with decades estimated to be needed for the regeneration of many of the woody shrubs that formerly covered extensive areas of the terrain.

The role and restoration of Sphagnum bogs

A notable and highly-valued feature of the Victorian alpine terrain, the Bogong High Plains in particular, is the concentration of wetlands and mossbeds. While the spatial extent of wetlands (peats or humified peats) on the Bogong High Plains is relatively low – occupying approximately 10 percent of the area of alpine and treeless subalpine vegetation (McDougall 1982) – they are generally rated highly in terms of their ecological and hydrological importance. They provide habitat for fauna, and, if the peat is saturated, will influence adjacent soils and vegetation. Though the hydrological role of *Sphagnum cristatum* bogs in catchment behavior is not entirely clear, it has long been speculated that they provide a temporary water store that dampens the hydrologic response exhibited in small upland catchments. This traditional thinking has largely been dispelled as more field evidence has been gathered (e.g., Wimbush 1970, Grover 2006), though there is scope for a more extensive examination of the issue. A second widespread view is that water passing through bogs is filtered (Wimbush 1970), and aerated (Legoe 1981). This hypothesis also awaits formal testing and evaluation.

Ecological restoration of burned bogs

As previously mentioned, many *Sphagnum* bogs were burned in the 2003 fires, some severely. An additional impact on these bogs was provided by cattle grazing, which for many years has been permitted during the summer months. In light of the damage caused by the fires, grazing permits were not renewed following the fires, and cattle are now excluded permanently from many of the areas that were burned . On the Bogong High Plains, ecological restoration works began several months post-fire by ecologists using techniques developed during the previous decade, and targetted the combined effects of the 2003 fires and longer-term cattle grazing. In relation to restoration work carried out by Parks Victoria (State Government land managers responsible for the Alpine National Park) , during the two consecutive summers following the fire, approximately 3100 weirs (small runoff detention structures) were placed across flow paths in 16 bogs that covered an area of 80 ha. Several different types of weirs were used, the most common of which were non-woven geotextile socks (270 gsm), filled with hardwood bark chips, and anchored to the ground with either wooden pegs or galvanized steel pins. Less commonly used were coir logs (1.2 m in length and 0.30 m diameter), and hessian bags filled with hardwood bark chips and wrapped in geotextile (K. Cosgriff, 2006 personal communication). The intention of this aspect of rehabilitation work was to counteract the loss of water from the moss beds, and reduce the risk of erosional scour creating free drainage from them. Figure 5 provides an example of a typical weir placed across the channel of a dissected bog. This particular weir has successfully resulted in upstream ponding, and the wetting of peat layers which would otherwise have become desiccated and subject to further deterioration.

In conjunction with Parks Victoria rehabilitation, significant rehabilitation and demonstration research sites have been created by La Trobe University, and the Department of Sustainability and Environment, Arthur Rylah Institute for Environmental Research (ARIER). Similar weir structures were used, with the addition of straw bales wrapped in geotextile. In combination with the weir installation programs, restoration of bogs in poor condition (with bare peat and channels associated with cattle tracks) has taken place using species most commonly found in bogs and observed post-fire regrowth. These include monocots (*Carex gaudichaudiani*, *Isolepis* sp., *Carpha nivicola*, and *Poa costiniana*), forbs (*Myriophyllum pedunculatum*, *Scaevola hookeri*, and *Pratia surrepens*), and the shrub *Baeckea gunniana*. *Poa costiniana* seed was also directly sown on areas of bare peat. Small-scale transplanting of *Sphagnum* has been used near some weirs (W. Papst, 2006 personal communication).

In addition, monitoring water quality and biogeochemical changes during bog restoration projects is taking place. This research has a primary focus on changes in peat-water chemical composition associated with the immediate hydration of bogs (Silvester 2006), as a result of those measures previously discussed which alter the local water table (Fig. 5). Other bog restoration work is being carried out in fire-affected bogs in New South Wales and the Australian Capital Territory, as outlined by Hope et al. (2005).

The main objective in placing weirs in fire-affected bogs is to reduce the amount of channelized flow (i.e., to curb runoff) and to spread water out

Fig. 5 Post-fire rehabilitation works on the Bogong High Plains, Victoria. Upper: a series of weirs constructed along an eroding drainage line. Lower: a weir retaining water in a burned area of moss beds.

across the bogs in order to increase the local water table, thus promoting the regeneration of the *Sphagnum* and associated bog species (both floral and faunal). Ecological restoration works on the Bogong High Plains have focused primarily on bog restoration, presumably due to the view that burned grasslands and heathlands have a much greater capacity than the *Sphagnum* bogs to regenerate after fires (i.e., that they are less sensitive to disturbance by fires). In addition, while cattle grazing in alpine areas are no longer permitted in the Australian Alps, having been suspended in Victoria following the 2003 fires, there is a long history of bog deterioration through grazing (e.g., Costin et al. 1959, Wahren et al. 1999). Cattle entering and drinking from bogs, and eating associated palatable species, is thought to create compaction and entrenchment, thus dissecting the bogs and resulting in a combination of physical damage and altered hydrology. Cattle may form tracks around the peat, thus deflecting drainage accessions (Ashton and Williams 1989). A reduction in the local water table as a result of entrenchment leads to the drying of the fibric and upper hemic layers of peat, thus disrupting the integrity of the bog surface and leaving it prone to further deterioration. As such, the pre-fire condition of *Sphagnum* bogs was generally viewed to be poor, and it is hoped that the combination of removed grazing pressure, natural post-fire regeneration, and active ecological restoration, will prevent further dysfunction and sustain a healthy level of ecosystem maintenance.

Case Study 3: Post-fire River Channel Rehabilitation and Management in Northeast Victoria

As noted earlier, heavy loads of sediment (sand to boulder size) were moved into and along streams in the northeast of Victoria following the 2003 fires, as a result of exceptional hillslope runoff and sediment delivery. Some of this sediment passed across cleared grazing land, and the overbank deposition of bed sediments was widespread, especially on flatter land adjacent to the main regional streams and rivers. Exceptional loads of coarse bedload were also carried by many streams draining steep, burned catchments.

None of the low-order tributary streams draining to rivers such as the Mitta Mitta is gauged. Therefore, the exceptionally coarse bedload carried during post-fire storms, and remaining in the channels, provides an opportunity for the indirect estimation of peak discharges and catchment specific discharges.

The required bedload sampling and stream cross-section surveying has been carried out on a range of streams in northeast Victoria (Schultz 2005). Peak flood stage can be estimated from sand drapes on the banks, and by the pulverisation of the bark on trees partially submerged in the flow, which is caused by the repeated impact of swiftly-moving bedload grains.

Using estimates of the flow velocity needed to move typical large cobbles and boulders derived from the bedload equation of Neill (1967), and multiplying this by the surveyed cross-sectional area, Schultz (2005) derived estimates of specific discharge in the range 10 to 50 $m^3\ s^{-1}\ km^{-2}$. There is considerable uncertainty attached to these estimates, especially because of the likelihood of dynamic shifts in bed level associated with rapid aggradation and scour, which cannot be reconstructed. However, given that rainfall intensities of up to 100 mm h^{-1}, sustained for 2 hours, were estimated for this region in intense post-fire thunderstorms (Ferguson et al. 2004), these specific yields are possible from a 2 km^2 catchment. These values rank among the highest published values for flash flooding, and this confirms the efficiency of runoff production in the steep and intensely burned terrain of the Victorian uplands. Table 1 includes some comparative estimates of peak specific catchment runoff, though these are from intense rains over unburned catchments. In the case of the very high yields of the burned catchments, it is possible that the runoff efficiency was boosted by temporary water repellency of the soils, as well as by the steep slopes and virtually complete removal of ground cover resulting from the high fire intensities. The lack of ground cover left hillslope runoff paths open and unobstructed, and undoubtedly increased both the volume and speed of surface runoff.

The high runoff intensities just mentioned, together with steep stream gradients (see Schultz 2005 for details) resulted in very high bedload fluxes.

Table 1 Selected published estimates of flash flood peak specific discharges ($m^3\ s^{-1}\ km^{-2}$) resulting from intense rainfall events

Location and date of flood event	Catchment size (km^2)	Nature of rainfall event	Peak specific discharge ($m^3\ s^{-1}\ km^{-2}$)	Source
Granada, Spain, 1973	116	600 m in 24 h	13.4	Cordón et al. (2003)
Biescas, Spain, 1996	18	250 mm in 6 h	~ 30.0	Anquetin et al. (2004)
Gard, France, 2002	1400	> 600 mm in 24 h	~ 30.0	Anquetin et al. (2004)
Coromandel–Waikato region, New Zealand, 2002	Typically 10–35 km^2	Commonly 60–100 mm in 24 h, but up to 215 mm in 24 h	10 – 17	Munro (2002)
Lian-Hwa-Chi #3 watershed, Taiwan, 2001	3.4 ha	371 mm in < 24 h; maximum of 103 mm in 1 h	16.7	Cheng et al. (2005)
Moores Run, Maryland, USA, 2003	9.1	59 mm in < 1 hour; peak 15-minute intensity of 102 mm h^{-1}	13.2 (but up to 25.2 at one sub-catchment gauge)	Smith et al. (2005)
Jamaica, 1998	1.32	> 400 mm in 2 days	19.2	Laing (2004)

No bedload monitoring stations exist in the,affected area, but clear morphological evidence of the flux of bedload cobbles and boulders was visible over wide areas. It was not uncommon to see stream-side accumulations of cobbles lodged against trees that had been growing in the riparian zone, but which became in-channel trees as channel enlargement occurred during post-fire flash-flooding events.

The Region

The fires affected a total of six Victorian river catchments the upper Murray Mitta, the Kiewa, the Ovens, the Mitchell, the Tambo and the Snowy (Environment Protection Authority of Victoria (EPA) 2004a). The upper Murray Mitta, the Kiewa, and the Ovens catchments are located in the northeast region of Victoria which is a part of Australia's Murray-Darling Basin system. While this region represents only 2 percent of the geographic area of the Murray-Darling Basin, these 3 river basins flow northwards contributing 38 percent of the total water in the Murray-Darling Basin system (North East Catchment Management Authority 2005). The Mitchell River, the Tambo River and the Snowy River flow in a southerly direction as independent coastal catchments in southeastern Victoria. We will consider representative riverine issues that arose in northeastern Victoria, which plays a vital role in providing water resources for southeastern Australia, and approximately 35 percent of which was burned in the 2003 bushfires (Ferguson et al. 2004).

Physical characteristics

The northeast region of Victoria is bounded by the Murray River (the state boundary with New South Wales) in the north and east, the Victorian Alps in the south and the Warby Ranges to the west. The fire affected what is often referred to as the Victorian high country, the high mountain country of the eastern highlands, which separate the coastal catchments from the northern catchments of the Murray Darling Basin. Victoria's highest peak, Mt. Bogong (1986 m), and other peaks including Mt Feathertop (1922 m), Mt. Hotham (1862 m), and Mt. Buffalo (1723 m) are all in the region. The region is dominated by mountains and plateaus of varying size occurring over a range of elevations from 350 m to over 1986 m (Rowe 1967, Rowe 1972). It includes broad, mature valleys with extensive terraces and fans, a feature of the northern part of the catchments. Due to the mountainous nature of much of the region, the climate varies considerably (Rowe 1967). Average annual precipitation ranges from 762 mm in the northern parts where elevations are low, to 2000 mm in the southwestern highlands where snow forms the major part of winter precipitation; yet the Omeo region is influenced by a strong rain shadow (Rowe 1967). The distribution of vegetation in the region is strongly influenced by temperature and by soil-moisture availability (Rowe 1967).

Investigations carried out by Ferguson et al. (2004) focused on two contrasting areas of hill country – regular hillslopes up to 20° on metamorphic rocks in the Buckland Valley and the more variable hillslopes on granitic rocks around Omeo in Mitta Mitta River headwaters (Table 2).

Table 2 Properties of the hillslopes and drainage networks in the Buckland River and Omeo region study areas

	Buckland River	Omeo region
Slope	mostly 25–35°, with valley floor < 50 m wide dominated by Buckland River channel	mostly >15° but upper slopes 25–35°. Piedmont-type lower slopes leading onto valley floor and waterways
Geology	Palaeozoic metamorphosed siltstones and sandstone	Locally granitic dominated, with patches of Palaeozoic metamorphosed siltstones and sandstones
Fire intensity	Locally variable – moderate to high	High across large areas, especially steeper forested upper slopes, lower intensity on farmland of lower slopes.
Drainage network contributing to trunk waterways	Typically second and third-order systems, draining areas up to several square km	Typically first and second-order systems, draining less than one square km.

Buckland River physiography

The upper Buckland River catchment area is composed of regionally consistent Ordovician sandstones and shales that have undergone low grade metamorphism. The regionally uniform nature of the rocks is a key factor controlling the relatively consistent nature of the landscape. Slopes are dominantly in the 25 to 35° range.

The drainage network in the upper Buckland River area is dendritic. The Buckland River in this area is a fourth-order waterway, and is fed at approximately 90° by a series of second- and third-order streams that drain up to a few square kilometres. Figure 6 (upper) illustrates the typical drainage network for this area. Soils in the area are gravel loams. Colluvial deposits up to two metres thick, formed of gravel in a silt/clay matrix were observed in first and second-order stream lines.

Omeo region physiography

The landscape of the Omeo area, in particular around Mt. Sam in the Livingstone Creek catchment, is more varied than the Buckland River area. The geology is dominated by a mixture of Palaeozoic granitic intrusives and Ordovician low grade metamorphic rocks. Slope angles vary, and are typically

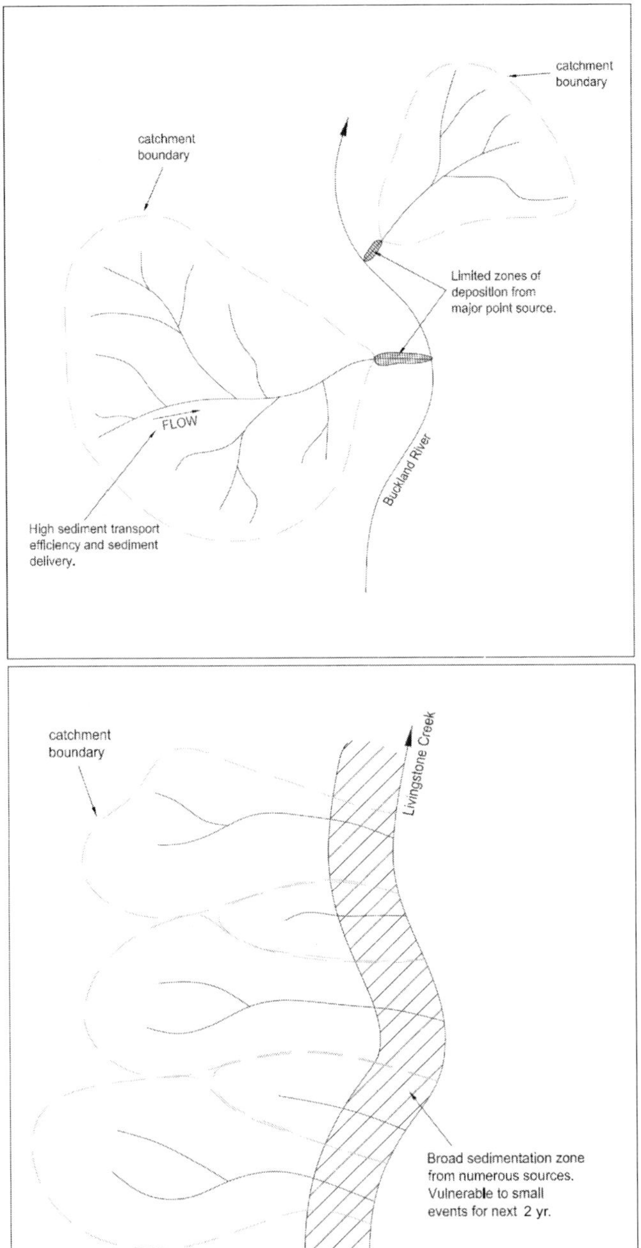

Fig. 6 Schematic diagrams showing the relationship between drainage network structure and debris flow deposition zones. Upper: Buckland River area. Lower: Livingstone Creek area, just north of Omeo.

greater than 15° in the mid and lower slopes, but up to 25 to 35° in the upper slopes. Soils are dominated by sand and sandy gravel, and are highly erosion-prone.

The drainage network in the Mt. Sam area is not dendritic. Third-order Livingstone Creek is fed by a series of first and second-order streams coming off the Mt. Sam massif (see Fig. 6, lower). Contributing catchment areas in these systems are typically in the 10 to 50 ha range.

Episodic intense rainfall

Shakesby et al. (2003) indicate that intense fire is capable of severely degrading soil cover and soil surfaces resulting in the promotion of major erosion. This change in cover at the catchment scale is reflected in changes in runoff, seen as increases in duration, frequency, and magnitude of channel hydrographs (Good 1973, O'Loughlin et al. 1982, Cornish and Binns 1987). Fire-affected slopes are also seen as an important control on the rates of soil erosion after intense rainfall (Good 1973, Brooker et al. 1993, Prosser and Williams 1998, English et al. 2005) yet the magnitude of erosion after fires is considered to be substantially dependent on the timing and characteristics of post-fire rain events (Hudson 1986, Wallbrink et al. 2004). Rainfall events undoubtedly trigger post-fire impacts, including the delivery of soil, ash, and other debris to rivers and waterways, affecting water flow and quality, leaving sediment and debris accumulations, and affecting stream biota including fish and invertebrates. But the volume, intensity, and spread of rainfall over the following months are always unknown, so that the magnitude of post-fire impacts in any bushfire event is hard to predict.

Summer in southeastern Australia is renowned for thunderstorms in addition to the hot, dry conditions already described. The first thunderstorm in the fire-affected regions of northeast Victoria was on 26 February 2003. Heavy rainfall was recorded on this date across much of the high country. According to local residents in the region, the storm had isolated pockets that resulted in rainfall intensity estimated to have been over 100 mm in less than an hour. Areas that were particularly affected were the Dingo Creek in the upper catchment of the Buckland River in the Ovens River basin, and various creeks and rivers (Bingo Munji Creek, Livingstone Creek, Cobungra River, Mitta Mitta River to name a few) in the Mitta Mitta catchment, which drains to the Murray River. The thunderstorm runoff carried very high loads of ash and according to many, the 'rivers ran black'. The rainfall in the catchment area of Dingo Creek resulted in a major sediment slug being carried with a flood pulse into the Buckland and then Ovens River. The Ovens River system is important to the region in that it provides water for stock and domestic use, irrigation, town water supply, and the lower reaches are listed as Heritage River. This event was well documented by Arthur Rylah Institute in an assessment of the impacts to aquatic species post fire. Ash, sediment, and fire debris washed into waterways after rainfall alters water chemistry and reduces dissolved oxygen levels causing stress and even death to fish and other biota. The study of the impacts of bushfires following a flash flood event in the catchment of the ovens river by the Environment Protection Authority of Victoria (EPA) (2004b) details the substantial runoff responses to the rainfall in Dingo Creek to the Ovens River where the flood pulse peaked at around 6,000 ML per day from the base flow of less than 500 ML per day (EPA 2004b). It was reported that turbidity in the Ovens River peaked at 70,000 NTU on February 27 at Myrtleford, when levels are usually below 10 NTU (EPA 2004b). The level of suspended solids is typically less than 6 mg L^{-1} in Ovens River; however, it peaked at 33,000 mg L^{-1} and Dissolved Oxygen (DO) concentrations dropped to 0.1 mg L^{-1} (EPA 2004b). These low DO levels were recorded at Tarrawingee, approximately 100 km downstream of Dingo Creek, and remained low for approximately 12 hours before commencing to rise again. It was observed that material deposited with the receding flood pulse was spread across the bed and banks infilling pool habitat and changing channel profiles (EPA 2005). An ongoing study of this sediment deposition found that 12 months later almost all of this material had been washed downstream.

Effects in low and higher order stream channels

Depositional and erosional processes on the sandy soils found on granitic rocks in the Omeo/Livingstone Creek area are similar to observations of the post-bushfire erosion process observed near Sydney by Humphreys (1985) and Paton et al. (1995), where post-fire organic debris, leaves, twigs, and bark, together with very large quantities of ash, were carried from the hill slopes as floating load followed by deposits of sand-sized bed load and finer suspended load. Similar dominant clast size is the key similarity between fires on the Triassic Hawkesbury Sandstone and the granitic rocks in the Livingstone Creek catchment. Small organic material such as ash, leaves, bark, and other post-fire residue is easily transported by flowing water (Wallbrink et al. 2004). Rocks, trees, and boulders are not generally thought to be so easily transported however this was a commonly observed load carried by runoff from the intense rain event in late February. These post-fire loads caused havoc for road authorities, filling drains and blocking culverts under roadways, and also for farmers, destroying fence lines and paddocks. The effects in low and higher order stream channels were extremely variable, depending on factors such as geology, the intensity of the fire, and the rainfall. Although the bushfires had many devastating consequences, the combined conditions of bushfires followed by intense rainfall resulted in fire-induced erosion, soil redistribution, and down gradient impacts, as well as presenting quite diverse challenges for management and rehabilitation.

After the first storm event and with subsequent localized rainfall events over the summer it was observed that the floating load was usually carried with the first rainfall in lower order streams. In contrast, higher order streams were receiving reoccurring large sediment loads, as lower order streams in different parts of the catchment received localized thunderstorm rainfall. Many sites in the Mitta Mitta river catchment were observed to continue to carry high sediment loads for the first two years after the fire. Where there was a continuous down gradient, drainage lines were observed to erode in response to increased flows. Where the gradient reduced due to confluence with larger waterways or where barriers to flow were encountered such as road culverts, the flow velocity would be reduced resulting in the deposition of bed and suspended loads. Small alluvial fans were a common site at the base of lower order streams at their confluences with trunk waterways such as the Livingstone Creek and Mitta Mitta River.

The expansion of alluvial fans following fires causes morphological changes including both increasing and decreasing gradients both upstream and downstream of fans (compare with Benda et al. 2003). This increased sediment storage in the fans was also associated with widening of floodplains and side channels and the creation of terraces. Where these fans were being created due to roadways, each rain event made it apparent, to the chagrin of the road authorities, that the road systems were unable to cope with the increased flows and sediment loads. The culverts on roads that were intersecting lower order streams on relatively steep slopes had become blocked, creating dams of sediment on the upstream side and on the downstream side the roads were being scoured away. Even road side drainage was affected. On gentle grades drains were being filled from hill slope erosion, reducing road drainage and on the steeper section of roads, drains were being scoured out, increasing in width and depth. Both scenarios created dangerous conditions for road users. Unfortunately, due to the mountainous terrain in most of the fire-affected region there was very little that could be done to control impacts on the road systems. In critical areas additional culverts were placed, down gradient road batters were reinforced to reduce scour, and maintenance crews were on standby to clear roads following rainfall events. Despite understanding that the work required was a result of natural forces, it was interesting to be told by crews undertaking this work that they found it very disheartening to return to the same location on a regular basis, as if they 'hadn't done their job properly the first time.'

In the Buckland River catchment, which also drains northward to the Murray River, erosion and associated sedimentation caused by high rainfall events in February 2003 differed greatly from those observed in the Livingstone Creek area just described. In the Buckland River area, all first-, second- and third-order stream lines were gullied out, with gullies typically up to 2.5 m deep and approximately 1 m wide. Gullies extend for many hundreds of meters up slope, often coming to within 100 m of the ridgeline, and multiple headcuts or steps up to 1.5 m high were observed in gullies on second and third order stream lines. Slope wash from adjacent slopes contributed sediment to gullies, but the total volume was probably less than that from the incision of the gullies themselves.

Debris flow deposition where the tributary streams joined the Buckland River was widespread. Debris flow deposits typically formed in the lower end of each tributary valley, where valley floor widths increase to above 20 m, and gradients drop below 15°. In the lower end of one third-order tributary, the zone of debris flow deposition extended for several hundred meters upstream of the Buckland River valley. The Buckland River valley itself is typically up to 50 m wide in this area, and debris flow deposits, particularly from third-order systems, in some cases covered the entire valley floor width, temporarily damming the Buckland River. It is estimated that at the margin of the Buckland River valley, the larger debris flow deposits are up to five m thick, decreasing rapidly into the valley floor center. The deposited material was dominantly gravel to cobble sized, with a silty/sandy matrix. Field inspections took place in October 2003, some nine months after the heavy rainfall event that washed large volumes of clay and silt sized particles into the Buckland River. Six days later these fine-grained suspended sediments caused great problems to the town of Wangaratta some 75 km downstream. The surface morphology of individual debris flow deposits was characterized by relatively flat gently dipping surfaces (see Fig. 7, upper) and linear lobes up to 1 m high and 5 m wide (see Fig. 7, lower) Coarse woody debris up to 50 cm in diameter and 5 m long has been deposited in the area. During the site inspection in October 2003, low flow channels were observed across all debris flow deposits. It is presumed that these formed after original debris flow deposition.

Physiographic controls over depositional processes

The markedly differing nature of depositional responses between the Mt. Sam/Livingstone Creek area and the upper Buckland River following post-fire rainfall is a clear function of different geology and drainage patterns. In the upper Buckland River areas of up to 4 km^2 contributed to individual debris flow deposits, through dendritic networks of second- and sometimes third-order. Debris flow deposition was focused into the lower end of tributary systems and the valley floor of the Buckland River valley. Deposition of coarse gravels was confined to the immediate confluence zone where tributary second and third order streamlines met the Buckland River valley, indicating that the combination of steep slopes and confinement were required to entrain such coarse clasts.

In contrast, the Mt. Sam area exhibited a broader array of depositional processes controlled by a more varied landscape and a dominantly sandy grain size. Numerous first- and second-order streamlines contributed sandy sediment to Livingstone Creek and low (<5°) hillslopes at the base of the Mt. Sam massif. Slopewash transported sand across bare slopes, with minor lobes deposited where flow concentration decreased below the transport threshold, most probably due to a decrease in slope angle. Flows in Livingstone Creek were competent to readily entrain the abundant sandy bedload carried in from

Fig. 7 Debris flow deposits in the Buckland River area. Upper: flat, gently dipping surface of debris flow deposit. Lower: panoramic view showing linear debris lobes.

tributary streamlines and direct slope connection, resulting in reworking and redeposition along the waterway.

Management planning

While observations in the region were providing general indications that rain events were the key trigger for impacts to riverine environments, the volume, intensity, and spread of rainfall over the following months was unknown so an approach to identify the high risk areas was required to facilitate management of the impacts on roads, streams, and property. The first step taken was a pilot study, which followed a risk analysis method to determine the dominant erosion and sedimentation process and the risks to stream values within the catchment area. The pilot study focused on two sub-catchments, the Omeo district and the upper Buckland River, as both these areas were affected by the thunderstorm event in February 2003 and were considered likely to be subject to further rainfall events. While the focus was on two subcatchment areas, the methodology of the pilot study could be readily applied on a catchment-wide basis. The outcome of the study was to produce a map of sediment sources and type, a risk method based on the sediment map and stream assets to identify risks and recovery priorities, recommendations for fire recovery management, and details of recommended trial rehabilitation works (Ferguson et al. 2004). Stream values were assessed under five major categories: channel form, in stream habitat, vegetation, water quality, and social and economic value. In prioritizing works, preference was given to protecting high values (assets) from the major identified threatening processes. Management works have focused on the following key areas in relation to waterways: erosion, sediment, and weeds. The focus of management for each of these key areas in relation to stream values is addressed below.

Management response to sediment deposition

A trial site for works that was a part of a pilot study (Ferguson et al. 2004) was an ephemeral waterway in the Mitta Mitta catchment and a tributary of the Livingstone Creek that flows into the Mitta Mitta River upstream of the Dartmouth dam. This site was chosen as a pilot site for several reasons: there was a high sediment load being mobilized into the Livingstone Creek which was causing blockage of the creek resulting in water quality concerns for the local community, the site was relatively easy to access, the landholders were keen to participate in the work, and the waterway intersected the Omeo Highway and the road culverts were blocked with sand and debris with each rainfall event, causing major difficulties for road users. It was important to recognize that the sedimentation was a natural result of the effect of fire on the sub-catchment and the episodic, intense rain events. The sediment load could not be stopped, regeneration of vegetation in the sub-catchment was expected to take some years, and the upper part of the catchment was steep and inaccessible to machinery. Thus, it was decided that the only possible management solution was to increase the natural recovery rate by assisting in the stabilization of the deposited sediments. The test site was a low-order waterway with a catchment area of 2 km^2, dropping almost 500 m in elevation over a waterway length that is just under 3.5 km. The changes to the catchment were resulting in an increase in rainfall runoff which was causing the upper half of the waterway to actively erode with some of the eroded material being deposited in the lower half of the waterway with the rest being mobilized into the trunk stream. Due to terrain constraints, no work could be done to prevent the erosion of the upper part of the waterway, and rainfall runoff would naturally diminish with the recovery of plant cover. The top part of the waterway, which lay within farm land, was fenced to exclude stock and revegetated to encourage natural recovery and protection of the stream system. The bottom part of the waterway where deposition was naturally occurring was chosen as an ideal place to put a series of pile fields to maintain flows yet to reduce sediment mobilization by increasing channel roughness. The pile fields were rows of hard wood timber piles with an average diameter of 300 mm, a length of 3 m, to be driven at intervals of 600 mm center-to-center spacings. The timber piles were salvage timber from the fires. They were driven with a pile driver (an excavator with a hydraulic hammer) into the bed and banks of the waterway until 500 mm above the deposited sediment or to refusal. They were placed approximately 600 mm apart from center-to-center of each pile, across the width of the waterway. A second row was placed within 100 mm to provide additional strength and resistance. A total of six sets of piles were placed in this waterway to reduce the sediment movement (Fig. 8). Since they were installed there have been several localized thunderstorm events in the catchment and to date the piles have held well. A low flow path has established in the waterway demonstrating that flows are still moving through the system as required however, sediment and debris movement has significantly reduced. The pile fields have been observed to have worked well in this situation, and as a result of their success they were used for three other waterways in the district where the site constraints were suitable to this type of work. With little rain in the district it is debatable that these piles have been really tested. Humphreys (1985) and Paton et al. (1995) argue that ground cover and resistance to erosion recover to pre-fire levels within three to six years after a fire. Similar patterns of reduction from initially high values of erosion were demonstrated by Croke et al. (1999) and Wallbrink and Croke (2002). This being the case, the fields have passed the crucial initial period as well as the first three critical years.

Another area where management works were undertaken focused on the Benambra Creek that had been dredged many years ago into a small channel. This creek was on a wide swampy floodplain and before people understood the ramifications of draining such areas, the central channel was dug. With the increased runoff post-fire, the creek system has had much greater flows than usual and bed and bank erosion was noticeably increasing. A survey was undertaken of the creek including a long profile of the bed and cross sections of the channel, to establish the locations of changes in grade along the bed. From this work a series of grade control structures were designed to stabilize the bed and reduce the grade and thus the rate of erosion. These works were designed to prevent the channel developing and becoming disconnected from the floodplain.

Weeds

The preceding examples of post-fire runoff and sediment delivery have illustrated something of the range of responses seen in the alpine and montane areas, and illustrated some of the rehabilitation techniques that are being deployed. Management of weed invasion in fire-affected areas was also a focus of the recovery program for catchments and waterways. Two particular weed species have been the focus of joint management of riverine

Fig. 8 Upper: Aerial photograph taken about one year after the 2003 fires, showing a tributary to Livingstone Creek, north of Omeo. The small tributary has delivered a load of sands and gravels sufficient to build a delta into Livingstone Creek, forcing flow against the east bank. Lower: View of the installation of piles in the tributary stream, designed to intercept and retain some of the coarse bedload material upstream of Livingstone Creek.

environments in the northeast of Victoria, the particular exotic species being willows (*Salix*) and English Broom (*Cytisis scoparisus*). Willows are not an indigenous species to Australia and were introduced as an erosion control tool on rivers and streams in the absence of immediate alternatives. Willow invasion of waterways has become a major stream management problem in Victoria, particularly in northeast Victoria. However, since the fires, willow species have been profiting from the lack of competition and have been found throughout the fire-affected area. In particular, many populations have been found colonizing the alpine wetlands described earlier, including those on the Bogong High Plains. Extensive work has been undertaken to map populations in the alpine region to ensure that populations are systematically controlled. Stem injecting of willows along major waterways has been undertaken where as in the alpine bogs with sensitive ecosystems the majority of the willows have been hand pulled, with the generous help of many community volunteers. The second invasive species prolifically colonizing riverine environments post-fire is English Broom. This particular species has been in the upper catchment of the Mitta Mitta River since historic mining times; however, it has been largely contained within a control area. Since the fires, with the movement of sediment within the catchment, the seeds have spread using waterways as vectors and the control area has been breached. English Broom is particularly difficult to control owing to the viability of the seeds, which remain viable for approximately 80 years. However, one of the positive outcomes of the fires was that the intensity of the fire in many areas is thought to have killed a large proportion of the viable Broom seed. This has presented an ideal management opportunity to control the Broom regrowth. However, with sediment loads moving through the catchment with rain events, road repair works, rehabilitation of containment lines, and other post-fire recovery management activities, the risk of spreading the remaining viable seed is quite high. Similar to the willows, mapping works have been undertaken to establish the extent of the Broom invasion, and staged works undertaken to restrict this plant to the original control area. With the alpine and sub-alpine terrain in fire-affected areas, the recovery of native understory and overstory vegetation is expected to take longer than areas of lower elevation. It is therefore expected that invasive weed species will require longer-term management until the native vegetation has become reestablished.

CONCLUSIONS

In this chapter, we have described some of the impacts of a major bushfire event (the 2003 bushfires) on a range of environments from high-elevation, treeless alpine terrain with alpine humus soils, to steep montane forests on erodible granitic soils. There was indeed a range of impacts on hydrologic and geomorphic processes that can be categorized using the scheme presented earlier. However, the magnitude of some impacts were outside the normal post-fire response range.

The steeper and drier terrain on granites north of Omeo almost certainly exhibited some surface runoff under pre-fire conditions. But following the fires, steep slopes in small catchments within this area exhibited extraordinarily high runoff yields. These slopes thus provide a clear instance of the *intensification* of the pre-existing process of hillslope runoff.

The extraordinary mobilization of bedload, given the small size of many of the catchments, was related to exceptionally high specific discharges, of >10 $m^3 s^{-1} km^{-2}$. The introduction of very coarse bedload, much of which remains in the stream channels, may be immobile until another severe fire event is followed by intense thunderstorms. Such accumulations therefore constitute the kind of evidence that, if it could be located in buried sediments and dated using residual charcoal fragments, might allow the pre-European chronology of severe bushfires to be extended.

Changes in geomorphic processes resulting from the fire, but with hillslope runoff response lagging, was exemplified by the changes in drop splash dislodgement of soil in the Omeo area. These changes in intensity of soil splash dislodgment had evidently arisen because the fire-adapted trees have regenerated by budding more rapidly than the understory and ground litter layers could redevelop to protect the soil surface. In this situation, the splash dislodgement would have become more intense with no external change in factors like rainfall or storm intensity. Additionally, post-fire hillslope runoff clearly delivered sediment to the drainage lines at rates that could not be matched by the ability of the stream to convey the material. For reasons that remain unknown, and which would be valuable to understand, conditions have now changed such that stream capacity exceeds the arriving hillslope sediment flux, and the channels are being scoured. However, the fact that the change from aggradation to degradation arose in the absence of any external trigger (such as an exception rainfall event), and after a lapse of several years, marks it as an instance of a complex response chain. This was not a direct response to fire, but rather an event that followed as one of a sequence of subsequent events. Potential mechanisms underlying the change from aggradation to degradation include recovery of hillslope ground cover, and the partial exhaustion of the hillslope sediment supply.

The soil infiltration data showed marked spatial variability. Low infiltration rates were associated with locations where post-fire runoff had resulted in surface scour. On the other hand, areas where hillslope sediment had accumulated, such as against stems or fallen woody debris, exhibited quite high infiltration rates. These spatial patterns seem likely to be quite different from any spatial patterns that might have existed before the bushfire, and thus provide an instance of *spatial rearrangement* that is not necessarily associated with a change in process intensity.

The rehabilitation programs undertaken in the various environments fall into two broad classes. In one category are attempts to accelerate the recovery of hydrologic and geomorphic processes toward pre-fire conditions. For example, to accelerate the recovery of damaged moss beds, weirs were installed to retain water within the wetland, foster growth of new moss cover, and return the system to a condition where it is once again self-supporting without management intervention. Another example of this class of rehabilitation is the cross-draining of fire control lines and the use of woody debris and salvaged foliage mats to provide protection and encourage re-colonization of the control lines by native vegetation. The control of coarse-grained stream sediments using pile fields represents a different class of rehabilitation related to the general improvement of ecosystem function. In any of these instances of rehabilitation work, the occurrence of a further bushfire event would complicate the analyses of the efficacy of the intervention works. In a similar sense, regional environmental change may cloud the assessment of the efficacy of the intervention works. This could arise through change in climatic conditions, including precipitation amount and seasonality, temperature, or humidity. But there may also be indirect effects of climate change on the fire regime itself, which may confound attempts to resolve the rate and course of post-fire recovery, as well as the efficacy and longevity of the effects of rehabilitation works.

Acknowledgments

The authors thank Mr. Cain Trist, Manager, Forest Estate Condition and Monitoring, Department of Sustainability and Environment, Victoria, Australia for providing a report on post-fire recovery relating to the 2003 bushfires.

References

Anquetin, S., J.D. Creutin, G. Delrieu, V. Ducrocq, E. Gaume, and I. Ruin [eds.]. 2004. Increasing the forecasting lead-time of weather driven flash floods. Laboratoire d'étude des Transferts en Hydrologie et Environment, Grenoble, France. *Online at* http://www.cnrm.meteo.fr/cyprim/global/documents_chantier_mediterranee/JRC_final_April_2004.pdf [accessed 16 Sep 2008].

Ashton, D.H. 1976. The development of even-aged stands of *Eucalyptus regnans* F. Muell. in Central Victoria. *Australian Journal of Botany,* 24: 397-414.

Ashton, D.H. and P.M. Attiwill. 2004. Tall open-forests. pp.157-196. *In* R.H. Groves [ed.]. *Australian Vegetation, 2nd edition*. Cambridge University Press, Cambridge, UK.

Ashton, D.H. and R.J. Williams. 1989. Dynamics of the sub-alpine vegetation in the Victorian region. pp. 143-168. *In* R. Good [ed.]. The Scientific Significance of the Australian Alps. Proceedings of the First Fenner Conference on the Environment. The Australian Alps National Parks, Weston, Australian Capital Territory, Australia.

Attiwill, P.M. 1994. Ecological disturbance and the conservative management of eucalypt forests in Australia. *Forest Ecology and Management,* 63: 301-346.

Australian Institute of Criminology. 2004. The cost of bushfires. Bushfire Arson Bulletin 2, 23 November 2004. Australian Institute of Criminology, Griffith, Australian Capital Territory, Australia. *Online at* http://www.aic.gov.au/publications/bfab/bfab002.pdf [accessed 16 Sep 2008].

Barlow, B.A. 2004. Phytogeography of the Australian region. pp. 3-35. *In* R.H. Groves [ed.]. *Australian Vegetation, 2nd edition.* Cambridge University Press, Cambridge, UK.

Benda, L., D. Miller, P. Bigelow, and K. Andras. 2003. Effects of post wildfire erosion on channel environments, Boise River Idaho. *Forest Ecology and Management,* 178: 105-119.

Berg, S.S. 2006. The Effects of Wildfire on Hydrological Behaviour and Soil Physical Properties in the Alpine Zone, Northeast Victoria. Ph.D. thesis, School of Geography and Environmental Science, Monash University, Melbourne, Australia.

Brooker, F.A., W.E. Dietrich, and L.M. Collins. 1993. Runoff and erosion after Oakland firestorm: expectations and observations. *California Geology,* 46: 159-173.

Cary, G.J. 2002. Importance of a changing climate for fire regimes in Australia. pp. 26-46. *In* R.A. Bradstock, J.E. Williams, and A.M. Gill [eds.]. *Flammable Australia. The Fire Regimes and Biodiversity of a Continent.* Cambridge University Press, Cambridge, UK.

Chappell, J.M.A. 1983. Thresholds and lags in geomorphologic changes. *Australian Geographer,* 15: 358–366.

Cheng, J.D., Y.C. Huang, H.L. Wu, J.L. Yeh, and C.H. Chang. 2005. Hydrometeorological and land use attributes of debris flows and debris floods during Typhoon Toraji, July 29-30, 2001 in central Taiwan. *Journal of Hydrology,* 306: 161-173.

Commonwealth Scientific and Industrial Research Organisation (CSIRO). 2003. Submission from CSIRO to House Select Committee on the Recent Australian Bushfires. No. 434. CSIRO, Clayton, South Victoria, Australia. *Online at* http://www.aph.gov.au/HOUSE/committee/bushfires/inquiry/subs/sub434.pdf [accessed 16 Sep 2008].

Cordón, E.R., Vigil E. Alvarez, Perez L. Garrido, F. Martin, Duran De Las Heras, J.J. Valsero, and R. Madrid García. 2003. The catastrophic flood event of the 18th of October, 1973, in the Rambla of Albuñol (La Rabita, Granada, Spain). *In* V.R. Thorndycraft, G. Benito, M. Barriendos, and M.C. Llasat [eds.]. Palaeofloods, Historical Floods and Climatic Variability: Applications in Flood Risk Assessment. Proceedings of the PHEFRA Workshop, Barcelona, 2002. EL Centro de Ciencias Medioambientales, Madrid, Spain.

Cornish, P.M. and D. Binns. 1987. Streamwater quality following logging and wildfire in a dry sclerophyll forest in southeastern Australia. *Forest Ecology and Management,* 22: 1-28.

Costin, A.B. 1989. The Alps in a global perspective. pp. 7-19. *In* R. Good [ed.]. The Scientific Significance of the Australian Alps. Proceedings of the First Fenner Conference on the Environment. The Australian Alps National Parks, Weston, Australian Capital Territory, Australia.

Costin, A.B., D.J. Wimbush, D. Kerr, and L.W. Gay. 1959. Studies in Catchment Hydrology in the Australian Alps. I. Trends in soils and Vegetation, Technical Paper No. 13. CSIRO, Division of Plant Industry, Melbourne, Victoria, Australia.

Croke, J., P. Hairsine, and P. Fogarty. 1999. Runoff generation and re-distribution in logged eucalyptus forests, south-eastern Australia. *Journal of Hydrology*, 216: 56-77.

Department of Sustainability and Environment (DSE). 2005. Report of the investigation into the future of cattle grazing in the Alpine National Park (Alpine Grazing Taskforce). Department of Sustainability and Environment, Melbourne, Victoria, Australia.

Department of Sustainability and Environment (DSE). 2006. Fire information reports. The state of Victoria, Australia, *Online at* http://www.dse.vic.gov.au [accessed 28 Aug 2008].

Department of Water Resources Victoria. 1989. Water Victoria: A Resource Handbook. Victorian Government Publishing Office (on behalf of the Department of Water Resources Victoria), Melbourne, Victoria, Australia.

Dodson, J.R. 2004. Quaternary vegetation history. pp. 37-56. *In* R.H. Groves [ed.]. *Australian Vegetation, 2nd edition*. Cambridge University Press, Cambridge, UK.

Downes, R.G. 1962. The Victorian High Plains – the environment and its use. *Proceedings of the Royal Society of Victoria*, 75: 339-347.

English, P., P. Wallbrink, G. Humphreys, R. Shakesby, S. Doerr, W. Blake, C. Chafer, and B. Vigneswaran. 2005. Impacts of water quality by sediments and nutrients released during extreme bushfires. Report 2: Tracer assessment of post-fire sediment and nutrient redistribution on hillslopes: Nattai National Park, NSW. CSIRO Land and Water, Australia.

Environment Australia 2001. State of the Environment 2001. Commonwealth of Australia, Australian State of the Environment Committee, Canberra, Australia.

Environment Protection Authority Victoria (EPA). 2004a. River Health: A snapshot of the effects of the 2003 bushfire. Environment Protection Authority, Melbourne, Victoria, Australia. *Online at* http://epanote2.epa.vic.gov.au/EPA/Publications.nsf/PubDocsLU/bushfire?OpenDocument. [accessed 21 Aug 2008].

Environment Protection Authority Victoria (EPA). 2004b. The impacts of bushfires following a flash flood event in the catchment of the Ovens River. Environment Protection Authority, Melbourne, Victoria, Australia. *Online at* http://epanote2.epa.vic.gov.au/EPA/Publications.nsf/PubDocsLU/ovens_catchment_report?OpenDocument. [accessed 21 Aug 2008].

Environment Protection Authority Victoria (EPA). 2005. A quantitative study of bushfire impacts on the Buffalo and King Rivers: first year findings. EPA Publication No. SR6, Environment Protection Authority Victoria, Melbourne, Victoria, Australia. *Online at* http://epanote2.epa.vic.gov.au/EPA/Publications.nsf/PubDocsLU/SR6?OpenDocument [accessed 21 Aug 2008].

Ferguson, R., C. Dwyer, T. Loffler, and R. Hardie. 2004. Fire Recovery Pilot Study: Buckland River and Omeo District Catchments, EarthTech Engineering Pty Ltd., Melbourne, Victoria, Australia.

Gill, A.M. 2005. Landscape fires as social disasters: an overview of 'the bushfire problem'. *Environmental Hazards*, 6: 65-80.

Gill, A.M. and P.C. Catling. 2002. Fire regimes and biodiversity of forested landscapes of southern Australia. pp. 351-369. *In* R.A. Bradstock, J.E. Williams, and A.M. Gill [eds.]. *Flammable Australia. The Fire Regimes and Biodiversity of a Continent.* Cambridge University Press, Cambridge, UK.

Good, R.B. 1973. A preliminary assessment of erosion following wildfires in Kosciusko National Park, NSW. *Journal of the Soil Conservation Service of NSW,* 29: 191-199.

Grover, S.P.P. 2006. Carbon and Water Dynamics of Peat Soils in the Australian Alps. Ph.D. thesis, School of Life Sciences, La Trobe University, Melbourne, Victoria, Australia.

Ham, J. 2003. An assessment of the impact of fire control lines on soil properties and the local environment: A case study of Bunyip State Park, Vic. Bachelor of Environmental Science (Honours) thesis, School of Geography and Environmental Science, Monash University, Victoria, Australia.

Hatten, J., D. Zabowski, G. Scherer, and E. Dolan. 2005. A comparison of soil properties after contemporary wildfire and fire suppression. *Forest Ecology and Management,* 220: 227-241.

Haydon, D.T., J.K. Friar, and E.R. Pianka. 2000. Fire-driven dynamic mosaics in the Great Victoria Desert, Australia. *Landscape Ecology,* 15: 373-381.

Hobgen, S. 2005. The effects of high intensity bushfire on sediment movement in a small steep catchment in North Eastern Victoria. Bachelor of Environmental Science (Honours) thesis, School of Geography and Environmental Science, Monash University, Melbourne, Victoria, Australia (unpublished).

Hope, G., J. Whinam, and R. Good. 2005. Methods and preliminary results of post-fire experimental trials of restoration techniques in the peatlands of Namadgi (ACT) and Kosciuszko National Parks (NSW). *Ecological Management & Restoration,* 6: 214-217.

Hudson, N.W. 1986. *Soil Conservation.* Revised reprint 1986, B.T. Batsford Limited, London, UK.

Humphreys, G.S. 1985. Bioturbation, rainwash and texture contrast soils. PhD thesis, School of Earth Sciences, Macquarie University, Sydney, Australia.

Jacobs, P. and I. Walker. 2003. Alpine fires 2003: rehabilitation and recovery of public land. Report *Online at* http://www.environment.sa.gov.au/biodiversity/pdfs/bushfire/jacobs_peter.pdf [accessed 16 sep 2008].

Kakadu Board of Management. 1999. Kakadu National Park: Plan of Management. Commonwealth of Australia, Department of Environment and Heritage. Australian Government Printing Office, Canberra, Australia Capital Territory, Australia.

Keeves, A. and D.R. Douglas. 1983. Forest fires in South Australia on 16 February 1983 and consequent future forest management aims. *Australian Forestry,* 46: 148-162.

Kershaw, A.P. 1994. Pleistocene vegetation of the humid tropics of north-eastern Queensland, Australia. *Palaeogeography, Palaeoclimatology, Palaeoecology,* 109: 399-412.

Kershaw, A.P., J.S. Clark, A.M. Gill, and D.M. D'Costa. 2002. A history of fire in Australia. pp. 3-25. *In* R.A. Bradstock, J.E. Williams, and A.M.Gill [eds.]. *Flammable Australia. The Fire Regimes and Biodiversity of a Continent.* Cambridge University Press, Cambrige, UK.

Kershaw, A.P., S. van der Kaars, P. Moss, B. Opdyke, F. Guichard, S. Rule, and C. Turney. 2006. Environmental change and the arrival of people in the Australian region. *Before Farming* Online journal 2006 (1): article 2. *Online at* http://www.waspress.co.uk/journals/beforefarming//journal_20061/abstracts/index.php [accessed 16 Sep 2008].

Laing, A.G. 2004. Cases of heavy precipitation and flash floods in the Caribbean during El Niño winters. *Journal of Hydrometeorology,* 5: 577-594.

Lawrence, R.E. 1999. The impact of land cover changes on hydrological sustainability in the Victorian alpine area. *Proceedings of the Royal Society of Victoria,* 111: ix-xxii.

Legoe, G.M. 1981. Peat and peatlands. A summary of research and literature. Occassional Paper 3, NSW Department of Environment and Climate Change, Sydney South, New South Wales, Australia.

Lindenmayer, D.B., R.B. Cunningham, C.F. Donnelly, and J.F. Franklin. 2000. Structural features of old-growth Australian montane ash forests. *Forest Ecology and Management,* 134: 189-204.

McCarthy, M.A., A.M. Gill, and D.B. Lindenmayer. 1999. Fire regimes in mountain ash forest: evidence from forest age structure, extinction models and wildlife habitat. *Forest Ecology and Management,* 124: 193-203.

McDougall, K.L. 1982. The Alpine Vegetation of the Bogong High Plains. Publication No. 357, Ministry for Conservation, Melbourne, Victoria, Australia.

McIntosh, P.D., M.D. Laffan, and A.E. Hewitt. 2005. The role of fire and nutrient loss in the genesis of the forest soils of Tasmania and southern New Zealand. *Forest Ecology and Management,* 220: 185-215.

Munro, A. 2002. The weather bomb 21 June 2002. Technical Report 2002/10. Environment Waikato, Hamilton, New Zealand.

Neave, M. and A.D. Abrahams. 2002. Vegetation influences on water yields from grassland and shrubland ecosystems in the Chihuahuan Desert. *Earth Surface Processes and Landforms,* 27: 1011-1020.

Neill, C.R. 1967. Mean-velocity criterion for scour of coarse uniform bed-material. pp. C6.1-C6.9. *In* Proceedings of the Twelfth Congress, International Association for Hydraulic Research, Colorado State University, Fort Collins, Colorado, USA.

North East Catchment Management Authority (NECMA). 2005. Annual Report 2004-2005. North East Catchment Management Authority, Wangaratta, Victoria, Australia.

O'Loughlin, E.M., N.P. Cheney, and J. Burns. 1982. The Bushranger Experiment: Hydrological response of a eucalypt catchment to fire. *In* E.M. O'Loughlin and L.J. Bren [eds.]. *The First National Symposium on Forest Hydrology.* The Institution of Engineers, Australia, National Conference Publication, 82: 132-138.

Paton, T.R., G.S. Humphreys, and P.B. Mitchell. 1995. *Soils: A new global view.* University College London Press, London, UK.

Pickering, C.M. and K. Barry. 2005. Size/age distribution and vegetative recovery of *Eucalyptus nihophila* (snowgum, Myrtaceae) one year after fire in Kosciuszko National Park. *Australian Journal of Botany,* 53: 517-527.

Prosser, I. and L. Williams. 1998. The impact of wildfire on runoff and erosion in native Eucalyptus forest. *Hydrological Processes,* 12: 251-265.

Rawson, R.P., P.R. Billing, and S.F. Duncan. 1983. The 1982-83 forest fires in Victoria. *Australian Forestry,* 46: 163-172.

Rieman, B.E., R.E. Gresswell, and M.K. Young. 2003. Editorial: Introduction to the effects of wildland fire on aquatic ecosystems in the western USA. *Forest Ecology and Management,* 178: 1-3.

Roberts, R.G., R. Jones, N.A. Spooner, M.J. Head, A.S. Murray, and M.A. Smith. 1994. The human colonization of Australia – optical dates of 53,000 and 60,000 years bracket human arrival at Deaf-Adder Gorge, Northern Territory. *Quaternary Science Reviews,* 13: 575-583.

Rowe, R.K. 1967. A study of the Land in the Victorian Catchment of Lake Hume. Soil Conservation Authority, Kew, Victoria, Australia.

Rowe, R.K. 1972. A study of the Land in the Catchment of the Kiewa River. Soil Conservation Authority, Victoria, Australia.

Russell-Smith, J., C. Yates, A. Edwards, G.E. Allan, G.D. Cook, P. Cooke, R. Craig, B. Heath, and R. Smith. 2003. Contemporary fire regimes of northern Australia, 1997-2001: change since Aboriginal occupancy, challenges for sustainable management. *International Journal of Wildland Fire,* 12: 283-297.

Schultz, B.K. 2005. Stream morphological response to high intensity fire in small montane catchments, North East Victoria. Bachelor of Environmental Science (Honours) thesis, School of Geography and Environmental Science, Monash University, Melbourne, Victoria, Australia.

Schumm, S.A. 1979. Geomorphic thresholds: the concept and its applications: *Transactions of the Institute of British Geographers,* 4: 485-515.

Shakesby, R.A., C.J. Chafer, S.H. Doerr, W.H. Blake, P.J. Wallbrink, G.S. Humphreys, and B.A. Harrington. 2003. Fire severity, water repellency characteristics and hydrogeomorphological changes following the Christmas 2001 Sydney Forest fires. *Australian Geographer,* 34(2): 147-175.

Silvester, E. 2006. Monitoring water quality and biogeochemical changes during peatland rehabilitation. A report to Parks Victoria. Department of Environmental Management and Ecology (DEME), La Trobe University, Albury-Wodonga, Australia.

Singh, G., A.P. Kershaw, and R. Clark. 1981. Quaternary vegetation and fire history in Australia. pp. 23-54. *In* A.M. Gill, R.A. Groves and I.R. Noble [eds.]. *Fire and Australian Biota.* Australian Academy of Science, Canberra, Australia.

Smith, J.A., A.J. Miller, M.L. Baeck, P.A. Nelson, G.T. Fisher, and K.L. Meierdiercks. 2005. Extraordinary flood response of a small urban watershed to short-duration convective rainfall. *Journal of Hydrometeorology,* 6: 599-617.

Turney, C.S.M., M.I. Bird, L.K. Fifield, R.G. Roberts, M.A. Smith, C.E. Dortch, R. Grün, E. Lawson, L.K. Ayliffe, G.H. Miller, J. Dortch, and R.G. Cresswell. 2001. Early human occupation at Devils' Lair, south-western Australia 50,000 years ago. *Quaternary Research,* 55: 3-13.

Wahren, H.A., R.J. Williams, and W.A. Papst. 1999. Alpine and subalpine wetland vegetation on the Bogong High Plains, south-eastern Australia. *Australian Journal of Botany,* 47: 165-188.

Wahren, H.A., W.A. Papst, and R.J. Williams. 2001. Early post-fire regeneration in subalpine heathland and grassland in the Victorian Alpine National Park, south-eastern Australia. *Australian Ecology,* 26: 670-679.

Wallbrink, P.J. and J. Croke. 2002. A combined rainfall simulator and tracer approach to assess the role of Best Management Practices in minimising sediment redistribution and loss in forests after harvesting. *Forest Ecology and Management,* 170(1-3): 217-232.

Wallbrink, P., P. English, C. Chafer, G. Humphreys, R. Shakesby, W. Blake, and S. Doerr. 2004. Impacts of water quality by sediments and nutrients released during extreme bushfires. Report 1: A review of the literature pertaining to the effect of fire on erosion and erosion rates, with emphasis on the Nattai catchment, NSW, following the 2001 bushfires. CSIRO Land and Water, Australia.

Williams, R.J. and D.H. Ashton. 1987. Effects of disturbance and grazing by cattle on the dynamics of heathland and grassland communities on the Bogong High Plains, Victoria. *Australian Journal of Botany*, 35: 413-431.

Williams, A.A.J., D.J. Karoly, and N.J. Tapper. 2001. The sensitivity of Australian fire danger to climate change. *Climatic Change*, 49: 171-191.

Williams, R.J. and A.B. Costin. 1994. Alpine and subalpine vegetation. pp. 467-500. *In* R.H. Groves [ed.]. *Australian Vegetation, 2nd edition*. Cambridge University Press, Cambridge, UK.

Wimbush, D.J. 1970. Hydrological Studies on *Sphagnum* Bogs in the Snowy Mountains, New South Wales. Masters thesis, Department of Botany, University of Sydney, Sydney, Australia.

19

Rehabilitation Strategies after Fire: The California, USA Experience

Peter M. Wohlgemuth[1]*, *Jan L. Beyers*[1] and *Ken R. Hubbert*[2]

Abstract

Emergency rehabilitation to mitigate the effects of flooding, accelerated erosion, and sedimentation that inevitably follows wildfire has been practiced in California, USA for nearly a century. However, California is a physically and culturally diverse area, and rehabilitation measures that work in one part of the state may not be appropriate in other regions. Rehabilitation philosophy can vary with different land management or hazard protection agencies, and may often reflect socio-political considerations as much as the resources or values at risk. Initial rehabilitation efforts in California focused on seeding burned hillsides and building engineering structures in the stream channels or at the mouths of canyons. Measures were refined, new techniques were developed, and the process was formalized in the 1960s and 1970s, with many agencies adopting Burned Area Emergency Response (BAER) programs. Current rehabilitation approaches still include treatments both on the hillslopes and in the stream channels, but also address the problems associated with wildland roads. Many lessons have been learned over the years, but many challenges remain, not the least of which is communicating the vast body of knowledge and experience both to the general public and to political decision-makers in order to make educated post-fire rehabilitation choices.

INTRODUCTION

During a two-week period of October 2003, nearly a dozen wildfires were burning simultaneously in southern California. Fanned by hot, dry *foehn*-type winds, known locally as Santa Ana's, this Fire Siege eventually consumed the

[1] *U.S. Department of Agriculture, Forest Service, Pacific Southwest Research Station, Riverside, California, USA.*
[2] *Hubbert and Associates, Apple Valley, California, USA.*
* *Corresponding author: Peter M. Wohlgemuth, U.S. Department of Agriculture, Forest Service, Pacific Southwest Research Station, 4955 Canyon Crest Drive, Riverside, California 92507 USA. Tel: 951 680 1538, Fax: 951 680 1501, e-mail: pwohlgemuth@fs.fed.us*

vegetation on over 290,000 hectares, making it the largest fire event in Californian history. The fires occurred in a variety of plant communities, including some forested lands, but primarily burned in chaparral shrublands and coastal sage scrub ecosystems. Fire suppression costs were in the tens of millions of dollars (US).

Even before the fires were out, interdisciplinary teams of resource specialists deployed to inventory the damage, assess the likely impacts of the coming winter storms, identify the values at risk from the inevitable flooding and accelerated erosion that would ensue, and recommend cost-effective mitigation treatments. These emergency rehabilitation measures are designed to protect human lives, public and private property, infrastructure (roads, bridges, pipelines, utility lines, communication sites, reservoirs, etc.), water quality, heritage and archaeology resources, and the habitat for threatened and endangered species of animals and plants. Every fire in the 2003 southern California Fire Siege was different in its location, topography, proximity to urban areas, and threats to values at risk, hence the recommended treatments varied accordingly. Cumulative treatment costs eventually exceeded US$7 million.

The foregoing example illustrates the problem of wildfire across California. Wildfire is commonplace – with the state experiencing numerous fires every year – and extremely large fires or complexes of many fires are not rare occurrences. Consequently, rehabilitation after fires is a highly-evolved management practice in California. However, California is a large and diverse area containing many different vegetation communities and resources in need of protection. Moreover, rehabilitation techniques have developed over time and continue to be refined. This chapter begins with a discussion of California as a diverse landscape, and then traces the historical development of burned area emergency rehabilitation practices throughout California, including the lessons learned and the future directions.

CALIFORNIA AS A DIVERSE LANDSCAPE

California covers a huge area – some 41,000 km^2 – larger than many countries (Fig. 1). Moreover, California is extremely heterogeneous, exhibiting much diversity in both physical and social characteristics. Consequently, post-fire rehabilitation strategies depend on the geography, the agency jurisdiction, and the resources or values at risk associated with a specific wildfire.

Physical Landscape

California covers 10 degrees of latitude that encompasses coastal environments, mountains, valleys and lowlands, and deserts (Fig. 1). California contains both the tallest mountain (Mt. Whitney – 4418 m) and the lowest elevation (Death Valley – 86 m below sea level) in the co-terminus

Fig. 1 Location map of California, USA with topography (courtesy of geology.com).

United States (excluding Alaska and Hawaii) (Oakeshott 1978). Although the state is incredibly diverse, there are some recognizable regional similarities across California. The northern portion of the state is more mountainous, receives more rainfall, and has more forest stands than its southern counterpart, which is generally less mountainous, drier, and dominated by brushfields and desert scrub vegetation. The contrast between the coastal strip and the inland section of California is even more pronounced, with greater rainfall and moderate temperatures along the coast promoting more mesic conditions that support luxurious vegetation growth and some unique environmental niches (Durrenberger and Johnson 1976).

The topography of California is dominated by coastal mountain ranges, a large central valley, and a large mountain mass in most of the eastern parts of the state (Fig. 1). The structural grain of this topography trends along a northwest-southeast axis. The lofty heights coupled with the large relative relief in the uplands reflect the ongoing tectonic activity at the junction of the Pacific and North America crustal plates (Hornbeck et al. 1983). The overall climate of California is classified as Mediterranean; characterized as having cool, moist winters and hot, dry summers. Temperatures approach 40°C in the summer but seldom drop below –10°C in the winter, except in the high mountains. Annual precipitation varies from 40 to 60 cm along the coast to 10 to 25 cm inland, with mountain ranges receiving 70 to 100 cm along orographic gradients. A section of the northwest coast can receive as much as 250 cm of rainfall annually (Hornbeck et al. 1983). While there are rivers and perennial streams throughout California (especially in the north), many of the canyons and valleys only support ephemeral watercourses for much of the year (primarily in the south).

Vegetation patterns in California mirror the trends in temperature and precipitation. Forests of pine (*Pinus*), fir (*Abies*), and oak (*Quercus*) with minor other hardwoods dominate the uplands, especially in the northern section of the state. Foothills and the lee side of many lower mountain chains are covered with chaparral brushfields. Grasslands were once prevalent in the valley bottoms and lowlands, but most of these areas have been converted to agriculture or urban centers. Sparse scrub – including cacti – and sagebrush covers most of the deserts and arid rangelands, especially in the southern part of the state (Hornbeck et al. 1983).

Erosion and sedimentation in California reflect both gravitational processes associated with steep topography and hydrologic processes resulting from seasonal winter rains. Dry ravel is the dry gravitational flow of soil material, and can be a major process by which sediment is delivered from the hillslopes to the stream channels (Rice 1974). Ravel is a ubiquitous erosion process in southern California, but also occurs on steep slopes in the north. Fire greatly accentuates the efficacy of dry ravel (Rice 1974).

Rainsplash and overland flow are the two hydrologic erosion processes that operate on hillside slopes. Raindrops can dislodge loose soil material and preferentially transport it downhill. Overland flow may concentrate into micro-channels that can entrain and transport sediment, creating rills or gullies, especially on steep slopes (DeBano et al. 1998). Both rainsplash and overland flow are negligible on unburned landscapes with a sufficient vegetation canopy. Fire consumes the protective vegetation and may alter the surface soil structure. This is typical of fire behavior in chaparral brushfields, but is also experienced in forested ecosystems, especially if there is a brush understory. The bare soil becomes very susceptible to erosion by hydrologic processes. Moreover, chemical changes in the soil following fire can produce a non-wettable or water repellent condition that restricts infiltration and promotes overland flow (DeBano 1981). This extra water flows from the hillsides to the adjacent stream channels, where it can mobilize sediment stored in dry ravel deposits. The slurry that is created can propagate down the stream channel as a debris flow with tremendous erosive power (Wells 1987).

Management Landscape

Although there are significant tracts of private property, most of the uplands of California are administered as public lands. Lands managed by the federal government include 7 National Parks, 9 National Monuments or National Recreation Areas, 18 National Forests, 6 military reservations, and scattered areas under the jurisdiction of the Bureau of Land Management. There are also many square kilometers of land administered by local Native American tribes. Lands managed by the State of California include 10 State Parks and several State Forests. Numerous county and municipal parks also cover upland areas.

Across the various federal, state, county, and municipal agencies that administer these lands, there is no uniform policy for post-fire rehabilitation. In keeping with their conservation preserve mandate, national and state parks tend to avoid ground-disturbing mitigation practices, unless there are high-value resources at risk, such as critical habitat for endangered species. They also avoid the use of non-native seeds or mulches that may introduce exotic weeds into pristine stands of native vegetation. Several federal agencies – the U.S. Department of Agriculture, Forest Service and the U.S. Department of Interior, Bureau of Land Management – have long-established programs of emergency response that include standing teams of resource specialists who consider all techniques when making their rehabilitation recommendations.

A major difference in agency philosophy is the contrast between land management and hazard protection. Land management agencies, at whatever governmental level, have long-range plans that direct recreation development, commodity extraction, and resource protection on the lands within their jurisdiction. Hence, post-fire rehabilitation would proceed only within the context of the existing land management planning. Hazard protection agencies, such as the California Department of Forestry and Fire Protection, which oversees fire suppression and post-fire rehabilitation on non-federal public and private lands, react to immediate emergencies and respond to specific events with mitigation as their sole focus. Thus conflicts can arise in recommending and implementing post-fire rehabilitation projects for fire incidents that cover multiple landowners.

Socio-political Landscape

Specific post-fire rehabilitation measures are usually dictated by on-site or downstream resources or values at risk from the flooding and accelerated erosion that typically follow wildfires. While resource specialists strive to make cost-effective treatment recommendations based on the best available science, often decisions reflect socio-political considerations.

Protection of human life is paramount when recommending emergency rehabilitation treatments. Measures to prevent threats to human communities are implemented first, often with redundant techniques. Early warning systems may be established to sound the alarm if rainfall in the burn area exceeds some critical threshold or if the creeks rise to some specified flood stage. Evacuation orders may also be issued if the area is deemed too dangerous for human occupation. Following the 2003 Old Fire in southern California, a storm cell stalled over the burned area. After several days of rain, a high-intensity burst occurred on Christmas Day, producing debris flows in the stream channels within the burned area. Tragically, 16 fatalities occurred in these flow events because people were unaware or ignored the repeated evacuation orders (Hubbert 2005).

Threats to private property and technological infrastructure are usually the main targets of rehabilitation measures. Homeowners and business proprietors downstream from a burned area may erect sandbag barriers or deflection structures to protect their property, but the most effective treatments try to control runoff and erosion at the source; in the burned area itself. Roads and bridges are critical values at risk, as they often provide the only access to isolated communities or remote communication sites. Pipelines, whether they carry water or petroleum products, also receive special consideration for rehabilitation treatment protection because of the potential for environmental disaster if flooding or erosion should damage the conduit. Similarly, utility lines warrant special protection. Flood control reservoirs and debris basins were built specifically to handle the floods generated from burned areas, but the hillsides and stream channels contributing to water supply reservoirs are usually treated to prevent siltation and the loss of storage capacity.

Loss of soil productivity is always an issue for maintaining ecosystem integrity, but it is of special concern in forest stands used for commercial timber production. Although burning liberates nutrients from the standing biomass and the forest floor, accelerated erosion and leaching can remove these chemical compounds before they can benefit stand regeneration. Rehabilitation treatments on hillside slopes attempt to retain valuable topsoil and prevent site degradation through unacceptable losses of critical nutrients, such as nitrogen and phosphorus (DeBano et al. 1998).

The soil and nutrients stripped off the burned hillsides are usually deposited in the stream channels at the bottom of the slopes. If these channels support surface runoff, the character of the water will be tremendously altered. Increased turbidity from the sediments and elevated solute concentrations of compounds from the ash can be lethal to aquatic organisms, especially fish populations important for food production and recreation. Post-fire sedimentation will also preferentially fill stream pools, further degrading fish habitat. Rehabilitation efforts strive to keep sediment from initially entering the stream channels, and then subsequently remove the entrained load, usually by a series of barrier structures.

The California floristic province is considered a hotspot of biodiversity (Myers et al. 2000) because of the large number of rare species and the ongoing habitat loss that affects them. As a result, consideration of rare species is now a part of most post-fire rehabilitation decisions within the state. For National Forest lands, rehabilitation treatments may be applied to prevent permanent impairment of critical habitat for threatened and endangered species (Forest Service Manual 2523.2), including measures such as planting of streamside willows for waterway shading, so long as the treatment(s) will be effective within two years. Areas known to harbor populations of rare plants may be excluded from seeding or mulch treatment prescriptions in order to prevent competition or other adverse effects. For example, after the Cedar Fire of 2003 in San Diego County, locations of suitable habitat for rare endemic annual plants were excluded from an aerial hydromulch application.

In extreme cases, where post-fire flooding or debris flows are expected to greatly alter habitat, populations of rare species may be collected and moved or retained in a zoo for later reintroduction. After the Old Fire in 2003, for example, mountain yellow-legged frogs were captured from a creek on the San Bernardino National Forest and taken to the Los Angeles zoo, where they thrived. Winter storms scoured out the stream channel where they had occurred. Amazingly, however, some frogs were found in the creek a year later, possibly washed down from tributary streams unaffected by the fire. Other treatments for rare species may include seeding grasses along stream banks to help reduce sediment movement into streams, as was done after the 2000 Manter Fire to protect populations of California golden trout.

Heritage or archeological resources often need protection following fires. The removal of plant cover is the primary concern due to increased accessibility and visibility of the cultural sites. The sites then become more susceptible to vandalism and artifact looting. However, managers can also perform surveys to identify new, undiscovered sites before vegetation recovery. Treatments in California usually consist of patrols, fencing, and road closures to discourage the above activities. Monitoring of the sites continues until vegetative groundcover recovers to the extent that visibility of the sites is obscured.

Wildfire can create ideal conditions for the growth and spread of invasive species. The bare mineral soil, lack of overstory shading, and readily available nutrients deposited in the ash layer after a fire provides an ideal seedbed for many non-native plants (Brooks and D'Antonio 2001). The role of fire in the spread of invasive plants has been receiving increased attention in recent years (e.g., Galley and Wilson 2001, Brooks et al. 2004). Fire disturbance has aided the spread of weed infestations that were already present before the fire, either in the seedbank or in proximity to the burn. In most cases, weed species that are found were likely present prior to the burning, but were released from competition following the fire. Over the years, pre-fire seed banks of invasive weed species have increased in quantity and area covered. Most of the invasive weeds are prolific seed producers and can remain viable for long periods of time. In California, fires that occur too frequently can also facilitate the conversion of native shrub and herbaceous vegetation into non-native grasslands (Keeley 2001).

Preventing the spread of invasive species into burned areas has become an important part of post-fire rehabilitation strategies in California. Measures taken can range from surveys to identify and remove new infestations to seeding native or short-lived non-native grasses in an attempt to out-compete invasive plants. However, it is still unknown whether this seeding can prevent the establishment of noxious weeds. Increased weed infestations are commonly observed in areas of fire suppression activities. Most of the incursions are associated with roadsides, bulldozer lines constructed to create fuel breaks, and drainages near human habitation (BAER Guidance Paper: Noxious and Invasive Weed Treatment, March 2004, unpublished data).

INITIAL REHABILITATION EFFORTS

The association between wildfire and subsequent flooding, accelerated erosion, and massive sedimentation has long been recognized throughout California, especially in the mountains along the southern coast. Observations of the phenomenon were reported in newspaper articles as far back as the late 1800s, and erosion control treatments following fires were suggested in the early part of the last century (Munns 1919). The event that galvanized public awareness and focused attention on the need for post-fire rehabilitation projects was the 1934 New Year's Day flood in La Crescenta, located in the foothills near Los Angeles. A high-intensity storm produced heavy rainfall on a freshly burned landscape and produced debris flows that scoured side tributaries to depths of up to 5 m and moved boulders the size of automobiles several kilometers from the mountain front (Kraebel 1934). The community experienced massive property damage and several people were killed. Initial rehabilitation efforts in the first half of the 20th century consisted of seeding the burned hillsides to produce a rapidly growing ground cover, and building engineering structures in the stream channels and at the mouths of the canyons to trap and remove sediment.

Foresters in southern California tried seeding burned-over hillside slopes with native shrubs as early as the 1920s to try to reduce post-fire erosion (Department of Forester and Fire Warden 1985). When they realized that shrub seeds germinated no earlier than natural regeneration, faster-growing non-native herbaceous species, such as Mediterranean mustards (*Brassica* L. spp.) were used (e.g., Gleason 1947). However, mustard seeds would subsequently be spread to downstream agricultural areas, where the plants were considered nuisance weeds by the local farmers. By the 1950s annual ryegrass (*Lolium multiflorum* Lam., native to Europe and Asia) had come to be regarded as the most effective and economical choice for charred California wildlands (Barro and Conard 1987). Ryegrass was relatively inexpensive, readily available, and could easily be applied from the air over wide areas.

One benefit of heavily seeding chaparral burns with ryegrass was that it tended to inhibit shrub seedling regeneration, which could be used to increase pasturage by type-converting dense shrub stands to annual grasses with occasional shrubs (see Schultz et al. 1955). Type-conversion of brushlands was also a goal of some prescribed fire programs on both private and public lands during this period, especially in the coast ranges and the foothills of the Sierra Nevada (Keeley 2001).

As in other areas, seeding forestlands after fire for soil retention and range improvement used species of predominantly non-native pasture grasses and forbs (see Chapter 11). Seed mixes generally contained annual grasses to provide quick cover, perennial grasses to establish long-term protection, and often legumes to add nitrogen to the soil (Ratliff and McDonald 1987).

Engineering solutions for post-fire rehabilitation and erosion control took the form of building barrier structures to trap and retain the transported debris once it had reached the stream channels. Channel checks are low dams (<2 m high) constructed in a series, spaced approximately 50 to 60 m apart. Although they could be made of any materials, often they were fabricated using wire cages filled with local rock material ranging in size from large cobbles to small boulders (Eaton 1932). These check dams could only trap and store modest amounts of sediment material, but they also served as drop structures to dissipate the energy of the flow, as well as grade control features to prevent channel incision and potential destabilization of the lateral banks and adjacent colluvial toeslopes (Eaton and Gillelen 1931). However, these check dams are time-consuming to plan and complete. As it is unrealistic to build a series of check dams on many creeks in the short period between the fire and flood-producing rains, they must be constructed in advance on streams within fire-prone areas.

Debris basins are large earthen or concrete settling reservoirs built at the mouth of major canyons where the streams flow onto the adjacent flatlands. As sediment-laden floods and/or debris flows enter the basin, the coarse load drops out in the low energy environment and becomes trapped. Relatively clear water then escapes over a spillway. After the event, the basin is cleaned out with heavy equipment so that storage capacity is sufficient for the next storm. This becomes problematic when a series of storms strikes over the course of several days. Moreover, these debris basins are very expensive and require considerable time to build. As with check dams, debris basins must be constructed in advance at the base of canyons within fire-prone areas. Many dozens of debris basins were built by local flood control districts throughout California between 1925 and 1960.

DEVELOPMENT OF THE REHABILITATION PROCESS

Nationally, the 1960s and early 1970s saw the preparation of the first official written reports on emergency postfire watershed rehabilitation. Funding for postfire rehabilitation treatments came from fire suppression accounts, emergency flood control programs, or funds appropriated for watershed restoration. After a Congressional inquiry on fiscal accountability, formal authority was provided for Burned Area Emergency Rehabilitation, now called Burn Area Emergency Response (BAER) programs in the federal interior and related agencies appropriation in 1974 (Robichaud et al. 2000). This authority integrated the evaluation of burn severity, funding request procedures, and treatment options. In the U.S. Department of Agriculture, Forest Service, a standardized Burned Area Report form was developed and used for this purpose. Interdisciplinary teams were assembled after major fires to assess the burned area, identify threats, and recommend treatments.

This was also a period of active investigation into the effectiveness of established and new post-fire rehabilitation treatments in California. Broadcast seeding of entire burned areas, usually with annual ryegrass, was commonplace (Barro and Conard 1987), and controversy over its use ensued, particularly in southern California chaparral (Gautier 1983). Engineering solutions to runoff and erosion threats were also employed. The contribution of roads to sediment movement and water channelization was recognized, and road treatments to reduce these effects were increasingly prescribed (Burroughs and King 1989).

Hillslope Treatments

After the 1960 Johnstone Peak fire on the San Dimas Experimental Forest in southern California, Forest Service researchers set up an extensive experiment testing different hillslope rehabilitation treatments in burned-over chaparral watersheds. Several seeding treatments, hillslope terracing, and various combinations were applied to small catchments, in which sediment deposition and runoff were measured. When the first winter after the fire proved to be one of the driest on record, with negligible grass establishment, the treatments were reapplied, and the next year the highest rate annual grass-seeded catchments recorded a 16 percent reduction in sediment production. Seeded grass cover was about 8 percent (Krammes and Hill 1963). The researchers noted that the seeded catchments had lower cover of native plants than the unseeded controls (Corbett and Green 1965). In contrast, the watersheds with the terrace treatment reduced sediment production by 60 percent compared to un-terraced catchments (Rice et al. 1965).

Contour trenches were also installed operationally after fires where flood control was a major concern. In the Sierra Nevada, their effectiveness was found to depend on the soils and geology of the affected area (DeByle 1970). However, terraces and trenches are radical ground disturbing practices, and remnants of these treatments that were applied in the 1960s in some cases can still be seen on the landscape today.

The effectiveness of annual ryegrass seeding for erosion control in chaparral ecosystems was questioned in the late 1970s and early 1980s. After fire in a chaparral, annual plants and shrub seedlings take advantage of the abundant light, space and soil nutrients suddenly available and fairly quickly occupy burned sites (Sampson 1944, Sweeney 1956, Keeley 1991). Seeded ryegrass was shown to displace those species, often without increasing ground cover or reducing erosion (Keeley et al. 1981, Gautier 1983, Taskey et al. 1989). The debate fueled research that extended into the 1990s (see next section).

Seeded annual ryegrass could also have negative effects on pine seedling establishment (Griffin 1982), as had been shown for other pasture grass species commonly used for post-fire seeding (Baron 1962). Massive wildfires in the Sierra Nevada, northern California, and southern Oregon in 1987 resulted in millions of dollars being spent on emergency rehabilitation treatments and brought greater scrutiny to the methods employed. A number of investigations of treatment effectiveness were reported at a symposium in 1988 (Berg 1989). Grass seeding was found to be highly variable in efficacy. Two papers attested to the effectiveness of straw mulch for erosion control (Gross et al. 1989, Miles et al. 1989), and others pointed out that post-fire logging, or 'salvage logging', could have beneficial effects on burned watersheds, especially when residual material, or 'slash', was left on the ground for soil protection (Barker 1989, Poff 1989). The symposium concluded with encouragement to develop better information to help managers make the best choices (MacDonald 1989).

Channel Treatments

Post-fire rehabilitation measures in stream channels continued to be dominated by engineering structures. More check dams were built in fire-prone watersheds, especially in the steep mountains of southern California (Ruby 1973). Besides these semi-permanent channel traps, land managers and protection agencies also began experimenting with temporary low dam structures made of logs or straw bales (Miles et al. 1989). Log dams, used in the forested northern part of the state, were fabricated from fire-killed trees taken directly off the burn site. Trees were felled, stacked to the desired height, secured with cables or bailing wire, and braced on the downstream face. It was of little consequence that these dams were semi-permeable, as long as they retained the coarser debris load. Straw bales – brought to the work site by trucks, helicopters, and/or hand crews – were used as large building blocks secured with fence posts and bailing wire to construct custom dams throughout California. Over time, both the logs and the straw would biodegrade, and the wedges of trapped sediment would slowly be released back into the channel system, and the longitudinal profile of the stream would return to a natural gradient (Miles et al. 1989).

Channel clearing was a popular rehabilitation treatment throughout California in the 1970s and 1980s (Barro et al. 1989). Clearing was done to prevent freshly toppled fire-killed trees (and any pre-fire downed material) from organizing into impromptu dams that would temporarily restrain the rising waters. The fear was that these makeshift dams would eventually burst, creating a large flood that would be dangerous to any downstream resources or values at risk. However, this desire to quickly and efficiently convey flood waters away from the burned area had to be balanced by the what often became the wholesale removal of the entire riparian ecosystem (Barro et al. 1989). Ironically, in northern California, post-fire measures in stream channels began to include intentionally placing pieces of large woody debris into the drainages to create stability points in order to preserve or create new fish habitat (O'Connor and Ziemer 1989).

Debris basins continued to be built at the canyon mouths of fire-prone watersheds to protect downstream communities from floods and/or debris flows. Over 100 of these structures now exist in Los Angeles County alone. Besides these carefully engineered structures, agencies also began to use temporary earthen catchment basins on smaller stream tributaries to protect high-value features, such as remote developments or critical habitat (USDA Forest Service 1992). Usually these earthen structures would not be maintained, and they would eventually be breached and revert back to a natural landscape after the post-fire emergency was over. Another new development to protect downstream human communities were deflection walls that, rather than creating a holding barrier, attempted to divert the sediment-laden flows away from property and infrastructure and back to the natural channels (Robichaud et al. 2000).

CURRENT REHABILITATION APPROACHES

From 1990 to the present, there has been a shift in the combinations of land, channel, road/trail, and protection/safety treatments that are selected by the BAER assessment teams. There has been a trend towards prescribing less hillslope treatments. Seeding has decreased in northern California, and is rarely prescribed in southern California. In southern California, aerial straw mulch and straw wattles have replaced seeding treatments. Straw mulching is also popular in northern California because of slower re-establishment of ground cover in the forested regions compared to the rapid recovery of southern California chaparral systems. Contour-felled logs, also known as log erosion barriers (LEBs), are used in northern California as a replacement for seeding where trees are available. Channel treatments are still being prescribed, although there is no clear consensus of their success or failure. More channel treatments are used in northern than in southern California because of fishery issues. BAER teams are not only treating to prevent excess sediment, but are also attempting to reestablish fish habitat as well. More recently, there has been greater emphasis on road and trail treatments, off-highway vehicle containment, protection of cultural resources, and noxious weed treatments. Additionally, the importance of monitoring the success or failure of these treatments has been recognized, and studies have been implemented to assess or quantitatively evaluate their performance.

Hillslope Treatments

The source of much of the sediment of concern to land managers and protection agencies in the post-fire environment originates on the hillside slopes. The following rehabilitation techniques attempt to retain the soil material on the hillslopes and delay its delivery to the adjacent stream channels.

Seeding

Routine broadcast use of annual ryegrass for post-fire stabilization seeding has decreased tremendously since 1990. Research studies that pointed out its lack of effectiveness for erosion control and detrimental impacts on native plants, especially annual fire-followers, contributed to this trend (Conard et al. 1995, Wohlgemuth et al. 1998, Beyers 2004). The U.S. Department of Agriculture, Forest Service rarely seeds at all in southern California, where unpredictable rains and generally good natural vegetation recovery do not make seeding cost-effective. Other agencies have tried native seed mixes, with generally low success (Keeley et al. 1995). Land managers recognize the dilemma posed by seeding in forested areas of northern California, where loss of site productivity due to erosion is a major concern but successful grass growth may suppress natural tree seedling establishment (Griffith 1998, Van der Water 1998). The U.S. Department of Agriculture, Forest Service now uses non-reproducing cereal grains for most post-fire seeding, both to reduce erosion and to reduce establishment by invasive non-native plants. Seeding is done only in carefully-targeted areas, such as above roads or streams critical to the survival of rare fish. However, even cereals may displace native plant species and suppress tree seedlings in the short term (Keeley 2004). For more discussion of post-fire grass seeding, see Chapter 11.

Contour-felled log erosion barriers

Contour-felled logs erosion barriers are more apt to be used in northern California because of the predominance of forested ecosystems. In southern California, most of the area consumed by fire is in chaparral shrublands where logs are scarce. The 2003 southern California wildfires burned through over 290,000 ha of diverse plant communities dominated by chaparral, but coniferous forests comprised only 5 percent of the total area burned (Keeley et al. 2004). In northern California, contour-felled logs have been minimally used and are not usually prescribed. Miles et al. (1989) monitored the effectiveness of the contour-felled logs on the Shasta-Trinity National Forest and found that the logs retained 3.6 to 15.1 Mg ha^{-1} of soil on site. They considered the sediment trapping efficiency low and costs of the treatment high (Robichaud et al. 2000). See Chapter 12 for further discussion.

Aerial straw mulch, hand-placed straw, and wood mulch

Due to its relatively low cost and history of success in reducing hillslope erosion, aerial straw mulching is one of the more popular hillslope erosion control treatment used in California. On the Grand Prix/Old Fire, the average cost including helicopter, personnel, straw, trucking, salary and per diem was US$1850 per hectare (Hubbert 2005, 2006). Miles et al. (1989) reported 13.5 to 22.5 Mg ha^{-1} reduction in soil erosion on the Shasta-Trinity National Forest when wheat straw was applied at 4.5 Mg ha^{-1}.

The use of certified weed-free rice straw has replaced the use of other non-certified straws such as wheat on most lands in California. The change to rice straw reflected the threat of invasive weed species that were being introduced in the applied straw. Rice straw is certified to be weed free, but monitoring by botanists after application have suggested otherwise. After the 2001 Darby Fire on the Stanislaus National Forest, 28 ha were aerial straw mulched with weed free rice straw and 25 ha of yellow starthistle (*Centaurea solstitalis*) were mapped the following year (Clines 2005). Yellow starthistle is of special concern as it is considered one of California's worst noxious weeds; infesting rangelands, pastures, hay fields, and orchards. In horses it can cause the fatal nervous disorder equine nigropallidal incephalomalacia, or 'chewing disease.'

Following the Grand Prix/Old Fire of 2003, land managers monitored both aerial straw mulch and hand-placed straw (Hubbert 2005, 2006). High winds contributed the most to straw mulch failure, either blowing the mulch offsite or piling the straw in deep clumps so that vegetation was suppressed. Poor application of the straw mulch also contributed to the failure of the material to provide cover. For best results, the large 330 kg hay bales were to be dropped from 60 m above the surface. However, because of the unevenness of the terrain, the bales were either dropped too low and did not break up completely resulting in piling or clumping; or were dropped too high resulting in uneven coverage and scattering beyond the projected treatment area. When applied correctly, straw mulch provides ground cover that reduces erosion and increases soil infiltration. Janicki and Grant (2002) noted that 330 kg bales did not break up as well as lighter 250 kg bales. Straw suppliers stated that bailing pressure, moisture content, and how fine the straw is chopped are factors determining how well the straw breaks up and spreads (Janicki and Grant 2002). Additionally, on steep slopes, mulch may increase downstream peak flows by artificially decreasing storage capacity (lowering evaporation and allowing greater infiltration), resulting in larger subsurface preferential flow. This may be an important factor in steep watersheds where soils are shallow and have developed on hard, unweathered bedrock.

Hand-placement of straw is more labor intensive than aerially applied straw, but usually results in better ground coverage. Good ground cover only holds true in areas that do not experience high winds. Most of the hand-applied straw was blown off-site in the wind-prone areas of the Grand Prix/ Old Fire (Hubbert 2005, 2006). In areas where the straw was blown off-site, high impacts of foot traffic during straw application disturbed the soil surface increasing the soil erosion hazard. Ground cover approached 70 percent in areas that did not experience strong winds.

Recent unpublished experimentation with wood mulches has shown it to be an effective treatment. If a tool is developed that helps apply the mulch in a cost effective way, it will gain in popularity, especially with the concerns of noxious plants in straw.

Aerial hydromulching and road hydromulch

Hydromulching combines a wood and paper fiber matrix with a non water-soluble binder, mixes the ingredients into a slurry, and applies the mixture by high-pressure nozzles or by a helicopter. The intent of the mulch treatment is to bind the loose surface soil together, reducing detachment and transport by rainsplash and overland flow, while still allowing infiltration across the landscape. Aerial hydromulching is a relatively new treatment in California and because of its expense (~US$4000 per ha) has been used sparingly. A total of US$320,000 was spent for aerial hydromulch on the Curve Fire that occurred on the Angeles National Forest in 2002. After the 2003 Cedar Fire, aerial hydromulch was applied to watersheds located above a residential community to help reduce flood peaks and sediment yield downstream. A total of 444 ha were treated at a total cost of US$1,650,000.

The use of road hydromulch (hydromulch applied through hoses from tanker trucks) has remained stable. Due to its expense and limited coverage range, it is mainly used to protect specific structures of high value. For example, following the Curve Fire, US$930,000 was spent protecting a state highway that provided access to the local mountains (Andressen 2002).

Straw wattles (fiber rolls)

The use of straw wattles has increased slightly in recent years. Straw wattles provide an alternative hillslope treatment when there are no trees available for contour-felling and seeding is undesirable. Implementation of the treatment is very labor intensive, and can be expensive if no volunteers are available. Although they weigh only 14 kg, they can be awkward and often take two people to transport them. With placement of the wattles, foot traffic is increased causing added disturbance to the soil surface.

Straw wattles were used following the 2003 Cedar Fire. Many of the fiber rolls were placed incorrectly across natural drainage positions. During the first year winter storm events, most of these failed (Hubbert 2005, 2006). Problems continued when the fiber rolls were repaired by backfilling the undercut portions with fresh hillslope material adjacent to the rolls. The next storm events removed this material as well, resulting in additional material being transported off-site (Hubbert 2005, 2006). Placement of the fiber rolls with their ends turned downslope also caused problems. Rills formed at the edge of many of the down-turned rolls, contributing to increased erosion the first year. This site experienced low rainfall, and therefore it was difficult to determine if the treatment was successful or not. Even after some intense rain events, there was little sediment accumulating behind the wattles (Hubbert 2005, 2006).

Channel Treatments

Sediment can be stored in stream channels for many years awaiting a significant flushing event, especially in ephemeral watercourses. In the post-fire environment, the extra water delivered from the burned hillsides swells the streams and can entrain and transport the stored sediment, threatening downstream infrastructure and habitat. The following treatments attempt to reduce the energy of the flowing water and retain some of the coarsest sediments behind barrier structures.

Straw bale check dams

Straw bale check dams have been used frequently in northern California, but much less so in southern California. Results have been mixed in evaluating the success or failure of the treatment. In a report by Miles et al. (1989), 1300 5-bale check dams were installed with only 13 percent failing the first year due to piping or undercutting. They considered straw bale check dams easy to install and highly effective (Robichaud et al. 2000). Collins and Johnston (1995) reported a 63 percent failure of 440 straw bale check dams 4 months following the 1991 Oakland Fire. After the Old Topanga Fire of 1993, Booker and Dietrich (1998) monitored straw check dams in 3 fire areas and reported that initially the dams had less than a 50 percent success ratio, with total failure by the 2nd year. They suggested that temporary structures should not be used in catchments with drainage areas greater than 1 ha. The straw bale dams failed because of piping, dam faces being undermined by flow over the structure, and destabilization of channel banks due to localized flow.

More recently in southern California monitoring of straw bale check dams was conducted following the Grand Prix/Old Fire (Hubbert 2005, 2006). After the Christmas Day storm of 2003, sediment completely filled to storage capacity all of the check dams. Once the check dams were filled with sediment, water was free to flow over the dams with no loss of energy. This resulted in severe downcutting and gully formation below the dams. Following the second winter of above average precipitation, all the straw bale check dams failed. In many cases, no signs of the straw bales could be found. Under these circumstances, the treatment was considered a failure.

Log check dams

Log check dams have not been used extensively in California as a treatment. Miles et al. (1989) reported a 15 to 30 yr life expectancy for the dams, but that they were expensive and labor intensive to install. After the 2003 Piru Fire, it was recommended by the BAER implementation team to construct log check dams in the lower and upper portions of a creek that drained into a water supply reservoir. The lower drainage section of log check dams immediately filled with sediment following the 2003 Christmas Day storm. The log check dams began to fail during rain events in February 2004. The check dams that failed at the sides resulted in further cutting of the bank resulting in bank erosion and more sediment contributed to the channel (Hubbert 2005, 2006). All of the check dams failed during the record-breaking rain events of the 2005 winter.

Engineering Techniques: Road, Stream Crossing, and Trail Treatments

Due to the rapidly expanding urban interface and greater encroachment into wildlands, post-fire road and stream crossing treatments have continued to increase in both quantity and costs in California. On the 2001 Star Fire (El Dorado National Forest), costs for road and trail treatments were US$112,000 out of a total BAER expenditure of US$190,100. Where road access was critical on the Piru Fire, costs for road treatments were US$2,136,500 out of a total of US$2,324,000. In southern California, a large portion of treatment money was spent on road debris removal (dry ravel accumulation on roads), debris basin cleanouts, and culvert cleanouts.

The use of storm patrols has also increased over the last five years. On the 2004 Fred's Fire (El Dorado National Forest), US$69,900 out of a total of US$116,000 was spent on storm patrols. Storm patrols were considered effective as pro-active maintenance operation to keep drainage structures free of sediment and debris. During the large storm events in the winter of 2004-2005, storm patrols were less successful in preventing drainage structure failures on the Piru Fire. This was mainly due to numerous landslides that blocked road access during the storms.

After the Piru Fire, concrete construction barriers were placed at low water crossings to add structural integrity and stability to the road surface. In some cases, low water crossings were considered more desirable than a culvert pipe because of its ability to pass a large amount of debris without 'plugging'. However, low water crossing can become flooded and cause delays, so are often used only on low traffic roads. By placing the barriers at a lower position below the road, it was believed that the vertical curve of the road could be lowered thus preventing flooding and failure. The barriers were originally placed too high with no spillway or low point. At first, water worked itself around the barriers. Subsequent storm events resulted in further erosion around the newly placed barriers, introducing up to a meter of new sediment. This was repaired by lowering and moving barriers nearer the bank, and creating a spillway toward the center. During the record setting rain events in the winter of the 2004-2005, all the barriers failed with some broken up and transported down stream for kilometers (Hubbert 2005, 2006).

Upgrading of culverts has continued to increase and has been considered a successful treatment. Trash rack installation has also increased and has been determined to be an effective treatment (Robichaud et al. 2000). Regrading rolling dips and repairing and replacing overside drains has continued to be an important and ongoing treatment. Rolling dips and overside drains are relatively inexpensive and fail often, but are considered to be effective in maintaining road access by removing water from the road. Of the road treatments prescribed on the Cedar Fire, more than 80 percent involved overside drains (Hubbert 2005, 2006).

The use of closure gates, removable pipe barriers, and fences that prevent public access has increased in recent years. Some closure gates are considered critical in protecting the public from rock fall, washouts, hazard trees, and flash flood events. Another important purpose of the gates is to limit public access of unauthorized off-highway vehicles (OHVs). The OHVs have become a major resource problem. On the Angeles and San Bernardino National Forests in southern California, thousands of hectares have been degraded from past OHV use because of the close proximity of 20 million people and 130 km of urban interface. After the Grand Prix/Old Fire, a total of 15 gates were installed. More than 5 km of fencing was installed bordering roads on the Cedar Fire (Hubbert 2005, 2006).

Trail repair and maintenance has seen an increase in BAER prescriptions since 1990. Due to increased use of the wildlands by a rapidly growing population base, keeping hiking trails open and accessible to the public has gained in importance. In addition, some high profile trails generate a large public response to their needs for repair.

No Treatment Alternative

Over the last decade, it has become apparent that the capability of wildlands to recover without treatments needs to be documented. In the majority of cases, wildfires do not devastate forest or rangeland ecosystems and eliminate sources of seed for desired tree and plant species. Wildfires do not sterilize soils. They do not delay or even preclude the reestablishment of plant cover, and they do not adversely impact the sustainability of ecosystems and the well-being of adjacent ecological communities. Moreover, only by monitoring no treatment areas can the effectiveness of treatments be evaluated. Future BAER assessment teams unfamiliar with local recovery periods would also benefit from this monitoring information that documents the natural recovery. Monitoring burned but untreated hillslopes provides data for future determinations of potential erosion risk that may be useful for future post-fire treatment decisions. This could be critical as both erosion and sediment potential define the emergency, based on the values at risk (Napper 2005).

LESSONS LEARNED AND FUTURE DIRECTIONS

After nearly a century of post-fire emergency rehabilitation in California, the practice has become fairly well-developed and there have been many lessons learned along the way. It was once standard operating procedure to seed every burned hectare and erect heroic engineering works on every stream channel to protect all onsite and downstream values at risk. Today this is seen as both unnecessary and unrealistic. We now understand that fire is an integral part of nearly all ecosystems, and that the consequences of wildfire are not necessarily devastating to the environment. There are still critical natural resources and human development that need extra protection in the aftermath of fire, but these need to be clearly identified and specifically targeted with environmentally sensitive measures. If there are no resources or values at risk, no treatments are needed. After the 2003 southern California Fire Siege, only about 2 percent of the over 290,000 ha burned were treated with some form of emergency rehabilitation measures (Hubbert 2005).

While human life and infrastructure are of paramount importance, emergency rehabilitation techniques must also be sensitive to environmental concerns. Grass seeding has been all but abandoned in California, as research has shown that rapidly growing non-native grasses do little to reduce erosion, provide no extra ground cover, and can be harmful to native species (Wohlgemuth et al. 1998). Radical ground disturbing measures whose impacts on the landscape can last much longer than the anticipated emergency are now discouraged. Engineering solutions can be prohibitively expensive and their protection results can not be guaranteed if the worst case scenario is realized and the design criteria are exceeded. Perhaps the most realistic protection measures from the consequences of post-fire flooding would be to redirect human development away from the canyon mouths and steeplands at the wildland/urban interface.

Although post-fire rehabilitation and restoration have been practiced for many decades, land management and hazard protection agencies have not always done a thorough job of monitoring the success and effectiveness of rehabilitation practices (GAO 2003). Both the U.S. Department of Agriculture, Forest Service (Robichaud et al. 2000) and the U.S. Department of Interior (Pyke and McArthur 2002) produced reports recommending that more research on and systematic monitoring and analysis of post-fire stabilization methods be conducted. Results of some of that monitoring have been described in this chapter and other chapters, and many other projects are being documented, often in internal agency reports. In the future these assessments need to consider not only the effectiveness of these practices but also their cost-effectiveness.

The practice of post-fire logging or 'salvage logging', often considered part of post-fire rehabilitation or restoration, has come under greater scrutiny as well (McIver and Starr 2000, Donato et al. 2006). The U.S. Department of Agriculture, Forest Service and the U.S. Department of Interior Joint Fire Science research program has recently (2006) awarded several grants dealing with the impacts of current and past post-fire logging, including projects in Sierra Nevada and northern California (see http://jfsp.nifc.gov for a list of past and current projects).

Perhaps the greatest challenge in the future to post-fire rehabilitation specialists lies in the realm of education. Vast knowledge and experience has accrued over the past century, but the communication of this information has lagged. The general public needs to be aware that the emergency is not over when the flames are doused, understand that – even seemingly removed from the burned area – there can be serious consequences of flooding and sedimentation in the post-fire environment, and appreciate the possibilities and costs of emergency rehabilitation. More importantly, land managers and political decision-makers need to receive objective information from the rehabilitation specialists in order to make educated post-fire restoration choices.

Acknowledgements

The authors thank all the tireless workers who have dedicated at least a portion of their professional careers to refining and improving post-fire emergency rehabilitation throughout the state of California over the past century.

References

Andressen, V. 2002. Curve Fire Burned-Area Report – FS-2500-8 from November, 2002. U.S. Department of Agriculture, Forest Service, Angeles National Forest, Acadia, California, USA.

Barker, P.F. 1989. Timber salvage operations and watershed resource values. pp. 1-2. *In* N.H. Berg [tech. coord.]. Proceedings of the Symposium on Fire and Watershed Management; October 26-28, 1988; Sacramento, California. General Technical Report PSW-109. U.S. Department of Agriculture, Forest Service, Pacific Southwest Forest and Range Experiment Station, Berkeley, California, USA.

Baron, F.J. 1962. Effects of different grasses on ponderosa pine seedling establishment. Research Note PSW-199. U.S. Department of Agriculture, Forest Service, Pacific Southwest Forest and Range Experiment Station, Berkeley, California, USA.

Barro, S.C. and S.G. Conard. 1987. Use of ryegrass seeding as an emergency revegetation measure in chaparral ecosystems. General Technical Report PSW-102. U.S. Department of Agriculture, Forest Service, Pacific Southwest Forest and Range Experiment Station, Berkeley, California, USA.

Barro, S.C., A.G. Campbell, and P.M. Wohlgemuth. 1989. Post-fire interactions between riparian vegetation and channel morphology and the implications for stream channel rehabilitation choices. pp. 51-53. *In* D.L. Abell [tech. coord.]. Proceedings of the California Riparian Systems Conference; September 22-24, 1988; Davis, California. General Technical Report PSW-110. U. S. Department of Agriculture, Forest Service, Pacific Southwest Forest and Range Experiment Station, Berkeley, California, USA.

Berg, N.H. [tech. coord.]. 1989. Proceedings of the Symposium on Fire and Watershed Management; October 26-28, 1988; Sacramento, California. General Technical Report PSW-109. U.S. Department of Agriculture, Forest Service, Pacific Southwest Forest and Range Experiment Station, Berkeley, California, USA.

Beyers, J.L. 2004. Postfire seeding for erosion control: Effectiveness and impacts on native plant communities. *Conservation Biology,* 18: 947-956.

Booker, F.A. and W.E. Dietrich. 1998. Landscape and Management Response to Wildfires. *In* California Burned Watersheds Erosion Study CDF No. 8CA38629, California Department of Forestry and Fire Protection, Sacramento, California, USA.

Brooks, M. and C. D'Antonio. 2001. The role of fire in promoting plant invasions. pp. 29-30. *In* M. Kelly [ed.]. Proceedings of the California Exotic Pest Plant Council Symposium. Vol. 6: 2000-2002. California Exotic Pest Plant Council, Berkeley, California, USA. Online at *http://www.cal-ipc.org/symposia/archive/pdf/18602.pdf* [accessed 23 Sep 2008].

Brooks, M.L., C.M. D'Antonio, D.M. Richardson, J.B. Grace, J.E. Keeley, J.M. DiTomaso, R.J. Hobbs, M. Pellant, and D. Pyke. 2004. Effects of alien plants on fire regimes. *BioScience,* 54: 677-688.

Burroughs, E.R., Jr. and J.G. King. 1989. Reduction of soil erosion on forest roads. General Technical Report INT-264. U.S. Department of Agriculture, Forest Service, Intermountain Research Station, Ogden, Utah, USA.

Clines, J. 2005. Preventing weed spread in contaminated hay and straw. pp. 4-6. *In* G. Skurka [ed.]. Proceedings California Invasive Plant Council Symposium, Volume 9: 2005. California Invasive Plant Council, Berkeley, California, USA. *Online at* http://www.cal-ipc.org/symposia/archive/pdf/2005%20Symposium%20proceedings.pdf [accessed 22 Aug 2008].

Collins, L.M. and C.E. Johnston. 1995. Effectiveness of straw bale dams for erosion control in the Oakland Hills following the fire of 1991. pp. 171-183. *In* J.E. Keeley and T. Scott [eds.]. *Brushfires in California Wildlands: Ecology and Resource Management*. International Association of Wildland Fire, Fairfield, Washington, USA.

Conard, S.G., J.L. Beyers, and P.M. Wohlgemuth. 1995. Impacts of post-fire grass seeding on chaparral ecosystems: What do we know and where do we go from here? pp. 149-161. *In* J.E. Keeley and T. Scott [eds.]. *Brushfires in California Wildlands: Ecology and Resource Management*. International Association of Wildland Fire, Fairfield, Washington, USA.

Corbett, E.S. and L.R. Green. 1965. Emergency revegetation to rehabilitate burned watersheds in southern California. Research Paper PSW-22. U. S. Department of Agriculture, Forest Service, Pacific Southwest Forest and Range Experiment Station, Berkeley, California, USA.

DeBano, L.F. 1981. Water repellent soils: A state-of-the-art. General Technical Report PSW-46. U.S. Department of Agriculture, Forest Service, Pacific Southwest Forest and Range Experiment Station, Berkeley, California, USA.

DeBano, L.F., D.G. Neary, and P.F. Ffolliott. 1998. *Fire's Effects on Ecosystems*. John Wiley & Sons, Inc., New York, New York, USA.

DeByle, N.V. 1970. Infiltration in contour trenches in the Sierra Nevada. Research Note INT-115. U.S. Department of Agriculture, Forest Service, Intermountain Forest and Range Experiment Station, Ogden, Utah, USA.

Department of Forester and Fire Warden. 1985. Emergency revegetation in Los Angeles County 1919-1984. County of Los Angeles, Fire Department, Los Angeles, California, USA.

Donato, D.C., J.B. Fontaine, J.L. Campbell, W.D. Robinson, J.B. Kauffman, and B.E. Law. 2006. Post-wildfire logging hinders regeneration and increases fire risk. *Science*, 311: 352.

Durrenberger, R.W. and R.B. Johnson. 1976. *California: Patterns on the Land*. Mayfield Publishing Company, Palo Alto, California, USA.

Eaton, E.C. 1932. San Gabriel project: Check dams. Los Angeles County Flood Control District Report. Los Angeles, California, USA.

Eaton, E.C. and F. Gillelen. 1931. Report on check dams. Los Angeles County Flood Control District Report. Los Angeles, California, USA.

Galley, D.E.M. and T.P. Wilson [eds.]. 2001. Proceedings of the invasive species workshop: The role of fire in the control and spread of invasive species. Fire Conference 2000: The First National congress on Fire Ecology, Prevention and Management. Miscellaneous Publication No. 11. Tall Timbers Research Station, Tallahassee, Florida, USA.

[GAO] General Accounting Office. 2003. Wildland fires: Better information needed on effectiveness of emergency stabilization and rehabilitation treatments. GAO-03-430. U.S. General Accounting Office, Washington, D.C., USA.

Gautier, C.R. 1983. Sedimentation in burned chaparral watersheds: Is emergency revegetation justified? *Water Resources Bulletin*, 19: 793-802.

Gleason, C.H. 1947. Guide for mustard sowing in burned watersheds of southern California. U.S. Department of Agriculture, Forest Service, California Region, San Francisco, California, USA.

Griffin, J.R. 1982. Pine seedlings, native ground cover, and *Lolium multiflorum* on the Marble-Cone burn, Santa Lucia Range, California. *Madroño*, 29: 177-188.

Griffith, R.W. 1998. Burned area emergency rehabilitation. pp. 4-7. *In* S. Gray [chairman]. Proceedings of the 19th Annual Forest Vegetation Management Conference: Wildfire Rehabilitation. University of California Cooperative Extension Forestry, Redding, California, USA.

Gross, E., I. Steinblums, C. Ralston, and H. Jubas. 1989. Emergency watershed treatments on burned lands in southwestern Oregon. pp. 109-114. *In* N.H. Berg [tech. coord.]. Proceedings of the Symposium on Fire and Watershed Management; October 26-28, 1988; Sacramento, California. General Technical Report PSW-109. U.S. Department of Agriculture, Forest Service, Pacific Southwest Forest and Range Experiment Station, Berkeley, California, USA.

Hornbeck, D., P. Kane, and D. Fuller. 1983. *California Patterns: A Geographical and Historical Atlas*. Mayfield Publishing Company, Palo Alto, California, USA.

Hubbert, K.R. 2005. Treatment effectiveness monitoring for southern California wildfires: 2003 to 2004. Report on file at U.S. Department of Agriculture, Forest Service, Pacific Southwest Research Station, Riverside, California, USA. *Online at* http://www.fs.fed.us/psw/publications/4403/BAEREffectivenessMonitoring SoCA.pdf [accessed 22 Aug 2008].

Hubbert, K.R. 2006. Treatment effectiveness monitoring for southern California wildfires: The 2nd and 3rd years 2004 to 2005 and 2005 to 2006. Report on file at U.S. Department of Agriculture, Forest Service, Pacific Southwest Research Station, Riverside, California, USA.

Janicki, A. and S. Grant. 2002. Heli-mulching on the Darby Fire: A Case Study. U.S. Department of Agriculture, Forest Service, Stanislaus National Forest, Sonora, California, USA.

Keeley, J.E. 1991. Seed germination and life history syndromes in the California chaparral. *The Botanical Review*, 57: 81-116.

Keeley, J.E. 2001. Fire and invasive species in Mediterranean-climate ecosystems of California. pp. 81-94. *In* Miscellaneous Publication No. 11. Tall Timbers Research Station, Tallahassee, Florida, USA.

Keeley, J.E. 2004. Ecological impacts of wheat seeding after a Sierra Nevada wildfire. *International Journal of Wildland Fire*, 13: 73-78.

Keeley, J.E., M. Carrington, and S. Trnka. 1995. Overview of management issues raised by the 1993 wildfires in southern California. pp. 83-89. *In* J.E. Keeley and T. Scott [eds.]. *Brushfires in California Wildlands: Ecology and Resource Management*. International Association of Wildland Fire, Fairfield, Washington, USA.

Keeley, J.E., C.J. Fotheringham, and M.A. Moritz. 2004. Lessons from the October 2003 wildfires in southern California. *Journal of Forestry*, 102: 26-21.

Keeley, S.C., J.E. Keeley, S.M. Hutchinson, and A.W. Johnson. 1981. Postfire succession of herbaceous flora in southern California chaparral. *Ecology*, 62: 1608-1621.

Kraebel, C.J. 1934. The La Crescenta flood. *American Forests*, 40: 251-254, 286-287.

Krammes, J.S. and L.W. Hill. 1963. "First aid" for burned watersheds. Research Note PSW-29. U.S. Department of Agriculture, Forest Service, Pacific Southwest Forest and Range Experiment Station, Berkeley, California, USA.

MacDonald, L. 1989. Rehabilitation and recovery following wildfires: A synthesis. pp. 141-144. *In* N.H. Berg [tech. coord.]. Proceedings of the Symposium on Fire and Watershed Management; October 26-28, 1988; Sacramento, California. General Technical Report PSW-109. U.S. Department of Agriculture, Forest Service, Pacific Southwest Forest and Range Experiment Station, Berkeley, California, USA.

McIver, J.D. and L. Starr. [tech. eds.]. 2000. Environmental effects of postfire logging: Literature review and annotated bibliography. General Technical Report PNW-

GTR-486. U.S. Department of Agriculture, Forest Service, Pacific Northwest Research Station, Portland, Oregon, USA.

Miles, S.R., D.M. Haskins, and D.W. Ranken. 1989. Emergency burn rehabilitation: cost, risk, and effectiveness. pp. 97-102. *In* N.H. Berg [tech. coord.]. Proceedings of the Symposium on Fire and Watershed Management; October 26-28, 1988; Sacramento, California. General Technical Report PSW-109. U.S. Department of Agriculture, Forest Service, Pacific Southwest Forest and Range Experiment Station, Berkeley, California, USA.

Munns, E.N. 1919. The control of flood water in southern California. *Journal of Forestry,* 17: 423-429.

Myers, N., R.A. Mittermeier, C.G. Mittermeier, G.A.G. da Fonseca, and J. Kent. 2000. Biodiversity hotspots for conservation priorities. *Nature,* 403: 853-858.

Napper, C. 2005. Why monitor the 'no treatment' areas within a wildland fire? *In* K.R. Hubbert. Treatment effectiveness monitoring for southern California wildfires: 2003 to 2004. Report on file at U.S. Department of Agriculture, Forest Service, Pacific Southwest Research Station, Riverside, California, USA. *Online at* http://www.fs.fed.us/psw/publications/4403/ BAER EffectivenessMonitoring SoCA.pdf [accessed 22 Aug 2008].

O'Connor, M.D. and R.R. Ziemer. 1989. Coarse woody debris ecology in a second-growth *Sequoia sempervirens* forest stream. pp. 165-171. *In* D.L. Abell [tech. coord.]. Proceedings of the California Riparian Systems Conference; September 22-24, 1988; Davis, California. General Technical Report PSW-110. U.S. Department of Agriculture, Forest Service, Pacific Southwest Forest and Range Experiment Station, Berkeley, California, USA.

Oakeshott, G.B. 1978. *California's Changing Landscapes.* McGraw-Hill Book Company, New York, New York, USA.

Poff, R.J. 1989. Compatibility of timber salvage operations with watershed values. pp. 127-140. *In* N.H. Berg [tech. coord.]. Proceedings of the Symposium on Fire and Watershed Management; October 26-28, 1988; Sacramento, California. General Technical Report PSW-109. U.S. Department of Agriculture, Forest Service, Pacific Southwest Forest and Range Experiment Station, Berkeley, California, USA.

Pyke, D.A. and T. McArthur. 2002. Emergency fire rehabilitation of BLM lands in the Intermountain West: Revegetation and monitoring. Interim Report to U.S. Department of Interior, Bureau of Land Management. U.S. Department of Interior, U.S. Geological Survey, Forest and Rangeland Ecosystems Science Center, Corvallis, Oregon, USA.

Ratliff, R.D. and P.M. McDonald. 1987. Postfire grass and legume seeding: what to seed and potential impacts on reforestation. pp. 111-123. *In* Proceedings of the Ninth Annual Forest Vegetation Management Conference. University of California Cooperative Extension Forestry, Redding, California, USA.

Rice, R.M. 1974. The hydrology of chaparral watersheds. pp. 277-334. *In* M. Rosenthal [ed.]. *Proceedings of the Symposium on Living with the Chaparral; March 30-31, 1973; Riverside,* California. The Sierra Club, San Francisco, California, USA.

Rice, R.M., R.P. Crouse, and E.S. Corbett. 1965. Emergency measures to control erosion after a fire on the San Dimas Experimental Forest, pp. 123-130. Miscellaneous Publication 970, U. S. Department of Agriculture, Forest Service, Washington, D.C., USA.

Robichaud, P.R., J.L. Beyers, and D.G. Neary. 2000. Evaluating the effectiveness of postfire rehabilitation treatments. General Technical Report RMRS-GTR-63. U.S. Department of Agriculture, Forest Service, Rocky Mountain Research Station, Fort Collins, Colorado, USA.

Ruby, E.C. 1973. Evaluation of check dams for sediment control. U.S. Department of Agriculture, Forest Service, Angeles National Forest, Arcadia, California, USA.

Sampson, A.W. 1944. Plant succession on burned chaparral lands in northern California. *California Agricultural Experiment Station Bulletin*, 685: 1-144.

Schultz, A.M., J.L. Launchbaugh, and H.H. Biswell. 1955. Relationship between grass density and brush seedling survival. *Ecology*, 36: 226-238.

Sweeney, J.R. 1956. Responses of vegetation to fire: A study of the herbaceous vegetation following chaparral fires. *University of California Publications in Botany*, 28: 143-249.

Taskey, R.D., C.L. Curtis, and J. Stone. 1989. Wildfire, ryegrass seeding, and watershed rehabilitation. pp. 149-161. *In* N.H. Berg [tech. coord.]. Proceedings of the Symposium on Fire and Watershed Management; October 26-28, 1988; Sacramento, California. General Technical Report PSW-109. U.S. Department of Agriculture, Forest Service, Pacific Southwest Forest and Range Experiment Station, Berkeley, California, USA.

U.S. Department of Agriculture, Forest Service. 1992. Forest Service Handbook 2509.13, Burned-Area Emergency Rehabilitation Handbook. U.S. Department of Agriculture, Forest Service, Washington, D.C., USA.

Van der Water, R. 1998. Post-fire riparian zone management: The Salmon River experience. pp. 25-40. *In* S. Gray [chairman]. Proceedings of the 19th Annual Forest Vegetation Management Conference: Wildfire Rehabilitation. University of California Cooperative Extension Forestry, Redding, California, USA.

Wells, W.G., II. 1987. The effects of fire on the generation of debris flows in southern California. *Reviews in Engineering Geology*, 7: 105-114.

Wohlgemuth, P.M., J.L. Beyers, C.D. Wakeman, and S.G. Conard. 1998. Effects of fire and grass seeding on soil erosion in southern California chaparral. pp. 41-51. *In* S. Gray [chairman]. Proceedings of the 19th Annual Forest Vegetation Management Conference: Wildfire Rehabilitation. University of California Cooperative Extension Forestry, Redding, California, USA.

20

Fire Disturbance Regimes, Ecosystem Recovery and Restoration Strategies in Mediterranean and Temperate Regions of Chile

J.J. Armesto[1,2]*, *M.A. Bustamante-Sánchez*[1,2], *M.F. Díaz*[1], *M.E. González*[4], *A. Holz*[3], *M. Nuñez-Avila*[4] and *C. Smith-Ramírez*[1,2]

Abstract

We provide an overview of the major causes and historical recurrence times of fire disturbance, both natural and anthropogenic, in Mediterranean-type vegetation and temperate forests of Chile (33 to 42° S), southwestern South America. For each vegetation type, we examine the main consequences of fire for species and ecosystems and describe post-fire regeneration and potentially effective strategies for ecological restoration. In both ecosystem types, human activities have greatly altered the historical disturbance regime. Increasing frequency of small-scale fires and massive plantations of fire-prone exotics have accelerated the degradation of Chilean Mediterranean vegetation, which exhibit low resilience to repeated fire disturbance. Active restoration of Mediterranean sclerophyllous vegetation is urgently needed given the magnitude of the loss of species, vegetation cover, and ecosystem services. Recurrent burning remains the major threat to restoration efforts. In temperate forests, historical fire regimes were characterized by low-frequency, large scale fires, but increasing bamboo cover and canopy opening due to logging are changing the susceptibility of forests to fire. Synchronous die-off of bamboo populations in the forest understory significantly increases fire probabilities. Several long-lived conifers have vegetative and reproductive traits that make them resistant to fire. Passive restoration is a viable strategy

---

[1] *Center for Advanced Studies in Ecology and Biodiversity, Pontificia Universidad Católica de Chile, Santiago, Chile.*
[2] *Insituto de Ecología y Biodiversidad, Universidad de Chile, Santiago, Chile.*
[3] *Department of Geography, University of Colorado, Boulder, Colorado, USA.*
[4] *Facultad de Ciencias Forestales (FORECOS), Universidad Austral de Chile, Valdivia, Chile.*
* *Corresponding author: J.J. Armesto, Center for Advanced Studies in Ecology and Biodiversity, Pontificia Universidad Católica de Chile. Alameda 340, P.C. 6513677, Santiago, Chile. Tel: +56 2 354 2610, Fax: +56 2 354 2621, e-mail: jarmesto@uchile.cl*

for these conifer-dominated, montane forests, but the harvest of dead timber is a serious threat to ecosystem recovery. Active restoration is needed in lowland evergreen forests, where the loss of canopy cover alters hydrology and raises the water table, slowing the process of tree invasion.

INTRODUCTION

Fire is an important factor effecting the vegetation and landscapes in Mediterranean and temperate regions around the world (Keeley 1989, Lepart and Debussche 1992, Cowling et al. 1996, Kitzberger and Veblen 1999, Jasinski and Payette 2005). However, the impact of fire on ecological processes and local biodiversity vary greatly among regions, with different implications for management and restoration. In some areas vegetation composition and structure, as well as plant life history traits, are largely dependent on the recurrence of wildfire, in other areas fire may be a novel and highly disrupting disturbance process introduced by humans. In the first case, the exclusion of fire, through vegetation management and fire prevention programs, can have negative consequences for species richness and ecosystem dynamics. Greater understanding of these fire-dependent systems has led managers to use controlled-fire programs to maintain biodiversity and ecosystem processes. Fire management programs can be used to promote vegetation recovery, conserve regional biodiversity, restore original vegetation, and prevent the occurrence of devastating fires associated with fuel buildup. In contrast, where wildfires are rare events in the lifetime of the dominant life forms, and people have increased fire frequency in recent times, there is a need to prevent or limit wildfire to enhance ecosystem recovery and thus maintain biotic diversity. In the latter ecosystems, recurrent fire may arrest successional change, promote invasion by fire prone, exotic species, and cause significant losses of local biodiversity which need to be addressed in restoration programs.

In this chapter, we provide a general overview of fire disturbance regimes and their historical role in both Mediterranean vegetation and the main types of temperate forests in Chile. These ecosystems display contrasting histories of fire impact and recovery processes. Our goals are 1) to establish the present characteristics of fire disturbance regimes, as modified by human activities, in contrast to the historical (pre-European) role of fire on the local ecosystems, as presently understood; 2) to examine major adaptations of plant species to tolerate or recover from distuıbance in each ecosystem type; 3) to assess what we know about ecosystem resilience properties, or the system's ability to recover its original composition, biodiversity and structure following fire; and 4) to discuss potentially effective strategies for post-fire restoration of biotic communities in each area, identifying situations where human interventions are necessary.

Fire in Mediterranean Sclerophyllous Vegetation

The analysis of fire frequency in Chilean matorral (Armesto et al. 1995, Montenegro et al. 2004) indicates that the number and extent of fires increased in the past 30 years, in correspondence with the increase in human population, the expansion of roads and urban areas. From the main causes of fire (Table 1), we can infer that the probability of fire increases with decreasing distances to roads and populated areas. The main causes of wildfire in central Chile between 1963 and 1998, as recorded by the Chilean Forest Service (CONAF), are road traffic (30 percent of fires) and clearing of vegetation for agriculture and forestry (20 percent). Lightening-related woodland fires are rare and represent less than one percent of recorded events (Table 1; Montenegro et al. 2004). Therefore, it can be estimated that any given area of the Chilean matorral now burns, as a result of human activities, with a frequency <20 years (Armesto et al. 1995). This frequency is similar to that documented for the fire-prone California chaparral, where many fire-related adaptations have been described (Hanes 1971, Keeley 1989).

Table 1 Major causes of wildfire in central Chile between 1963 and 1998 (data from Armesto et al. 1995, Montenegro et al. 2004)

Causes	% incidence
Roads and transportation-related	24
Clearing of land for agriculture or forestry	28
Accidental, recreation-related	20
Military exercises	12
Lightning	<1

In central Chile, fire has increased in recurrence and extent since the Spanish settlement (Armesto et al. 1995, Montenegro et al. 2004). This does not mean that aboriginal fires did not exist (e.g., Heusser 1983) as charcoal records indicate, but that they must have been limited to near settlements. However, the rapid expansion of farmland after European settlement, which relied on the use of fire to clear vegetation, changed the physiognomy of landscapes and the abundance of different life forms (Armesto and Gutiérrez 1978). In addition, firewood harvest and making of charcoal were common activities during the 19th and early 20th centuries (Fuentes and Muñoz 1995).

The absence of fire-induced traits in Chilean sclerophyllous vegetation strongly indicates that fire disturbance did not play a major role in these ecosystems in the absence of humans (Table 2). This contrasts with life history characteristics of plants from other Mediterranean-climate areas of the world where wildfires have been a permanent selective force (Keeley 1995). Massive soil seed banks, highly responsive to fire, which are characteristic of the California chaparral (Parker and Kelly 1989) are absent from Chilean sclerophyllous vegetation, particularly among shrub species (Muñoz and

Table 2 Post-fire responses of herbs, shrubs and trees in the Chilean Mediterranean vegetation compared to California chaparral (modified from Montenegro et al. 2004 and Armesto et al. 1995). ++ Common response; + less frequent or rare, 0 absent.

Plant response	Chile	California
Annual pyrophytes	0	++
Resprouting from buried bulbs	+	++
Resprouting from lignotuber or roots	++	++
Fire-stimulated flowering	0	+
Smoke-induced germination	0	++
Heat-induced germination	0	++
Large dormant seed banks	0	++
Seed release from cones or fruits	0	+
Large fleshy-fruited, bird dispersed seeds	++	+

Fuentes 1989, Jiménez and Armesto 1992). Seeds of Chilean matorral species do not germinate in response to smoke or high temperatures (Muñoz and Fuentes 1989). Smoke stimulates seed germination in the California chaparral and South African fynbos (Brown 1993), but it has no effect on Chilean shrub seeds (Muñoz and Fuentes 1989). Few, if any, annual plants germinate after fire in Chilean sclerophyllous vegetation (Armesto and Gutiérrez 1978, Montenegro et al. 2004). Chilean Mediterranean shrubs also lack serotinous fruits which release their seeds in response to fire as found in numerous South African and Australian species (Cowling 1996).

Seed dispersal syndromes differ greatly between Chilean sclerophyllous shrubs and California and Australian species (Hoffmann and Armesto 1995), particularly because of the high frequency of large, avian-dispersed fleshy fruits in Chile and the commonness of gravity and ant dispersal in the other two floras. These differences in dispersal syndromes suggest that Mediterranean sclerophyllous vegetation in central Chile historically had a predominantly closed cover (Hoffman and Armesto 1995), where directed-seed dispersal by birds (Weeny 2001) could have been important to reach a limited number of open sites suitable for seed germination. In fire-prone Mediterranean environments, such long-distance dispersal mechanisms may be less necessary because high fire recurrence creates open areas with a shorter frequency than the life span of dominant shrub species. This idea is further supported by the fact that Chilean myrtaceous shrubs from sclerophyllous vegetation bear fleshy fruits, whereas their family relatives in the sclerophyllous flora of Australia (e.g., *Eucalyptus*) have dry, gravity-dispersed fruits (Hoffmann and Armesto 1995).

Consequently, we argue that anthropogenic fire had a profound effect on floristic diversity and function of Mediterranean ecosystems in Chile than in other Mediterranean areas of the world. One of the main consequences has been a change in the general physiognomy of landscapes. In central Chile, vegetation cover has changed from largely closed woodland, with scattered openings, as can still be seen in some sheltered ravines, to an open mosaic of discrete shrub patches separated by open spaces dominated by herbaceous vegetation (Fuentes et al. 1984, 1986, Armesto and Pickett 1985). This patchy vegetation structure has major consequences for the dynamics of vegetation. Shrub regeneration is now strongly driven by nursing effects of shrubs, serving as recruitment sites, and is severely limited by increased animal grazing in open areas between shrubs (Holmgren et al. 2000). In addition, the distribution and composition of herbaceous vegetation in the Chilean matorral has been strongly modified, favoring a number of fire-tolerants such as *Pasithaea coerulea* and *Alstroemeria* spp., which can resprout from buried bulbs or roots (Montenegro et al. 2004) and facilitate the spread of European weeds. The widespread effects of fire in the Chilean Mediterranean landscapes may have also changed plant species richness, although evidence of local extinction is limited and not clearly attributable to fire disturbance. Nevertheless, population sizes are declining due to the inability of some species to recover from repeated wildfire, thus affecting the genetic diversity of central Chilean shrubs, which have low seed production and are obligate outbreeders (Arroyo and Uslar 1993).

Ecological Restoration Strategies

From the discussion above, it seems likely that over broad areas of central Chile, lack of propagules, limited sprouting ability, intense grazing, and fire recurrence are strong inhibitors of ecosystem recovery (Armesto et al. 1995). Site and landscape considerations must be taken into account when deciding where interventions are necessary to facilitate regeneration. To overcome the main limitations to shrub establishment, it may be necessary to provide mechanisms to enhance seed dispersal inputs and protect regeneration from grazing and fire. For practical reasons, the paucity of seed sources may be considered the first limitation to overcome (Fig. 1). At the landscape level, it could be useful to identify significant remnants that can serve as seed sources for adjacent open areas. Such areas could be focal points to carry out site-specific interventions and to prevent grazing and fire. This makes it highly relevant to preserve small remnants of sclerophyllous vegetation (Armesto et al. 2002) that may be important reservoirs of propagules and species for rehabilitation.

Passive restoration

Although degraded woodlands in central Chile have only limited ability to recover spontaneously from fire (Jiménez and Armesto 1992, Armesto et al. 1995, Segura et al. 1998, Holmgren et al. 2000, Montenegro et al. 2004), several conditions must be met to facilitate recovery. The disturbing agents must be removed, propagule sources must exist at the site or in a nearby area as a source of new colonists, soils must remain reasonably intact, and aggressive weeds must be controlled if the original community is to re-establish. When

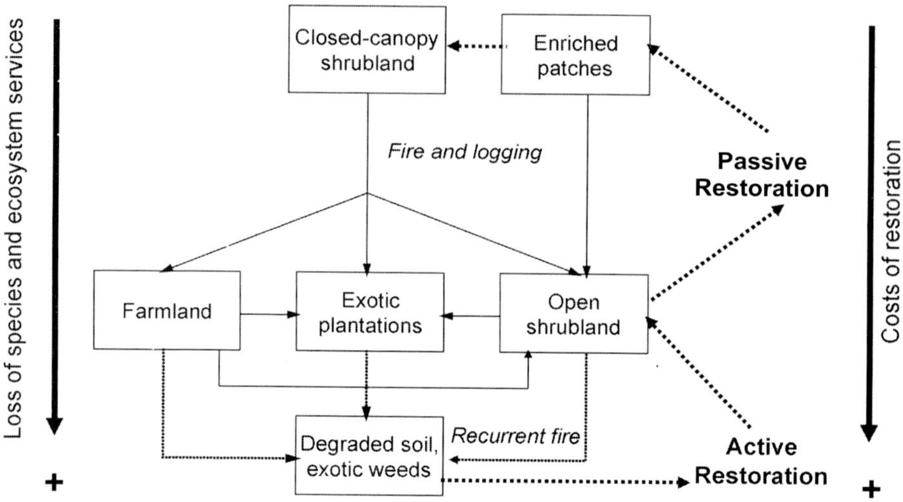

Fig. 1 Main transitions associated with land use and fire disturbance in Chilean Mediterranean vegetation. Return times to the closed-canopy stage are unknown, but increase to the bottom of the diagram. Biodiversity and ecosystem services are enhanced at the top of the diagram. Active restoration may require the occurrence of a 'window of opportunity' derived from a wet phase of ENSO (see text).

these conditions are met, **passive restoration** can be achieved by protecting the site from fire and grazing and allowing natural regeneration to proceed (Fig. 1).

As fire is a widespread and frequent anthropogenic disturbance in central Chile today, fire prevention, detection, and community education are a pre-requisite for spontaneous regeneration or successful restoration. Local community involvement in fire prevention is one of CONAF's (Chilean Forest Service) main goals, because people are responsible for most fires in central Chile (Table 1). Fires also affect pine and eucalyptus plantations, which cover nearly one million hectares, especially on the coastal mountains of central Chile. The expansion of exotic plantations of these fire-prone species in recent decades (Armesto et al. 2001) may have contributed to enhance fire disturbance to native vegetation, as fires initiated in plantations often extend to neighboring native shrublands. Fire was also used as a tool to eradicate native shrubs for planting exotic timber trees, a method that was stimulated by Government subsidies to plantations on non-forested land (Lara and Veblen 1993).

Passive restoration is therefore a viable strategy for only a limited number of locations in central Chile, especially applicable in areas where remnants of native vegetation are still found and can provide abundant seed input. However, such efforts must necessarily be accompanied by effective fire prevention policies and local community education to be successful, given the long periods necessary for the recovery of shrub cover.

Active restoration

The area of degraded vegetation in central Chile has increased in recent decades, due to the recurrence of fire and grazing (CONAF-CONAMA-BIRF 1999). Chances for spontaneous regeneration (passive restoration) are considerably reduced by the prevalence of multiple disturbance agents, including fire, livestock grazing, seedling predation by European rabbits, expansion of Eurasian weeds, and leaf litter extraction for horticultural proposes, among other uses (Fuentes and Muñoz 1995, Lienlaf 1996). Accordingly, for most areas in central Chile, **active restoration** is required, which implies direct human intervention. The design of effective manipulations depends on a deeper understanding of the ecosystem under management and definition of ecologically and socially valid restoration targets (Armesto et al. 2007).

To ensure long-term restoration and to speed up the process of vegetation recovery, it may be necessary to reintroduce certain key woody species known to facilitate seed input and improve conditions for seedling establishment (Armesto and Pickett 1985). Through *enrichment planting* local genetic stocks of trees whose populations are low in number can be propagated and planted in degraded areas, or secondary forests, to increase their population size and long-term viability. This requires the careful choice of a variety of genetic sources within the study area to prevent in-breeding and improve the likelihood that seedlings planted can tolerate climatic fluctuations. This is especially important given that many tree and shrub species in the Chilean matorral are obligate outbreeders (Arroyo and Uslar 1993).

Degraded areas generally present unfavorable conditions (summer stress, presence of herbivores, absence of pollinators, nutrient-poor soils, etc.) that limit recruitment of both natural and planted seedlings. Experimental shrub seedlings planted in open gaps on the Andean foothills around Santiago were unable to survive even when protected from herbivores (Fuentes et al. 1986). Careful selection of suitable pioneer species may be advised. Drought resistant species, such as *Baccharis*, *Quillaja*, and *Lithrea* seem likely candidates (Armesto et al. 1995). On more mesic sites, long-term observations indicate that matorral can regenerate more easily (Fuentes et al. 1984), even in burned sites. Shrub seedlings can survive in relatively large open gaps, especially when the herbaceous layer is removed (Holmgren et al. 2000). Therefore, active restoration through enrichment planting will have better results in humid sites, such as south-facing slopes and riparian habitats.

To persist through the Mediterranean summer drought, transplanted seedlings should be able to take advantage of infrequent wet periods. It has recently been suggested that the restoration of degraded semiarid ecosystems might be achieved effectively by adjusting grazing practices depending on El Niño Southern Oscillation (ENSO) rainfall variability (Holmgren and Scheffer 2001). In normal years, because of extremely dry summers, herbivore control alone will not ensure seedling establishment. On the other hand, high rainfall during winter and spring of El Niño years alone is not enough to enhance matorral recovery, because herbivory remains a crucial problem. Accordingly, grazing control during the wet phases of ENSO may stimulate recovery of woodlands, by promoting a shift between two alternative stable states (Holmgren and Scheffer 2001). Several observations support the idea that long-lasting recovery of woody vegetation in semiarid ecosystems depends on controlling grazing pressure during wet years, associated with positive phases of ENSO (e.g., Austin and Williams 1988, Gutiérrez et al. 1997, Holmgren et al. 2001). Nonetheless, the applicability of this model to restoration practices in Chilean matorral remains to be demonstrated.

To enhance woody seedling establishment, beginning from intermediate successional stages, with the presence of pioneer shrubs such as *Acacia caven* or *Baccharis* spp. may be most effective (Armesto and Pickett 1985, Fuentes et al. 1986, Armesto et al. 1995), as facilitative interactions play a major role in Mediterranean ecosystems. Shrubs can have a net positive effect on recruitment of other species (Bertness and Callaway 1994, Callaway and Walker 1997, Brooker and Callaghan 1998). One of the main nursing effects is an improvement of water relations of seedlings (Holmgren et al. 1997). Lower air and soil temperatures and increased soil water content under nurse shrubs reduce thermal and water stress on seedlings during dry periods (e.g., Valiente-Banuet and Ezcurra 1991, Aguiar and Sala 1994). However, even in mesic sites of central Chile few naturally established seedlings have been recorded (Armesto and Pickett 1985, Holmgren et al. 2000). Consequently, seed availability, rather than differential seedling emergence and survival, may be limiting colonization after wildfires. Chilean sclerophyllous vegetation presents an important number of species with fleshy fruits dispersed by birds (Hoffmann et al. 1989, Hoffmann and Armesto 1995). Thus, artificial or natural bird perches can be used as a cost-effective method to increase seed availability (Handel 1997). A degraded or burned area containing resprouting shrubs and scattered remnant (perch) trees is likely to receive greater avian seed dispersal than a treeless area (Debussche & Isenmann 1994, Verdu & García-Fayos 1996, Pausas et al. 2006). Accordingly, intense and recurrent fire over much of central Chile produce open conditions that are difficult to revert without intervention because of limited seed input.

Synthesis of Recommendations

Mediterranean vegetation in central Chile has been greatly modified over the past two centuries by the introduction of a high frequency of fires, caused primarily by people and rarely by natural ignition sources. The recurrence of wildfire is enhanced by rapid fuel buildup, as arboreal species change their growth form to multiple-stemmed, low shrubs with many dead stems due to crown shading (J.J. Armesto, unpublished observation). In addition, massive plantations of fire-prone exotic trees, such as *Pinus radiata* and *Eucalyptus* spp., further enhance fire spread and impact native vegetation found in areas surrounding the plantations. In this scenario, which also includes a large number of native and introduced grazers and high inter-annual variability in rainfall, restoration and long-term rehabilitation efforts are logistically difficult, expensive and have an uncertain outcome. Accordingly, restoration assays should be carried out at small scales, as pilot projects, including a major component of education to describe the advantages of rehabilitation for local communities (e.g., reduction of the threat of landslides). In addition, efforts must take advantage of windows of opportunity provided by rainfall variability (Holmgren and Scheffer 2001) in order to combine passive and active restoration strategies.

Fire and Ecosystem Recovery in Temperate Forests

Temperate rain forests cover much of southern Chile (Alaback 1991) and are generally considered an ecosystem relatively free of frequent wildfires (Armesto et al. 2001). However, several forest types in the temperate and sub-Antarctic region are dominated by tree species that may have a long history of association with fire disturbance, as indicated by their life cycles and morphology. On the other hand, the probability of fire increases eastwards, as one moves over the Andes towards the drier forest-steppe boundary (Kitzberger et al. 1997), and finally humans have used fire extensively to clear land for farming and ranching, thus altering the historical disturbance regimes (Armesto et al. 2001). Some forest fires have changed the landscape in ways that cannot be easily or rapidly restored without human intervention (Díaz and Armesto 2007). Other forests are apparently more resilient to fire disturbance, such as *Araucaria* forests in the Andes (Gonzalez et al. 2005), and passive restoration is a viable strategy. In what follows, we discuss the historical role of fire disturbance for the major temperate forest types in Chile and examine our current understanding of ecosystem resilience and rehabilitation of fire-damaged areas in each forest type. We also discuss how humans have altered fire disturbance regimes since pre-colonial times.

Fire History of *Araucaria–Nothofagus* Forests

Araucaria araucana (locally known as Araucaria) is a long-lived evergreen conifer of great cultural and scientific significance. In Chile, about 95 percent of the *Araucaria* forests are found in the Andes at an altitude of 1000 to 1600 m, between 37° 30′ S and 39° 40′ S. Two small additional populations are located on the coastal range between 37° 40′ S and 38° 40′ S (CONAF-CONAMA-BIRF 1999). Araucaria form pure or mixed stands with deciduous or evergreen species of *Nothofagus*. *Araucaria* forest cover, estimated to be 500,000 ha by 1550′s, has been reduced by almost 50 percent due to logging and fires, mostly during the 20th century (Lara et al. 1999).

Fire has played a key role in the ecology of *Araucaria–Nothofagus* ecosystems. Although fire disturbances have been typically dismissed as a natural factor controlling the dynamics of these temperate ecosystems (González 2002), charcoal particles in pollen records for at least the 44,000 yr BP (Heusser 1984) suggest that natural fires may have influenced the landscape prior to the arrival of Native American populations (Dillehay 1997). Also, the presence of fire-adapted species indicates that fire has been a significant factor in the history of these ecosystems. *Araucaria araucana* and associated *Nothofagus* species are well recognized for several adaptations to cope with fire (Table 3). *Araucaria araucana*, in particular, is well adapted to withstand fires because of its unusually thick bark (>10 cm thick), protected terminal buds, and its ability to resprout after fire (Veblen 1982a, Burns 1993). If fire does not kill the upper branches of Araucaria trees, the thick bark protects them from surface fires and thus regeneration takes place under the original canopy. In addition, *Nothofagus* species are known to resprout vigorously or to colonize massively after stand-replacing fires (Veblen et al. 1992a, González et al. 2005).

Table 3 Fire adaptations or tree strategies to resist, respond to, or recover from fire in *Araucaria–Nothofagus* forests in Chile and Argentina

Species	Capacity of resprouting	Thick and resistant bark	Massive recruitment from seeds	Cones resistant to fire
N. antarctica	X			
N. nervosa	X	X	X	
N. pumilio			X	
N. obliqua	X	X	X	
N. dombeyi			X	
A. araucana	X	X		X

Source: Veblen 1982, Burns 1993, Veblen et al. 1996.

Fire has been a ubiquitous disturbance in the mountain landscape dominated by *Araucaria–Nothofagus* forests over at least the past millennia. These forest ecosystems have been shaped by a mixed-severity fire regime including stand-replacing and less severe surface fires. Large and more infrequent high severity events are commonly followed by a distinct regeneration pulse of *Nothofagus* species along with a few *Araucaria* trees established close to remnant mature trees of *Araucaria* (González et al. 2005). In contrast, more frequent and low-severity fires typically result in none or little mortality of canopy trees and limited establishment of new individuals. Less frequent crown fires characterize the fire regime of more mesic, old-growth pure *Nothofagus pumilio* forests and mixed *Araucaria–Nothofagus* forests. On the other hand, for *Araucaria–Nothofagus* forests in drier, warmer environments, fire return times are shorter and could be less destructive (González et al. 2005).

Human activity has significantly altered the fire disturbance regime in the Araucarian region. Our initial evaluation indicates that, compared with the pre-European period (pre-1883), fire frequency increased following Euro-Chilean settlement in the late 19th century (González et al. 2005). Large tracts of subalpine forests were intentionally burned by new settlers in an effort to clear land for livestock ranching. Thus, pre-settlement fire intervals were largely compared with the settlement period (González et al. 2005). The more frequent but, in some habitats, less severe fire regime recorded during the 20th century was mostly the result of the extensive burning practiced by cattle ranchers (González et al. 2005).

For most of the major habitat types in the Valdivian Ecoregion, fire regimes established during the Holocene have been severely altered during the past 200 years. In the Araucarian region, human-altered fire regimes have strongly changed the structure, composition and function of these ecosystems, threatening the conservation of biodiversity. For example, increased fire frequency has changed the average stand age (and patch size) in the landscape, which is now mainly dominated at stand scale by young, post-fire *Nothofagus* stands, of 50 to 150 years old (c. 500,000 ha; CONAF-CONAMA-BIRF 1999). Large areas, originally covered by old-growth forests, now have a sparse tree cover or are densely dominated by bamboo species (Donoso 1993a, González et al. 2002; see below). Thus, the novel disturbance patterns threaten the persistence of native plant and animal populations, especially species restricted to old-growth habitat, such as the Magellanic Woodpecker and the southern Spotted Owl (Díaz et al. 2005). Consequently, human-induced changes in fire regime have resulted in current forest ecosystem structures and functions outside their historical range of variability (Morgan et al. 1994).

Restoration strategies

Restoration efforts strive to restore natural species composition and stand structures and, perhaps more importantly, natural processes that maintain the ecological integrity, resilience, and sustainability of ecosystems (White and Walker 1997, Landres et al. 1999). Therefore, understanding the role of fire is essential to develop socially and ecologically acceptable solutions for restoration.

In the case of *Araucaria–Nothofagus* forests, understanding their historic fire regime is fundamental for sound ecological restoration and fire hazard management efforts. Dendroecological reconstructions of fire regimes (González et al. 2005) for forests in mesic habitats are mostly characterized by long intervals between catastrophic fires. Therefore, ecological restoration efforts should focus on returning this ecological process to its natural and historic range of variability. For this purpose, frequent anthropogenic fires must be suppressed or controlled from large portions of the landscape, and furthermore, grazing and salvage felling from burned sites should be reduced or stopped (González 2005, González and Veblen 2007). The latter processes curtail forest recovery and promote the invasion of sites by exotic weeds. Ecologically sound restoration approaches should aim not only to replicate and reestablish particular stand reference conditions but also to restore the natural variability and resilience characteristic of *Araucaria–Nothofagus* forests.

Fire in Alerce (*Fitzroya cupressoides*) Forests

Alerce forests cover about 260,000 ha in Chile and 22,600 ha in Argentina, mainly in the Andes, but also in more limited coastal and lowland locations in the Chilean Lake District (39 to 40° S). Alerce forests have endured an intense and continuous exploitation since the 16th century to the present. Unfortunately, logging in many cases was associated with human-set fires to clear the brush, as the thick bark protects the timber of large Alerces from surface fires. In the accessible lowlands of the Chilean Lake District, several thousand hectares of Alerce were burned by settlers and loggers in the mid-19th century and the stands are almost completely vanished (Fraver et al. 1999, Silla et al. 2002). Most stands on the coastal range of the Lake District have also been subjected to catastrophic fires (Veblen and Ashton 1982, Armesto et al. 1995) and hundreds of hectares are now covered by standing dead trees. Only Andean forests remain largely intact, and affected by small scale fires (Donoso 1993a). Thus, anthropogenic fire has been a key factor in the history of Alerce forests during the Holocene.

Incidence of fire

On the coastal range, which runs parallel to the Pacific Ocean with maximum elevations of 1000 m, fires have been more extensive than in the Andes. Dating of fires in two stands on the coastal range using dendrochronology (Lara et al. 1999) shows that *Fitzroya* trees were able to survive fires which occurred between 1397 and 1943. Massive re-colonization of these sites by Alerce took place 5 to 17 years after the fire. Several burned stands in the same area maintain few scattered or small groups of live Alerces and few green stands also remain in the vicinity serving as seed sources (C. Smith-Ramírez, personal communication). Many fires are more than a century old. Historians, such as Molina et al. (2006), report the occurrence of large fires by the end of the 19th century, presumably set by indigenous people to open grazing grounds. A large fire was recorded in 1960 after a large earthquake that affected the Chilean Lake District.

Further south along the coastal range, a higher fire frequency has been associated with episodic droughts (Molina et al. 2006). In the summer of 1987-88, 11000 ha of native forest, including Alerce, were burned presumably by people (Lara et al. 1999). Ten years later, during the dry summer of 1997-98, 9000 ha were burned, including 2339 ha of Alerce (Neira and Díaz 1998). These fires were associated with years of lower precipitation and warm ENSO episodes in central Chile (Lara et al. 2003). Dendrochronological records from 345 trees (Wolodarsky et al. 2005) suggest that human-set fires increased in frequency following a 1976 Chilean decree that banned the commercial harvest of live, but not dead Alerce. In the coastal range of Chiloé, southern limit of the distribution of *Fitzroya* along the coast, Urrutia (2002) dated fire scars of trees, recording large fires in 1111, 1454, 1664, and 1868. A fire burned several commercially logged stands around 1930.

In conclusion, large forest fires repeatedly affected the coastal populations of Alerce. Most of these fires have been associated with human activities, such as logging and clearing of grazing grounds, and most of them also respond to dry periods when burning can be more severe and extensive. Although the frequency of fire in Alerce forests in pre-Columbian times remains unknown, we can assume that it was lower although not entirely absent. Lightning fires do occur in Alerce stands (Armesto et al. 1995) and although rare, the long life span of Alerce (>3000 years) increases the chances of large catastrophic fires during dry episodes coupled to fuel buildup. The thick, fire-resistant bark of Alerce could be a response to a pre-historic fire disturbance regime.

Potential for recovery of burned forest stands

Abundant regeneration of *Fitzroya* has been documented within recently burned forest stands in the coastal range of Chiloé and the Chilean Lake District (Veblen and Ashton 1982, Smith-Ramírez 2007, Páez and Armesto, unpublished data, Smith-Ramírez et al., unpublished data). Smith-Ramírez (2007) showed that, in contrast to open, burned stands, old-growth Alerce forests, not recently affected by fire, in coastal mountains have little or no regeneration of *Fitzroya*. Similar lack of regeneration has been recorded in undisturbed Andean stands. In contrast, regeneration of *Fitzroya* in 17 stands burned in the past 100 years or less showed high sapling densities (>10 thousand trees per hectare). Smith-Ramírez (2007) argues that the occurrence of fire seems to have a stimulating effect on Alerce regeneration. This effect seems to be greater than the negative impact of logging. In logged stands regeneration varied from none to abundant (Smith-Ramírez 2007).

In contrast to the coastal range, Lara (1991) found that logging and fire had negative impacts on Alerce regeneration in two areas of Andean forests. In these disturbed areas, Alerce was replaced by fast-growing angiosperms, such as *Nothofagus betuloides, N. nitida, Drimys winteri*, and other broad-leaved trees (Lara 1991).

A few stands remain in the lowlands of the Chilean Lake District (39 to 40° S). Silla et al. (2002) found that 63 percent of saplings (<50 cm tall) in these stands originated from vegetative growth. Regeneration from seed varied between 1 and 99 percent depending on the site. This result shows that *Fitzroya* is able to resprout after fire disturbance.

Restoration of Alerce forests

From the results discussed above it seems likely that Alerce stands affected by fire may be able to regenerate passively from seed and vegetative growth in both coastal and lowland sites. The major threat to these regenerating stands is the high frequency of anthropogenic fire, because of the long time it takes for the saplings to reach maturity. From the historical data, it seems that fire disturbance was not an unusual event for *Fitzroya* forests, much as in the case of *Araucaria* forests in the Andes. In both cases, regeneration could take place as long as fire frequency remained low. With growing human presence over the past two centuries, including logging and fire, recurrent fires may limit the regeneration of *Fitzroya*. This effect is more negative if we consider *Fitzroya*'s low seed viability and variable seed production (Donoso 1993b). Successful plantations of Alerce, propagated from cuttings, in the lowlands of the Lake District (Lara et al., unpublished data) are indicative of the ability of this long-lived conifer to be used in reclamation programs. The salvage of dead logs and snags, which is the only legal form of harvesting Alerce timber in Chile, can be strongly harmful to the regeneration process by eliminating substrates that facilitate tree recruitment and removing massive quantities of nutrients from the ecosystem.

Human-induced Fire and Restoration of *Pilgerodendron* Forests

Pilgerodendron uviferum (D. Don) Florin (Cupressaceae) (locally known as Cipres de las Guaitecas) is a long-lived conifer, endemic to southern Chile and adjacent areas of Argentina. This conifer extends its range over 1600 km (39° 36′ to 54° 20′ S) reaching the southern tip of South America (Moore 1983), and occupying elevations from sea-level to 1200 m. *Pilgerodendron* occurs typically in wet, poorly drained sites with high annual rainfall (>2500 mm), often found in association with *Sphagnum* bogs, where it forms pure stands (Cruz and Lara 1981). On favorable sites the species can reach diameters of up to 1.1 m, heights >20 m and ages >500 years (Szeicz et al. 2000). Regeneration is largely from seed, but it may also resprout (Cruz and Lara 1981).

Palynological studies in Chiloé Island (42° 30′ S), southern Chile, show that stands of *Pilgerodendron* were a patchy component of lowland forests in late glacial and early post-glacial times (Villagrán 1991). In this region, *Pilgerodendron* declined in abundance later in the Holocene as cold-sensitive Valdivian rainforest taxa expanded their range to higher latitudes (Villagrán 1991). Although the relationship between the past distribution of *P. uviferum* and that of other evergreen rainforest taxa is unknown, it is likely that *Pilgerodendron* populations became fragmented and restricted to sites where evergreen broad-leaved tree species were less successful, such as poorly-drained soils in depressions (Armesto et al. 1995).

Exploitation of *Pilgerodendron*, such as that of Alerce, began during the pre-colonial period, but increased greatly after European settlement in the late 19th century (Cruz and Lara 1981). Due to the quality of its timber (Roig and Boninsegna 1991), populations were rapidly decimated, leading to the present mosaic of forest fragments and pastures observed in northern Chiloé Island (Willson and Armesto 1996). Local people frequently used fire to gain access to logging areas, as *Pilgerodendron* bark is highly resistant to fire. Historical records document numerous and extensive fires in the 1940's and 50's which affected thousands of hectares of *Pilgerodendron* forests. Since 1973, this conifer is listed on Appendix I of CITES (Convention on International Trade in Endangered Species) both in Chile and Argentina. In addition, *Pilgerodendron* has been classified as range restricted or scattered in Argentina and vulnerable according to the IUCN categories of threat in Chile (Farjon and Page 1999).

Restoration strategies and field assays

It is now widely appreciated that understanding patterns of genetic variation is of critical importance to the conservation and restoration of threatened taxa (Ennos 1998). Using random amplified polymorphic DNA (RAPD) markers, Allnutt et al. (2003) showed that *Pilgerodendron* populations differed greatly in genetic diversity, with some post-fire remnant populations characterized by very low genetic variability. In addition, a high degree of genetic differentiation was recorded among populations (Allnut et al. 2003). The high degree of genetic differentiation recorded for *Pilgerodendron* suggests that transfer of germplasm between populations should be avoided, to ensure that the genetic material is adapted to local site conditions (Ennos 1998).

Restoration of fire-degraded and declining populations of *Pilgerodendron* is not only an urgent need, but it also represents an opportunity for the recovery of products and services that have been highly degraded by human impact. It is widely appreciated that fragmentation and isolation of populations can result in reduced genetic variation, which in turn may reduce reproduction or survival and thereby population viability (Sherwin and Moritz 2000).

The landscape of northern Chiloé Island has shifted from nearly all forested to a large fraction of land covered by pastures and secondary shrublands, where tree regeneration is limited (see below). Most shrublands originated from logging or burning of forests that included *Pilgerodendron* among other evergreen trees. In this context, a study of genetic diversity of *Pilgerodendron* and a subsequent restoration experiment was conducted in northern Chiloé Island, in Senda Darwin Biological Station (SDBS, 42° S), 20 km north of Ancud. The population and genetic structure of *Pilgerodendron uviferum* was studied in five degraded, post-fire, lowland stands (35 to 65 years old), located in the vicinity of SDBS, to assess their conservation status and potential for recovery. Genetic diversity assessment (Gabriel et al., unpublished data) revealed that stands were largely composed of genetically distinct individuals, indicating that resprouting was not a prevalent response to fire. These stands were used as source populations for restoration.

We evaluated the rehabilitation potential of *Pilgerodendron* in wet secondary shrublands, particularly testing for the effects of *Sphagnum* substrate on tree growth and survival. In Senda Darwin Biological Station, we set up four plots in each of two secondary shrubland areas, one with >60 percent *Sphagnum* cover and one without *Sphagnum* (<10 percent cover). *Pilgerodendron* saplings were planted every 4 m within 28 × 28 m plots (n = 49 trees per plot). Plants were obtained from cuttings collected from three remnant populations as described above, rooted and maintained in a greenhouse for two years before planting. Four years after planting (2005), a repeated measures ANOVA showed significant effects of substrate ($p<0.001$) and population of origin ($p<0.001$) on tree growth. We conclude that the presence of *Sphagnum* significantly slowed the growth of *Pilgerodendron* in secondary shrublands and that the genetic origin of plants used in assays is a critical factor for successful rehabilitation.

Both demographic and genetic evidence support the hypothesis that *Pilgerodendron* is able to recover from fire disturbance. Possibly this capacity is a legacy from the long history of population reduction and expansion during Glacial periods in southern Chile (Villagrán 1991). *Pilgerodendron* has a demonstrated ability to grow in the close vicinity of ice sheets and moraine fields in southern Chile (Szeicz et al. 2000) where disturbances are recurrent. Thus, it seems reasonable to believe that fire control programs, to remove this threat from degraded areas of northern Chiloé, will allow this slow-growing conifer to reproduce and persist. As its current restriction to marginal habitats where soil conditions are strongly unfavorable for tree growth, ecological restoration of *Pilgerodendron* should not just be limited to areas where remnant stands are found today. Restoration programs are likely to be more successful in open areas of low agricultural output (e.g., waterlogged, nutrient-poor, and/or shallow soils), which are widespread in northern Chiloé Island. These areas frequently remain unused by people due to the high costs of drainage and fertilization, and their reclamation with *Pilgerodendron*, obtained from local seed sources or cuttings, may add value to these currently unproductive lands.

Fire in Evergreen Forest Types

Fire disturbance and logging of forests in south-central Chile greatly increased during the 20th century (Lara et al. 1996, Aravena et al. 2002). These changes may have triggered switches from evergreen forests to waterlogged, open shrublands in poorly drained lowland sites. These humid sites become later invaded by *Sphagnum* moss, which creates conditions unfavorable for re-colonization of trees.

In Chiloé Island, Chile (42° S), tree cutting for timber and human set fires to clear land for livestock and farming have transformed the originally continuous forest cover into scattered forest fragments surrounded by extensive zones of bushes and anthropogenic meadows. Shrublands and meadows often exhibit seasonal water-logging (Papic, 2000). Compared with forest fragments, the water content of soils is always greater in adjacent secondary shrublands, even during summer months (Díaz et al. 2007). The presence of remnant snags and logs, showing fire scars and charcoal, in these open shrublands indicates that they were originally forested, but have remained treeless for at least 50 to 60 years. Changes in vegetation cover can have profound effects on the hydrology of an ecosystem because the water balance of a site is strongly dependent on two main factors, canopy cover and soil drainage. Canopy removal by logging or burning stands alters the hydrological cycle and may raise the water table in degraded woodlands that receive high annual rainfall. Several studies have shown an elevated water table level following deforestation (Ruprecht and Schofield 1991, Roy 1998).

Rain can follow three pathways upon entering a forested ecosystem: a) evaporation from foliage and soil surfaces to the atmosphere (evapotranspiration, ET), b) infiltration to deeper soil layers, including lateral transfer across geological strata, and c) runoff and stream flow. During the growing season, ET is an important regulator of hydrological fluxes of forested basins, including stream and river flow (Bormann and Likens 1979, Likens and Bormann 1986, Aber and Federer 1992, Larcher 1995, Jarvis et al. 1997, Kimball et al. 1997, White et al. 1998). As chemical flows are intimately linked to the hydrologic cycle, disturbances that alter the hydrologic flows can have consequences for nutrient budgets of terrestrial and aquatic ecosystems (Likens et al. 1978, Godoy et al. 1999, Jackson et al. 2001, Salmon et al. 2001, Perakis and Hedin 2002). Changes in ET that occur after disturbances altering the cover of a basin (Swift et al. 1975, Huber et al. 1985, Whitehead and Kelliher 1991) may affect local hydrology including overland and deep water flows. In Chiloé forest, we documented that 53 percent of the incoming water returns to the atmosphere via ET (Díaz et al. 2007). More than half of this value represents interception or direct evaporation from the wet canopy. Water not intercepted by the canopy reaches the soil as throughfall or stemflow. Both of these components combined account for up to 67 percent of total precipitation (Díaz et al. 2007).

After a fire, trees are replaced by shrubs and ET decreases to just nine percent of the incoming rainfall, mainly because of a marked reduction in foliage interception surfaces and canopy transpiration (Díaz et al. 2007). Consequently, most of the rain falls directly on the soil and net precipitation (water that reaches the soil) increases to 91 percent. This situation drastically changes soil infiltration and the depth of the water table in disturbed forest sites. Burned sites dominated by shrubs in Chiloé have a shallower water table than forests (Díaz et al. 2007). Over the entire year, the average difference in water table depth between shrublands and adjacent forest was 35 cm. In shrublands the water table reached the surface during winter months.

Consequences for ecosystem recovery and restoration

Hydrologic changes associated with anthropogenic fire and/or logging of forests have spread rapidly through the rural landscape of southern Chile, especially during the past decades (Willson and Armesto 1996, Armesto et al.

2001), with broad consequences for successional processes, tree regeneration, and restoration practices. Recent studies (Papic 2000, Díaz and Armesto 2007) postulate an interruption of successional processes because of water-logging of soils. This alteration of the water table level would be difficult to reverse, at least for a long period until drainage of the sites changes and canopy cover recovers to its pre-disturbance condition. Tree invasion of these post-fire shrublands is severely limited (Díaz and Armesto 2007), presumably because seedlings of most pioneer tree species cannot colonize seasonally water-logged sites. If this is the case, forest regeneration is primarily limited by hydrology following the loss of forest cover, leading to dominance by shrubs and mosses. Tree invasion of shrublands may also be delayed by low seed inputs and lack of woody detritus that provide sites for seed germination and establishment (R. Jaña, unpublished data). Accordingly, sites disturbed by fire and logging suffer a reversion of succession towards a stage dominated by *Sphagnum* mosses that retard tree establishment. Such conditions may be analogous to those of the early Holocene, following deglaciation in lowland areas, when evergreen trees colonized poorly drained, fluvio-glacial deposits and *Sphagnum*-dominated moorlands (Heusser 1984, Villagrán 1988, 1991).

Díaz and Armesto (2007) proposed the semi-deciduous pioneer tree *Embothrium coccineum* (Proteaceae) as a suitable candidate for restoration of tree cover in waterlogged sites, because this species is capable of becoming established directly over *Sphagnum* cushions, thereby avoiding the effects of the elevated water table. *Embothrium* can grow rapidly on top of *Sphagnum* cushions, and slowly contribute to change the hydrologic balance of the site. In addition, successional processes in these wet, post-fire shrublands could be enhanced by the maintenance of woody residues that may serve as establishment sites for trees (Papic 2000). As many tree species of evergreen rain forests regenerate on downed logs (Christie and Armesto 2003), the presence of remnant logs at the disturbed site provides elevated substrates for tree establishment, above the water table. Survival of planted seedlings was greater on logs than on soil in a wet secondary shrubland (Papic 2000). We recently initiated a field study to assess the ability of native trees to invade a seasonally water-logged meadow providing artificial woody substrates. To enhance seed input, it may be necessary to add artificial perches to the experimental design, because meadows receive limited seed deposits from avian-dispersed, fleshy-fruited species, which dominate the temperate forest flora (Jaña et al., manuscript in preparation). Once broad-leaved, evergreen trees become established, the normal hydrologic cycle of a forest should slowly be recovered, and succession could lead to forest with a deeper table.

Bamboo Cycle and Fire Disturbance in Chilean Temperate Forests

Bamboos are important components of the forest understory in many tropical (Widmer 1997, Gratzer et al. 1999, Marod et al. 1999, Tabarelli and Mantovani 2000, Banana and Tweheyo 2001, Franklin and Bowman 2003) and temperate forests (Taylor et al. 1995, Abe 2002, Narukawa and Yamamoto 2002, Goto 2004). In temperate forests of southern South America, several bamboo species in the genus *Chusquea* have been characterized as keystone species because their presence inhibits tree regeneration (Veblen 1979, Veblen et al. 1981, Veblen 1982b, Armesto and Figueroa 1987, Donoso 1993a) and their life-cycles influence fire disturbance regimes (Veblen et al. 2003).

The rapid growth rates, large sizes, and high culm (stem) densities are salient features that allow *Chusquea* species to dominate the forest understory from sea level to the alpine tree line at ca. 1,200 m in south-central Chile (Veblen 1982b, Veblen et al. 1983, Donoso 1993a). Bamboo cover is especially abundant within tree-fall gaps and at sites recently disturbed by fire or logging (González et al. 2002). Although most *Chusquea* species are highly shade-intolerant, bamboos create shaded conditions that affect tree regeneration, in particular that of the relatively shade-intolerant *Nothofagus* species (Holz and Veblen 2006). Given the rarity of bamboo flowering events, there is little understanding of the relation between synchronous bamboo flowering and die-off on tree regeneration dynamics in south-central Chile. Recent studies document that a window of opportunity is created after synchronous flowering and die-off of bamboo, which releases resources for the growth and establishment of tree seedlings and saplings (González et al. 2002, Holz and Veblen 2006).

In addition to its influence on tree regeneration dynamics in southern South American forests, the life cycle of bamboo affects fire regimes in temperate rain forests by modifying fuel quality and accumulation. Even during the vegetative phase *Chusquea* species are highly flammable (González et al. 2005). Due to their tall culms, bamboos can drive the fire from the understory to the forest canopy (i.e., *C. culeou*; Veblen et al. 1992a) and, moreover, bamboos resprout vigorously after fires. However, in the presence of livestock, the fuel loads and culm heights of *C. culeou*-dominated understories are greatly reduced (Veblen et al. 1992b, Relva and Veblen 1998).

Synchronous die-off of *Chusquea* increases the availability of dry litter and withering culms in the understory, which in turn enhance forest flammability for ca. four to five years after bamboo flowering (Veblen et al. 2003), even in the humid understories of mesic Andean forests. Throughout the last century, widespread fires resulted from the combination of warm and dry climatic conditions and bamboo flowering events (Veblen et al. 2003, González et al. 2005). From 1890 to 1915, it is believed that a *Chusquea* species—most likely *C. culeou*—flowered massively in the mesic Argentinean and Chilean forests (San Martin 1940, Tortorelli 1947, González 1986), coinciding with some extremely dry years (Villalba 1990, Villalba et al. 1997, Villalba et al. 1998), and resulting in a peak of forest burning by European settlers (Tortorelli 1947, Veblen et al. 2003). During a two year period (1940 to 1942), *C. culeou* also flowered massively in the Chilean and Argentinean Andes (González Cangas 1998, Veblen et al. 2003), coinciding with a severe summer drought (Veblen et al.

2003) that resulted in extensive intentional burning during 1943 and 1944 (Tortorelli 1947).

However, widespread fires that coincided with extreme droughts also occurred during periods when *Chusquea* species were believed to be in their vegetative phase in the region (e.g., 1997 to 1998). Additionally, not every bamboo flowering event has resulted in widespread fires. Thus, five to six years after massive flowering of *C. montana* in Andean sub-alpine forests (Holz and Veblen 2006) and of *C. culeou* in montane forests, fire has not been recorded. These observations suggest that even though leaf litter and culms are extremely flammable and favor the spread of understory fires, ignition sources and, most importantly, regional climatic conditions need to be conducive for widespread fires to occur.

One of the most fascinating and frequently discussed aspects of bamboo ecology is their life cycle. Many—but not all—bamboo species are monocarpic or semelparous plants which means that they flower once in their life and subsequently seed and die off. The synchronous flowering, seeding, and subsequent death of large populations of bamboo can extend over thousands of hectares of forest understory, and may occur at decadal or longer intervals depending on the species and population (Gadgil and Prasad 1984, Nelson 1994, Makita et al. 1995, Franklin 2004). In the case of southern South American forests, some *Chusquea* species flower synchronously over two to three years at intervals from about 17 to over 70 years and die massively thereafter over areas of hundreds-to-thousands of square kilometers (Veblen 1982, Pearson et al. 1994, González and Donoso 1999, González et al. 2002, Holz and Veblen 2006). For instance, it is believed that *C. culeou* and *C. quila* flower in cycles of about 60 to 70 years (González et al. 2002).

Several hypotheses have been proposed to explain this peculiar flowering behavior, including internal genetic traits in response to seed predators (Janzen 1976), physiological factors (Gielis et al. 1997, Nadgauda et al. 1997), climatic fluctuations such as ENSO (Jaksic and Lima 2003), and fire recurrence (Keeley and Bond 1999, Keeley and Bond 2001). Given that it is difficult to determine the exact flowering cycle of most bamboo species due to its low frequency and that most historically documented evidence has been hearsay, each of the above hypotheses have weaknesses. For instance, the fire hypothesis suggests that lightning-ignited wildfire has led to the synchronized mast flowering and recruitment of bamboo (Keeley and Bond 1999). However, it is not clear how bamboo seeds on the ground or recently established bamboo seedlings would have a higher chance to succeed in the presence of a fire. A favored hypothesis explains the flowering cycle as controlled by internal genetic traits, rather than by external climatic signals, and occurring as a response to seed predators (Janzen 1976). This hypothesis assumes that predators will be quickly satiated by an unexpected abundance of seeds, resulting in a time lag between the peak of bamboo seed production and the increase of predator population, which in turn results in successful germination of the majority of seeds before the predator population reaches its peak. However, it has been argued that the selective influence of seed predators on bamboo flowering is unlikely due to the differences in the life spans of bamboo species and predators (Pearson et al. 1994).

Control of bamboo fires

Although we are not aware of past or ongoing restoration projects in temperate bamboo forests in the southern Andes, past chronicles (Rothkugel 1916) and current research (see review by Veblen et al. 2003) provide valuable information that can be used to guide conservation or restoration efforts. Examples from northern Patagonia on the eastern slopes of the Andes illustrate what could happen to forested ecosystems when altered by humans beyond their historical range of variability. As on the western slopes of the southern Andes, settlers on the eastern slopes intentionally burned large areas of the landscape from the mid and late 1800s to the early 1900s. However, on the eastern side, repeated fires gradually transformed mesic *Nothofagus dombeyi*-dominated forests into secondary shrub/bamboo-dominated communities (Kitzberger and Veblen 1999, Veblen et al. 1999, Mermoz et al. 2005). Moreover, these fire-induced changes promoted positive feedbacks favoring the dominance of fast-growing species with vigorous post-fire sprouting, at the expense of slower growing, seeding species (Kitzberger and Veblen 1999, Veblen et al. 1999). Following these massive fire disturbance events, a combination of reduced human-set fires, increasing pressure from fuel-reducing livestock, and promotion of restoration via the suppression of fires derived from any ignition source (i.e., natural or human-related), has resulted in an abnormal decrease in fire frequency. This, in turn, has resulted in a tendency for tree-dominated communities to increase to historically peak levels at the expense of grasslands and shrublands (Veblen and Lorenz 1988, Kitzberger and Veblen 1999).

Today, *Austrocedrus chilensis* forests at the steppe-forest ecotone and sub-mesic forest patches at higher elevation are denser, more continuous, more flammable, and more susceptible to high severity fires (Veblen et al. 1992a). Tree invasion of the fire-prone *Austrocedrus* steppe has led to greater fuel buildup, which is altering fire regimes. Sites that historically supported only surface fires are now susceptible to larger stand-replacing fires (Veblen et al. 1992a, Kitzberger and Veblen 1999). Although this evidence should be interpreted with caution, when extrapolated to wetter forests on the Chilean side of Andes, it illustrates the unexpected ecosystem consequences of both the control of human-set fires and restoration practices. Thus, alteration of the historical range of variability of ecosystems may develop into unwanted results.

Synthesis of Recommendations

Different forest types in the temperate rain forest region of southern South America have contrasting histories of fire in pre-colonial times, but our review of the literature indicates that fire disturbance has historically been a significant component of disturbance regimes in most forests and certainly at present influencing the structure and dynamics of forests and landscapes within the region. Consequently, the examination of factors determining fire susceptibility, recurrence, and recovery in various types of native forests is of major importance for elaborating restoration plans. The conifers *Araucaria*, *Fitzroya*, *Pilgerodendron*, and *Austrocedrus*, because of their longevity, rates of fuel buildup, and chances of natural ignition by lightening, have been subjected to a long history of fire that predates the arrival of Native Americans and Europeans. They all have vegetative and reproductive traits that make them resistant to fire and, to different extents, capable of resprouting after a moderate severity fire. In the case of conifer-dominated forests, passive restoration based on seed sources remaining at the burned site or in its vicinity seems to be a viable strategy for ecosystem recovery within reasonable time frames. Nevertheless, salvage logging may be quite harmful to the regeneration potential of all these forests. This problem is especially serious in the case of *Fitzroya*, because of legal extraction of high volumes of dead timber from burned sites.

Changes in disturbance regimes, as a consequence of human intervention to either prevent or set fires in these temperate forests may have serious consequences for ecosystem dynamics, such as greater fuel buildup and greater recurrence of large-scale fire. Accordingly, better understanding of the differences between past and present fire regimes is necessary in most forest types. In addition, the role of bamboo cover in maintaining or altering fire regimes in rain forests is still poorly understood. The increase in bamboo cover due to logging and fire may be a positive feedback on fire return times in secondary forests, thus changing their structure and composition. The maintenance of the old-growth condition in forests seems jeopardized by increased fire frequency. Finally, the hydrologic balance of some evergreen forests on poorly drained glacio-fluvial deposits may be drastically altered by fire, returning the land to a postglacial successional stage, where flooding and *Sphagnum* invasion delay the process of colonization by tree species. We are conducting field assays to enhance tree invasion in seasonally waterlogged meadows, by providing woody substrates for tree establishment and perches to facilitate the input of avian-disseminated seed.

Acknowledgements

A part of this work was funded by Fondecyt grant 1501-0001 (to CASEB), by the Millennium Scientific Initiative, and by Reforlan (International Cooperation Program of the European Union-Framework 6). M.E.G. thanks International Foundation for Science (IFS D/3124-2) and Dirección de Investigación de la Universidad Austral de Chile (DID- S-2005-08).

References

Aber J.D. and C.A. Federer. 1992. A generalized, lumped-parameter model of photosynthesis, evapotranspiration and net primary production in temperate and boreal forest ecosystems. *Oecologia,* 92: 463-474.

Abe, M., J. Izaki, H. Miguchi, T. Masaki, A. Makita, and T. Nakashizuka. 2002. The effects of *Sasa* and canopy gap formation on tree regeneration in an old beech forest. *Journal of Vegetation Science,* 13: 565-574.

Aguiar M.R. and O.E. Sala. 1994. Competition, facilitation, seed distribution and the origin of patches in a Patagonian steppe. *Oikos,* 70: 26-34.

Alaback, P.B. 1991. Comparative ecology of temperate rainforests of the Americas along analogous climatic gradients. *Revista Chilena de Historia Natural,* 64: 399-412.

Allnutt, T.R., A.C. Newton, A. Premoli, and A. Lara. 2003. Genetic variation in the threatened South American conifer *Pilgerodendron uviferum* (Cupressaceae), detected using RAPD markers. *Biological Conservation,* 114: 245-253.

Aravena J.C., M.R. Carmona, C.A. Pérez, and J.J. Armesto. 2002. Changes in tree species richness, stand structure and soil properties in a successional chronosequence of forest fragments in northern Chiloé Island, Chile. *Revista Chilena de Historia Natural,* 75: 339-360.

Armesto, J.J. and J.R. Gutiérrez. 1978. El efecto del fuego en la estructura de la vegetacion de Chile central. *Anales del Museo de Historia Natural,* 11: 43-48.

Armesto, J.J. and J. Figueroa. 1987. Stand structure and dynamics in the temperate rain-forests of Chiloé Archipelago, Chile. *Journal of Biogeography,* 14: 367-376.

Armesto, J.J. and S.T.A. Pickett. 1985. A mechanistic approach to the study of succession in the Chilean matorral. *Revista Chilena de Historia Natural,* 58: 9-17.

Armesto, J.J., C. Villagrán, J.C. Aravena, C. Pérez, C. Smith-Ramírez M. Cortés, and L.O. Hedin. 1995. Conifer forests of the Chilean coastal range. pp. 156-170. *In* R.S. Hill and N. Enright [eds.]. *Ecology of the Southern Conifers.* Melbourne University Press, Melbourne, Australia.

Armesto, J.J., C. Papic, and P. Pliscoff. 2002. Relevancia de las pequeñas áreas silvestres para la conservación de la biodiversidad en el bosque nativo. *Ambiente & Desarrollo,* 18(1): 60-70.

Armesto, J.J., R. Rozzi, and J. Caspersen. 2001. Temperate forests of North and South America. pp. 223-249. *In* F.S. Chapin, O.E. Sala, and E. Huber-Sannwald [eds.]. *Global Biodiversity in a Changing Environment. Scenarios for the 21st Century* Springer, New York, New York, USA.

Armesto, J.J., P.E. Vidiella, and H.E. Jiménez. 1995. Evaluating causes and mechanisms of succession in the Mediterranean regions of Chile and California. pp.418-434. *In* M.T.K. Arroyo, P.H. Zedler, and M.D. Fox [eds.]. *Ecology and Biogeography of Mediterranean Ecosystems in Chile, California, and Australia.* Springer-Verlag, New York, New York, USA.

Armesto, J.J., S. Bautista, E.D. Val, B. Ferguson, X. García, A. Gaxiola, H. Godinez-Álvarez, G. Gann, F. López-Barrera, R. Manson, M. Núñez-Ávila, C. Ortiz-Arrona, P. Tognetti, and G. Williams-Linera. 2007. Towards an ecological restoration network: reversing land degradation in Latin America. *Frontiers in Ecology and the Environment,* 5(4): W1-W4. *Online at* http://www.esajournals.org/doi/full/10.1890/1540-9295%282007%295%5Bw1%3ATAERNR%5D2.0.CO%3B2 [accessed 23 Sep 2008].

Arroyo, M.T.K. and P. Uslar. 1993. Breeding systems in a temperate mediterranean-type climate montane sclerophyllous forest in central Chile. *Botanical Journal of the Linnean Society,* 111: 83-102.

Austin, M.P. and O. Williams. 1988. Influence of climate and community composition on the population demography of pasture species in semi-arid Australia. *Vegetatio,* 7: 43-49.

Banana, A.Y. and M. Tweheyo. 2001. The ecological changes of *Echuya afromontane* bamboo forest, Uganda. *African Journal of Ecology,* 39: 366-373.

Bertness, M.D. and R.M. Callaway. 1994. Positive interactions in communities. *Trends in Ecology and Evolution,* 9: 191-193.

Bormann, F.H. and G.E. Likens. 1979. *Pattern and Process in a Forested Ecosystem.* Springer-Verlag, New York, New York, USA.

Brooker, R.W. and T.V. Callaghan. 1998. The balance beween positive and negative interactions and its relationship to environmental gradients: a model. *Oikos,* 81: 196-207.

Brown, N.A.C. 1993. Promotion of germination of fynbos seeds by plant-derived smoke. *New Phytologist,* 123: 575-584.

Burns, B. 1993. Fire-induced dynamics of *Araucaria araucana–Nothofagus antarctica* forest in the southern Andes. *Journal of Biogeography,* 20: 669-685.

Callaway, R.M. and L.R. Walker. 1997. Competition and facilitation: a synthetic approach to interactions in plant communities. *Ecology,* 78: 1958-1965.

Christie, D.A. and J.J. Armesto. 2003. Regeneration microsites and tree species coexistence in temperate rain forests of Chiloé Island, Chile. *Journal of Ecology,* 91: 776-784.

CONAF-CONAMA-BIRF. 1999. Catastro y evaluación de recursos vegetacionales nativos de Chile. CONAF-CONAMA, Santiago, Chile.

Cowling, R.M., P.W. Rundel, B.B. Lamont, M.K. Arroyo, and M. Arianoutsou. 1996. Plant diversity in Mediterranean-climate regions. *Trends in Ecology and Evolution,* 11: 363-366.

Cruz, G. and A. Lara. 1981. Tipificación, cambio de estructura y normas de manejo para Ciprés de las Guaytecas (*Pilgerodendron uvifera* (D. Don) Florin) en la Isla Grande de Chiloé. Thesis, Universidad de Chile, Santiago, Chile.

Debussche, M. and P. Isenmann. 1994. Bird-dispersed seed rain and seedling establishment in patchy mediterranean vegetation, *Oikos,* 69: 414-426.

Donoso, C. 1993a. Bosques templados de Chile y Argentina. Variación, Estructura y Dinámica. Editorial Universitaria S.A., Santiago, Chile.

Donoso, C. 1993b. Producción de semillas y hojarasca de las especies del tipo forestal alerce (*Fitzroya cupressoides*) de la Cordillera de la Costa de Valdivia, Chile. *Revista Chilena de Historia Natural,* 66: 53-64.

Díaz, I.A., J.J.Armesto, S. Reid, K.E. Sieving, and M.F. Willson. 2005. Linking forest structure and composition: avian diversity in successional forests of Chiloé Island, Chile. *Biological Conservation,* 123: 91-101.

Díaz M.F. and J.J. Armesto. 2007. Limitantes físicos y bióticos de la regeneración arbórea en matorrales sucesionales de la Isla Grande de Chiloé, Chile. *Revista Chilena de Historia Natural,* 80: 13-26.

Díaz, M.F., S. Bigelow, and J.J. Armesto. 2007. Alteration of the hydrologic cycle due to forest clearing and its consequences for rainforest succession. *Forest Ecology and Management,* 244: 32-40.

Dillehay, T. 1997. *Monte Verde, a Late Pleistocene settlement in Chile. Vol. 2: The Archaeological Context and Interpretation*. Smithsonian Institution Press, Washington, D.C., USA.

Ennos, R.A. 1998. Genetic constraints on native woodland restoration. pp.27-34. *In* A.C. Newton and P. Ashmole [eds.]. *Native Woodland Restoration in Southern Scotland: Principles and Practice*. Borders Forest Trust, Ancrum, Jedburgh, UK.

Farjon, A. and C.N. Page [Compilers]. 1999. *Conifers—Status Survey and Conservation Action Plan*. IUCN/SSC Conifer Specialist Group Report, The IUCN Publication Services Unit, Cambridge, UK.

Franklin, D.C. 2004. Synchrony and asynchrony: observations and hypotheses for the flowering wave in a long-lived semelparous bamboo. *Journal of Biogeography*, 31: 773-786.

Franklin, D.C. and D. Bowman. 2003. Bamboo, fire and flood: regeneration of *Bambusa arnhemica* (Bambuseae : Poaceae) after mass-flowering and die-off at contrasting sites in monsoonal northern Australia. *Australian Journal of Botany*, 51: 529-542.

Fraver, S., M.E. González, F. Silla, and A. Lara. 1999. Composition and structure of remnant *Fitzroya cuppressoides* forests of southern Chile´s Central Depression. *Journal of the Torrey Botanical Society*, 126: 49-57.

Fuentes, E.R., R.D. Otaiza, M.C. Alliende, A.E. Hoffmann, and A. Poiani. 1984. Shrub clumps of the Chilean matorral vegetation: Structure and possible maintenance mechanisms. *Oecologia*, 64: 405-411.

Fuentes, E.R., A.J. Hoffmann, A. Poiani, and M.C. Alliende. 1986. Vegetation change in large clearings: patterns in the Chilean matorral. *Oecologia*, 68: 358-366.

Fuentes, E.R. and M. Muñoz. 1995. The human role in changing landscapes in central Chile: Implications for intercontinental comparisons. pp.401-417. *In* M.T.K. Arroyo, P.H. Zedler, and M.D. Fox [eds.]. *Ecology and Biogeography of Mediterranean Ecosystems in Chile, California, and Australia*. Springer-Verlag, New York, New York, USA.

Gadgil, M. and S.N. Prasad. 1984. Ecological determinants of life-history evolution of two Indian bamboo species. *Biotropica*, 16: 161-172.

Gielis, J., P. Goetghebeur, and P. Debergh. 1997. Morphological and biochemical aspects of flowering in bamboos – the development of model systems. pp.179-186. *In* G.P. Chapman [ed.]. *The Bamboos*. Academic Press, London, UK.

Godoy R., C. Oyarzún, and J. Bahamondes. 1999. Flujos hidroquímicos en un bosque de *Nothofagus pumilio* en el Parque Nacional Puyehue, sur de Chile. *Revista Chilena de Historia Natural*, 72: 579-594.

González Cangas, Y. 1998. Memoria histórica y saber cotidiano: validación del conocimiento en el florecimiento de la *Chusquea quila* en el sur de Chile (X Región). Thesis. Facultad de Educación y Humanidades, Universidad de la Frontera, Temuco, Chile.

González, J.G. 1986. Villa Rica: historia inédita. Municipalidad de Villarica, Villarica, Chile.

González, M.E. 2002. *Araucaria-Nothofagus* forests in south-central Chile. Ph.D. dissertation. Department of Geography, University of Colorado, Boulder, Colorado, USA.

González, M.E. 2005. Fire history data as reference information in ecological restoration. *Dendrochronologia*, 22: 149-154.

González, M.E. and C. Donoso. 1999. Seed and litter fall in Chusquea quila (Poaceae: Bambusoideae), after synchronous flowering in south-central Chile. *Revista Chilena de Historia Natural,* 72: 169-180.

González, M.E. and T.T. Veblen. 2006. Climatic influences on fire in *Araucaria araucana-Nothofagus* forests in the Andean cordillera of south-central Chile. *Ecoscience,* 13: 342-350.

González, M.E. and T.T. Veblen. 2007. Wildfire in *Araucaria araucana* forests and ecological considerations about salvage logging in areas recently burned. *Revista Chilena de Historia Natural,* 80: 243-253.

González, M.E., T.T. Veblen, C. Donoso, and L. Valeria. 2002. Tree regeneration responses in a lowland *Nothofagus*-dominated forest after bamboo dieback in South-Central Chile. *Plant Ecology,* 161: 59-73.

González, M.E., T.T. Veblen, and J.S. Sibold. 2005. Fire history of *Araucaria–Nothofagus* forests in Villarrica National Park, Chile. *Journal of Biogeography,* 32: 1187-1202.

Goto, Y. 2004. Early post-fire vegetation regeneration in *Larix kaempferi* artificial forests with an undergrowth of *Sasa senanensis*. *Ecological Research,* 19: 311-321.

Gratzer, G., P.B. Rai, and G. Glatzel. 1999. The influence of the bamboo *Yushania microphylla* on regeneration of *Abies densa* in central Bhutan. *Canadian Journal of Forestry Research,* 29: 1518-1527.

Gutiérrez, J.R., P.L. Meserve, S. Herrera, L.C. Contreras, and F.M. Jaksic. 1997. Effects of small mammals and vertebrate predators on vegetation in the Chilean semiarid zone. *Oecologia,* 109: 398-406.

Handel, S.N. 1997. Role of plant-animal mutualisms in the design and restoration of natural communities. pp. 11-132. *In* K.M. Urbanzka, N.R. Webb, and P.J. Edwards [eds.]. *Restoration Ecology and Sustainable Development*. Cambridge University Press, Cambridge, UK.

Hanes, T.L. 1971. Succession after fire in the chaparral of southern California. *Ecological Monographs,* 41: 27-52.

Heusser, C.J. 1966. Late-Pleistocene pollen diagrams from the Province of Llanquihue, Southern Chile. *American Philosophical Society Proceedings,* 110: 269-305.

Heusser, C.J. 1983. Quaternary pollen record from laguna Tagua Tagua, Chile. *Science,* 219: 1429-1432.

Heusser, C.J. 1984. Late-Glacial-Holocene Climate of the Lake District of Chile. *Quaternary Research,* 22: 77-90.

Hoffmann, A.J. and J.J. Armesto. 1995. Modes of seed dispersal in the Mediterranean regions in Chile, California, and Australia. pp. 289-310. *In* M.T.K. Arroyo, P.H. Zedler, and M.D. Fox [eds.]. *Ecology and Biogeography of Mediterranean Ecosystems in Chile, California and Australia*. Springer-Verlag, New York, New York, USA.

Hoffmann, A.J., S. Teillier, and E.R. Fuentes. 1989. Fruit and seed characteristics of woody species in Mediterranean-type regions of Chile and California. *Revista Chilena de Historia Natural,* 62: 43-60.

Holmgren, M., M. Scheffer, and M.A. Huston. 1997. The interplay of facilitation and competition in plant communities. *Ecology,* 78: 1966-1975.

Holmgren, M., A.M. Segura, and E.R. Fuentes. 2000. Limiting mechanisms in the regeneration of the Chilean matorral. *Plant Ecology,* 147: 49-57.

Holmgren, M. and M. Scheffer. 2001. El Niño as a window of opportunity for the restoration of degraded arid ecosystems. *Ecosystems,* 4: 151-159.

Holmgren, M., M.E. Ezcurra, J.R. Gutiérrez, and G.M.J. Mohren. 2001. El Niño effects on the dynamics of terrestrial ecosystems. *Trends in Ecology and Evolution,* 16: 89-84.

Holz, C.A. and T.T. Veblen. 2006. Tree regeneration responses to *Chusquea montana* bamboo die-off in a subalpine *Nothofagus* forest in the southern Andes. *Journal of Vegetation Science,* 17: 19-28.

Huber, A., C. Oyarzún, and A. Ellies. 1985. Balance hídrico en tres plantaciones de Pinus radiata. II: Humedad del suelo y evapotranspiración. *Bosque,* 6: 74-82.

Jackson, R.B., S.R. Carpenter, C.N. Dahm, D.M. McKnight, R.J. Naiman, S.L. Postel, and S.W. Running 2001. Water in a changing world. *Ecological Applications,* 11: 1027-1045.

Jaksic, F.M. and M. Lima. 2003. Myths and facts on ratadas: Bamboo blooms, rainfall peaks and rodent outbreaks in South America. *Australian Ecology,* 28: 237-251.

Janzen, D.A. 1976. Why bamboos wait so long to flower. *Annual Review of Ecology and Systematics,* 7: 347-391.

Jarvis P.G., J.M. Massheder, S.E. Hale, J.B. Moncrieff, M. Rayment, and S.L. Scott. 1997. Seasonal variation of carbon dioxide, water vapor, and energy exchanges of a boreal black spruce forest. *Journal of Geophysical Research,* 102: 28953-28966.

Jasinski, J.P.P. and S. Payette. 2005. The creation of alternative stable states in the southern Boreal forest, Quebec, Canada. *Ecological Applications,* 75: 561-583.

Jiménez, H.E. and J.J Armesto. 1992. Soil seed bank of disturbed sites in Chilean matorral: its importance in early secondary succession. *Journal of Vegetation Science,* 3: 579-586.

Keeley J.E. 1993. Smoke-induced flowering in the fire-lily *Cyrtanthus ventricosus*. *South African Journal of Botany,* 59: 638.

Keeley J.E. 1995. Seed germination patterns in fire-prone Mediterranean-climate regions. pp.239-273. *In* M.T.K. Arroyo, P.H. Zedler, and M.D. Fox [eds.]. *Ecology and Biogeography of Mediterranean Ecosystems in Chile, California and Australia.* Springer-Verlag, New York, New York, USA.

Keeley J.E. [ed.]. 1989. *California Chaparral: Paradigms Reexamined.* Museum of Natural History, Los Angeles, California, USA.

Keeley, J.E. and W.J. Bond. 1999. Mast flowering and semelparity in bamboos: the bamboo fire cycle hypothesis. *American Naturalist,* 154: 383-391.

Keeley, J.E. and W.J. Bond. 2001. On incorporating fire into our thinking about natural ecosystems: a response to Saha and Howe. *American Naturalist,* 158: 664-670.

Kimball, J.S., M.A. White, and S.W. Running. 1997. BIOME-BGC simulations of stand hydrologic processes for BOREAS. *Journal of Geophysical Research,* 102: 29043-29051.

Kitzberger, T. and T.T. Veblen. 1999. Fire-induced changes in northern Patagonian landscapes. *Landscape Ecology,* 14: 1-15.

Kitzberger, T., T.T. Veblen, and R. Villalba. 1997. Climatic influences on fire regimes along a rain forest-to-xeric woodland gradient in northern Patagonia, Argentina. *Journal of Biogeography,* 24: 35-47.

Landres, P.B., P. Morgan, and F.J. Swanson. 1999. Overview of the use of natural variability concepts in managing ecological systems. *Ecological Applications,* 9: 1179-1188.

Lara, A. 1991. The dynamics and disturbance regimes of *Fitzroya cupressoides* forest in South-Central Andes of Chile. Ph.D. Thesis, Department of Geography, University of Colorado, Boulder, Colorado, USA.

Lara, A. and T.T. Veblen. 1993. Forest plantations in Chile: A successful model? pp.118-139. *In* A. Mather [ed.]. *Afforestation Policies, Planning and Progress.* Belhaven Press, London, UK.

Lara, A., C. Donoso, and J.C. Aravena. 1996. La conservación del bosque nativo de Chile: problemas y desafíos. pp.335-362. *In* J.J. Armesto, C.Villagrán, and M.T.K. Arroyo [eds.]. *Ecología de los bosques nativos de Chile.* Editorial Universitaria, Santiago, Chile.

Lara, A., S. Fraver, J.C. Aravena, and A. Wolodarsky-Franke. 1999. Fire and dynamics of *Fitzroya cupressoides* (alerce) forests of Chile's cordillera Pelada. *Ecoscience,* 6: 100-109.

Larcher, W. 1995. *Physiological Plant Ecology,* Third Edition. Springer-Verlag, Berlin, Germany. pp. 215-275.

Lepart, J. and M. Debussche. 1992. Human impact on landscape patterning: Mediterranean examples. pp. 76-106. *In* A.J. Hansen and F. di Castri [eds.]. *Landscape Boundaries Consequences for Biotic Diversity and Ecological Flows.* Springer-Verlag, New York, New York, USA.

Lienlaf, E.F. 1996. Estimación de la producción de hojarasca y su pérdida mediante la extracción de 'tierra de hojas' en formaciones esclerófilas de la Reserva Nacional Río Clarillo. Thesis, Facultad de Ciencias Agrarias y Forestales, Universidad de Chile, Santiago, Chile.

Likens, G.E. and F.H. Bormann. 1996. *Biogeochemistry of a Forested Ecosystem. 2nd edition.* Springer-Verlag, New York, New York, USA.

Likens, G.E., F.H. Bormann, R.S. Pierce, and W.A. Reiners. 1978. Recovery of a deforested ecosystem. *Science,* 199: 492-496.

Makita, A., H. Makita, and A. Nishiwaki. 1995. Mass flowering of *Sasa kurilensis* to the south of Lake Towada, northern Japan, in 1995. *Bamboo Journal,* 13: 34-41.

Marod, D., U. Kutintara, C. Yarwudhi, H. Tanaka, and T. Nakashisuka. 1999. Structural dynamics of a natural mixed deciduous forest in western Thailand. *Journal of Vegetation Science,* 10: 777-786.

Mermoz, M., T. Kitzberger, and T.T. Veblen. 2005. Landscape influences on occurrence and spread of wildfires in Patagonian forests and shrublands. *Ecology,* 86: 2705-2715.

Molina, R., M. Correa, C. Smith-Ramierez, and A. Gainza. 2006. *Alerceros Huilliches de la Cordillera de la Costa de Osorno.* Andros, Santiago, Chile.

Montenegro, G., R. Ginocchio, A. Segura, J.E. Keeley, and M. Gomez. 2004. Fire regimes and vegetation responses in two Mediterranean-climate regions. *Revista Chilena de Historia Natural,* 77: 455-464.

Moore, D.M. 1983. *Flora of Tierra del Fuego.* Anthony Nelson, Shropshire, UK.

Morgan, P., G.H. Aplet, J.B. Haufler, H.C. Humphries, M.M. Moore, and W.D. Wilson. 1994. Historical range of variability: a useful tool for evaluating ecosystem change. *Journal of Sustainable Forestry,* 2: 87-111.

Muñoz, M. and E.R. Fuentes. 1989. Does fire induce shrub germination in the Chilean matorral? *Oikos,* 56: 177-181.

Nadgauda, R.S., C.K. John, M.S. Joshi, V.A. Parasharami, and A.F. Mascarenhas. 1997. Application of in vitro techniques for bamboo improvement. pp. 163-177. *In* G.P. Chapman [ed.]. *The Bamboos.* Academic Press, London, UK.

Narukawa, Y. and S. Yamamoto. 2002. Effects on dwarf bamboo (*Sasa* sp.) and forest floor microsites on conifer seedling recruitment in a subalpine forest, Japan. *Forest Ecology and Management,* 163: 61-70.

Neira, E. and S. Díaz. 1998. Superficie afectada por incendios en bosques de alerce y siempreverde en el año 1998 para el fundo esperanza norte y sur Lote A, cordillera de la Costa – Llanquihue. Informe Técnico. Universidad Austral de Chile – WWF, Valdivia, Chile.

Nelson, B.W. 1994. Natural forest disturbance and change in the Brazilian Amazon. *Remote Sensing Reviews,* 10: 105-125.

Papic, C. 2000. Regeneración de plántulas arbóreas sobre material leñoso en descomposición en un bosque sucesional de Chiloé, Chile. Master's Thesis, Facultad de Ciencias, Universidad de Chile, Santiago, Chile.

Parker, V.T. and V.R. Kelly. 1989. Seed banks in California chaparral and other mediterranean climate shrublands. *In* M.A. Leck, V.T. Parker, and R.L. Simpson [eds.]. *Ecology of Soil Seed Banks.* Academic Press, New York, New York, USA.

Pausas, J.G., A. Bonet, F.T. Maestre, and A. Climent. 2006. The role of the perch effect on the nucleation process in Mediterranean semi-arid oldfields. *Acta Oecologica,* 29: 346-352.

Pearson, A.K., O.P. Pearson, and I.A. Gomez. 1994. Biology of the bamboo *Chusquea culeou* (Poaceae, Bambusoideae) in southern Argentina. *Vegetatio,* 111: 93-126.

Perakis, S.S. and L.O. Hedin. 2002. Nitrogen loss from unpolluted South American forests mainly via dissolved organic compounds. *Nature,* 415: 416-419.

Relva, M.A. and T.T. Veblen. 1998. Impacts of introduced large herbivores on *Austrocedrus chilensis* forests in northen Patagonia, Argentina. *Forest Ecology and Management,* 108: 27-40.

Roig, F.A. and J. Boninsegna. 1991. Estudios de crecimiento radial, basal, en altura y de las condiciones climáticas que afectan el desarrollo de *Pilgerodendron uviferum. Revista Chilena de Historia Natural,* 64: 53-63.

Rothkugel, M. 1916. *Los Bosques Patagónicos.* Ministerio de Agricultura, Buenos Aires, Argentina.

Roy, V. 1998. Impacts of the modification of the hydrological regime by clearcutting and draining forested wetlands on growth of natural and planted seedlings. Ph.D Dissertation, Université Laval, Québec, Canada.

Ruprecht J.K. and N.J. Schofield. 1991. Effects of partial deforestation on hydrology and salinity in high salt storage landscapes. I. Extensive block clearing. *Journal of Hydrology,* 129: 19-38.

Salmon, C.D., M.T. Walter, L.O. Hedin, and M.G. Brown. 2001. Hydrological controls on chemical export from an undisturbed old-growth Chilean forest. *Journal of Hydrology,* 253: 69-80.

San Martin, F. 1940. *El Paso de la Villarica.* Talleres de Sebastián de Amorrortu e Hijos, Buenos Aires, Argentina.

Sherwin, W.B. and C. Moritz. 2000. Managing and monitoring genetic erosion. pp.9-34. *In* A.G. Young and G.M. Clarke [eds.]. *Genetics, Demography and Viability of Fragmented Populations.* Cambridge University Press, Cambridge, UK.

Szeicz, J.M., A. Lara, S. Díaz, and J.C. Aravena. 2000. Dendrochronological studies of *Pilgerodendron uviferum* in southwestern South America. pp. 245-270. *In* F.A. Roig [ed.]. *Dendrocronología en América Latina.* EDIUNC, Mendoza, Argentina.

Silla F., S. Fraver, A. Lara, T. Allnut, and A.C. Newton. 2002. Regeneration and stand dynamics of *Fitzroya cupressoides* (Cupressaceae) forests of southern Chile's Central Depression. *Forest Ecology and Management*, 165: 213-224.

Segura, A.M., M. Holmgren, J.J. Anabalón, and E.R. Fuentes. 1998. The significance of fire intensity in creating local patchiness in the Chilean matorral. *Plant Ecology*, 139: 259-264.

Smith-Ramírez, C. 2007. Regeneration of the threatened conifer *Fitzroya cupressoides* after indigenous and non-indigenous timber harvesting in southern Chilean forests. *Forest Ecology and Management*, 248: 193-201.

Swift L.W., W.T. Swank, J.B. Mankin, R.J. Luxmoore, and R.A. Goldstein. 1975. Simulation of evapotranspiration and drainage from mature and clear-cut deciduous forests and young pine plantation. *Water Resources Research*, 11: 667-673.

Tabarelli, M. and W. Mantovani. 2000. Gap-phase regeneration in a tropical montane forest: the effect of gap structure and bamboo species. *Plant Ecology*, 148: 149-155.

Taylor, A.H., Q. Zisheng, and J. Liu. 1995. Tree regeneration in an *Abies faxoniana* forest after bamboo dieback, Wang Lang Natural Reserve, China. *Canadian Journal of Forest Research*, 25: 2034-2039.

Tortorelli, L.A. 1947. *Los Incendios de Bosques en la Argentina*. Ministerio de Agricultura, Buenos Aires, Argentina.

Urrutia, R. 2002. Desarrollo de una cronología de anchos de anillos para Alerce (*Fitzroya cupressoides*) y reconstrucción de la historia de incendios en el área de Abtao, Parque Nacional Chiloé, X Región.Thesis Ingeniero Forestal. Pontificia Universidad Católica de Chile, Santiago, Chile.

Valiente-Banuet, A. and E. Ezcurra 1991. Shade as a cause of the association between the cactus *Neobuxxaumia tetetzo* and the nurse plant *Mimosa luisana*. *Journal of Ecology*, 79: 961-971.

Veblen, T.T. 1979. Structure and dynamics of *Nothofagus* forests near timberline in south-central Chile. *Ecology*, 60: 937-945.

Veblen, T.T. 1982a. Regeneration patterns in *Araucaria araucana* forests in Chile. *Journal of Biogeography*, 9: 11-28.

Veblen, T.T. 1982b. Growth patterns of *Chusquea* bamboos in the understory of Chilean *Nothofagus* forests and their influences in forest dynamics. *Bulletin of the Torrey Botanical Club*, 109: 474-487.

Veblen, T.T. and D.H. Ashton. 1982. The regeneration status of *Fitzroya cupressoides* in the Cordillera Pelada, Chile. *Biological Conservation*, 23: 141-161.

Veblen, T.T., C.Z. Donoso, F.M. Schlegel, and B.R. Escobar. 1981. Forest dynamics in south-central Chile. *Journal of Biogeography*, 8: 211-247.

Veblen, T.T., T. Kitzberger, and A. Lara. 1992a. Disturbance and forest dynamics along a transect from Andean rain forest to Patagonian shrubland. *Journal of Vegetation Science*, 3: 507-520.

Veblen, T.T., T. Kitzberger, E. Raffaele, and D.C. Lorenz. 2003. Fire History and Vegetation Changes in Northern Patagonia. pp. 265-295. *In* T.T. Veblen, W.L. Baker, G. Montenegro, and T.W. Swetnam [eds.]. *Fire Regimes and Climatic Change in Temperate Ecosystems of the Western Americas*. Springer-Verlag, New York, New York, USA.

Veblen, T.T., T. Kitzberger, R. Villalba, and J. Donnegan. 1999. Fire history in northern Patagonia: The role of humans and climatic variation. *Ecological Monographs*, 69: 47-67.

Veblen, T.T. and D.C. Lorenz. 1988. Recent vegetation changes along the forest/steppe ecotone in northern Patagonia. *Annals of the Association of American Geographers*, 78: 93-111.

Veblen, T.T., M. Mermoz, C. Martin, and T. Kitzberger. 1992b. Ecological impacts of introduced animals in Nahuel Huapi National Park, Argentina. *Conservation Biology*, 6: 71-83.

Veblen, T.T., F.M. Schlegel, and J.V. Oltremari. 1983. Temperate broad-leaved evergreen forest of South America. pp. 5-31. *In* J.D. Ovington [ed.]. *Temperate Broad-leaved Evergreen Forests*. Elsevier, Amsterdam, The Netherlands.

Verdu, M. and P. Garcia-Fayos. 1996. Nucleation process in a Mediterranean bird-dispersed plant. *Functional Ecology*, 10: 275-280.

Villagrán, C. 1988. Late quaternary vegetation of southern Isla Grande de Chiloé, Chile. *Quaternary Research*, 30: 304-314.

Villagrán, C. 1991. Historia de los bosques templados del sur de Chile durante el Tardiglacial y Postglacial. *Revista Chilena de Historia Natural*, 64: 447-460.

Villalba, R. 1990. Climatic fluctuations in northern Patagonia during the last 1000 years as inferred from tree-ring records. *Quaternary Research*, 34: 346-360.

Villalba, R., J.A. Boninsegna, T.T. Veblen, A. Schmelter, and S. Rubulis. 1997. Recent trends in tree-ring records from high elevation sites in the Andes of northern Patagonia. *Climate Change*, 36: 425-454.

Villalba, R., E.R. Cook, G.C. Jacoby, R.D. D 'Arrigo, T.T. Veblen, and P.D. Jones. 1998. Tree-ring based reconstructions of precipitation in Patagonia since A.D. 1600. *Holocene*, 8: 677-692.

Waring, R.H. and W.H. Schlesinger. 1985. *Forest Ecosystems: Concepts and Management*. Academic Press, Oxford, UK. pp. 94-120.

Wenny, D.G. 2001. Advantages of seed dispersal: A re-evaluation of directed dispersal. *Evolutionary Ecology*, 3: 51-74.

White, J.D., S.W. Running, P.E. Thornton, R.E. Keane, K.C. Ryan, D.B. Fagre, and C.H. Key. 1998. Assessing simulated ecosystem processes for climate variability research at Glacier National Park, USA. *Ecological Applications*, 8: 805-823.

White, P.S. and J.L. Walker. 1997. Approximating nature's variation: selecting and using reference information in restoration ecology. *Conservation Biology*, 5: 338-349.

Whitehead, D. and F.M. Kelliher. 1991. A canopy water balance model for a *Pinus radiata* stand before and after thinning. *Agricultural and Forest Meteorology*, 55: 109-126.

Widmer, Y. 1997. Life history of some *Chusquea* species in old-growth oak forest in Costa Rica. pp. 17-31. *In* G.P. Chapman [ed.]. *The Bamboos*. Academic Press, Oxford, UK.

Willson, M.F. and J.J. Armesto. 1996. The natural history of Chiloé: on Darwin's trail. *Revista Chilena de Historia Natural*, 69: 149-161.

Wolodarsky-Francke A., A. Lara, R. Reyes, P. Romero, E. Cuq, and J.C. Cisternas. 2005. Determinación de áreas geográficas y volúmenes de madera de alerce (*Fitzroya cupressoides*) muerto antes y después de 1976, en las comunas de Fresia y Purranque, X Región. Informe proyecto Lahuén (CONAF-Universidad Austral de Chile), Santiago, Chile.

21

Summary and Remarks

Peter R. Robichaud[1*] and *Artemi Cerdà*[2]

INTRODUCTION

This publication, *Fire Effects on Soils and Restoration Strategies*, provides a global view of current knowledge of fire effects on soil and potential mitigation of those effects for use by scientists, policy-makers, land managers, and students. This final chapter will summarize the collected expertise of the contributing authors, and identify the themes that link the chapters within each section—*Section I: Fire Effects on Soil* (Chapters 2 to 9); *Section II: Rehabilitation and Restoration Strategies* (Chapters 10 to 14); *and Section III: Regional Strategies* (Chapters 15 to 20).

For over 400 million years fire has been affecting the atmosphere, climate, and evolution of terrestrial ecosystems. In Chapter 1, A. Scott reviews the evidence for the existence of wildfire "in deep time, before man," (420 to 400 million years ago) and suggests that wildfires increased with atmospheric oxygen levels during the Carboniferous and Permian eras (350 to 250 my). As long as fires have existed, fire-induced changes in soil properties have impacted ecosystem functioning through changes in system flora and fauna, nutrient cycling, hydrological activity, erosion and sedimentation processes, and gaseous exchange with the atmosphere. Today, the dramatic changes brought about by wildfire might be of less concern if these fires were rare occurrences with few human impacts; however, just the opposite is true. Not only do wildland fires occur frequently, but these fires and post-fire effects impact the health, safety, and resources of many people. In addition, the number and severity of wildland fires are predicted to continue increasing as human impacts on the land and global climate changes accelerate. Recognition of these trends have made wildland fires and post-fire effects important natural resource concerns throughout the world.

[1] *Peter R. Robichaud, U.S. Department of Agriculture, Forest Service, Rocky Mountain Research Station, Moscow, Idaho, USA.*

[2] *Artemi Cerdà, Departament de Geografia. Universitat de València, Valencia, Spain.*

* *Corresponding author: Peter R. Robichaud, U.S. Department of Agriculture, Forest Service, Rocky Mountain Research Station, 1221 S. Main St., Moscow, Idaho 83843, USA. Tel: 1 208 883 2349, Fax: 1 208 883 2318, e-mail: probichaud@fs.fed.us*

SECTION I

Section I provides a review of the current understanding of fire effects on soil. The chapters divide this topic by types of fire impacts on soil, such as geomorphic processes (Chapter 2), biological effects (Chapter 5), and soil erosion after fire (Chapter 6). However, many of these topics overlap. For example, the effects of fire on nutrient cycles (Chapter 8) is a subset of the chemical effects of fire on soil (Chapter 4), and fire-induced soil water repellency (Chapter 7) has significant influence on post-fire infiltration (Chapter 3). Since each chapter reflects the research and current understanding of its authors, a range of perspectives is apparent, especially when comparing these overlapping topics.

There is strong agreement throughout Section I that the magnitude of fire-induced changes in soil properties and the length of time needed for soil to recover to its pre-fire functionality are dependent on soil burn severity. The degree of soil burn severity is dependent on the peak temperatures reached during the fire and duration of those temperatures within the soil. Generally, prescribed fires result in less soil heating (i.e., lower burn severity) than wildland fires, and as a result, fire effects are less severe.

In Chapter 2, Moody and Martin state that elevated soil temperatures increase the availability of sediment for erosion by lowering thresholds of soil movement by removing the protective cover over the soil. With the occurrence of high intensity post-fire rainfall, floods can occur and result in eroded hillslopes, incised channels, and deposited sediment across the burned landscape. These post-fire erosion and deposition processes are caused in part by reductions in soil infiltration rates. As discussed in Chapter 3 (Cerdà and Robichaud), fires generally reduce the infiltration capacity of soils, causing more surface flow. Immediately after a fire, the highly absorbent ash layer may absorb some or all rainfall for some storms; however, this water holding capacity quickly disappears as the ash is washed downslope or blown away leaving exposed mineral soil. Reduction in infiltration rates after fires can be attributed to loss of vegetation and litter cover, soil sealing, changes in soil structure, and fire-induced soil water repellency. In Chapter 7, Doerr et al. discuss fire effects on soil water repellency for both naturally water repellent soils and on soils with little or no water repellency prior to burning. Compounds with hydrophobic properties exist in the organic material found on the forest floor and in the organic soil horizon. During combustion, these compounds can be volatized and condense on soil particles beneath the soil surface, thereby inducing or intensifying soil water repellency.

Post-fire increases in runoff and hillslope erosion (up to three orders of magnitude above pre-fire rates) reflect the changes in hydrological functioning of burned soil. In Chapter 6, D. Scott et al. describe how the consumption of vegetation and forest floor material increases overland flow velocities (due to loss of water storage capacity and reduced surface roughness) which then has greater erosivity and sediment transport capacity. The result is both increased overland flow and erosion. The erosional response to wildfires is highly variable and reflects the inherent erosivity of the site—both the physiographic factors, such as drainage and slope characteristics that may influence erosion energy and concentration times of surface water runoff, and rainfall characteristics, such as intensity and duration. In addition, the erosional response is impacted by the extent and degree of soil burn severity. Because of vegetative and soil changes, some forms of gravity-driven erosion (dry ravel and mass failures) and wind erosion may also increase after fires.

Fire-induced changes in physical, chemical, and biological soil properties contribute to lower post-fire soil infiltration and higher erosion rates. In Chapter 4, Úbeda and Outeiro describe the effects of combustion and soil heating that cause decreases in soil organic matter content, aggregate stability, porosity, cation exchange capacity, and water storage capacity. Fire also increases soil bulk density, erodibility, water repellency, pH, and soil nutrient availability (short-term). In Chapter 5, Mataix-Solera et al. discuss fire effects specific to soil biology. Fire affects soil microbes directly through heating, where near-surface soil temperatures during fires are often sufficient to kill soil microorganisms, and indirectly through the physical and chemical changes in soil. Thus, soil burn severity is a strong predictor of fire effects on soil biological properties. The various classes of microorganisms respond to and recover from fire differentially, which leads to shifts in population ratios between groups of soil microbes. For example, fungi are more sensitive to heating than bacteria and actinomycetes and larger proportions of the fungal community die during a fire. Changes in soil properties, such as an increase in soil pH due to ash, can also influence the observed post-fire increase in the bacteria/fungi ratio. General post-fire recovery of burned soil is highly dependent on recovery of the soil biota as they directly impact soil chemical and physical properties and soil system functioning.

Raison et al. (Chapter 8) examines fire effects on the nutrient cycle. During fires soil nutrients may be removed by direct combustion and volatilization, and once the fire has ended by leaching or erosion of ash and soil. Fire effects, such as soil heating, can also transform soil nutrients to less usable chemical forms, thereby affecting decomposer activity and nutrient uptake by vegetation. Changes in nutrient uptake and turnover can, in turn, cause changes in vegetation composition, structure, and growth rates. In a continuing cascade of cause and effect, changes in vegetation impact hydrology, soil temperature regimes, and may trigger vegetation succession. Because the nutrient cycle is area-specific, local understanding of the relationship between fire and nutrient cycling is needed to guide post-fire management.

The disappearance (combustion) and charring of vegetation is, perhaps, the most dramatic visual fire effect; however fire is a natural process and post-fire recovery includes re-vegetation of the burned landscape. Vegetation recovery mechanisms determine ecosystem resilience, or the capacity to return to pre-fire conditions, and generally ecosystems submitted to frequent fires tend to be more resilient. In Chapter 9, Lloret and Zedler use the *fire regime-resilience coupled model* to discuss post-fire successional trajectories (for example, a typical trajectory will have species that were not dominant in the pre-fire conditions replaced by shade-tolerant species). However, under certain conditions, such as climate change, species change through planting, and/or fire regime modification by management, this coupling is disrupted and may result in decreased fire resilience, which may push succession trajectories toward new communities. For example, fire suppression strategies may lead to the loss of early successional resilient species, while too frequent fires may result in forests giving way to more open communities, such as savannah and grassland.

SECTION II

It is well-established that wildland fires are natural events that are part of succession for most terrestrial ecosystems, and post-fire recovery of burned landscapes does not require human intervention. Nonetheless, post-fire restoration strategies are the focus of an ever-increasing amount of international research and natural resource management activity and funding. This relatively new focus is the result of the expansion of the human population and human activities into fire-prone areas around the globe. With human life and safety, managed natural resources, built infrastructure, and private homes and buildings at risk from both wildland fires and the secondary effects of those fires, a 'hands-off' natural recovery process is rarely accepted as appropriate public policy. In most cases, there is considerable public pressure to do as much as possible to suppress wildfires, mitigate the secondary effects (increased runoff, flooding, and erosion) of those fires, and accelerate the natural recovery process. Consequently, after many fires, land managers often decide to use stabilization and rehabilitation strategies to reduce erosion and sedimentation and to protect important downstream structures and resources as well as the soil resource itself.

Post-fire stabilization and rehabilitation strategies are generally expensive and do not always provide the expected benefits. Some treatments have unintended consequences such as the introduction of invasive weeds through the use of agricultural straw mulch. In addition, quantification of post-fire treatment effectiveness requires expensive, long-term research efforts that are site-specific. In *Section II* (Chapters 10 to 14) contributing authors examine common post-fire restoration strategies—their effectiveness, limitations, and on-going developments. The research studies described in these chapters reveal key parameters of the most successful strategies, which are leading researchers, engineers, and entrepreneurs to develop new, improved treatment materials and application processes.

Chapter 10 (Robichaud) provides an overview of the post-fire treatments that are used immediately after fire suppression to reduce potential runoff, flooding, erosion, and debris flows to stabilize the burned area, as well as the treatments designed for short-term rehabilitation and long-term restoration and recovery. Stabilization treatments for hillslopes, roads, and channels are reviewed along with some longer-term treatments used for ecological restoration. In addition, Robichaud examines the parameters that inform post-fire treatment decisions, such as the use of runoff and erosion prediction models, post-fire assessment of values at risk for damage or loss, and cost-benefit analysis based on known and predicted treatment effectiveness. Decisions of where and when to apply post-fire rehabilitation treatments are as important, if not more important, as treatment choices. The process of treatment justification can, in the end, support a *no treatment option* as the best, most cost-effective decision. The high cost of post-fire rehabilitation treatments makes effectiveness monitoring important as it is necessary to determine if treatments are functioning as desired, compare treatment effectiveness, and identify conditions that enhance or limit treatment effectiveness.

Given that burned hillslopes are a major source of post-fire sediments and the cost for treating large hillslope areas is high, there has been more recent research on the development and evaluation of hillslope treatments than on other types of post-fire treatments. Three chapters (Chapters 11, 12, and 13) of *Section II* provide in-depth reviews of the three major types of hillslope treatments—seeding, erosion barriers, and mulching.

Broadcast seeding of grasses, usually from aircraft, is the oldest and most common post-fire treatment because rapid establishment of vegetation has been regarded as the most cost-effective method to improve water infiltration and hold soil on burned hillslopes. In Chapter 11, Beyers reviews the use of post-fire seeding not only to stabilize burned slopes, but also to prevent invasion of burned areas by noxious weeds, replace undesirable annual grasses on burned rangelands, and re-establish desirable vegetation including shrub and tree species. Enhancement of post-fire ground cover (especially in the first year after the fire when it is most needed) by a seeded species depends on the amount and timing of growing-season rainfall, protection of the seed from predators and desiccation, and resting the area from grazing until seeded species and natural regeneration are well-established. These favorable conditions are often not met resulting in a mixed record of treatment effectiveness for erosion reduction and rangeland rehabilitation. Seeded grasses can displace native herbaceous plants in post-fire succession, and high grass cover can limit recruitment of tree and shrub seedlings on burned sites. The increasing use of native seed for post-fire treatment has focused on noxious weed reduction and rangeland rehabilitation rather than potential erosion control.

The use of erosion barriers, which provide mechanical barriers to overland flow, promote infiltration, and trap sediment, for post-fire rehabilitation treatment is reviewed by Robichaud (Chapter 12). Erosion barriers have been used in the United States for the past 30 years, particularly contour-felled logs installed after fires on forest land. However, recent data from several studies, where sediment from treated and untreated sites was compared, indicate that post-fire erosion barrier treatment effectiveness is less than expected. Although erosion barriers can be effective for low- to moderate-intensity rainfall events, their effectiveness is greatly reduced for high-intensity rainfall events. Their effectiveness is also dependent on the quality of the installation and decreases over time as the sediment storage areas above the barriers become filled and the barrier can no longer trap mobilized sediment. The impact of these research results has been a reduction in the use of erosion barriers for post-fire hillslope stabilization, especially when high-intensity rainfall is likely.

In contrast, effectiveness studies of straw mulches being used to mitigate erosion from burned hillslopes generally have been positive, and the use of mulches (as well as the development of new mulches and application methods) for post-fire treatment is expanding. In Chapter 13, Bautista et al. review the use of post-fire mulch treatments that are applied to reduce the effects of raindrop impact and overland flow and reduce soil loss. Mulch immediately increases ground cover on the treated areas and is most effective during the critical first post-fire year when erosion is likely to be the greatest. A wide range of mulch materials (e.g., agricultural straw, wood strands, wood shreds) have been shown to be effective in reducing post-fire erosion, making mulch one of the most effective post-fire rehabilitation treatments available. However, the main disadvantage of post-fire mulch treatments is its high cost, which limits its use to burned areas where erosion potential is high and threatens to damage or destroy valuable resources. In addition, mulching can introduce non-native plants to areas where it is applied. The use of certified 'weed-free' straw or straw from plants that will not grow in the burned area (e.g., rice straw applied on burned upland forest) can reduce this risk. Mulches made from shredded or manufactured forest materials generally do not contribute to the spread of invasive weeds. The rate and uniformity of mulch application impacts its effectiveness. If mulch is applied too thinly it does not provide the ground cover needed to reduce erosion, and thick mulch can suppress the growth of vegetation—an advantage if the suppressed vegetation is invasive weeds, but a disadvantage if vegetative recovery or seeded species are inhibited. Developments of new mulch materials (hydromulch, paper mulches, polyacrylamide, etc.) and application methods (helimulching) are continually being developed, and the use of mulches composed of local, site-specific forest materials (wood shreds, pine needles, shredded debris from post-fire logging, etc.) becomes more attractive as a way to reduce transportation and manufacturing costs.

Most post-fire stabilization and short-term rehabilitation treatments are used to mitigate the post-fire effects on physical ecosystem components, such as soil, water, and hydrologic processes. Long-term rehabilitation and restoration activities are more often focused on biotic components of the ecosystem—natural recovery of native communities and habitat, maintenance of biodiversity, and the critical need for disturbance habitats for some biota. In Chapter 14, Vallejo et al. explain that long-term post-fire restoration not only aims to restore ecosystem structure and function, but also to recover ecosystem fire resilience and reduce future fire propagation potential. This commitment needs restoration strategies that promote secondary succession towards more mature, more resilient plant communities at a landscape scale. Field-tested strategies for long-term ecological restoration include plant species selection, seeding of woody plants, development of quality nursery stock, site preparation or soil amendment and fertilization; however, the complex interaction of different environmental conditions make the results of every restoration project unique and difficult to compare. Given the uncertainties of environmental conditions and the myriad of interactions, adaptive management principles should be applied throughout long-term post-fire restoration projects. Improving landscape and ecosystems quality (biodiversity, resilience, structure, function, etc.) and reducing wildfire propagation may justify the expense of long-term post-fire restoration projects to land management agencies, but the less-than-obvious results and long-term nature of these projects make it difficult to gain and maintain public support for these efforts.

SECTION III

Each of the six chapters in *Section III* focuses on current post-fire rehabilitation and restoration efforts in a specific region and/or country (Australia, Canada, Chile, Portugal, and two areas of the western United States) where wildfires have significant impact on the land and people who live there. In each chapter the regional climate, topography, soils, and dominant ecosystem(s) are discussed, and although wildfire effects may be similar, they often vary by magnitude across regions. The sense of urgency to respond to various wildfire effects also varies regionally, and descriptions of post-fire rehabilitation and restoration are markedly different among the countries represented. From the chapters in *Section III* it appears that, except for the western US, post-fire hillslope stabilization treatments are used infrequently, and are generally restricted to rehabilitation of the fire lines created during fire suppression efforts. The main deterrents to broader use of stabilization treatments appears to be the large cost of hillslope treatments and the logistical difficulties of applying them. The chapters from Australia (Chapter 18), Canada (Chapter 17), Chile (Chapter 20), and Portugal (Chapter 15) focus on longer-term restoration of natural habitat and, in some cases, timber and/or grazing productivity. The two chapters from the US (Chapters 16 and 19) focus on immediate post-fire stabilization and rehabilitation efforts to mitigate the fire effects on soil, water, and hydrologic processes. Little mention of long-term restoration efforts, such as tree planting, land-use restrictions, and riparian rehabilitation are given in these two chapters despite their extensive use in the United States. Thus, it is important to recognize that each chapter in *Section III* provides a limited view of post-fire rehabilitation—not only in terms of regional applicability, but also in terms of the complete rehabilitation/ restoration process that may be used within the region or country.

The focus of both Chapters 16 and 19 is stabilization treatments applied immediately after fire suppression to reduce potential increases in runoff, flooding, erosion, and debris flows in Colorado and California, USA. In Chapter 16, MacDonald and Larsen confirm that erosion rates from areas of the Colorado Front Range burned at high severity (5 to 10 Mg ha^{-1} yr^{-1} for the first two to three years after burning) are significantly greater than from areas burned at moderate or low severity and that it typically takes three to five years for hillslope-scale sediment yields to decline to near-background levels. Because post-fire changes in soil, water, and hydrological functioning pose significant risk for flooding, water quality, and other resources, there is public pressure and land agency support for immediate post-fire hillslope stabilization treatments. Given that post-fire sediment yields are most closely associated with the amount of surface cover and rainfall erosivity, the most effective post-fire rehabilitation treatments are those that immediately provide surface cover, such as straw mulching. Other hillslope treatments, such as seeding, seeding combined with scarification, hydromulching, contour-felled log erosion barriers, and direct application of polyacrylamide to the soil surface, have been used with varying degrees of success; however, straw mulch has been the most cost-effective treatment to date. Although Wohlgemuth et al. approach the topic of post-fire rehabilitation in California (Chapter 19) from a broader historical and state-wide perspective, like Chapter 16, the discussion of rehabilitation efforts is focused on mitigation of flooding, erosion, and sedimentation. The post-fire rehabilitation efforts include hillslope stabilization treatments (seeding, mulching, etc.) with evaluations of success similar to those in the Colorado Front Range. In addition, the discussion in Chapter 19 includes use of engineered structures in the stream channels or at the mouths of canyons to mitigate debris flows and treatments designed to protect wildland roads and trails from post-fire damage.

Dunkerly et al. (Chapter 18) discuss the large-scale of Australian wildland fires where more than one million hectare have been consumed in a single fire event, and its role as a driver of ecosystem change. These large fires incur enormous financial costs for suppression and loss of assets and primary production. To illustrate current post-fire rehabilitation efforts, Dunkerly et al. examined the impacts of lightening-ignited forest fires that occurred in 2003 and burned for over 60 days, sweeping across large areas of New South Wales and Victoria. Following these fires, the protection of waterways was a high priority for the post-fire rehabilitation work. The 9,000 km of fire control lines were rehabilitated by covering them with earth and vegetation residues, installation of cross-drains, and some replanting of vegetation. The loss of canopy and ground cover in montane forests and alpine environments, resulted in increased surface runoff and widespread erosion. Post-fire high-intensity rain storms caused exceptionally large inputs of coarse sediments, as well as ash and finer sediments, into area waterways. Rehabilitation included sediment control devices along many roads and culverts to limit the coarse sediment inputs into these regional streams and rivers. Significant areas of moss and peat were burned, and rehabilitation included installation of weirs on seepage outlets to retain water and promote moss regrowth.

Canada has a wide range of fire landscapes, and the steeper mountainous areas of the western provinces have the greatest risks for large post-wildfire erosion. Curren and D. Scott (Chapter 17) describe current post-fire assessment and treatments being used in western Canada. Until recently post-fire rehabilitation efforts focused primarily on the fire access roads and fire lines constructed while fighting the fire, along with some post-fire broadcast seeding and reforestation efforts. Forest fires in western Canada have increased in frequency and severity during the past decade, and the consequences of climate variation, such as increased areas of dead timber (fuels) due to dramatic proliferation of mountain pine beetle populations, has increased the potential for severe wildfires. Some provincial land management agencies are implementing post-fire risk assessment procedures, similar to those used in the US, for making stabilization and rehabilitation treatments decisions, which include rehabilitation of burned hillslopes where potential increased erosion poses a risk to valued resources. The use of prescribed fires to restore 'natural disturbance regimes' and reduce the severity of future forest fires are also being recommended.

Ferreira et al. (Chapter 15) also emphasizes the need to use prescribed burning to reduce the risk of large, high-severity wildfires on managed forest lands in Portugal. These forest lands consist largely of non-native eucalyptus plantations that provide a significant economic benefit to the country and are highly prone to fire. Since 1980 the mean annual area burned in Portugal (which has the highest fire incidence in Europe) slightly exceeds one percent of the total area of the country. Recognizing that soil erosion and soil nutrient loss is greater immediately after a fire, the authors suggest that it is necessary to identify burned areas for rehabilitation and implement cost-effective mitigation strategies. However, reintroduction of prescribed fire is strongly advocated by these authors as a mechanism for reducing the number and severity of forest fires in Portugal. Not only could prescribe fires reduce the severity of future fires, but extensive use of prescribed fires would create natural fuel breaks and slow their progression.

In contrast to Portugal, where rehabilitation and restoration efforts are aimed at restoring and/or protecting the managed timber resource, Arnesto et al. (Chapter 20) discuss potential restoration techniques that could support the resurgence of Chile's natural (and unique) forest ecosystems. Much of the Chilean Mediterranean and temperate forests have been replaced by fire-prone exotics that provide a more marketable timber resource. Fire frequency in these timber stands has increased and many of these fires spread from the timber into surrounding natural areas causing further degradation of the natural vegetation, which exhibit low resilience to repeated fire disturbance. In the temperate forests, increases in bamboo cover and open canopy (due to logging) are also increasing fire frequency. Passive restoration is a viable strategy for these conifer-dominated montane forests, but that makes the harvest of dead timber a serious threat to ecosystem recovery. Active restoration is needed in central Chile to re-establish Mediterranean vegetation in higher elevations and evergreen forests in the lowland, where the loss of canopy cover alters hydrology, raising the water table and slowing the process of tree invasion.

FUTURE RESEARCH DIRECTIONS

Research on forest fires, hydrology, engineering, fire ecology, climate change, aquatic ecology, forestry, and many other fields informs land management decisions related to post-fire stabilization, rehabilitation, and long-term restoration. Progress in all these fields will continue to improve the quality of these management decisions and effectiveness of interventions. Given the high cost of post-fire treatments, it is essential to maximize treatment benefits through selective treatment use in areas that pose the greatest threat to valued resources. Two areas of research that may improve the cost-effective use of post-fire rehabilitation treatments are the continued refinement of 1) remote sensing techniques to assess fire effects and the spatial continuity of soil burn severity and 2) techniques to estimate the potential threat posed by the post-fire landscape, such as predictive post-disturbance erosion models. Improvements in both of these research areas will directly impact post-fire assessment and the targeted use of post-fire treatments where and when they can be most effective.

Continued research is also needed in the areas of fire effects and the post-fire environment, long-term treatment effects, and new treatment effectiveness. Although much research has been done on fire effects, little is known about the effects of ash on infiltration, soil sealing, or on water quality once it moves into the stream channels. Research on the fate of ash requires an immediate response after fires as ash is easily moved by both wind and water. Long-term research is also needed. Monitoring of post-fire treatments, if it is done at all, rarely extends beyond three years and the long-term effects of treatments are largely unknown. As new rehabilitation treatment materials and techniques are developed, it is important to evaluate both their individual effectiveness as well as the effectiveness of potential treatment combinations.

In many areas, natural ecosystems have been replaced with tree plantations that provide a more consistent and economically viable timber resource (e.g., Portugal and Chile, Chapters 15 and 20, respectively). However, some scientists and land managers are concerned that the resilience and biodiversity of the natural systems are a valued resource that is endangered by the current land use management efforts to maximize timber production. Given that these managed timber stands often have different fire regimes than the natural ecosystems they replaced, a management decision to restore natural land cover will involve changes in the way post-fire rehabilitation and restoration are done. Researchers, land managers, and policy makers will need to work together to determine where and when such transitions should and can be made.

The research needs discussed above are examples, not an exhaustive list, and other researchers and land managers would likely exemplify research needs a bit differently. Nonetheless, there is agreement that additional research will improve our understanding of post-fire effects and provide more and better management options for effective post-fire rehabilitation and restoration.

The impacts of wildfires and post-wildfire effects will continue to demand a human response to protect lives, property, and valued resources well into the future. Through their contributions to this compendium, the authors submit that this response should be science-based and reflect our collective knowledge.

Index

F

G

H

I

L